FUNDAMENTALS OF BIOGEOGRAPHY

Second Edition

Fundamentals of Biogeography presents an accessible, engaging, and comprehensive introduction to biogeography, explaining the ecology, geography, history, and conservation of animals and plants. Starting with an outline of how species arise, disperse, diversify, and become extinct, the book examines how environmental factors (climate, substrate, topography, and disturbance) influence animals and plants; investigates how populations grow, interact, and survive, and how communities form and change; and explores the connections between biogeography and conservation.

The second edition has been extensively revised and expanded throughout to cover new topics and revisit themes from the first edition in more depth. Illustrated throughout with informative diagrams and attractive photographs, and including guides to further reading, chapter summaries, and an extensive glossary of key terms, *Fundamentals of Biogeography* clearly explains key concepts in the history, geography, and ecology of life systems. In doing so, it tackles some of the most topical and controversial environmental and ethical concerns, including species overexploitation, the impacts of global warming, habitat fragmentation, biodiversity loss, and ecosystem restoration.

Fundamentals of Biogeography presents an appealing introduction for students and all those interested in gaining a deeper understanding of key topics and debates within the fields of biogeography, ecology, and the environment. Revealing how life has been and is adapting to its biological and physical surroundings, Huggett stresses the role of ecological, historical, and human factors in fashioning animal and plant distributions, and explores how biogeography can inform conservation practice.

Richard John Huggett is a Reader in Geography at the University of Manchester.

ROUTLEDGE FUNDAMENTALS OF PHYSICAL GEOGRAPHY SERIES

Series Editor: John Gerrard

This new series of focused, introductory textbooks presents comprehensive, up-to-date introductions to the fundamental concepts, natural processes, and human/environmental impacts within each of the core physical geography sub-disciplines. Uniformly designed, each volume contains student-friendly features: plentiful illustrations, boxed case studies, key concepts and summaries, further reading guides, and a glossary.

FUNDAMENTALS OF BIOGEOGRAPHY
Second Edition

Richard John Huggett

Routledge
Taylor & Francis Group

LONDON AND NEW YORK

First edition published 1998

Second edition 2004
by Routledge
2 Park Square, Milton Park, Abingdon,
Oxfordshire OX14 4RN

Simultaneously published in the USA and Canada
by Routledge
29 West 35th Street, New York, NY 10001

Routledge is an imprint of the Taylor & Francis Group

© 1998, 2004 Richard John Huggett

Typeset in Garamond and Futura by Florence Production Ltd, Stoodleigh, Devon
Printed and bound in Great Britain by Bell & Bain Ltd, Glasgow

British Library Cataloguing in Publication Data
A catalogue record for this book is available from the British Library

Library of Congress Cataloging in Publication Data
Huggett, Richard J.
Fundamentals of biogeography/Richard John Huggett. – 2nd ed.
p. cm. – (Routledge fundamentals of physical geography series)
Includes bibliographical references and index.
1. Biogeography. I. Title. II. Series.
QH84.H84 2004
578′.09 – dc22
2003027028

ISBN 0–415–32346–0 (hbk)
ISBN 0–415–32347–9 (pbk)

for my family

CONTENTS

SERIES EDITOR'S PREFACE

We are presently living in a time of unparalleled change and when concern for the environment has never been greater. Global warming and climate change, possible rising sea-levels, deforestation, desertification, and widespread soil erosion are just some of the issues of current concern. Although it is the role of human activity in such issues that is of most concern, this activity affects the operation of the natural processes that occur within the physical environment. Most of these processes and their effects are taught and researched within the academic discipline of physical geography. A knowledge and understanding of physical geography, and all it entails, is vitally important.

It is the aim of this *Fundamentals of Physical Geography Series* to provide, in five volumes, the fundamental nature of the physical processes that act on or just above the surface of the earth. The volumes in the series are *Climatology*, *Geomorphology*, *Biogeography*, *Hydrology*, and *Soils*. The topics are treated in sufficient breadth and depth to provide the coverage expected in a *Fundamentals* series. Each volume leads into the topic by outlining the approach adopted. This is important because there may be several ways of approaching individual topics. Although each volume is complete in itself, there are many explicit and implicit references to the topics covered in the other volumes. Thus, the five volumes together provide a comprehensive insight into the totality that is Physical Geography.

The flexibility provided by separate volumes has been designed to meet the demand created by the variety of courses currently operating in higher education institutions. The advent of modular courses has meant that physical geography is now rarely taught, in its entirety, in an 'all-embracing' course but is generally split into its main components. This is also the case with many Advanced Level syllabuses. Thus students and teachers are being increasingly frustrated by the lack of suitable books and are having to recommend texts of which only a small part might be relevant to their needs. Such texts also tend to lack the detail required. It is the aim of this series to provide individual volumes of sufficient breadth and depth to fulfil new demands. The volumes should also be of use to sixth form teachers where modular syllabuses are becoming common.

The volumes have been written by higher education teachers with a wealth of experience in all aspects of the topics they cover and a proven ability in presenting information in a lively and interesting way. Each volume provides a comprehensive coverage of the subject matter using clear text divided into easily accessible sections and subsections. Tables, figures, and photographs are used where appropriate as well as boxed case studies and summary notes. References to important previous studies and results are included but are used sparingly to avoid overloading the text. Suggestions for further reading are also provided. The main target readership is introductory level undergraduate students of physical geography or environmental science, but there will be much of interest to students from other disciplines and it is also hoped that sixth form teachers will be able to use the information that is provided in each volume.

John Gerrard

AUTHOR'S PREFACE TO
THE SECOND EDITION

The first edition of *Fundamentals of Biogeography* was published in 1998. Since that time, the subject has moved on, with some interesting developments. Other textbooks have appeared. *Biogeography: An Ecological and Evolutionary Approach* (2000) by Cox and Moore is now in its sixth edition and it has been joined by Brown and Lomolino's *Biogeography*, second edition (1998), and MacDonald's *Biogeography: Introduction to Space, Time and Life* (2003). After having read these books and having taught the material in *Fundamentals of Biogeography* for several years, I felt that some rearrangement, especially in the early chapters and in the final chapter would be beneficial. The key changes are the division of the book into four parts: Introducing Biogeography, Ecological Biogeography, Historical Biogeography, and Conservation Biogeography.

Part I consists of four chapters dealing with the nature of biogeography, basic biogeographical processes, and distributions. A chapter on speciation, diversification, and extinction is new.

Part II is long. It starts with a chapter on habitats, environments, and niches. Four chapters follow that cover environmental factors: climate, substrate, topography, and disturbance. This expansion allows a much fuller treatment of this material than in the first edition. The remaining chapters in this part examine populations, interacting populations, communities (including a new section on the theory of island biogeography), and community change.

Part III tackles the history of organisms. Three chapters consider dispersal and diversification in the distant past (a largely new chapter), vicariance in the distant past, and past community change (with much new material).

Part IV explores the application of biogeography to conservation issues, focusing on conserving species and populations, and conserving communities and ecosystems. These two chapters replace the final chapter in the first edition and are sharply focused on conservation biogeography.

The effect of these major changes, plus some updating of examples and ideas, should be to give the book an even tighter, more logical, and better-balanced structure, and to offer students better value for money.

Once again, I should like to thank many people who have made the completion of this book possible: Nick Scarle for revising many of the first edition diagrams and drawing the many new ones; Andrew Mould for having the good sense to realize that the book needed a refreshing overhaul and expansion; Chris Fastie, Rob Whittaker, Stephen Sarre, Karen A. Poiani, and Pat Morris for letting me re-use their photographs; and Francisco L. Pérez, Stefan Porembksi, and Cam Stevens for supplying me with fresh ones; Clive Agnew and other colleagues in the School of Geography at Manchester University for lending their support for writing a textbook in a research-driven climate; Derek Davenport for endless discussions on all manner of things; and, as always, my wife and family for letting me spend so much time in front of the PC.

Richard John Huggett
Poynton
December 2003

AUTHOR'S PREFACE TO THE FIRST EDITION

Biogeography means different things to different people. To biologists, it is traditionally the history and geography of animals (zoogeography) and plants (phytogeography). This, historical biogeography, explores the long-term evolution of life and the influence of continental drift, global climatic change, and other large-scale environmental factors. Its origins lie in seventeenth-century attempts to explain how the world was restocked by animals disembarking from Noah's ark. Its modern foundations were laid by Charles Darwin and Alfred Russel Wallace in the second half of the nineteenth century. The science of ecology, which studies communities and ecosystems, emerged as an independent study in the late nineteenth century. An ecological element then crept into traditional biogeography. It led to analytical and ecological biogeography. Analytical biogeography considers where organisms live today and how they disperse. Ecological biogeography looks at the relations between life and the environmental complex. It used to consider mainly present-day conditions, but has edged backwards into the Holocene and Pleistocene.

Physical geographers have a keen interest in biogeography. Indeed, some are specialist teachers in that field. Biogeography courses have been popular for many decades. They have no common focus, their content varying enormously according to the particular interests of the teacher. However, many courses show a preference for analytical and ecological biogeography, and many include human impacts as a major element. Biogeography is also becoming an important element in the growing number of degree programmes in environmental science. Biogeography courses in geography and environmental science departments are supported by a good range of fine textbooks. Popular works include *Biogeography: Natural and Cultural* (Simmons 1979), *Basic Biogeography* (Pears 1985), *Biogeography: A Study of Plants in the Ecosphere* (Tivy 1992), and *Biogeography: An Ecological and Evolutionary Approach* (Cox and Moore 1993), the last being in its fifth edition with a sixth in preparation.

As there is no dearth of excellent textbooks, why is it necessary to write a new one? There are at least four good reasons for doing so. First, all the popular texts, though they have been reissued as new editions, have a 1970s air about them. It is a long time since a basic biogeography text appeared that took a fresh, up-to-date, and geographically focused look at the subject. Second, human inter-action with plants and animals is now a central theme in geography, in environmental science, and in environmental biology. Existing textbooks tackle this topic, but there is much more to be said about application of biogeographical and ecological ideas in ecosystem management. Third, novel ideas in ecology are guiding research in biogeography. It is difficult to read articles on ecological biogeography without meeting metapopulations, heterogeneous landscapes, and complexity. None of these topics is tackled in existing textbooks. They are difficult topics to study from research publications because they contain formidable theoretical aspects. Nevertheless, it is very important that students should be familiar with the basic ideas behind them. First- and second-year undergraduates can handle them if they are presented in an informative and interesting way that avoids excessive mathematical formalism. Fourth, environmentalism in its glorious variety has mushroomed into a vast interdisciplinary juggernaut. It impinges on biogeography to such an extent that it would be inexcusably remiss not to let it feature in a substantial way. It is a facet of biogeography that geography students find fascinating. Without doubt, a biogeography textbook for the next millennium should include discussion of environmental and ethical concerns about such pressing issues as species exploitation, environmental degradation, and biodiversity. However, biogeography is a vast subject and all textbook writers adopt a somewhat individualistic viewpoint. This book is no exception. It stresses the role of ecological, geographical, historical, and human factors in fashioning animal and plant distributions.

I should like to thank many people who have made the completion of this book possible. Nick Scarle patiently drew all the diagrams. Sarah Lloyd at Routledge bravely took yet another Huggett book on board. Several people kindly provided me with photographs. Rob Whittaker and Chris Fastie read and improved the section on vegetation succession. Michael Bradford and other colleagues in the Geography Department at Manchester University did not interrupt my sabbatical semester too frequently. Derek Davenport again discussed all manner of ideas with me. And, as always, my wife and family lent their willing support.

Richard John Huggett
Poynton
December 1997

ACKNOWLEDGEMENTS

The author and publisher would like to thank the following for granting permission to reproduce material in this work:

The copyright of photographs remains held by the individuals who kindly supplied them (please see photograph captions for individual names); Figure 4.10 from Figure in *The Mammals of Britain and Europe* by A. Bjärvall and S. Ullström (London and Sydney: Croom Helm) © 1986, reproduced with kind permission of Croom Helm; Figure 6.8 after Figures from the MONARCH Report, reproduced by kind permission of Pam Berry; Figure 7.3 after distribution maps of meadow oat-grass and wavy hair-grass from *New Atlas of the British and Irish Flora: An Atlas of the Vascular Plants of Britain, Ireland, the Isle of Man and the Channel Islands* by C. D. Preston, D. A. Pearman, and T. D. Dines (Oxford: Oxford University Press) © 2002, reproduced by kind permission of Oxford University Press; Figure 7.11 after Figure 4 from I. S. Downie, J. E. L. Butterfield, and J. C. Coulson (1995), Habitat preferences of sub-montane spiders in northern England (*Ecography* 18, 51–61) © 1995, reproduced by kind permission of Blackwell Publishing Ltd; Figure 10.12 after distribution maps kindly supplied by Duncan Halley and reproduced with his permission; Figure 11.1 after Figure on p. 28 of 'Ecological chemistry' by L. P. Brower (*Scientific American* 220 (February), 22–9) © 1969, reproduced by kind permission of *Scientific American*; Figure 12.20 after Figures 1, 3, and 4 from 'A species-based theory of insular biogeography' by M. V. Lomolino (*Global Ecology & Biogeography* 9, 39–58) © 2000, reproduced by kind permission of Blackwell Science Ltd; Figure 13.5 after Figure 9.1 from 'European expansion and land cover transformation', pp. 182–205, by M. Williams, in I. Douglas, R. J. Huggett, and M. E. Robinson (eds) *Companion Encyclopedia of Geography: The Environment and Humankind* (London: Routledge) © 1996, reproduced with kind permission of Routledge; Figures 14.3 and 14.4 after Figures 8e and 10 from 'The Great American Interchange: an invasion induced crisis for South American mammals' by L. G. Marshall (1981), in M. H. Nitecki (ed.) *Biotic Crises in Ecological and Evolutionary Time*, pp. 133–229 (New York: Academic Press) © 1981, reproduced by kind permission of L. G. Marshall; Figure 14.5 after Figure 5 from *Splendid Isolation: The Curious History of South American Mammals*, by G. G. Simpson (New Haven, CT, and London: Yale University Press) © 1980, reproduced by kind permission of Yale University Press; Figure 17.2 after Figure 1 from 'Minimum viable population and conservation status of the Atlantic Forest spiny rat *Trinomys eliasi'* by D. Brito, and M. de Souza Lima Figueiredo (*Biological Conservation* 122, 153–8) © 2003, reproduced with permission from Elsevier.

Note

Every effort has been made to contact copyright holders for their permission to reprint material in this book. The publishers would be grateful to hear from any copyright holder who is not here acknowledged and will undertake to rectify any errors or omissions in future editions of this book.

PART I INTRODUCING BIOGEOGRAPHY

WHAT IS BIOGEOGRAPHY?

Biogeographers study the geography, ecology, and evolution of living things. This chapter covers:

- ecology – environmental constraints on living
- history and geography – time and space constraints on living

---

Biogeographers address a misleadingly simple question: why do organisms live where they do? Why does the speckled rangeland grasshopper live only in short-grass prairie and forest or brushland clearings containing small patches of bare ground? Why does the ring ouzel live in Norway, Sweden, the British Isles, and mountainous parts of central Europe, Turkey, and southwest Asia, but not in the intervening regions? Why do tapirs live only in South America and southeast Asia? Why do the nestor parrots – the kea and the kaka – live only in New Zealand?

Two groups of reasons are given in answer to such questions as these – ecological reasons and historical-cum-geographical reasons.

ECOLOGY

Ecological explanations for the **distribution** of organisms involve several interrelated ideas. First is the idea of *populations*, which is the subject of *analytical biogeography*. Each species has a characteristic life history, reproduction rate, **behaviour**, means of dispersal, and so on. These traits affect a population's response to the environment in which it lives. The second idea concerns this biological response to the **environment** and is the subject of *ecological biogeography*. A population responds to its physical surroundings (**abiotic** environment) and its living surroundings (**biotic** environment). Factors in the abiotic environment include such physical factors as temperature, light, **soil**, geology, topography, fire, water, and air currents; and such chemical factors as oxygen levels, salt concentrations, the presence of toxins, and acidity. Factors in the biotic environment include competing species, parasites, diseases, predators, and humans. In short, each species can tolerate a range of environmental factors. It can only live where these factors lie within its tolerance limits.

The speckled rangeland grasshopper

This insect (*Arphia conspersa*) ranges from Alaska and northern Canada to northern Mexico, and from California to the Great Plains. It lives at less than 1,000 m elevation in the northern part of its range and up to 4,000 m in the southern part. Within this extensive latitudinal and altitudinal range, its distribution pattern is very patchy, owing to its decided preference for very specific habitats (e.g. Schennum and Willey 1979). It requires short-grass prairie, or forest and brushland openings, peppered with small pockets of bare ground. Narrow-leaved grasses provide the grasshopper's food source. It needs the bare patches to perform its courtship rituals. Dense forest, tall grass **meadows**, or dry scrubland fail to meet these ecological and behavioural needs. Roadside meadows and old logged areas are suitable and subject to slow **colonization**. Moderately grazed pastures are also suitable and support large populations.

Even within suitable habitat, the grasshopper's low vagility (the ease with which it can spread) limits its distribution. This poor ability to spread is the result of complex social behaviour, rather than an inability to fly well. Females are rather sedentary, at least in mountain areas, while males make mainly short, spontaneous flights within a limited area. The two sexes together form tightly knit population clusters within areas of suitable habitat. Visual and acoustic communication displays hold the cluster together.

Ring ouzel

A mix of ecology and history may explain the biogeography of most species. The ring ouzel or 'mountain blackbird', which goes by the undignified scientific name of *Turdus torquatus* (Box 1.1), lives in the cool temperate climatic zone, and in the alpine equivalent to the cool temperate zone on mountains (Figure 1.1). It likes cold climates. During the last **ice age**, the heart of its range was probably the Alps and Balkans. From here, it spread outwards into much of Europe, which was then colder than now. With climatic warming during the last 10,000 years, the ring ouzel has left much of its former range, surviving only in places that are still relatively cold because of their high latitude or altitude. Even though it likes cold conditions, most ring ouzels migrate to less severe climates during winter. The north European populations move to the Mediterranean while the alpine populations move to lower altitudes.

HISTORY

Historical-cum-geographical explanations for the distribution of organisms involve two basic ideas, both of which are the subject of *historical biogeography*. The first idea concerns *centres-of-origin* and **dispersal** from one place to another. It argues that species originate in a particular place and then spread to other parts of the globe, if they should be able and willing to do so. The second idea considers the importance of geological and climatic changes splitting a single population into two or more isolated groups. This idea is known as **vicariance** biogeography. The following case studies illustrate these two basic biogeographical processes.

Tapirs

The tapirs are close relatives of the horses and rhinoceroses. They form a family – the Tapiridae. There are four living species, one of which dwells in southeast Asia and three in central and South America (Plate 1.1). Their present distribution is thus broken and poses a problem

Box 1.1

WHAT'S IN A NAME? CLASSIFYING ORGANISMS

Everyone knows that living things come in a glorious diversity of shapes and sizes. It is apparent even to a casual observer that organisms appear to fall into groups according to the similarities between them. No one is likely to mistake a bird for a beetle, or a daisy for a hippopotamus. Zoologists and botanists classify organisms according to the similarities and differences between them. Currently, five great kingdoms are recognized – prokaryotae (monera), protoctista, plantae, fungi, and animalia. These chief subdivisions of the kingdoms are phyla. Each phylum represents a basic body plan that is quite distinct from other body plans. This is why it is fairly easy, with a little practice, to identify the phylum to which an unidentified organism belongs. Amazingly, new phyla are still being discovered (e.g. Funch and Kristensen 1995).

Organisms are classified hierarchically. Individuals are grouped into species, species into genera, genera into families, and so forth. Each species, genus, family, and higher-order formal group of organisms is called a **taxon** (plural **taxa**). Each level in the hierarchy is a taxonomic category. The following list shows the classification of the ring ouzel:

Kingdom: Animalia (animals)
Phylum: Chordata (chordates)
Subphylum: Vertebrata (vertebrates)
Class: Aves (birds)
Subclass: Neornithes ('new birds')
Superorder: Carinatae (typical flying birds)
Order: Passeriformes (perching birds)
Suborder: Oscines (song birds)
Family: Muscicapidae (thrush family)
Subfamily: Turdinae (thrushes, robins, and chats)
Genus: *Turdus*
Species: *Turdus torquatus*

Animal family names always end in -idae, and subfamilies in -inae. Dropping the initial capital letter and using -ids as an ending, as in felids for members of the cat family, gives them less formal names. Plant family names end in -aceae or -ae. The genus (plural genera) is the first term of a binomial: genus plus species, as in *Turdus torquatus*. It is always capitalized and in italics. The species is the second term of a binomial. It is not capitalized in animal species, and is not normally capitalized in plant species, but is always italicized in both cases. The specific name signifies either the person who first described it, as in *Muntiacus reevesi*, Reeve's muntjac deer, or else some distinguishing feature of the species, as in *Calluna vulgaris*, the common (= vulgar) heather. If subspecies are recognized, they are denoted by the third term of a trinomial. For example, the common jay in western Europe is *Garrulus glandarius glandarius*, which would usually be shortened to *Garrulus g. glandarius*. The Japanese subspecies is *Garrulus glandarius japonicus*. In formal scientific writing, the author or authority of the name is indicated. So, the badger's full scientific name is *Meles meles* L., the L. indicating that Carolus Linnaeus (1707–78) first described the species. The brown hare's formal name is *Lepus europaeus* Pallas, which shows that Peter Simon Pallas first described it (in 1778). In this book, the authorities will be omitted because they confer a stuffy feel. After its first appearance in each chapter, the species name is abbreviated by reducing the generic term to a single capital letter. Thus, *Meles meles* becomes *M. meles*.

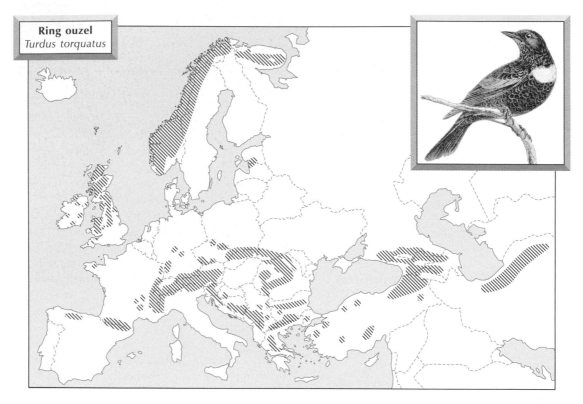

Figure 1.1 The breeding distribution of the ring ouzel (*T. torquatus*).
Sources: Map adapted from Cramp (1988); picture from Saunders (1889)

for biogeographers. How do such closely related species come to live in geographically distant parts of the world? Finds of fossil tapirs help to answer this puzzle. Members of the tapir family were once far more widely distributed than at present (Figure 1.2). They lived in North America and Eurasia. The oldest fossils come from Europe. A logical conclusion is that the tapirs evolved in Europe, which was their centre-of-origin, and then dispersed east and west. The tapirs that went northeast reached North America and South America. The tapirs that chose a southeasterly **dispersal route** moved into southeast Asia. Subsequently, probably owing to climatic change, the tapirs in North America and the Eurasian homeland went extinct. The survivors at the trop-

ical edges of the distribution spawned the present species. This explanation is plausible, but it is not watertight – it is always possible that somebody will dig up even older tapir remains from somewhere else. The incompleteness of the fossil record dogs historical biogeographers and dictates that they can never be fully confident about any hypothesis.

Nestor parrots

The nestor parrots (Nestorinae) are endemic to New Zealand. There are two species – the kaka (*Nestor meridionalis*) and the kea (*N. notabilis*) (Plate 1.2). They are closely related and are probably descended from a 'proto-kaka' that reached

Plate 1.1 Central American or Baird's tapir (*Tapirus bairdi*), Belize.

Photograph by Pat Morris.

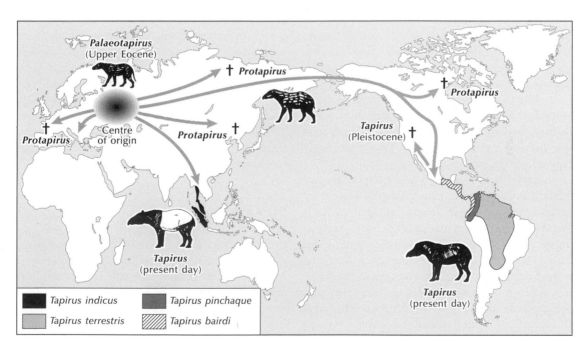

Figure 1.2 Tapirs: their origin, spread, and present distribution.

Source: Adapted from Rodríguez de la Fuente (1975)

(a)

(b)

Plate 1.2 Nestor parrots. (a) Kaka (*N. meridionalis*).
(b) Kea (*N. notabilis*).
Photographs by Pat Morris.

New Zealand during the **Tertiary period** (see Appendix). Then, New Zealand was a single, forest-covered island. The proto-kaka became adapted to forest life. Late in the Tertiary period, the northern and southern parts of New Zealand split. North Island remained forested and the proto-kakas there continued to survive as forest parrots, feeding exclusively on vegetable matter and nesting in tree hollows. They eventually evolved into the modern kakas.

South Island gradually lost its forests because mountains grew and climate changed. The proto-kakas living on South Island adjusted to these changes by becoming 'mountain parrots', depending on alpine shrubs, insects, and even **carrion** for food. They forsook trees as breeding sites and turned to rock fissures. The changes in the South Island proto-kakas were so far-reaching that they became a new species – the kea. After the **Ice Age**, climatic amelioration promoted some reforestation of South Island. The kakas dispersed across the Cook Strait and colonized South Island. Interaction between North and South Island kaka

populations is difficult across the 26 km of ocean. In consequence, the South Island kakas have become a subspecies. The kaka and the kea are now incapable of interbreeding and they continue to live side by side on South Island. The kea has never colonized North Island, probably because there is little suitable habitat there. The biogeography of the nestor parrots thus involves **adaptation** to changing environmental conditions, dispersal, and vicariance events.

SUMMARY

Ecology (including behaviour), history, and geography determine the distribution of organisms. Most species distributions result from a combination of all these factors, but biogeographers tend to specialize in ecological aspects (ecological biogeography) or historical aspects (historical biogeography). Ecological biogeographers are interested in the effects of environmental factors in constraining species ranges, and in the role of past environmental changes in shaping species ranges. Some historical biogeographers are interested in finding centres-of-origin and dispersal routes of various groups of organism; others prefer to interpret biogeographical history through vicariance (range-splitting) events.

ESSAY QUESTIONS

1 What is ecological biogeography?

2 What is historical biogeography?

FURTHER READING

Brown, J. H. and Lomolino, M. V. (1998) *Biogeography*, 2nd edn. Sunderland, MA: Sinauer Associates.
An excellent text covering virtually all aspects of biogeography in rich detail.

Cox, C. B. and Moore, P. D. (2000) *Biogeography: An Ecological and Evolutionary Approach*, 6th edn. Oxford: Blackwell.
Already a classic. A deservedly popular textbook.

George, W. (1962) *Animal Geography*. London: Heinemann.
Written before the revival of continental drift, but worth a look.

MacDonald, G. (2003) *Biogeography: Introduction to Space, Time and Life*. New York: John Wiley & Sons.
An excellent, up-to-date textbook with lots of examples.

Pears, N. (1985) *Basic Biogeography*, 2nd edn. Harlow: Longman.
A good, if dated, text on ecological biogeography.

Tivy, J. (1992) *Biogeography: A Study of Plants in the Ecosphere*, 3rd edn. Edinburgh: Oliver & Boyd.
An enjoyable introductory textbook with a chapter on historical biogeography.

BIOGEOGRAPHICAL PROCESSES I

SPECIATION, DIVERSIFICATION, AND EXTINCTION

Species appear, flourish, and disappear. This chapter covers:

- How new species arise
- How species diversify
- Why species become extinct

GENETIC INFORMATION: THE BASIS OF EVOLUTIONARY CHANGE

New species arise through the process of **speciation**. The nature of speciation and its causes are fiercely debated. A key element is a store of genetic information that is mutable. The debate largely focuses on the relative importance of the various sources of change; for example, is geographical isolation the primary driving force or is it ecological differentiation? This section will examine these causes of change, having first outlined the nature of stores of genetic material.

Genes, genomes, and gene pools

Genetic information provides a blueprint, so to speak, for making an organism. The **genome** is the total genetic information stored in an organism (or organelle or cell). Genetic information is stored in **chromosomes** within cells. In prokaryotes (bacteria and cyanobacteria), deoxyribonucleic acid (DNA) forms a coiled structure called a nucleoid. In eukaryotes (all other organisms), chromosomes are made of DNA and protein. The *karyotype* is the number, size, and shape of the chromosomes in a somatic cell arranged in a standard manner. The human karyotype has 46 chromosomes.

DNA is found in mitochondria as well as in cell nuclei. Mitochondrial DNA (mtDNA) is a closed, circular molecule. (Mitochondria are vitally important organelles that are the site of cell respiration and produce energy-rich molecules of adenosine triphosphate (ATP).) As a rule, specific genes in mtDNA are present in all taxa, be they toad, tortoise, titmouse, or tiger. Because change within mtDNA occurs about five to ten times faster than in nuclear DNA, mtDNA 'records' recent demographic events: in short, it is a repository of phylogenetic history. Techniques of molecular biology allow this 'history' to be unlocked and have enabled a new branch of biogeography – **phylogeography** – to emerge (e.g. Avise 2000), though the role of this subject in biogeography is debated (see Ebach *et al.* 2002; Ebach and Humphries 2003).

A **gene** is a specific piece of genetic information – a unit of inheritance passed from generation to generation by the gametes (eggs and spermatozoa). It consists of a length of DNA at a particular site or **locus** on a chromosome. Some genes have alternative forms or **alleles** at the same locus, which differ in one or more bases. For instance, in humans, two alleles exist at a locus controlling eye colour: one allele determines blue eyes and the other brown eyes. Multiple alleles may occur at some loci, though there are seldom more than ten. Species with at least two discrete genetic variants are *polymorphic species* and display **polymorphism**. Alleles of one gene may be identical or may differ at the same loci in a genotype. Where two alleles received from both parents are identical, the condition is **homozygous** and the individual is a *homozygote*; where they differ, the condition is **heterozygous** and the individual a *heterozygote*. Dominant alleles affect the phenotype in a heterozygote and a homozygote, whereas only in a homozygote do recessive alleles affect the phenotype. In a randomly mating population, the average homozygosity is a measure of gene identity and the average heterozygosity is a measure of gene diversity.

The **genotype** is the genetic make-up of an individual, the sum of all its genes. It contrasts with its **phenotype**, which is its form, physiology, and way of life; in other words, the sum of its characteristics. The phenotype of an organism changes throughout its life, but its genotype remains the same, except for occasional **mutations**. The genotype determines the range of phenotypes that may develop, and the environment determines the actual phenotypes that do develop. For example, in California, USA, populations of Hansen's cinquefoil (*Potentilla glandulosa hanseni*) living at different altitudes have the same genotype but different phenotypes.

A *gene pool* is the totality of the genes or genotypes of all individuals within a 'reproductive community' at a given time. A reproductive community is a community comprising individuals that reproduce sexually (as opposed to asexually) and mate with each other. A *panmictic population*, in which matings occur at random, is the smallest reproductive community. A **deme** is a local population of interbreeding individuals (although it could also be a population of individuals that reproduce asexually).

A **species** is the most inclusive reproductive community. Before the 1990s, it was a mainstay of evolutionary theory that the genetic discontinuities between species are absolute because *isolating mechanisms* prevent sexually reproducing organisms of different species from interbreeding. Many biologists now question this view and call into doubt the reality of species as fully isolated entities (e.g. Mallet 2001; Wu 2001). They see speciation as a process in which groups of organisms gradually become genealogically distinct, rather than a discontinuity that affects all genes at once. This means that boundaries between species are fuzzier than was once thought. In addition, some biologists now feel that total reproductive isolation is not the best definition of a species, a view bolstered, for example, by studies revealing the transfer of genes between closely related fruit fly (*Drosophila*) species (Wu 2001).

Some species display considerable geographical variability in form and possess distinct geographical **races** and ecological races, which are often designated as subspecies. *Geographical races* either live side by side and may grade into each other, or else live separately (that is, have a disjunct distribution). An example is the human species that has marked geographical racial variations – Amerindian, Polynesian, Asiatic, European, African, and so on. *Ecological races* have different genotypes and phenotypes, are interfertile with numerous intergrades in zones of contact, and live together (have sympatric distributions). Species bearing geographical and ecological races are *polytypic* (contain several types). Polytypism is the variability between populations or groups and is often expressed as several races or subspecies. It is not the same as polymorphism, which is the variability within populations. However, polymorphisms may furnish a store of genetic raw material from which polytypisms evolve.

Changing genes

Evolution occurs in populations, not in individuals. It requires changes in DNA, which may arise from processes of gene mutation and chromosomal change. Potential changes in a gene pool arise from several processes: mutation (which creates new alleles and alters chromosomes), genetic drift, gene flow, natural selection, and geographical variation.

Mutation

Mutations are changes in hereditary materials. As a rule, information encoded in a DNA sequence reproduces dependably during replication. Occasionally, 'errors' occur so that the nucleotide sequence in parent and daughter DNA molecules differ. *Gene* (or *point*) *mutation* involves the altering of the DNA sequence of a gene and the passing of the new nucleotide sequence to the offspring. *Chromosome mutation* (or *aberration*) involves the changing of the number of chromosomes, or the number or arrangement of genes in a chromosome (Table 2.1).

Genetic drift

Chance changes in the frequency of alleles, occurring as interbreeding populations exchange genetic material, produces *genetic drift*. Normally, genetic drift is a weak force of genetic change with very little effect on a large population. It may have a significant effect in small populations produced by a few individuals colonizing a new habitat (founder populations), by remnants of a population becoming marooned in refugia, by population crashes causing **bottlenecks** (sudden reductions in genetic diversity), and by metapopulation patches (p. 171) with limited inter-patch

Table 2.1 A classification of chromosome mutations

Change in	Mechanism	Description
Number of genes	Deficiency or deletion	Chromosome loses a segment of DNA containing one or more genes
	Duplication	A DNA segment of one or more genes occurs twice or more in a set of chromosomes
Location of genes	Inversion	Location of a block of genes inverted within a chromosome
	Translocation	Location of a block of genes changes in a chromosome
Number of chromosomes	Fusion	Two non-homologous chromosomes fuse into one
	Fission	A chromosome splits into two
	Aneuploidy	One or more chromosomes of the normal set is either missing or present in excess
	Haploidy and polyploidy	The number of sets of chromosomes is other than two. Most organisms are diploid (have two sets of chromosomes in their somatic cells (but one set in their gametic cells). Some organisms are normally haploid, that is, have one set of chromosomes. Organisms with more than two sets of chromosomes are polyploid, a condition common in many species in some groups of plants

Source: Based on discussion in Dobzhansky *et al.* (1977, 57–8)

gene flow. Genetic drift is possible in large populations where a few individuals monopolize breeding. An example is red deer stags that run harems.

Gene flow

This is the movement of genetic information between different parts of an interbreeding population. High gene flow keeps the gene pool 'well stirred'; low gene flow may lead to genetic divergence in different parts of the population.

Natural selection

According to the latest thinking of some biologists, **natural selection** is a primary driving force of speciation and may be more potent than allopatry (the geographical isolation of populations). The argument is that divergent natural selection, which fine-tunes phenotypes to local environments, may outweigh gene flow, leading to further divergence, and so forth, until speciation is accomplished (Dieckmann and Doebeli 1999; Via 2001).

Selection tests the genetic foundation of individuals, acting directly on the phenotype and indirectly on the genotype. It may be directional, stabilizing, or disruptive (Figure 2.1).

Directional or *progressive selection* drives a unidirectional change in the genetic composition of a population, favouring individuals with advantageous characteristics bestowed by a gene or set of genes (Figure 2.1a). It may occur when a population adapts to a new environment, or when the environment changes and a population tracks the

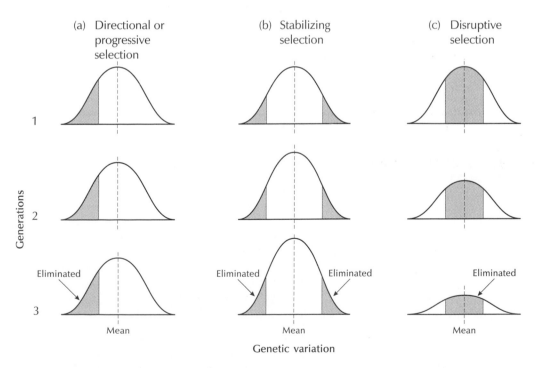

Figure 2.1 Directional, stabilizing, and disruptive selection.
Source: After Grant (1977)

changes. The response of the melanic (dark) form of the peppered moth (*Biston betularia*) when confronted with industrial soot illustrates the first case (Box 2.1). Anita Malhotra and Roger S. Thorpe (1991) believed that they demonstrated selection in action by manipulating natural populations of the Dominican lizard (*Anolis oculatus*). They translocated several ecotypes (a race adapted to local ecological conditions) of the lizard into large experimental enclosures, and monitored them over two months. The rate of survival and non-survival depended on the degree of similarity between the ecological conditions of the enclosure site and the original habitat. The changes in size of the white-tailed deer (*Odocoileus virginianus*) during the Holocene epoch, which appear to track environmental changes, exemplify the second type of directional selection (Purdue 1989).

Stabilizing selection occurs when a population is well adapted to a stable environment. In this case, selection weeds out the ill-adapted combinations of alleles and fixes those of intermediate character (Figure 2.1b). Stabilizing selection is omnipresent and probably the most common mode of selection. The peppered moth's response to industrial pollution illustrated directional selection; but the moth population also displays stabilizing selection. Before industrialization altered the moth's environment, stabilizing selection winnowed out the rare melanic mutants, a situation that probably prevailed for centuries.

Disruptive or *diversifying selection* favours the extreme types in a polymorphic population and eliminates the intermediate types, so encouraging polymorphism (Figure 2.1c). Disruptive selection under experimental conditions occurs in the common fruit fly (*Drosophila melanogaster*) (e.g. Thoday 1972). In nature, at least three situations may promote disruptive selection (Grant 1977, 98–9). The first situation is where well-differentiated polymorphs have a strong selective advantage over poorly differentiated polymorphic types, as in sexually dimorphic species, where males and females that possess distinct secondary sexual characters have a better chance of mating and reproduction than intermediate types (intersexes, homosexuals, and so on). The second situation is where a polymorphic population occupies a heterogeneous habitat. The polymorphic types could be specialized for different subniches in the habitat. This may occur in the sulphur butterfly (*Colias eurytheme*), the females of which species are

Box 2.1

NATURAL SELECTION IN ACTION: INDUSTRIAL MELANISM IN THE PEPPERED MOTH

The peppered moth (*B. betularia*) in Britain is a small, drab moth with three colour phases — whitish, brownish mottled, and dark (almost black) — that exist in British populations. In Victorian insect collections, the lighter and mottled types predominate. In the late nineteenth century, the rare dark phase began to make more appearances, until by the 1920s the species was nearly all black. H. Bernard Kettlewell and his colleagues correlated the change in predominate colour to the industrialization of the Manchester area (e.g. Kettlewell 1973, 131–51). Soot and sulphur discharged from mills and factories blackened tree trunks and killed mottled tree lichens. The lighter and mottled moths stood out clearly on the trunks and were easy pickings for predatory birds, while the darker moths became hard to see. By 1986, a study of 1,825 moths collected throughout Great Britain showed that the area dominated by the black moths is steadily shrinking towards the north-east corner of the country.

polymorphic for wing colour, with one gene controlling orange and white forms. At several localities in California, the white form has an activity peak in the morning and late afternoon, and the orange form has a peak of activity around midday, indicating that the polymorphic types have different temperature and humidity preferences. The third situation occurs when a plant population crosses two different ecological zones. Under these circumstances, different adaptive characteristics may arise in the two halves of the population and persist despite interbreeding. This appears to be the case for whitebark pine (*Pinus albicaulis*), a high-montane species living at and just above the **treeline** in the Californian Sierra Nevada (Clausen 1965). On the mountain slopes up to the timberline, the population grows as erect trees; above the timberline, it grows as a low, horizontal, elfinwood form. The arboreal and elfinwood populations are contiguous and cross-pollinated by wind, as witnessed by the presence of some intermediate individuals.

Geographical and ecological variation

The phenotypes and genotypes of many species display geographical and ecological variations. Two types of geographical variation arise: continuous geographical variation and disjunct geographical variation. Large and continuous populations, such as the human population and populations of many forest trees and plains grasses, have polymorphic local populations within them with characteristic balances of polymorphic types fashioned by gene flow and various kinds of selection. In moving through such a population, the frequencies of the polymorphic type shift little by little. The allele frequencies for human blood groups exhibit this pattern.

Many species adapt to conditions in their local environment, and especially to gradual geographical changes in climate across continents. Such adaptation is often expressed in the phenotype as a measurable change in size, colour, or some other trait. The gradation of form along a

climatic gradient is called a *cline* (Huxley 1942). Clines result from local populations developing tolerances to local conditions, including climate, through the process of natural selection (see Saloman 2002). Commonly observed clines of pigmentation, body size, and so on have generated a set of biogeographical rules (Box 2.2). Morphological clines may evolve very swiftly. In 1852, the house sparrow (*Passer domesticus*) was introduced into the eastern USA from England. Fifty years later it had already developed geographical variation in size and colour. Today it is smallest along the central Californian coast and southeast Mexico, and is largest on the Mexican Plateau, the Rocky Mountains, and the northern Great Plains. The clines in house sparrows that have evolved in North America resemble the clines found in Europe (Johnston and Selander 1971). The American robin (*Turdus migratorius*) displays similar geographical variation in size and shape to the house sparrow (Aldrich and James 1991): it is small in the southeastern USA and along the central Californian coast, and large in the Rocky Mountains and associated high plains. The European wild rabbit (*Oryctolagus cuniculus*), introduced in eastern Australia a little over a century ago, already displays clinal variation in skeletal morphology. The rapidity of clinal evolution revealed by these, and other, examples has been reproduced using genetic models of populations that show, even in the presence of gene flow, that clines can develop within a few generations (Endler 1977).

Disjunct geographical races evolve where a population is discontinuous, comprising a set of island-like, spatially separate subpopulations. For instance, fineflower gilia (*Gilia leptantha*) lives in openings of montane pine forests in southern California. The forests occur at middle and high elevations on mountain ranges separated by many miles of unforested lowlands. In consequence, the plant has a disjunct geographical distribution and four disjunct geographical races (Grant 1977, 166).

Box 2.2

BIOGEOGRAPHICAL RULES

Most biogeographical rules were established during the nineteenth century when it was observed that the form of many warm-blooded animal species varies in a regular way with climate.

Gloger's rule

Proposed by Constantin Wilhelm Lambert Gloger in 1833, this rule states that races of birds and mammals in warmer regions are more darkly coloured than races in colder or drier regions. Or, to put it another way, birds and mammals tend to have darker feathers and fur in areas of higher humidity. This is recognized as a valid generalization about clines of **melanism**. A credible explanation for it is that animals in warmer, more humid regions require more pigmentation to protect them from the light. Gloger's rule was first observed in birds, but was later seen to apply to mammals such as wolves, foxes, tigers, and hares. It has also been observed in beetles, flies, and butterflies. Given that colour variation shows a concordance of pattern in birds that have vastly different competitors, diets, histories, and levels of gene flow, some common physiological adaptation seems likely.

Bergmann's rule or the size rule

Established by Carl Bergmann in 1847, this rule states that species of birds and mammals living in cold climates are larger than their **congeners** that inhabit warm climates. It applies to a wide range of birds and mammals. Bergmann believed many species conform to the rule because big animals have a thermal advantage over small ones in cold climates: as an object increases in size, its surface area becomes relatively smaller (increasing by the square) than its volume (increasing by the cube).

Examples of Bergmann's rule, and exceptions to it, abound. In central Europe, the larger mammals, including the red deer (*Cervus elaphus*), roe deer (*Capreolus capreolus*), brown bear (*Ursus arctos*), fox (*Vulpes vulpes*), wolf (*Canis lupus*), and wild boar (*Sus scrofa*), increase in size towards the northeast and decrease in size towards the southwest. In Asia, larger tigers (*Panthera tigris*) tend to occur at higher latitudes. Species that decline to obey Bergmann's rule include the capercaillie (*Tetrao urogallus*), which is smaller in Siberia than in Germany. Also, many widespread Eurasian and North American bird species are largest in the highlands of the semi-arid tropics (Iran, the Atlas Mountains, and the Mexican Highlands), and not in the coldest part of their range. The geographical variation of size in some vertebrate and invertebrate poikilotherms ('cold-blooded') species conforms to Bergmann's rule (e.g. Lindsey 1966).

The explanation of Bergmann's rule is the subject of much argument, but climate does appear to play a leading role (see Yom-Tov 1993). This is borne out by Frances C. James's (1970) study of bird size and various climatic measures sensitive to both temperature and moisture (wet-bulb temperature, vapour pressure, and absolute humidity) in the eastern USA. She found that wing length, a good surrogate of body size, increased in size northwards and westwards from Florida in the following species: the hairy woodpecker (*Dendrocopos villosus*), downy woodpecker (*D. pubescens*), blue jay (*Cyanocitta cristata*), Carolina chickadee (*Parus carolinensis*), white-breasted nuthatch (*Sitta carolinensis*), and eastern meadowlark (*Sturnella magna*). In all cases, there was a tendency for larger (or longer-winged) birds to extend southwards in the Appalachian Mountains, and for smaller (or shorter-winged) birds to extend northwards in the Mississippi River valley. In the downy woodpecker, female white-breasted nuthatches, and female blue jays, relatively longer-winged birds tended to extend southwards into the interior highlands of Arkansas, and relatively shorter-winged birds to extend northwards into other river valleys. These subtle relations between clinal size variation and topographic features indicated that the link between the two phenomena might involve precise adaptations to very minor climatic gradients. The variation in wing length in these bird species correlated most highly with those variables, such as wet-bulb temperature, which register the combined effects of temperature and humidity. This suggested that size variation depends on moisture levels as well as temperature. James reasoned that a relationship with wet-bulb temperature and with absolute humidity in ecologically different species strongly suggests that a common physiological adaptation is involved. Absolute humidity nearly determines an animal's ability to lose heat: any animal with constant design will be able to unload heat more easily if it has a higher ratio of respiratory surface to body size. This new twist to Bergmann's rule bolsters some aspects of Bergmann's original interpretation about thermal budgets. Climate tends to be cooler, and therefore drier, at high altitudes and latitudes. This accounts for the fact that many clines of increasing size parallel increasing altitude and latitude. Additionally, size tends to increase in arid regions irrespective of altitude and latitude, and widespread species tend to be largest in areas that are high, cool, and dry. James concluded that, if the remarkably consistent pattern of clinal size variation in breeding populations of North American birds represents an adaptive response, then 'Bergmann's original rationale of

thermal economy, reinterpreted in terms of temperature and moisture rather than temperature alone, still stands as a parsimonious explanation' (James 1991, 698).

An observed body-mass decline in several resident bird species in Israel since 1950 presents an interesting demonstration of Bergmann's rule in operation (Figure 2.2). Yoram Yom-Tov (2001) found that the body mass of the yellow-vented bulbul (*Pycnonotus xanthopygos*), house sparrow (*Passer domesticus*), Sardinian warbler (*Sylvia melanocephala*), and graceful prinia (*Prinia gracilis*) showed significant declines during the second half of the twentieth century. Minimum summer temperatures in Israel rose by an average of 0.26°C per decade over the same period.

Allen's rule or the proportional rule

Joel A. Allen's (1877) rule extends Bergmann's rule to include protruding parts of the body, such as necks, legs, tails, ears, and bills. Allen found that protruding parts in wolves, foxes, hares, and wild cats are shorter in cooler regions. Like large body-size, short protruding parts help to reduce the surface area and so conserve heat in a cold climate. The jackrabbit (subgenus *Macrotolagus*), which lives in the southwestern USA, has ears one third its body length; in the common jackrabbit (*Lagus campestris*), which ranges from Kansas to Canada, the ears are the same length as the head. Another observation conforming to Allen's rule is that such mammals as bats, which have a large surface area for their body mass, are found chiefly in the tropics. Allen's rule has been observed in poikilotherms as well as homeotherms.

Guthrie's or Geist's rule

This is a modern biogeographical rule based on the observation that the seasonal amount of food

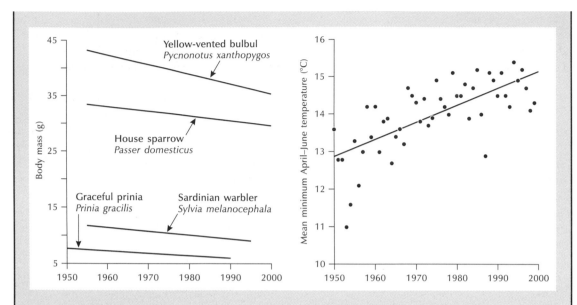

Figure 2.2 Body-mass decline in several resident bird species in Israel since 1950.
Source: Adapted from Yom-Tov (2001)

available influences body size in large mammals. Proposed by R. Dale Guthrie (1984) and Valerius Geist (1987), the basis of this rule is that animals in areas of high seasonal food abundance can achieve a greater proportion of their potential annual growth and therefore develop bigger bodies.

Differentiation into ecological races provides a third kind of intraspecies variation pattern. Many forms of ecological race include altitudinal races in montane species, host races in insects, and seasonal races in organisms with demarcated breeding systems.

SPECIES BIRTH: SPECIATION

Processes of speciation

Speciation is the production of new species. It demands mechanisms for bringing new species into existence and mechanisms for maintaining them and building them into cohesive units of interbreeding individuals that maintain some degree of isolation and individuality. There is a threshold at which *microevolution* (evolution through adaptation within species) becomes *macroevolution* (evolution of species and higher taxa). Once this threshold is traversed, evolutionary processes act to uphold the species' integrity and fine-tune the new species to its niche: gene flow may smother variation; unusual genotypes may be less fertile, or may be eliminated by the environment, or may be looked over by would-be mates. Various mechanisms may thrust a population through the *speciation threshold*, each being associated with a different model of speciation: allopatric speciation, peripatric speciation, stasipatric speciation, and sympatric speciation (Figure 2.3). Evolutionary biologists argue over the effectiveness of each type of speciation (e.g. Losos and Glor 2003).

Allopatric speciation

Geographical isolation reduces or stops gene flow, severing genetic connections between once interbreeding members of a continuous population. If isolated for long enough, the two daughter populations will probably evolve into different species. This mechanism is the basis of the classic model of **allopatric** ('other place' or geographically separate) **speciation**, as propounded by Ernst Mayr (1942) who called it *geographical speciation* and saw geographical subdivision as its driving force. Mayr recognized three kinds

of allopatric speciation: strict allopatry without a population bottleneck; strict allopatry with a population bottleneck; and extinction of intermediate populations in a chain of races:

1 *Strict allopatry without a population bottleneck* occurs in three stages. First, the original population extends its range into new and unoccupied territory. Second, a geographical barrier, such as a mountain range, forms and splits the population into two; and genetic modification affects the separate populations to the extent that if they come back into contact, genetic isolating mechanisms will prevent their reproducing. The model includes cases where species extend their range by traversing an existing barrier, as when birds cross the sea to colonize an island.

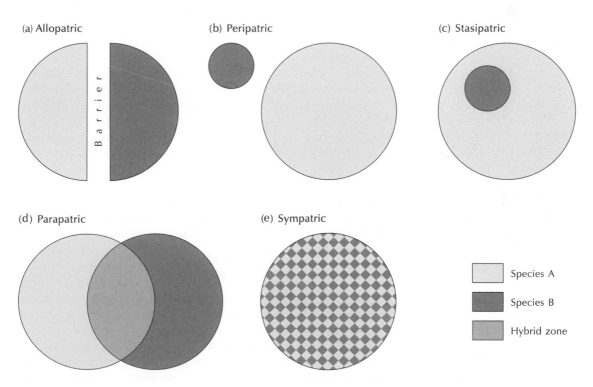

Figure 2.3 Types of speciation: allopatric speciation, peripatric speciation, stasipatric speciation, parapatric, and sympatric speciation.

2 In *strict allopatry with a population bottleneck*, a small band of founder individuals (or even a single gravid female) colonizes a new area. Mayr argued that the founding population would carry but a small sample of the alleles present in the parent population, and the colony would have to squeeze though a genetic bottleneck; he called this the *founder effect*. However, from the 1970s onwards, it became apparent that the bottleneck might not be as tight as originally supposed, even with just a few founding intervals (White 1978, 109). Some 10 to 100 founding colonists may carry a substantial portion of the alleles present in the parent population. Even a single gravid female colonist, providing she is heterozygous at 10–15 per cent of her gene loci and providing an equally heterozygous male (largely of different alleles) fertilized her, could carry a considerable amount of **genetic variability**. Admittedly, a newly founded colony will initially be more homozygous and less polymorphic than the parent population. If the colony should survive, gene mutation should restore the level of polymorphism, probably with new alleles with new allele frequencies.

3 Two species of European gull – the herring gull (*Larus argentatus*) and the lesser black-backed gull (*L. fuscus*) – exemplify the extinction of intermediate populations in a *chain of species* (Figure 2.4). These species are the terminal members of a chain of *Larus* subspecies encircling the north temperate region. Members of the chain change gradually but the end members occur sympatrically in northwest Europe without **hybridization**.

Vicariance events and dispersal-cum-founder events may drive allopatric speciation (Figure 2.5). Two species of North American pines illustrate vicariance speciation. Western North American lodgepole pine (*Pinus contorta*) and eastern North American jack pine (*P. banksiana*) evolved from a common ancestral population that the advancing Laurentide Ice Sheet split asunder some 500,000 years ago. The colonization of the Galápagos **archipelago** from South America by an ancestor of the present giant tortoises (*Geochelone* spp.), probably something like its nearest living relative the Chaco tortoise (*G. chilensis*), is an example of a dispersal and founder event. The giant tortoises are all the same species on the Galápagos – *G. elephantopus* – but there are 11 living subspecies and 4 extinct subspecies. Five subspecies occur on Isabela and the rest occur on different islands.

Once populations are isolated and become differentiated, they may then stay isolated and never come into contact again. In this case, it may take a long time for reproductive isolation to occur and it is difficult to know when the two populations are different species. If contact is re-established, perhaps because the barrier disappears, then three things may happen: (1) the populations may not interbreed, or fail to produce fertile offspring, reproductive isolation is complete, and speciation has occurred. This happened with the kaka and kea in New Zealand. (2) The two populations may hybridize, but the hybrids may be less fit than the offspring of the within-populations matings. *Reinforcement* is the process that selects within-population matings, and *isolating mechanisms* are traits that evolve to augment reproductive isolation. (3) The two populations may interbreed comprehensively, producing fertile fit hybrids, so that the populations merge and the differentiation is diluted, and eventually disappears.

Peripatric speciation

This is a subset of allopatric speciation. **Peripatric speciation** occurs in populations on the edge (perimeter) of a species range that become isolated and evolve divergently to create new species. A small founding population is often involved. An excellent example of this is the paradise kingfishers (*Tanysiptera*) of New Guinea (Mayr 1942). The main species, the common paradise kingfisher (*T. galatea galatea*), lives on the

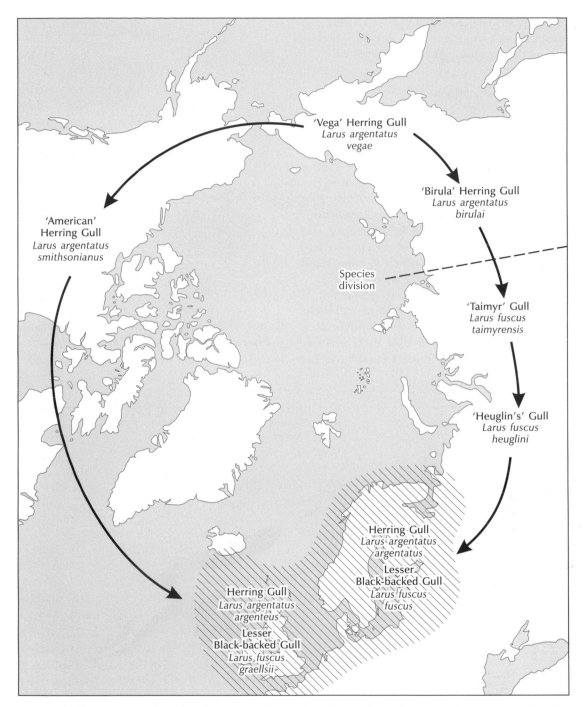

Figure 2.4 Chain or ring of species of gull around the North Pole. Change between subspecies is gradual but the two end members live sympatrically in northwest Europe.

main island. The surrounding coastal areas and islands house a legion of morphologically distinct races of the paradise kingfishers (Figure 2.6).

Parapatric speciation

Parapatric (abutting) **speciation** is the outcome of divergent evolution in two populations living geographically next to each other. The divergence occurs because local adaptations create genetic gradients or clines. Once established, a cline may reduce gene flow, especially if the species is a poor disperser, and selection tends to weed out hybrids and increasingly pure types that wander and find themselves at the wrong end of the cline. A true hybrid zone may develop, which in some cases, once reproductive isolation is effective, will disappear to leave two adjacent species. An example is the main species of the house mouse in Europe (Hunt and Selander 1973). A zone of hybridization separates the light-bellied eastern house mouse (*Mus musculus*) of eastern Europe and the

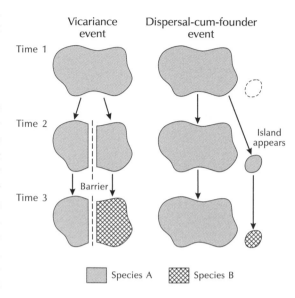

Figure 2.5 The chief drivers of allopatric speciation: vicariance events and dispersal-cum-founder events.
Source: Adapted from Brown and Lomolino (1998)

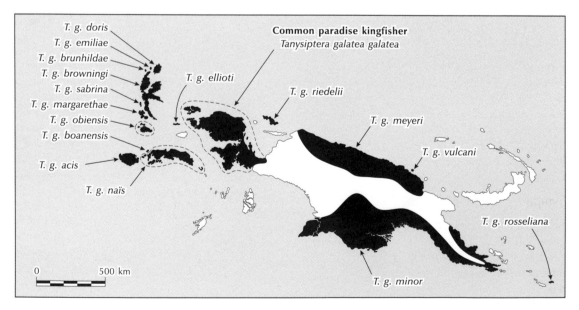

Figure 2.6 Races of paradise kingfishers (*Tanysiptera*) on New Guinea and surrounding islands.
Source: Partly adapted from Mayr (1942)

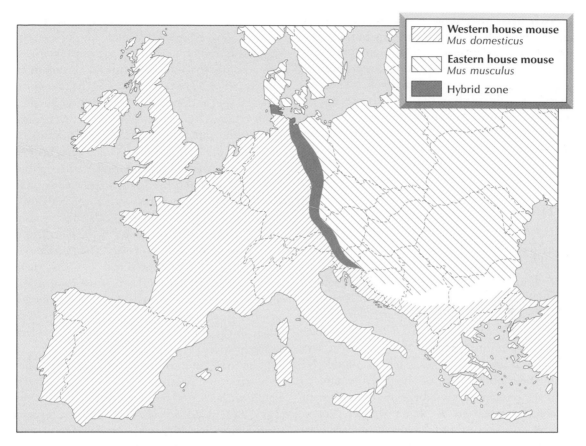

Figure 2.7 The distribution of the light-bellied eastern house mouse (*M. musculus*) of eastern Europe and the dark-bellied western house mouse (*M. domesticus*) of western Europe (which some authorities treat as a subspecies of *M. musculus*) and the zone of hybridization between them.

Source: Adapted from distribution maps in Mitchell-Jones *et al.* (1999)

dark-bellied western house mouse (*M. domesticus*) of western Europe (Figure 2.7).

Sympatric speciation

Sympatric speciation occurs within a single geographical area and the new species overlap – there is no spatial separation of the parent population. Separate genotypes evolve and persist while in contact with each other. Once deemed rather uncommon, new studies suggest that parapatric and sympatric speciation may be a potent process of evolution (e.g. Via 2001).

Several processes appear to contribute to sympatric speciation (p. 14). Disruptive selection favours extreme phenotypes and eliminates intermediate ones. Once established, natural selection encourages reproductive isolation through *habitat selection* or *positive assertive mating* (different phenotypes choose to mate with their own kind). Habitat selection in insects may have favoured sympatric speciation and account for much of the large diversity of that group. *Competitive selection*, a variant of disruptive selection, favours phenotypes within a species that avoid intense competition and clears out intermediate types.

Two irises that grow in the southern USA appear to result from sympatric speciation. The giant blue iris (*Iris giganticaerulea*) grows in damp meadows, while its close relative, the copper iris (*I. fulva*), grows in drier riverbanks. The two species hybridize, but the hybrids are not as successful in growing on either the dry or very wet sites as the pure forms and usually perish. So, the inability of hybrids to survive restricts the gene flow between the two species.

Stasipatric speciation

Stasipatric speciation occurs within a species range owing to chromosomal changes. Chromosomal changes occur through: (1) a change in chromosome numbers, or (2) a rearrangement of genetic material on a chromosome (an inversion) or a transferral of some genetic material to another chromosome (a translocation). *Polyploidy* doubles or more the normal chromosome component, and **polyploids** are often larger and more productive than their progenitors. Polyploidy is rare in animals but appears to be a major source of sympatric speciation in plants: 43 per cent of dicotyledon species and 58 per cent of monocotyledon species are polyploids. Stasipatric speciation seems to have occurred in some western house mouse (*M. domesticus*) populations (see p. 22). In Europe, the normal karyotype for the species contains 20 sets of chromosomes. Specimens with 13 sets of chromosomes were first discovered in southeast Switzerland in the Valle di Poschiavo. At first, these were classed as a new species and designated *M. poschiavinus*, the tobacco mouse. Later, specimens from other alpine areas of Switzerland and Italy (as well as from northern Africa and South America) also had non-standard karyotypes. Surprisingly, all the populations showed no morphological or genetic differences other than differences in their karyotypes and all belonged to *M. domesticus*.

Subterranean mole rats of the *Spalax ehrenbergi* complex living in Israel provide an outstanding example of a species' overall adaptive response to climate (Nevo 1986). These mole rat populations comprise four morphologically identical incipient chromosomal species (with **diploid** chromosome numbers 2n = 52, 54, 58, and 60). The four chromosomal species appear to be evolving and undergoing ecological separation in different climatic regions: the cool and humid Galilee Mountains (2n = 52), the cool and drier Golan Heights (2n = 54), the warm and humid central Mediterranean part of Israel (2n = 58), and the warm and dry area of Samaria, Judea, and the northern Negev (2n = 60). All the species are adapted to a subterranean ecotype: they are little cylinders with short limbs and no external tail, ears, or eyes. Their size varies according to heat load, presumably so that there is only a small risk of overheating under different climates: large individuals live in the Golan Heights; smaller ones in the northern Negev. The colour of the mole rats' pelage ranges from dark on the heavier black and red soils in the north, to light on the lighter soils in the south. The smaller body size and paler pelage colour associated mainly with 2n = 60 helps to mitigate against the heavy heat load in the hot steppe regions approaching the Negev desert. The mole rats show several adaptations at the physiological level. **Basal metabolic rates** decrease progressively towards the desert. This minimizes water expenditure and the chances of overheating. More generally, the combined physiological variation in basal metabolic rates, non-shivering heat generation, body-temperature regulation, and heart and respiratory rates, appears to be adaptive at both the mesoclimatic and microclimatic levels, and both between and within species, so contributing to the optimal use of energy. Ecologically, territory size correlates negatively, and population numbers correlate positively, with productivity and resource availability. Behaviourally, activity patterns and habitat selection appear to optimize energy balance, and differential swimming ability appears to overcome winter flooding, all paralleling the climatic origins of the different species. In summary, the incipient species are reproductively isolated to varying degrees, representing different adaptive systems that can be viewed genetically, physiologically, ecologically, and behaviourally. All are adapted to climate, defined by humidity and temperature regimes, and ecological speciation is correlated with the southward increase in aridity stress.

Lineages and clades

Several terms pertain to the study of species in the past (Figure 2.8). A *lineage* is single line of descent. One speaks of the human lineage and the reptilian lineage. **Extinction** is the termination of a lineage and marks the end of the line – a lineage that failed to survive to the present. Extinction is the eventual fate of all species, as discussed later in this chapter. In the fossil record, *speciation* is the branching of lineages. In other words, it marks the point where a single line of descent splits into two lines that diverge from their common ancestor. It occurs when a part of a population becomes reproductively isolated from the remaining populations of the established species by any of the mechanisms mentioned in the previous section. It is well nigh impossible to reconstruct the isolating mechanisms involved in fossil populations, although evidence may sometimes exist for geographical isolation.

Fossil assemblages traced through time reveal clades. A **clade** is a cluster of lineages produced by repeated branching (speciation) from a single lineage. The branching process that generates clades is *cladogenesis*. The clade Elephantinae comprises two extinct genera – *Primelephas* and *Mammuthus* – and two living genera – *Elephas* (modern Asian elephants) and *Loxodonta* (modern African elephants). Evidence from mtDNA suggests that *Elephas*, *Loxondonta*, and *Mammuthus* started to differentiate in the Late **Miocene epoch** around 5.6–7.0 million years ago. This supports fossil

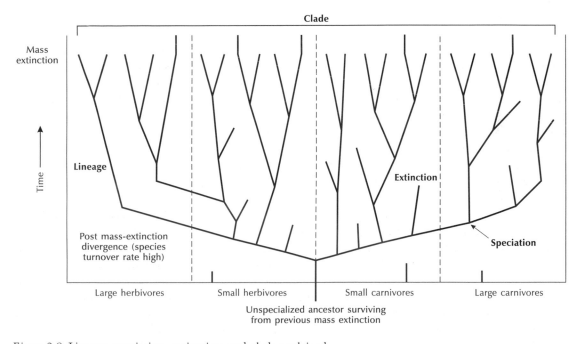

Figure 2.8 Lineage, speciation, extinction, and clade explained.

finds since 1980: the oldest known *Loxodonta* specimen comes from Baringo, Kenya, and Nkondo-Kaiso, Uganda, between 7.3 and 5.4 million years ago; and the oldest known *Elephas* comes from Lothagam, Kenya, about 6–7 million years ago (Tassy and Debruyne 2001).

Evolution

The fossil record permits the inference of evolutionary changes in organisms. Evolution that generates new species to create a clade or a group of clades is *cladogenesis* or *phylogenetic evolution*. Phylogeny is the origin of the branches and the 'trees of life' and encompasses the relationships between all groups of organisms as seen in the genealogical links between ancestors and descendants.

Phylogenetic evolution requires an answer to a difficult question – how do species relate to each other? *Cladistics* is a method of biological classification that attempts to find phylogenetic relationships by constructing branching diagrams based on shared derived characters (synapomorphies). Willi Henning (1966), a German entomologist, laid down its basic principles. Henning defined relationships using branching diagrams, which he saw as evolutionary trees. He contended that only shared derived characters (*synapomorphies*) betray a close common ancestry, and shared primitive characters (*symplesiomorphies*), inherited from a remote common ancestor, are irrelevant or misleading when seeking phylogenetic relationships. In addition, he recognized characters unique to any one group (*autapomorphies*). To illustrate these ideas, consider Figure 2.9, which represents the phylogenetic relationships of the New World monkeys. The vertical bars show synapomorphies. For instance, bimanual locomotion is a derived character shared by the woolly monkey, woolly spider monkey, and spider monkey. The three shades of circles show autapomorphies, that is, characters unique to a group. Nocturnality is an autapomorphy of the owl monkey, and tool use an autapomorphy of the capuchin.

Hennig set his ideas in an evolutionary framework – his branching diagrams are *evolutionary trees* with an implicit time dimension and with forks marking the splitting of ancestral species. The diagram of the New World monkey relationships (Figure 2.9) may be seen in this way. However, it is possible to look at branching diagrams in a more general way that has no evolutionary connotations. They can be seen as *cladograms* with no timescale and the nodes simply imply shared characters (synapomorphies). The differences between evolutionary trees and cladograms may appear minor, but they are hugely important (Patterson 1982). A cladogram is a summary the pattern of character distributions among taxa, in which the nodes are shared characters and the lines are immaterial, the relationships being expressible as a Venn diagram (Figure 2.10). An evolutionary tree is a summary of pattern plus a summary of the historical process of descent with modification (evolution) that created the pattern of characters, in which the nodes are real (if not also identifiable) ancestors, the forks are speciation events, and the lines are lineages of descent by modification.

Phylogenetic evolution contrasts with *phyletic evolution* (*anagenesis*, *chronospeciation*), in which an established species slowly changes into another species within the same lineage. The new species produced in this way go by a variety of names: *chronospecies*, *palaeospecies*, and *evolutionary species*. The recognition of chronospecies is arbitrary and subjective: it is assumed that, in Europe, *Mammuthus primigenius* evolved by phyletic evolution from *M. armeniacus*, which evolved by phyletic evolution from *M. meridionalis*, but it is somewhat arbitrary where the dividing lines between the chronospecies are placed. When a chronospecies changes into a new form, *pseudoextinction* (sometimes called *phyletic extinction*) occurs. Thus *M. meridionalis* became pseudoextinct when it evolved into *M. armeniacus*.

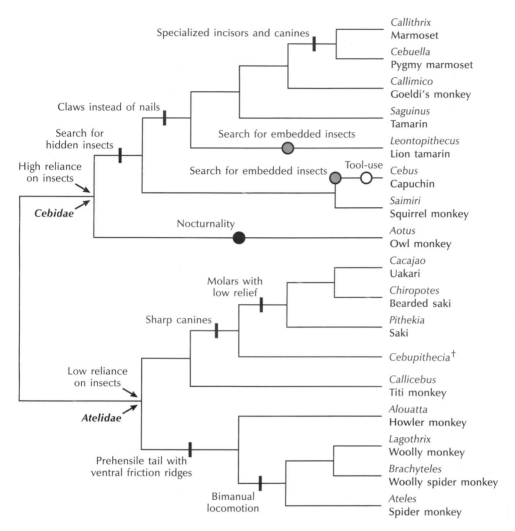

Figure 2.9 Phylogenetic relationships in the New World monkeys.
Source: Adapted from Horovitz and Meyer (1997)
† extinct

Evolution works through adaptation, a process in which natural selection shapes a character for current use. Occasionally, a *key innovation* is thrown up that permits species to exploit a lifestyle novel to that taxon or, sometimes, new to any form of life. Such key innovations may help to trigger **adaptive radiation** (see p. 29). Innovations may be brand new or they can be exaptations. *Exaptations* are characters acquired from ancestors that are co-opted for a new use. An example is the blue-tailed gliding lizard (*Holaspis guentheri*) from tropical Africa that has a flattened head, which allows it to hunt and hide in narrow crevices beneath bark. The flattened head also allows it to glide from tree to tree. The head flattening was originally an

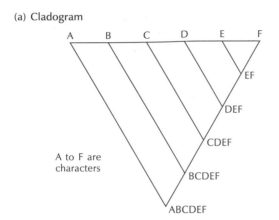

(a) Cladogram

A to F are characters

(b) Venn diagram

(c) Written form

$$(A(B(C(D(EF)))))$$

Figure 2.10 Relationships between hypothetical species expressed as (a) a cladogram, (b) a Venn diagram, and (c) in written form. All species share characters A to F.

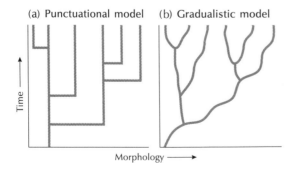

(a) Punctuational model   (b) Gradualistic model

Time

Morphology

Figure 2.11 Punctuational and gradualistic macroevolution.

adaptation to crevice use and was later co-opted for gliding (an exaptation) (Arnold 1994).

A big debate surrounds the question of evolutionary patterns shown by clades and groups of clades. Two extreme cases arise: (1) all evolution is concentrated in speciation (the branching of lineages) and (2) all evolution takes place within lineages (Figure 2.11). The first case leads to the *punctuational model* of evolution, with most major evolutionary transitions occurring at speciation events. The second cases give rise to the *gradualistic model* of evolution, with most evolution occurring as phyletic change and rapid divergent speciation playing a minor role.

SPECIES GROWTH: DIVERSIFICATION

Ecological diversification

This is the fate of nearly all newly formed species. Species are similar to each other after a speciation event, but are likely to diverge when exposed to different environments with different selective pressures. The biogeographical equivalent of the competitive exclusion principle (p. 192) states that species within similar niches have non-overlapping geographical distributions, whereas species that coexist in the same area and habitat tend to use significantly different resources. A striking example of this is *sibling* or *cryptic species* that are genetically distinct but very close in ecology and morphology. Sibling species commonly display abutting, but non-overlapping (parapatric) **geographical ranges**, numerous examples coming from animals and plants. Several species of pocket gophers (of the genera *Thomomys* and *Geomys*) have ranges that come into contact in North America but do not overlap. Where species ranges do overlap, then there are normally big differences in resource use achieved through slight differences in niches (seen in form, physiology, and behaviour) that evolve through character displacement (p. 198).

Adaptive radiation

Adaptive radiation is the diversification of species to fill a wide variety of ecological niches, or the 'rise of a diversity of ecological roles and attendant adaptations in different species within a lineage' (Givnish and Sytsma 1997, xiii). It is one the most important processes bridging ecology and evolution. It occurs when a single ancestor species diverges, through recurring speciation, to create many kinds of descendant species that become or remain sympatric. These species tend to diverge to avoid interspecific competition. Even when allopatric species are generated, some divergence still occurs as the allopatric species adapt to different environments.

Examples of adaptive radiation are legion. Darwin's finches (Geospizinae) on the Galápagos Islands are a famous example (Figure 2.12). A single ancestor, possibly the blue-black grassquit

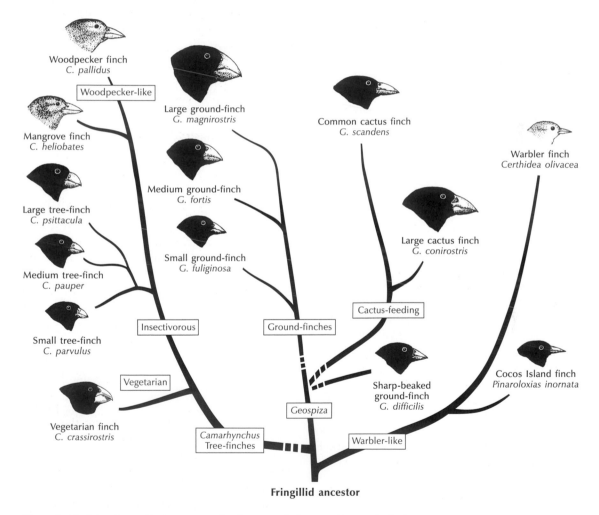

Figure 2.12 Adaptive radiation seen as beak size and shape in Darwin's finches.
Sources: Adapted from Grant (1986) and Lack (1947)

(*Volatinia jacarina*), colonized the archipelago from South America around 100,000 years ago. Allopatric speciation resulting from repeated episodes of colonization and divergence within the island group created 5 genera and 13 species. The beaks of the different species match their diet – seed eaters, insect eaters, and a bud eater.

The Hawaiian Islands have fostered adaptive radiations. The Hawaiian honeycreepers (Drepanidinae) were originally thought to have radiated from a single ancestral seed-eating finch from Asia to give 23 species in 11 genera. It is now known that many more species formed part of the radiation in the recent past, with 29–33 recorded in historical times and 14 as subfossil remains. The radiation produced seed eaters, insect eaters, and nectar eaters, all with appropriately adapted beaks. The Hawaiian silversword alliance, described as the most remarkable example of adaptive radiation in plants, displays an extreme and rapid divergence of form and physiology. The common ancestor of the silversword alliance, which split from Californian tarweeds about 13–15 million years ago, arrived in Hawaii some 4–6 million years ago. It has produced a wide range of plants that spans almost the complete gamut of environmental conditions found on Hawaii, with an altitudinal range from 75 to 3,750 m. The forms include acaulescent (stemless) or short-stemmed, monocarpic (flowering and bearing fruit only once before dying) or polycarpic (producing flowers and fruit several times in one season) rosette plants; long-stemmed, monocarpic or polycarpic rosettes plants; trees, shrubs, and sub-shrubs; mat plants, cushion plants; and lianas.

Lemurs in Madagascar are the product of an adaptive radiation in primates that began with the arrival of a common ancestor some 50 million years ago (Tattersall 1993). At least 45 species lived in the recent past, around 2,000 years ago when humans first arrived on the island; some 33 survive today in 14 genera (Table 2.2) The true lemurs comprise five arboreal (tree-living), vegetarian species that eat fruits, flowers, and leaves. Sportive lemurs are nocturnal and move mainly by jumps. Mouse lemurs (*Microcebus*) are small (up to 60 g), run like rodents, and eat insects as well as fruits. The indri and sifakas (*Propithecus*) are large animals (up to 1 m long). The aye-aye specializes in prising insects larvae from tree bark and fills the niche of woodpeckers. At least 15 species of subfossil lemur species in 8 or more genera reveal the 'big' end of the radiation. *Archaeolemur* lived on the ground and was about the size of a female baboon. The 77-kg *Megaladapis* was arboreal with a niche similar to that of a koala. At 60 kg, *Palaeopropithecus* was a sloth-like tree-dweller.

Non-radiative and non-adaptive radiation

Not all adaptations are radiative and not all are adaptive. *Non-radiative 'radiation'* occurs when vacant niche space permits a sort of ecological release involving diversification but not speciation within a lineage. An example is 'o'hia lehua (*Metrosideros polymorpha*). This Hawaiian tree species is very diverse and has a wide range of forms. It occupies bare lowlands to high bogs, occurs as a small shrub on young lava flows and as a good-sized tree in a canopy of mature forest. But it is ascribed to a single species despite such a rich variety of forms.

Non-adaptive radiation occurs where radiation is associated with no clear niche differentiation. It may occur when radiations have occurred allopatrically in fragmented habitats. For instance, on Crete, land snails of the **genus** *Albinaria* have diversified into a species-rich genus with little niche differentiation. All species occupy roughly the same or only a narrow range of habitats, but rarely do any two *Albinaria* species live in the same place.

Convergent evolution and parallel evolution

Convergent evolution is the process whereby different species independently evolve similar traits as a result of similar environments or

Table 2.2 Living lemurs

Genus	Number of species in genus	Species
Woolly lemurs (*Allocebus*)	2	Hairy-eared dwarf lemur (*Allocebus trichotis*)
		Eastern woolly lemur (*Avahi laniger*)
Western woolly lemur (*Avahi*)	1	Western woolly lemur (*Avahi occidentalis*)
Dwarf lemurs (*Cheirogaleus*)	2	Greater dwarf lemur (*Cheirogaleus major*)
		Fat-tailed dwarf lemur (*Cheirogaleus medius*)
Aye-ayes	1	Aye-aye (*Daubentonea madagascariensis*)
		[One extinct species]
True lemurs (*Eulemur*)	5	Crowned lemur (*Eulemur coronatus*)
		Brown lemur (*Eulemur fulvus*)
		Black lemur (*Eulemur macaco*)
		Mongoose lemur (*Eulemur mongoz*)
		Red-bellied lemur (*Eulemur rubriventer*)
Gentle lemurs (*Hapalemur*)	3	Golden bamboo lemur (*Hapalemur aureus*)
		Grey gentle lemur (*Hapalemur griseus*)
		Broad-nosed gentle lemur (*Hapalemur simus*)
Indris (*Indri*)	1	Indri (*Indri indri*)
Ring-tailed lemurs (*Lemur*)	1	Ring-tailed lemur (*Lemur catta*)
Sportive lemurs (*Lepilemur*)	7	Grey-backed sportive lemur (*Lepilemur dorsalis*)
		Milne-Edwards' sportive lemur (*Lepilemur edwardsi*)
		White-footed sportive lemur (*Lepilemur leucopus*)
		Small-toothed sportive lemur (*Lepilemur microdon*)
		Weasel sportive lemur (*Lepilemur mustelinus*)
		Red-tailed sportive lemur (*Lepilemur ruficaudatus*)
		Northern sportive lemur (*Lepilemur septentrionalis*)
Mouse lemurs (*Microcebus*)	4	Grey mouse lemur (*Microcebus murinus*)
		Pygmy mouse lemur (*Microcebus myoxinus*)
		Golden-brown mouse lemur (*Microcebus ravelobensis*)
		Red mouse lemur (*Microcebus rufus*)
Mirza	1	Coquerel's dwarf lemur (*Mirza coquereli*)
Phaner	1	Fork-crowned dwarf lemur (*Phaner furcifer*)
Sifakas (*Propithecus*)	3	Diademed sifaka (*Propithecus diadema*)
		Tattersall's sifaka (*Propithecus tattersalli*)
		Verreaux's sifaka (*Propithecus verreauxi*)
Ruffed lemurs (*Varecia*)	1	Ruffed lemur (*Varecia variegata*)
		[One recently extinct species]

selection pressures. For instance, sharks and bony fish, whales and dolphins, and extinct sea-going reptiles (ichthyosaurs) all evolved streamlined, torpedo-shaped bodies for cutting through water.

Parallel evolution (or **parallelism**) refers to changes in two closely related stocks that differ in minor ways and that both go through a similar series of evolutionary changes. It is similar to convergence, except that in convergence the original species are from very different stocks, unlike the stocks in parallel evolution, which are similar to start with. Marsupials and placental mammals are a case in point, though sometimes thought of as a case of convergent evolution.

SPECIES DEATH: EXTINCTION

Extinction is the doom of the vast majority of species (or genera, families, and orders); it is the rule, rather than the exception. A **local extinction** or **extirpation** is the loss of a species or other taxon from a particular place, but other parts of the gene pool survive elsewhere. The American bison (*Bison bison*) is now extinct over much of its former range, but survives in a few areas (p. 182). A **global extinction** is the total loss of a particular gene pool. When the last dodo died, its gene pool was lost forever. Supraspecific groups may suffer extinctions. An example is the global extinction of the sabre-toothed cats, one of the main branches of the cat family. A **mass extinction** is a catastrophic loss of a substantial portion of the world's species. Mass extinctions stand out in the fossil record as times when the extinction rate runs far higher than the background or normal extinction rate (p. 333). Some 99.99 per cent of all extinctions are normal extinctions.

The 'life-expectancy' of species varies between different groups. The fossil record suggests that mammal genera last about 10 million years, with primate genera enduring only 5 million years. Individual species survive even less time, something around 1 to 2 million years for complex animals. On the other hands, 'living fossils' appear to have persisted for ages with little change. Examples are the horseshoe crab (*Limulus* spp.), a relative of the spiders, which has lived and changed little for at least 300 million years; **cycads**, which are 'living fossil' plants surviving from the **Mesozoic era** (p. 64); and the ginkgo, which is remarkably similar to specimens that lived around 100 million years ago (Zhou and Zheng 2003). Probably the most famous 'living fossils' are the coelacanths – *Latimeria chalumnae* was found in 1938, *L. menadoensis* in 1998, and an as yet unnamed species in 2000. Coelacanths have persisted nearly unchanged for 70 million years.

Periods of rapid climatic change, sustained volcanic activity, and asteroid and comet impacts seem to cause mass extinctions. Normal extinctions depend on many interrelated factors that fall into three groups – biotic, evolutionary, and abiotic.

Biotic factors

Most biotic factors of extinction are **density-dependent factors**. This means their action depends upon population size (or density). The larger the population, the more effective is the factor. Density-dependent factors are chiefly biotic in origin. They include factors related to biotic properties of individuals and populations (body size, niche size, range size, population size, generation time, and dispersal ability) and factors related to interactions with other species (competition, disease, parasitism, predation).

Biotic properties

Body size, niche size, and range size all affect the probability of extinction. As a rule, large animals are more likely to become extinct than small animals. Smaller animals can probably better adapt to small-scale habitats when the environment changes. Large animals cannot so easily find suitable habitat or food resources and so find it more difficult to survive. Specialist species with narrow niches are more vulnerable to extinction than are generalists with wide niches.

Small populations are more prone to extinction through chance events, such as droughts, than are large populations. In other words, there is safety in numbers. Tropical birds living in patches of Amazon forest show that populations of 50 or more are about 5 times less likely to go extinct locally than are populations of 5 or fewer. Species with rapid generation times stand more chance of dodging extinction. Good dispersers are better placed to escape extinction that poor dispersers, as are species with better opportunities for dispersal. In addition, a species with a large gene pool may be better able to adapt to environmental changes than species with a small gene pool.

Geography can be important – widespread species are less likely to go extinct than species with restricted ranges. This is because restricted range species are more vulnerable to chance events, such as a severe winter or drought. In a widespread species, severe events may cause local extinctions but are not likely to cause a global extinction. This generalization is borne out by defaunation experiments on red mangrove (*Rhizophora mangle*) islands in the Florida Keys, USA, where the insects, spiders, mites, and other terrestrial animals were exterminated with methyl bromide gas (see Simberloff and Wilson 1970). Analysis of the data revealed that the probability of invertebrate extinctions decreased with the number of islands occupied (Hanski 1982). It should be pointed out that a widespread distribution is not a guarantee of extinction avoidance. The passenger pigeon (*Ectopistes migratorius*) and the American chestnut (*Castanea dentata*) were abundant with widespread distributions in eastern North America in the nineteenth century, but suffered *range collapse* and extinction in the case of the passenger pigeon (p. 181) and near extinction in the case of the American chestnut (p. 151) within 100 years. The bison (*B. bison*) (p. 182), trumpeter swan (*Olor buccinator*), whooping crane (*Grus americana*), and sandhill crane (*G. canadensis*), also once widely distributed in North America, have suffered range collapses.

Widespread species also appear to be less at risk than restricted species to mass extinctions. For instance, extinction rates of marine bivalves and gastropods that lived along the Atlantic and Gulf coastal plains of North America in the late **Cretaceous period** increased with species range (Jablonski 1986).

Biotic interactions

Competition can be a potent force of extinction. Species have to evolve to outwit their competitors, and a species that cannot evolve swiftly enough is in peril of becoming extinct.

Virulent **pathogens**, such as viruses, may evolve or arrive from elsewhere to destroy species. The fungus *Phiostoma ulmi*, which is carried mainly by the Dutch elm beetle (*Scolytus multistriatus*), causes Dutch elm disease. Starting in the Netherlands, Dutch elm disease spread across continental Europe and into the USA during the 1920s to 1940s, ravaging the elm populations. After a decline in Europe (but not in the USA), it re-emerged as an even more virulent form (described as a new species – *Ophiostoma novo-ulmi*) in the mid-1960s to affect Britain and most of Europe.

Predators at the top of **food chains** are more susceptible to a loss of resources than are herbivores lower down. A chief factor in the decline of tigers is not habitat loss or poaching, but a depletion of the **ungulate** prey base throughout much of the tigers' range (Karanth and Stith 1999).

Island mammal, bird, and reptile populations are especially vulnerable to all sorts of competitive and predatory **introduced species** (Table 2.3). Since 1600 (and up to the late 1980s), 113 species of birds have become extinct. Of this total,

Table 2.3 Recorded extinctions of mammals, birds, and reptiles, 1600 to 1983

Taxon	Mainland[a]	Island[b]	Ocean	Total extinctions	Approximate number of species in taxon	Percentage of taxon lost
Mammals	30	51	4	85	4,000	2.1
Birds	21	92	0	113	9,000	1.3
Reptiles	1	20	0	21	6,300	0.3

Source: Adapted from Reid and Miller (1989)
Notes: a Landmasses 1 million km² (the size of Greenland) or larger. b Landmasses less than 1 million km²

21 were on mainland areas and 92 on islands (Reid and Miller 1989). In many cases, numerous species of sea birds survive only on outlying islets where introduced species have failed to reach. The story for mammals and reptiles is similar.

Evolutionary factors

Several evolutionary changes may, by chance, lead to some species being more prone to extinction than others. *Evolutionary blind alleys* arise when a loss of genetic diversity during evolution fixes species into modes of evolutionary development that become lethal. A species may evolve on an island and not possess the dispersal mechanisms to escape if the island should be destroyed or should experience climatic change. Some species may become overspecialized through adaptation and fall into *evolutionary traps*. Faced with environment change, overspecialized species may be unable to adapt to the new conditions, their over-specialization serving as a sort of evolutionary straitjacket that keeps them 'trapped'. An interesting upshot of this idea is that species alive today must be descendants of non-specialized species. Behavioural, physiological, and morphological complexity, as varieties of specialization, also appear to render a species more prone to extinction. Simple species – marine bivalves for example – survive for about 10 million years, whereas complex mammals survive for 3 million years or less.

Abiotic factors

Abiotic factors of extinction are usually **density-independent factors**, which means that they act uniformly on populations of any size. Density-independent factors tend to be physical in origin – climatic change, sea-level change, flooding, asteroid and comet impacts, and other catastrophic events. These factors often produce fluctuations in population size that can end in extinction. Take the example of the song thrush (*Turdus philomelos*). This bird lives throughout the British Isles except Shetland (Venables and Venables 1955). It was absent from Shetland in the nineteenth century but established a colony on the island in 1906, breeding near trees, which were scarce in Shetland, the largest group being planted in 1909. By the 1940s, about 24 of breeding pairs inhabited the island. The severe winter of 1946–7 reduced the population to some three or four pairs from then until 1953. Somewhere between 1953 and 1969 the Shetland's song thrushes died out.

Abiotic factors are usually implicated in mass extinctions. However, several researchers stress the potential role of diseases as drivers of mass extinctions. Lethal pathogens carried by the dogs, rats, and other animals associated with migrating humans may have caused the **Pleistocene epoch** mass extinctions (MacPhee and Marx 1997) (p. 325). Similarly, it is possible that the terminal Cretaceous extinction event might have resulted from changes of palaeogeography, in which land connections created by falling sea-levels allowed massive migrations from one landmass to another, leading to biotic stress in the form of predation and disease:

> The shallow oceans drained off and a series of extinctions ran through the saltwater world. A monumental immigration of Asian dinosaurs streamed into North America, while an equally grand migration of North American fauna moved into Asia. In every region touched by this global intermixture, disasters large and small would occur. A foreign predator might suddenly thrive unchecked, slaughtering virtually defenseless prey as its population multiplied beyond anything possible in its home habitat. But then the predator might suddenly disappear, victim of a disease for which it had no immunity. As species intermixed from all corners of the globe, the result could only have been global biogeographical chaos.
>
> (Bakker 1986, 443)

SUMMARY

The creation of new species, their diversification, and ultimate extinction, all of which take place in an ever-changing environment, are the root processes behind biogeographical patterns. Changes in gene pools occasioned by mutation, genetic drift, gene flow, and natural selection drive the formation of new species and subspecies. In adapting to local environments, some species produce clines – geographical variations in particular characteristics. The mechanisms of speciation are complex and open to considerable debate. Biologists recognize several types of speciation – allopatric, peripatric, parapatric, sympatric, and stasipatric. Looked at over geological timescales, speciation is the branching of lineages (lines of descent), with extinctions marking the end of lineages. Clades are clusters of lineages formed by repeated branching or speciation events, a process called cladogenesis or phylogenetic evolution. Speciation within a lineage is phyletic evolution and produces chronospecies. Species diversification involves the initial separation of a new species driven by ecological factors and adaptive radiation, a key evolutionary process by which species diverge to take different ecological roles. Convergent evolution and parallel evolution arise from species experiencing the same environmental pressures in geographically separate regions coming to look alike. Extinction occurs locally and regionally, when it is an extirpation, and globally. Global extinction is the ultimate fate of all species. Extinction occurs because of biotic, evolutionary, and abiotic factors. Biotic factors include body size, range size, population size, dispersal ability, competition, disease, and predation. Evolutionary factors are a question of luck – some species during their evolution happen to acquire a characteristic that leaves them in evolutionary blind alleys or traps. Abiotic factors include climatic change, sea-level change, asteroid impacts, and other catastrophic events.

ESSAY QUESTIONS

1 **To what extent is geographical isolation necessary for the creation of new species?**

2 **How does adaptive radiation link ecological processes with evolutionary processes?**

3 **Why are some species more susceptible to extinction than others?**

FURTHER READING

Howard, D. J. and Berlocher, S H. (eds) (1998) *Endless Forms: Species and Speciation*. New York: Oxford University Press.

A collection of essays on the theme of generating new species and diversity. Not easy but rewarding.

Lawton, J. H. and May, R. M. (eds) (1995) *Extinction Rates*. Oxford: Oxford University Press.

Lots of recent figures on extinction rates.

Schilthuizen, M. (2001) *Frogs, Flies, and Dandelions: Speciation – the Evolution of New Species*. New York: Oxford University Press.

An engaging and informative introduction to the subject. Ideal for beginners.

Schluter, D. (2000) *The Ecology of Adaptive Radiation* (Oxford Series in Ecology and Evolution). Oxford: Oxford University Press.

Looks at the ecological causes of adaptive radiation.

Stearns, S. C. and Hoekstra, R. (2000) *Evolution: An Introduction*. Oxford: Oxford University Press.

A first-rate introduction to the subject.

3

BIOGEOGRAPHICAL PROCESSES II

DISPERSAL

Organisms, even sedentary ones, have a propensity to disperse. Individuals roam into new areas, either as adults or as eggs and seeds, and establish colonies. This chapter covers:

- How organisms spread
- How humans aid and abet the spreading

GETTING AROUND: THE MOVEMENT OF ORGANISMS

Dispersal is a vast subject that has long occupied the minds of ecologists and biogeographers (p. 293). All organisms can, to varying degrees, move from their birthplaces to new locations. Terrestrial mammals can walk, run, dig, climb, swim, or fly to new areas. The adults of higher plants and some aquatic animals are **sessile** (rooted to one spot), but are capable of roving large distances in their early stages of development. Organisms *disperse* when they move to, and attempt to colonize, areas outside their existing range. Some species travel huge distances on an annual basis to avoid harsh conditions, to feed, or to mate. Such *seasonal migrations* do not involve the colonization of new areas outside the species range and do not count as dispersal, and nor do episodic *irruptions* of populations, such as the irruptions of the desert locust (*Schistocera gregaria*) that swarms northwards from its central African core.

The stage in the life cycle of an organism that does the dispersing is a **propagule**. In plants and fungi, a propagule is the structure that serves to reproduce the species – seed, spore, stem, or root cutting. In animals, a propagule is the smallest number of individuals of a species able to colonize a new area. Depending upon the biological and behavioural needs of the species, it is a fertilized egg, a mated female, a male and a female, or a group of individuals.

Dispersal

Organisms disperse. They do so in at least three different ways (Pielou 1979, 243):

1 *Jump dispersal* is the rapid transit of individual organisms across large distances, often across inhospitable terrain. The jump takes less time than the life-span of the individual involved. An insect carried over sea by the wind is an example.

2 *Diffusion* is the relatively gradual spread or slow penetration of populations across hospitable terrain. It takes place over many generations. Species that expand their ranges little by little are said to be diffusing. Examples include the American muskrat (*Ondatra zibethicus*), spreading in central Europe after a Bohemian landowner introduced five individuals in 1905 and now inhabiting Europe in many millions (Elton 1958). Another example is the nine-banded armadillo (*Dasypus novemcintus*) that has spread, and is still spreading, from Mexico to the southeastern USA (Figure 3.1) (Taulman and Robbins 1996).

3 *Secular migration* is the spread or shift of a species that takes place very slowly, so slowly that the species undergoes evolutionary change while it is taking place. By the time population arrives in a new region, it will differ from the ancestral population in the source area. South American members of the family

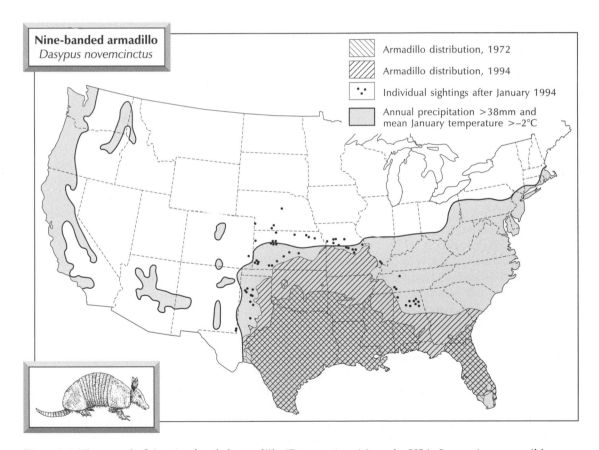

Figure 3.1 The spread of the nine-banded armadillo (*D. novemcinctus*) into the USA. Its maximum possible range is constrained by rainfall and temperature.

Source: Adapted from Taulman and Robbins (1996)

Camelidae (the camel family) – the llama (*Lama glama*), vicuña (*L. vicugna*), guanaco (*L. guanicoe*), and alpaca (*L. pacos*) – are examples. They are all descended from now extinct North American ancestors that underwent a secular migration during the **Pliocene epoch** over the then newly created Isthmus of Panama.

Agents of dispersal

Species may disperse by *active* movement (digging, flying, walking, or swimming), or by *passive* carriage. Physical agencies (wind, water, landmasses) or biological agencies (other organisms, including humans) bring about passive dispersal. These various modes of transport are given technical names – anemochore for wind dispersal, thalassochore for sea dispersal, hydrochore for water dispersal, anemohydrochore for a mixture of wind and water dispersal, and biochore for hitching a ride on other organisms.

Anemochores include many plant species that have seeds designed for wind transport. Some animals are anemochores: young black widow spiders (*Latrodectus mactans*) spin long strands of web that catch the wind and carry the small spiders many kilometres. Many other insects are borne aloft and carried great distance by the wind. Some carabid beetles only 3 mm in length can fly up to 100 miles (Erwin 1979). Ancestors of many of the native spiders, mites, and insects on the Hawaiian Islands made the air trip from Asia, Australia, and North America, which are 3,000 to 4,000 kilometres away. *Hydrochores* include the adults, larvae, and eggs of many aquatic organisms. The coconut palm (*Cocos nucifera*) is a striking *thalassochore*. The thick husk and shell of the coconut keep the seed afloat and safeguard it from sea water as it drifts for long periods in ocean currents. When a coconut beaches on a tropical island, it may germinate and grow into a mature tree. The tiny plumed seeds of the aspen are *anemohydrochores*, being capable of dispersal by wind or water. *Anemochory* is very useful for plants living in floodplains and on

islands. *Biochores* include *zoochores* (dispersed by animals) and *anthropochores* (dispersed by humans). Zoochores travel as seeds on fur, feathers, or clothing (exo-zoochory), as is the case with cleavers (*Galium aparine*), the fruits of which are spherical with hooked bristles and adhere to fur and clothing. Alternatively, they are deliberately moved and stored as seeds by a herbivore, as in the case of acorns collected and secreted by a squirrel. Or else they may pass as seeds though their digestive system of a herbivore that eats their fruits (endozoochory). The efficacy of organisms as agents of dispersal is surprising. A recent study, carried out in the Schwäbische Alb, southwest Germany, shows just how effective sheep are at spreading populations of wild plants by dispersing their seeds (Fischer *et al.* 1996). A sheep was specially tamed to stand still while it was groomed for seeds. Sixteen searches, each covering half of the fleece (it was difficult to search all parts of the animal), produced 8,511 seeds from 85 species. The seeds were a mixture of hooked, bristled, and smooth forms. They included sweet vernal-grass (*Anthoxanthum odoratum*), large thyme (*Thymus pulegioides*), common rock-rose (*Helianthemum nummularium*), lady's bedstraw (*Galium verum*), and salad burnet (*Sanuisorba minor*).

Figure 3.2 shows the means by which colonists were carried to Rakata, which lies in the Krakatau Island group, after the volcanic explosion of 1883. Notice that sea-dispersed or thalassochore species – most of which live along strandlines – are rapid colonizers. The anemochores comprise three ecological groups. The very early colonists are mostly ferns, grasses, and composites (members of the Compositae), which are common in early pioneer habitats. Forest ferns, orchids, and Asclepiadaceae (milkweed, butterfly flower, and wax plant family) dominate a second group. These second-phase colonists require conditions that are more humid. Numerically, most of them are **epiphytes**. The third group consists of seven primarily wind-dispersed trees. Animal-dispersed (zoochore) organisms are the slowest to colonize. Birds and bats mainly carry them.

Over geological timescales, drifting continents may ship whole **faunas** and **floras** across oceans. A continental block acts like a gigantic raft and serves as a kind of *Noah's ark*; it may also carry a cargo of fossil forms, and in this regard is like a *Viking funeral ship* (McKenna 1973) (p. 313).

Good and bad dispersers

Dispersal abilities vary enormously. This is evident in records of the widest known ocean gaps crossed by various land animals, either by flying, swimming, or on rafts of soil and **vegetation** (Figure 3.3). Bats and land birds, insects and spiders, and land molluscs form the 'premier league' of transoceanic dispersers. Lizards, tortoises, and rodents come next, followed by small carnivores. The poorest dispersers are large mammals and freshwater fish.

Not all large mammals are necessarily inept at crossing water. It pays to check their swimming proficiency before drawing too many biogeographical conclusions from their distributions. Fossil elephants, mostly pygmy forms, are found on many islands: San Miguel, Santa Rosa, and Santa Cruz, all off the Californian coast; Miyako and Okinawa, both off China; Sardinia, Sicily, Malta, Delos, Naxos, Serifos, Tilos, Rhodes, Crete, and Cyprus, all in the Mediterranean Sea; and Wrangel Island, off Siberia. Before reports on the proficiency of elephants as swimmers (D. L. Johnson 1980), it was widely assumed that elephants must have walked to these islands from mainland areas, taking advantage of former **land bridges** (though vicariance events are also a possibility). As elephants could have swum to the islands, new explanations for the colonization of the islands are required. Tigers, too, have a surprisingly high degree of mobility in water. They can swim for up to 29 km across rivers or 15 km across the sea (Kitchener 1999).

Supertramps are ace dispersers. They move with ease across ocean water and reproduce very rapidly, setting up thriving colonies. They were first recognized on the island of Long, off New Guinea

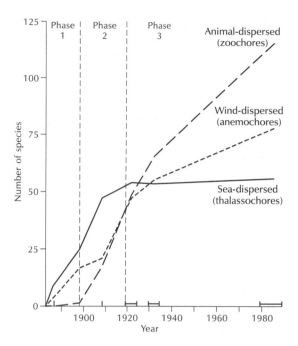

Figure 3.2 The means by which the spermatophyte flora reached Rakata, in the Krakatau island group, from 1883 to 1989. The collation periods are indicated on the horizontal axis. Human-introduced species are excluded.

Source: Adapted from Bush and Whittaker (1991)

(Diamond 1974). Long was devegetated and defaunated about two centuries ago by a volcanic explosion. The diversity of bird species is now far higher than would be expected (Figure 3.4). Of the 43 species present on the island, 9 were responsible for the high density. These were the supertramps. They specialize in occupying islands too small to maintain stable, long-lasting populations, or islands devastated by catastrophic disturbance – volcanic eruptions, **tsunamis**, or **hurricanes**. Competitors that can exploit resources more efficiently and that can survive at lower resource abundances eventually oust the supertramps from these islands. In the plant kingdom, the dandelion (*Taraxacum officinale*) is a supertramp. An **annual**, it tolerates a wide range of climates and thrives in a variety of soil conditions, matures and sets in one

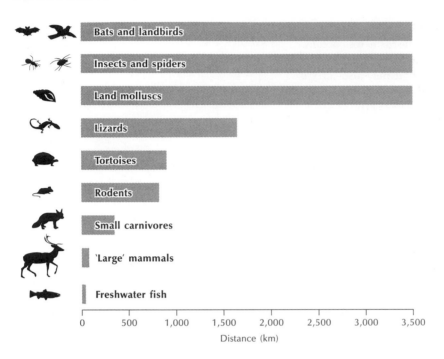

Figure 3.3 The widest ocean gaps crossed by terrestrial animals. The distances are extremes and probably not typical of the groups.
Source: Adapted from Gorman (1979)

growing season, and has small seeds with plumes that are readily broadcast by the wind or caught up in fir and clothing. The dandelion occurs on all continents save Antarctica.

Dispersal routes

The ease and rate at which organisms disperse depend on two things: the topography and climate of the terrain over which they are moving and the wanderlust of a particular species. Topography and climate may impose constraints upon dispersing organisms. Obviously, organisms disperse more easily over hospitable terrain than over inhospitable terrain. Obstacles or **barriers** to dispersal may be classified according to the 'level of difficulty' in crossing them. George Gaylord Simpson (1940) suggested three types:

1 'Level 1' barriers are **corridors** – routes through hospitable terrain that allow the unhindered passage of animals or plants in both directions.

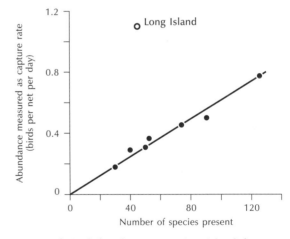

Figure 3.4 Bird abundances on various islands between New Guinea and New Britain. Abundance was measured by the daily capture rate in nets. Netting yields, which reflect total population densities, increase with the local number of species. Long Island is exceptional because population density is abnormally high for an island of its size. This high density is due largely to the presence of supertramp species.
Source: Adapted from Diamond (1974)

2 'Level 2' barriers are **filter routes**. An example is a land bridge combined with a climatic barrier that bars the passage of some migrants. The Panamanian Isthmus is such a route since it filters out species that cannot tolerate tropical conditions.

3 'Level 3' barriers are **sweepstakes routes**. They reflect the fact that, as in gambling, there are always a small number of winners compared with losers. In biology, the winners are those few lucky individuals that manage to survive a chance journey by water or by air and succeed in colonizing places far from their homeland.

Simpson's terms apply primarily to connections between continents. A more recent scheme, though redolent of Simpson's, is more applicable to connections between islands and continents and islands with other islands (E. E. Williams 1989). It recognizes five types of connection:

1 *Stable land bridges*. These are 'filter routes' in Simpson's sense. Faunas are free to move in both directions.

2 *Periodically interrupted land bridges*. These are akin to stable land bridges, but there are two differences. First, water gaps periodically interrupt access in both directions. Second, faunas on one or both sides may suffer extinction, owing to the loss of area during times of separation.

3 *'Noah's arks'*. These are fragments of lithospheric plates carrying entire faunas with them from one source area to another (see p. 313). Ordinarily, transport on Noah's arks is one way.

4 *Stepping-stone islands*. These are a fairly permanent or temporary series of islands separated by moderate to small water gaps. Traffic could be two-way, but often goes from islands of greater diversity to islands of lesser diversity.

5 *Oceanic islands*. These are situated a long way from the mainland and they receive 'waifs'. They are similar to stepping-stone islands, but there are fewer arrivals arriving at much longer time intervals. They are not true sweepstakes routes because the chances of arriving depends on species characteristics – some species have a much better chance of arriving than others do.

Of course, for terrestrial animals, crossing land is not so difficult as crossing water, and many large mammals have dispersed between biogeographical regions. Given enough time, probably no barrier is insurmountable:

One morning [in Glacier Park], dark streaks were observed extending downward at various angles from saddles or gaps in the mountains to the east of us. Later in the day, these streaks appeared to be much longer and at the lower end of each there could be discerned a dark speck. Through binoculars these spots were seen to be animals floundering downward in the deep, soft snow. As they reached lower levels not so far distant, they proved to be porcupines. From every little gap there poured forth a dozen or twenty, or in one case actually fifty five, of these animals, wallowing down to the timberline on the west side. Hundreds of porcupines were crossing the main range of the Rockies.

(W. T. Cox 1936, 219)

LIFE ON THE MOVE: DISPERSAL IN ACTION

Dispersal undoubtedly occurs at present but it is normally difficult to observe. The chief problem in detecting dispersal in action is that detailed species distributions are scarce. Most instances of organisms moving to new areas probably pass unnoticed – is a new sighting an individual that has moved in from elsewhere, or is it an individual that was born in the area but not seen before? Despite these problems, there are several amazing cases of present-day dispersal resulting from human introductions. People accidentally or purposely take introduced species to new areas, so aiding and abetting the spread of many species. An example of a deliberate introduction is the coypu (*Myocastor coypus*), brought from South America to Britain in the 1930s for its fur (nutria). Numerous escapes occurred and it established itself in two areas: at a sewage farm near Slough, where a colony lived from 1940 to 1954, and in East Anglia, with a centre in the Norfolk Broads (Lever 1979). A concerted trapping programme seems to have eradicated the coypu from Britain (Gosling and Baker 1989). An accidental introduction was the establishment of the ladybird *Chilocorus nigritus* in several Pacific islands, northeast Brazil, west Africa, and Oman after shipment from other areas (Samways 1989).

Successful, half-successful, and failed introductions

Not all introductions survive; some gain a foothold but progress little further; others go rampant and swiftly colonize large tracts of what is to them uncharted territory. A dispersing organism will fail if it cannot colonize a new location. Dispersal ability does not necessarily equate with colonizing ability. Some hundred species of birds from Asia and Europe arrive in North America every year but do not set up permanent populations. **Environmental factors** that may hinder colonization include adverse physical conditions and unfavourable biotic conditions. Tropical plants and animals that disperse to high latitudes are unlikely to survive the colder climate. Of the biotic factors that stand to impede colonization, competition ranks high. A colonist is unlikely to oust a superior competitor.

Failed dispersers include several species that were unsuccessfully taken to New Zealand — bandicoots, kangaroos, racoons, squirrels, bharals, gnus, camels, and zebras.

Amphibians and reptiles in Ireland appear to be *reluctant dispersers*. Just four species of amphibians and reptiles live in Ireland, compared with twelve on the British mainland. The species are the natterjack toad (*Bufo calamita*), the common

newt (*Triturus vulgaris*), the common or viviparous lizard (*Lacerta vivipara*), and the common frog (*Rana temporaria*). The common frog was introduced into ditches in University Park, Trinity College Dublin, in 1696. It still flourishes there today, and has spread to the rest of Ireland. So, why have only three species of amphibians and reptiles colonized Ireland? One explanation is that other newts and toads did establish bridgeheads, but they died out because they were unable to sustain large enough colonies for successful invasion. The present distribution of the natterjack toad in Ireland, which is restricted to a small part of Kerry and shows no signs of spreading, lends this view some support.

The European starling (*Sturnus vulgaris*) is a fine example of a *rampant disperser*. This bird has successfully colonized North America, South Africa, Australia, and New Zealand. Its spread in North America was an indubitable ecological explosion – within 60 years it had colonized the entire USA and much of Canada. There were several 'false starts' or failed introductions during the nineteenth century when attempts to introduce the bird in the USA failed. For example, in 1899, 20 pairs of starlings vanished after their release in Portland, Oregon. Then in April 1890, 80 birds were released in Central Park, New York, and in March the following year a further 80 birds were released. Within 10 years, the European starling was firmly established in the New York City area. From that staging post, it expanded its range very rapidly, colonizing some 7,000,000 km^2 in 50 years (Figure 3.5). The speed of dispersal was due to the irregular migrations and wanderings of non-breeding 1- and 2-year-old starlings. Adult birds normally use the same breeding ground year after year and do not colonize new areas. The roaming young birds frequented faraway places. Only after 5 to 20 years of migration between the established breeding grounds and the new sites, did the birds take up permanent residence and set up new breeding colonies. For example, the European starling was first reported in California in 1942. It first nested

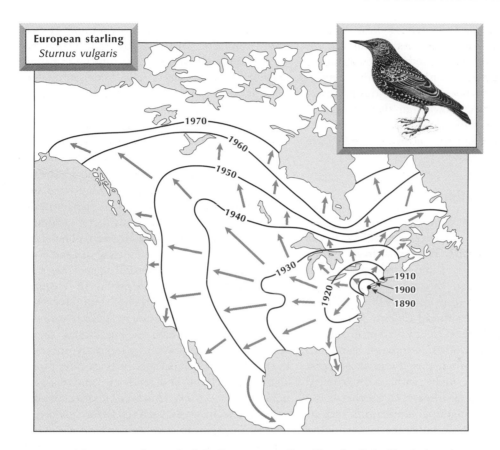

Figure 3.5 The westward spread of the European starling (*S. vulgaris*) in North America.
Sources: Map adapted from Kessel (1953) and Perrins (1990); picture from Saunders (1889)

there in 1949. Large-scale nesting did not occur until after 1958.

The success of many introductions is beyond question. Indeed, it is ironical that humans are brilliantly successful in the unintentional extermination of some species, mainly through habitat alteration and fragmentation, but hopelessly unsuccessful in the purposeful eradication of introduced species that have become pests. Invasion success depends on the interaction between the invader and the community it is invading. Predicting the fate and impact of a specific introduced species is very difficult (Lodge 1993). For instance, domestic and wild European rabbits have been liberated on islands all over the world (Flux and Fullagar 1992). The outcomes of these introductions range from, on the one hand, utter failure to, on the other, rabbit densities so high that islands lose their vegetation and soil. Some rabbit populations have survived remarkably adverse conditions for up to 100 years and then become extinct.

There are places where alien animals and plants are of major economic and conservation significance. A prime example is New Zealand (Atkinson and Cameron 1993). Plant introductions have averaged 11 species per year since European settlement in 1840, and weeds are increasingly altering distinctive landscapes. Many introduced animals act as disease vectors or threaten native biota. Recent studies of introduced wasps show adverse effects on honey-eating and insectivorous birds. Introduced possums prey on eggs and nestlings of native birds, damage native forests, and transmit bovine tuberculosis.

A few examples drawn from the animal and fungal kingdoms will serve to illustrate the rates of spread and the impact that invaders may have on native communities.

Animal introductions

The American mink (*Mustela vison*) is a medium-sized mustelid carnivore with a long body. It was introduced to British fur farms from North America in the 1930s. Some individuals escaped and soon established themselves in the wild. Mink is now found in many parts of the Britain and will continue to spread (Figure 3.6a). As a carnivore, it has had a rather different impact on native wildlife than other introduced mammals. A crucial question is whether the mink occupies a previously vacant niche with an anticipated mild overall ecological impact, or whether it is a species that is endangering competitors such as the otter (*Lutra lutra*) and prey species (including fish stocks). A recent survey helped to resolve this issue, and drew five conclusions (Birks 1990). First, the mink is little threat to the otter, although it exacerbates otter decline in areas where the otter is already **endangered**. Second, the mink has probably aided the decline of water vole (*Arvicola terrestris*) in some localities (see Carter and Bright 2003). Third, if the mink should have had any effects on waterfowl, then these have not been translated into widespread population declines. Fourth, there are grave potential risks of introducing mink to offshore islands. Fifth, the mink is not having a serious overall impact on fish stocks, at least in England and Wales.

The muntjac deer (*Muntiacus reevesi*) is small, standing about 50 cm at the shoulder (Plate 3.1). Its small size allows it to live in copses, thickets, neglected gardens, and even hedgerows (N.

Chapman *et al.* 1994). Following the first releases from Woburn, Bedfordshire, in 1901, the numbers of free-living Reeves' muntjac in Britain remained low until the 1920s, when populations were largely confined to the woods around Woburn, and possibly also around Tring in Hertfordshire. However, in the 1930s and 1940s there were further deliberate introductions in selected areas some distance from Woburn. Consequently, the subsequent spread of Reeves' muntjac was from several foci (Figure 3.6b). The spread in the second half of the twentieth century was aided by further deliberate and accidental releases. By these means, new populations continued to establish themselves outside the main range. Thus, the natural spread has been much less impressive than previously assumed; even in areas with established populations, it takes a long time for muntjac deer to colonize the entire available habitat. The natural rate of spread is probably about 1 km a year, comparable to that of other deer species in Britain. It prefers arable land classes and tends to shun marginal upland land classes. However, long-established populations in areas such as Betws-y-Coed in Wales show that muntjac deer may persist in low numbers in atypical habitats.

Plant introductions

In Britain, floods disperse Japanese knotweed (*Reynontria japonica*) **rhizomes**; and Indian balsam (*Impatiens glandulifera*), introduced from the Himalayas as a garden plant, has spread along riverbanks. In the western USA, two exotic species – tamarisk or salt cedar (*Tamarix hispida rubra*) and Russian olive (*Elaeagnus angustifolia*) – have colonized widely along rivers and cause serious problems in local ecosystems. The tamarisk is a very invasive shrub-tree introduced from Eurasia and planted across the western USA by government agencies in the early 1900s in an effort to control soil erosion. The Russian olive is a native tree of Europe and Asia introduced to North America by settlers some 150 years ago,

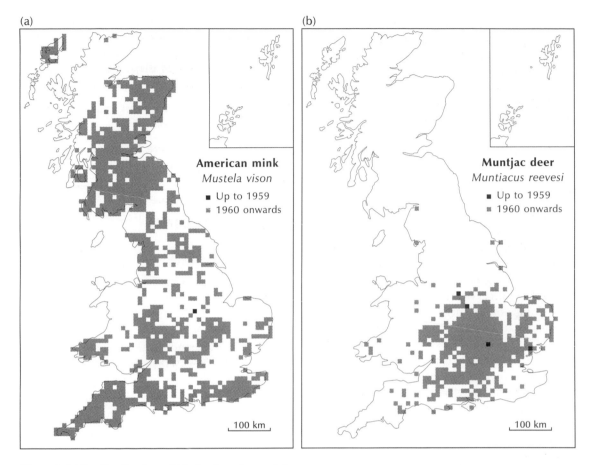

(a)   (b)

American mink
Mustela vison

■ Up to 1959
▨ 1960 onwards

100 km

Muntjac deer
Muntiacus reevesi

■ Up to 1959
▨ 1960 onwards

100 km

Figure 3.6 The distribution of (a) the American mink (*M. vison*) and (b) the muntjac deer (*M. reevesi*) in Britain.

Source: Adapted from H. R. Arnold (1993)

Plate 3.1
Muntjac deer
(*M. reevesi*).
Photograph by
Pat Morris.

since when it has colonized widely along rivers where it is having a severe impact on native birds and fish (e.g. Dixon and Johnson 1999). Similarly, purple loosestrife (*Lythrum salicaria*), originally introduced into North America from Europe in the early 1800s as an ornamental plant and a contaminant of ship ballast, spread via waterborne commerce to invade many wetlands, causing severe problems in some areas (Thompson *et al.* 1987; Mullin 1998).

SUMMARY

Organisms disperse. They may do so actively (by walking, swimming, flying, or whatever) and passively (carried by wind, water, or other organisms). Dispersal ability varies enormously throughout the living world. All organisms are capable of limited dispersal; most are mediocre dispersers; some, such as the supertramps, are expert dispersers. Ease of passage along dispersal routes varies from almost effortless to nigh on impossible. Dispersal occurs at present, much of it resulting from human introductions. Introduced species display all degrees of success from out-and-out failure to brilliant success. Some cause severe environmental problems.

ESSAY QUESTIONS

1 What principles of dispersal emerge from studies of newly colonized islands?

2 To what extent are alien species a threat to global biodiversity?

3 To what extent have humans 'homogenized' the world biota?

FURTHER READING

Cox, G. W. (1999) *Alien Species in North America and Hawaii: Impacts on Natural Ecosystems.* Washington, DC: Island Press.
An excellent introduction to the problem of invasive species.

Cronk, Q. C. B. and Fuller, J. L. (1995) *Plant Invaders: The Threat to Natural Ecosystems.* London: Chapman & Hall.
Covers problems caused by alien plants.

Elton, C. S. (1958) *The Ecology of Invasions by Animals and Plants.* London: Chapman & Hall.
A little classic gem. Still worth reading.

BIOGEOGRAPHICAL PATTERNS

DISTRIBUTIONS

All species and other groups of organisms have a particular geographical range or distribution. This chapter covers:

■ regional differences in faunas and floras
■ kinds of distributions
■ geographical range size and shape

GEOGRAPHY: BIOGEOGRAPHICAL REGIONS

Different places house different kinds of animals and plants. This became apparent as the world was explored. In 1628, in his *The Anatomy of Melancholy*, Robert Burton wrote:

> Why doth Africa breed so many venomous beasts, Ireland none? Athens owls, Crete none? Why hath Daulis and Thebes no swallows (so Pausanias informeth us) as well as the rest of Greece, Ithaca no hares, Pontus [no] asses, Scythia [no] wine? Whence comes this variety of complexions, colours, plants, birds, beasts, metals, peculiar to almost every place?
>
> (Burton 1896 edn: Vol. II, 50–1)

Regional differences in the distribution of species became increasingly manifest as explorers discovered new lands. In the mid-eighteenth century, George Leclerc, Compte de Buffon (1707–1788) studied the then known tropical mammals from the Old World (Africa) and the New World (central and South America). He found that they had not a single species in common. Later comparisons of African and South American plants, insects, and reptiles evinced the same pattern.

By the nineteenth century, it was clear that the land surface was divisible into biogeographical regions, each of which carries a distinct set of animals and a distinct set of plants. Augustin-Pyramus de Candolle considered plants and identified *areas of endemism*, that is botanical regions, each possessing a certain number of plants peculiar to them. He listed 20 such botanical regions or areas of endemism in 1820, and by 1838 had added another score, bringing the total to 40. In 1826, James Cowles Prichard, a zoologist, distinguished seven regions of mammals: the Arctic region, the temperate zone, the equatorial regions, the Indian isles, the Papuan region, the Australian region, and the extremities of America and Africa. William Swainson modified this scheme in 1835, by taking account of the 'five recorded varieties of humans', to give five regions: the European (or Caucasian) region, the Asiatic (or Mongolian) region, the American region, the Ethiopian (or African) region, and the Australian (or Malay) region.

The seminal work of an English ornithologist, Philip Lutley Sclater, and the eminent English biogeographer and naturalist, Alfred Russel Wallace, eclipsed the ideas of Prichard and Swainson on animal distributions. Using bird distributions, Sclater (1858) recognized two basic divisions (or 'creations', as he termed them) – the Old World (Creatio Paleogeana) and the New World (Creatio Neogeana) – and six regions. The Old World he divided into Europe and northern Asia, Africa south of the Sahara, India and southern Asia, and Australia and New Guinea. The New World he divided into North America and South America. Sclater's schema prompted a flurry of papers by English-speaking zoologists, including Thomas Henry Huxley and Joel Asaph Allen, each of whom promulgated his own favoured geographical classification. In his *The Geographical Distribution of Animals* (1876), Alfred Russel Wallace reviewed the competing systems, arguing persuasively in favour of adopting Sclater's six regions, or realms as Wallace dubbed them. Sclater's system and Wallace's minor amendments to it provided a nomenclature that survives today (Figure 4.1). Later suggestions were minor variations on the Sclater–Wallace

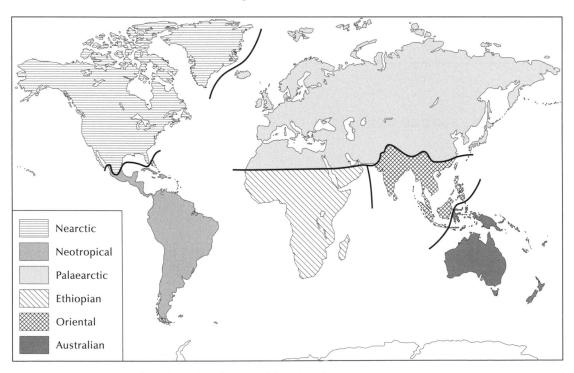

Legend:
- Nearctic
- Neotropical
- Palaearctic
- Ethiopian
- Oriental
- Australian

Figure 4.1 The Sclater and Wallace classification of faunal regions.

Table 4.1 Biogeographical regions and subregions, as defined by Alfred Russel Wallace

Region	Subregion
Palaeogaea (Old World)	
Palaearctic	North Europe
	Mediterranean
	Siberia
	Manchuria (or Japan)
Ethiopian	East Africa
	West Africa
	South Africa
	Madagascar
Oriental	Hindustan (or central India)
	Ceylon (Sri Lanka)
	Indo-China (or Himalayas)
	Indo-Malaya
Australian	Austro-Malaya
	Australia
	Polynesia
	New Zealand
Neogaea (New World)	
Neotropical	Chile
	Brazil
	Mexico
	Antilles
Nearctic	California
	Rocky Mountains
	Alleghenies
	Canada

Source: After Wallace (1876)

theme. Sclater and Wallace identified six regions – Nearctic, Neotropical, Palaearctic, Ethiopian, Oriental, and Australian. Together, the Nearctic and Palaearctic regions form Neogaea (the New World), while other regions form Palaeogaea (the Old World). Wallace's contribution was to identify subregions, four per region, which correspond largely to de Candolle's botanical regions (Table 4.1). Indeed, the nineteenth-century classification of biogeographical regions was essentially an attempt to group areas of endemism into a hierarchical classification according to the strengths of their relationships. It should be noted that C. Barry Cox (2001), in a study of biogeographical regions, considers the names Neotropical, Nearctic, and Palaearctic to be cumbersome and unnecessary, favouring South American, North American, and Eurasian as plain alternatives.

It is unanticipated and noteworthy that the distributions of species with good dispersal abilities, including plants, insects, and birds, tend to fall within traditional zoogeographical regional boundaries. The avifaunas of North America and Europe contain several families and many genera that the two regions do not share, even though dispersal across the north Atlantic and Pacific Oceans by 'accidental visitors' occurs every year. Even long-distance migrant bird taxa tend to stay either in the eastern hemisphere or in the western hemisphere, where they migrate between high and low latitudes, and appear ill disposed to disperse east–west between continents.

Mammal regions

Of the six faunal regions delineated by Sclater and Wallace, the *Palaearctic* or *Eurasian region* is the largest. It includes Europe, northern Africa, the Near East, and much of Asia (but not the Indian subcontinent or southeast Asia). Its mammal fauna is quite rich with some 40 families. Only two of these families are endemic to the Palaearctic region – the blind mole rats (Spalacidae) and the Seleviniidae, represented by one species, the dzhalman, which is a small insectivorous rodent (Table 4.2).

The *Nearctic* or *North American region* encompasses nearly all the New World north of tropical Mexico. Its fauna is diverse and includes families with a largely tropical distribution, such as the sac-winged or sheath-tailed bats (Emballonuridae), vampire bats (Desmodontidae), and javelinas or peccaries (Tayassuidae), and largely boreal families, such as the jumping mice (Zapodidae), beavers (Castoridae), and bears (Ursidae). Only two Nearctic families are endemic to the region (Table 4.2): the Aplodontidae, which contains

Table 4.2 Endemic mammal families in the faunal regions

Faunal region	Number of endemic families	Names of endemic families
Eurasian (Palaearctic)	2	Blind mole rats (Spalacidae); dzhalman (Seleviniidae)
North American (Nearctic)	2	Mountain beaver or sewellel (Aplodontidae); pronghorn antelope (Antilocapridae)
South American (Neotropical)	27	Solenodons (Solenodontidae); West Indian shrews (Nesophontidae); New World monkeys (Cebidae); marmosets (Callithricidae), caeonolestids or marsupial mice (Caenolestidae); monito del monte or 'monkey of the mountains' (Microbiotheriidae); anteaters (Myrmecophagidae); sloths (Bradypodidae); degus, coruros, and rock rats (Octodontidae); tuco-tucos (Ctenomyidae); spiny rats (Echimyidae); rat chinchillas (Abrocomidae); hutias and coypus (Capromyidae); chinchillas and viscachas (Chinchillidae); agouties (Dasyproctidae); pacas (Cuniculidae); pacarana (Dinomyidae); guinea-pigs and their relatives (Caviidae); capybaras (Hydrochoeridae); quemi and its allies (Heptaxodontidae)[a]; bulldog bats (Noctilionidae); New World leaf-nosed bats (Phyllostomidae); moustached bats, ghost-faced bats, and naked-backed bats (Mormoopidae); vampire bats (Desmondontidae), funnel-eared bats (Natalidae); smoky or thumbless bats (Furipteridae); disk-winged bats (Thyropteridae)
Ethiopian	15	Giraffes (Giraffidae); hippopotamuses (Hippopotamidae)[b]; aardvark (Orycteropodidae); tenrecs (Tenrecidae); the Old World sucker-footed bats (Myzopodidae); lemurs (Lemuridae); woolly lemurs (Indriidae); aye-ayes (Daubentoniidae); golden moles (Chrysochloridae); otter shrews (Potamogalidae); scaly-tailed squirrels (Anomaluridae); the spring hare or Cape jumping hare (Pedetidae); cane rats (Thryonomydiae); the rock rat or dassie rat (Petromyidae); African mole rats (Bathyergidae)
Oriental	5	Spiny dormice (Platacanthomyidae); tree shrews (Tupaiidae); tarsiers (Tarsiidae); flying lemurs or colugos (Cynocephalidae); Kitti's hog-nosed bat or bumblebee bat (Craseonycteridae)
Australian	19	Echidnas or spiny anteaters (Tachyglossidae); platypus (Ornithorhynchidae); marsupial 'mice' and 'cats' (Dasyuridae); Tasmanian wolf (Thylacinidae); numbat or banded anteater (Myrmecobiidae); marsupial mole (Notoryctidae); bandicoots and bilbies (Peramelidae); burrowing bandicoots (Thylacomyidae); spiny bandicoot and mouse bandicoot (Peroryctidae); striped possum, Leadbeater's possum, and wrist-winged gliders (Petauridae); feathertail gliders (Acrobatidae); pigmy possums (Burramyidae); brush-tailed possums, cuscuses, scaly-tailed possums (Phalangeridae); ringtail possums and great glider (Pseudocheiridae); kangaroos and wallabies (Macropodidae); rat kangaroos, potoroos, and bettongs (Potoroidae); koalas (Phascolarctidae); wombats (Vombatidae); noolbender or honey possum (Tarsipedidae)

Notes: a Recently extinct. b Those living on the Lower Nile are technically in the Eurasian region

one species, the mountain beaver or sewellel (*Aplondontia rufa*), and the Antilocapridae, which also contains one species, the pronghorn antelope (*Antilocapra americana*). Two other families are almost endemic: the pocket gophers (Geomyidae) live in North America, central America, and northern Colombia; and the kangaroo rats and pocket mice (Heteromyidae) live in North America, Mexico, central America, and north-western South America.

The *Neotropical* or *South American region* covers all the New World south of tropical Mexico. It boasts some 27 endemic families of mammals, including 12 **caviomorph rodent** families and 7 bat families (Table 4.2).

The *Ethiopian region* encompasses Madagascar, Africa south of a somewhat indeterminate line running across the Sahara, and a southern strip of the Arabian peninsula. It has about 15 endemic families, almost as many as the Neotropical region, including 2 shrew families (golden moles and otter shrews) and 5 rodent families (Table 4.2). Two other families – the elephant shrews (Macroscelididae) and gundis (Ctenodactylidae) – live only in Africa, but range into the north of the continent, which is part of the Palaearctic region.

The *Oriental region* covers India, Indo-China, southern China, Malaysia, the Philippines, and Indonesian islands as far east as Wallace's line. It has just five endemic families (Table 4.2): spiny dormice (Platacanthomyidae), tree shrews (Tupaiidae), tarsiers (Tarsiidae), flying lemurs or colugos (Cynocephalidae), and one endemic bat family – the Craseonycteridae – represented by a single species known as Kitti's hog-nosed bat or bumblebee bat (*Craseonycteris thonglongyai*), which was discovered in Thailand in 1973.

The *Australian region* includes mainland Australia, Tasmania, New Guinea, Sulawesi, and many small Indonesian islands. It possesses some 19 endemic families of mammals (Table 4.2).

Applying modern methods of numerical classification to mammal distributions brings out the similarities and differences of biogeographical regions. Using multidimensional scaling to data on the distribution of 115 mammal families (wholly marine families and the human family were omitted) in Wallace's 24 subregions, Charles H. Smith (1983) delineated similar regions to those in the Sclater–Wallace scheme, but significant differences emerged. In Smith's system, there are four regions – Holarctic, Latin American, Afro-Tethyan, and Island – and ten subregions (Figure 4.2). The Holarctic region comprises the Nearctic and the Palaearctic subregions; the Latin American region comprises the Neotropical and Argentine subregions; the Afro-Tethyan region comprises the Mediterranean, Ethiopian, and Oriental subregions; and the Island region comprises the Australian, the West Indian, and Madagascan subregions. Each subregion is as unique as it can be when compared with all other subregions. Several features of Smith's system are intriguing. First, it reveals a close similitude between the mammal families of the Ethiopian and Oriental regions. Second, it includes the Mediterranean subregion within the Ethiopian region, thus excluding it from the Palaearctic region. Third, it promotes Madagascar and the West Indies to distinct island subregions, removing them from the Ethiopian region and the Neotropical region, respectively.

Table 4.3 shows the regional richness and endemicity of mammal families in Smith's regions and subregions. Of the 115 mammal families used in the analysis, 43 (37 per cent) are endemic to subregions. The lowest subregional endemicity occurs in the Palaearctic subregion, with no endemic families, and the highest in the Neotropical subregion, with nine endemic families. Smith's analysis also indicated that the Nearctic, Palaearctic, Mediterranean, and Oriental subregions have high affinities with the faunas of other subregions, whereas the Argentine and Australian subregions have low affinities with the faunas of other subregions. Furthermore, the nature of the Neotropical, Argentine, Ethiopian, Australian, West Indian, and Madagascan faunas reflects the effects of isolation or inaccessibility (or both).

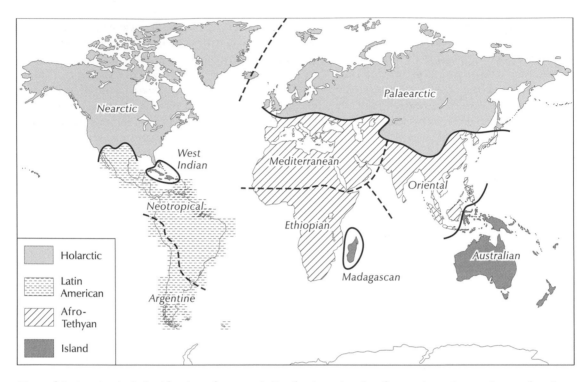

Figure 4.2 A numerical classification of mammal distributions showing four main regions and ten subregions. *Source*: Adapted from C. H. Smith (1983)

Table 4.3 Mammalian families in Smith's faunal regions

Mammal region	Number of families	Number of endemic families	Percentage of endemic families
Holarctic	36	6	17
Latin America	48	20	42
Afro-Tethyan	65	29	45
Island	35	15	43

Floral regions

In *The Geography of the Flowering Plants* (1974), British botanist Ronald Good summarized the distribution of living **angiosperms** by adapting a scheme devised by Adolf Engler during the 1870s. Good delineated six major *floral regions*, though he styled them 'kingdoms': the Boreal region, the Palaeotropical region, the Neotropical region, the Australian region, South African (Cape) region, and the Antarctic floral region. Each of these comprises a number of subregions (Good called them regions), of which there are 37 in total (Figure 4.3). (A similar set of floral 'kingdoms' was delineated by Armen L. Takhtajan in 1986.) The Boreal floral region spans North America and Asia, which share many families, including the birches, alders, hazels, and hornbeams (Betulaceae), mustard (Cruciferae), primrose (Primulaceae), and buttercup (Ranunculaceae). Six subregions are recognized: the Arctic and Subarctic, eastern Asia, western and central Asia, the Mediterranean, Euro-Siberia, and North America. The Palaeotropical region covers most of Africa, the Arabian peninsula, India, southeast Asia, and parts of the western and central Pacific. The subregions are not firmly agreed but Malesia, Indo-Africa, and Polynesia are commonly recognized. The Malesian subregion is exceptionally rich in forms with about 400 endemic genera. Madagascar, which is part of the Indo-African subregion but sometimes taken as a separate region, has 12 endemic families and 350 endemic genera. The

Neotropical region covers most of South America, save the southern tip and a southwestern strip, central America, Mexico (excepting the dry northern and central sections), and the West Indies and southern extremity of Florida. It is gloriously rich floristically, housing 47 endemic families and nearly 3,000 endemic genera. The

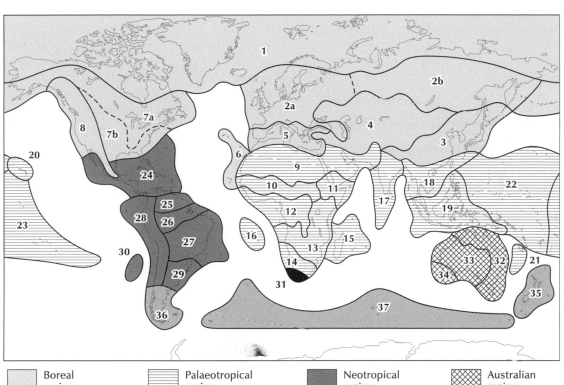

▨ **Boreal region**	▨ **Palaeotropical region**	▨ **Neotropical region**	▨ **Australian region**
1. Arctic and Sub-arctic	9. African–Indian Desert	24. Caribbean	32. N. and E. Australian
2. Euro-Siberian	10. Sudanese Park Steppe	25. Venezuela and Guiana	33. S. W. Australian
a. Europe	11. N. E. African Highland	26. Amazon	34. C. Australian
b. Asia	12. W. African Rain-forest	27. South Brazilian	
3. Sino-Japanese	13. E. African Steppe	28. Andean	▨ **Antarctic region**
4. W. and C. Asiatic	14. South African	29. Pampas	
5. Mediterranean	15. Madagascar	30. Juan Fernandez	35. New Zealand
6. Macaronesian	16. Ascension and St Helena		36. Patagonian
7. Atlantic North American	17. Indian	■ **South African region**	37. S. Temperate Oceanic Islands
a. Northern	18. Continental S. E. Asiatic		
b. Southern	19. Malaysian	31. Cape	
8. Pacific North American	20. Hawaiian		
	21. New Caledonia		
	22. Melanesia and Micronesia		
	23. Polynesia		

Figure 4.3 The 6 floral regions and 37 subregions mapped by Good.

Source: Adapted from Good (1974)

Cape region of South Africa is, for its small size, rich in plants with 11 endemic families and 500 endemic genera. The Australian region is highly distinct with 19 endemic families, 500 endemic genera, and over 6,000 flowering-plant species. The Antarctic region has a curious geography and includes a coastal strip of Chile and the southern tip of South America, the Antarctic and sub-Antarctic islands, and New Zealand. The sub-Antarctic subregion (southern Chile, Patagonia, and New Zealand) carries a distinctive flora involving some 50 genera, of which the southern beech (*Nothofagus*) is a characteristic element.

Regional similarities and differences

Comparisons and contrasts between taxa

The world's regional mammal faunas interconnect with each other in complex ways, as do the world's regional floras. Connections at the species level are weak, except between the Eurasian and North American regions, but some regions share genera and families. Each biogeographical region possesses two groups of families: those that are endemic or peculiar to the region, and those that are shared with other regions. Although no agreed system of naming shared taxa (species, genera, families, or whatever) exists, a useful scheme suggests that taxa shared between two biogeographical regions are *characteristic*, taxa shared between three or four biogeographical regions are *semi-cosmopolitan*, and taxa shared between five or more biogeographical regions are *cosmopolitan* (see p. 59). Links between regions are suggested by a mixing of some faunal or floral elements. A Malesian floral element is present in the tropical rainforests of northeastern Queensland, Australia. Antarctic and Palaeotropical floras mingle in South Island of New Zealand, Tasmania, and the Australian Alps. The strong affinity of the Ethiopian and Oriental faunal regions is reflected in a number of shared families: bamboo rats (Rhizomyidae), elephants (Elephantidae), rhinoceroses (Rhinocerotidae), chevrotains (Tragulidae), lorises and pottos (Lorisidae), galagos or bush-babies (Galagonidae), apes (Pongidae), and pangolins or scaly anteaters (Manidae).

Faunal and floral regions compared

The major floral regions and the major mammal regions are roughly congruent, but there are important differences between them. First, owing to the superior dispersal ability of some plants compared with terrestrial mammals, the floral regions tend to be less clear-cut than do the faunal regions. Second, although the boreal floral region is equivalent to the combined Eurasian and North American faunal regions (the Holarctic region), the North American floral subregion differs from the Nearctic faunal region in that it does not occupy all of Florida or Baja California. The Palaeotropical floral region is equivalent to the combined Ethiopian and Oriental faunal regions or a large part of Smith's Afro-Tethyan region, excluding the Mediterranean, which groups floristically with the Boreal region. The Australian floral region approximately corresponds with the Australian faunal region, though the dividing line with the Asian region lies between Australia and New Guinea, rather than further west as in the case of animals. Indeed, it is puzzling that the flora of New Guinea is Palaeotropical while its fauna is Australian. The Neotropical floral region broadly matches the Neotropical faunal region, but the floral Neotropical region, unlike the faunal Neotropical region, takes in Baja California and the southern end of Florida. The Cape floral region, which occupies the southern tip of Africa, bears no equivalent faunal region. The Antarctic floral region, like the Cape floral region, possesses no faunal counterpart, includes southern South America and New Zealand, and some of its members are found in Tasmania and southeastern Australia.

Transitional zones and filters

Various kinds of barrier, determined mainly by climate, mountains, and water gaps, separate the chief faunal and floral regions. Two water gaps – the Bering Strait and the Norwegian Sea, both of which experience cold climates – separate the North American region from the Eurasian region. A narrow land-link (the Isthmus of Panama), which replaced an earlier water gap, acts as a filter between North America and South America, with arid conditions lying north of the land-link in Mexico. The Sahara desert divides the Palaearctic region from the Ethiopian region. The Ethiopian region is insulated from the Oriental region by arid lands in southwest Asia and the Arabian peninsula. The Himalayas and their eastward extensions create a formidable barrier between the Oriental region and the Palaearctic region. In the region sometimes called Wallacea, a series of water gaps hinders movement between the Oriental region and the Australian region.

The borders between biogeographical regions are passable with varying levels of ease or difficulty. Seldom do the environmental conditions in the border areas allow unhampered access between regions. An open border once existed between Alaska and Siberia when, during the Pleistocene epoch, there was a dry-land connection across what is now the Bering Strait. Other borders tend to act as filters and prevent the passage of some species from one biogeographical region to another. In many cases, the border area is transitional as the fauna or flora of one biogeographical region intermixes with the fauna or flora of an adjacent biogeographical region. Two cases will illustrate these points.

Wallacea

The famous zoogeographical transition zone between Lydekker's line and Wallace's line is sometimes called Wallacea (Figure 4.4). Oriental and the Australian faunas grade into one another in a large area of Wallacea. The faunas of both these regions thin out across the transition zone. *Wallace's line*, which passes between Bali and Lombok and along the Makassar Strait between

Borneo and Sulawesi, marks the easternmost extension of a wholly Oriental fauna. A few Oriental species (shrews, civets, pigs, deer, and monkeys) have colonized Sulawesi and Bali, but they are genetically distinct from their relatives in the Oriental region. A very few Oriental species, all of which might have been introduced, occur on the islands as far east as Timor, but no Oriental species live beyond that point. *Lydekker's line*, which passes between the Australian mainland and Timor and between New Guinea and Seram and Halmahera, follows the edge of Australia's continental shelf (the Sahul Shelf). It marks the westernmost limit of a wholly Australian fauna. A few Australian species live on some small islands a little to the west, and as far west as Sulawesi and Lombok. *Weber's line* runs west of the Moluccas and east of Timor, and marks places with an equal mix of Oriental and Australian species. It is taken by some authorities as the dividing line between the Oriental and Australian faunas. However, the search for a hard-and-fast dividing line in such a patently transitional region seems pointless.

The Isthmus of Panama

South America presently joins North America, but, for most of the last 65 million years or so, it was an island-continent. Once during that time, from about 40 to 36 million years ago, a land connection with North America may have existed as a chain of islands. From 30 million to 6 million years ago, South America remained a colossal island and mammals had no possibility of interaction with other faunal regions. Even as recently as 6 million years ago, the Bolivar Trough connected the Caribbean Sea with the Pacific Ocean and deterred the passage of animals. By 3 million years ago, a land connection – the Panamanian land bridge – had developed that supplied a gateway for faunal interchange between North and South America. A flood of mammals simply walked into South America. The passage was two-way and known as the Great

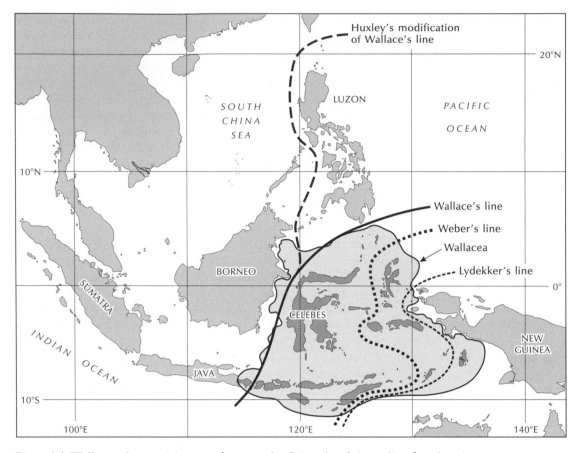

Figure 4.4 Wallacea: the transition zone between the Oriental and Australian faunal regions.

American Interchange (p. 314). Today, the Panamanian Isthmus is a biogeographical filter.

COSMOPOLITAN AND PAROCHIAL: PATTERNS OF DISTRIBUTION

All species, genera, families, and so forth have a geographical range or distribution. Distributions range in size from a few square metres to almost the entire terrestrial globe. The physical environment, the living environment, and history determine their boundaries. They tend to follow a few basic patterns – large or small, widespread or restricted, continuous or broken.

Large or small, widespread or restricted

An *endemic* species lives in only one place, no matter how large or small that place should be. A species can be endemic to Australia or endemic to a few square metres in a Romanian cave. A *pandemic* species lives in all places. The puma or cougar (*Felis concolor*), for example, is a pandemic species because it occupies nearly all the western New World, from Canada to Tierra del Fuego (Plate 4.1). It is also an endemic species of this region because it lives nowhere else. *Cosmopolitan* species inhabit the whole world, though not necessarily in all places. It is possible for a

Plate 4.1 Puma or mountain lion (*F. concolor*).
Photograph by Pat Morris.

Micro-endemic species

Some species have an extremely restricted or *micro-endemic* distribution, living as a single population in a small area. The Devil's Hole pupfish (*Cyprinodon diabolis*) is restricted to one thermal spring issuing from a mountainside in southwest Nevada, USA (Moyle and Williams 1990). The Amargosa vole (*Microtus californicus scirpensis*) inhabits freshwater marshes along a limited stretch of the Amargosa River in Inyo County, California, USA (Murphy and Freas 1988). The rare black hairstreak butterfly (*Strymonidia pruni*) is confined to a few sites in central England and in central and eastern Europe.

Endemic plant families

Two restricted and endemic flowering-plant families are the Degeneriaceae and the Leitneriaceae (Figure 4.6). The Degeneriaceae consists of a single tree species, *Degeneria vitiensis*, that grows on the island of Fiji. The Leitneriaceae also consists of a single species – the Florida corkwood (*Leitneria floridana*). This **deciduous** shrub is native to swampy areas in the southeastern USA where it is used as floats for fishing nets.

Cosmopolitan plant families

Two widespread flowering-plant families are the sunflower family (Compositae or Asteraceae) and the grass family (Poaceae or Graminae) (Figure 4.7). The Compositae is one of the largest families of flowering plants. It contains around 1,000 genera and 25,000 species. It is found everywhere except the Antarctic mainland, though it is poorly represented in tropical rain forests. The grass family comprises about 650 genera and 9,000 species, including all the world's cereal crops (including rice). Its distribution is worldwide, ranging from inside the polar circle to the equator. It is the chief component in about one fifth of the world's vegetation. Few plant formations lack grasses; some (**steppe**, **prairie**, and **savannah**) are dominated by them.

cosmopolitan species to occur in numerous small localities in all continents.

As a rule, pandemic or cosmopolitan species have widespread distributions, whereas endemic species have restricted distributions. Several small to medium mammal species in Europe have restricted and endemic distributions, including the broom hare (*Lepus castroviejoi*), the Pyrenean pine vole (*Microtus gerbei*), the Balkan snow vole (*Dinaromys bogdanovi*), and the Romanian hamster (*Mesocricetus newtoni*) (Figure 4.5). The capybara (*Hydrochoerus hydrochaeris*), the largest living rodent, is endemic to South America. It is also a pandemic, ranging over half the continent. The distinction between widespread and restricted often rests on the occurrence or non-occurrence of species within continents or biogeographical regions (Table 4.4).

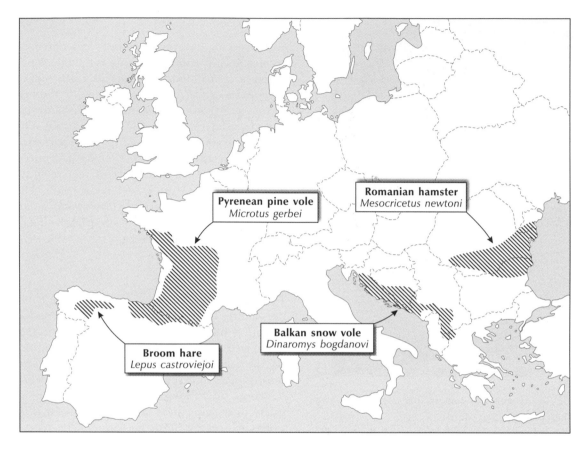

Figure 4.5 Broom hare (*L.. castroviejoi*), Pyrenean pine vole (*M. gerbei*), Balkan snow vole (*D. bogdanovi*), and Romanian hamster (*M. newtoni*) – four species with an endemic and restricted distribution.
Source: Adapted from Mitchell-Jones *et al.* (1999)

Zonal climatic distributions

Some animal and plant distributions follow climatic zones (Figure 4.8). Five relatively common zonal patterns are *pantropical* (throughout the tropics), *amphitropical* (either side of the tropics), *boreal* (northern), *austral* (southern), and *temperate* (middle latitude). The sweetsop and soursop family (Annonaceae), consisting of about 2,300 trees and shrub species, is pantropical, though centred in the Old World tropics. The sugarbeet, beetroot, and spinach family (Chenopodiaceae) consists of about 1,500 species, largely of perennial herbs, widely distributed either side of the

tropics in saline habitats. The arrowgrass family (Scheuchzeriaceae) comprises a single genus (*Scheuchzeria*) of marsh plants that are restricted to a cold north temperate belt, and are especially common in cold *Sphagnum* bogs. The poppy family (Papaveraceae) has some 250 species of mainly herbaceous annuals or perennials that are confined largely to the north temperate zone.

Continuous or broken

Plant and animal distributions have a third basic pattern – they tend to be either *continuous* or else

Table 4.4 Species classed according to range size and cosmopolitanism

Range size	Degree of cosmopolitanism			
	Endemic (peculiar)	Characteristic (shared between two biogeographical regions)	Semi-cosmopolitan (shared between three or four biogeographical regions)	Cosmopolitan (shared between five or more biogeographical regions)
Microscale	Black hairstreak butterfly (*Strymonidia pruni*) Amargosa vole (*Microtus californicus scirpensis*) Devil's Hole pupfish (*Cyprinodon diabolis*)	*Friesea oligorhopala* – a collembolan species found in Europe, Tripoli (Libya), Malta, Bahía Blanca (Argentina), and Santiago (Chile)	Skua (*Stercorarius skua*) – a coastal bird of Antarctica, southern South America, Iceland, and the Faeroes	Stenotypic ornamental plants
Mesoscale	California vole (*Microtus californicus*) Romanian hamster (*Mesocricetus newtoni*)	Rose pelican (*Pelecanus onocrotalus*) – a bird shared by central Asia and southern Africa	Olivaceus warbler (*Hippolais pallida*) – a passerine bird shared by southern Europe, the Near East, and scattered places in the Ethiopian region	Cormorant (*Phalacrocorax carbo*) – a bird from the Palaearctic, Oriental, Ethiopian, Australian, and Nearctic regions
Macro- and megascale	Capybara (*Hydrochoerus hydrochaeris*) – a pandemic	Puma or cougar (*Felis concolor*) – a pandemic of the New World	Black heron (*Nycticorax nycticorax*) – a pandemic bird of South America, North America, Africa, and Eurasia	Human (*Homo sapiens*) – a pandemic cosmopolite

Source: Adapted from Rapoport (1982)

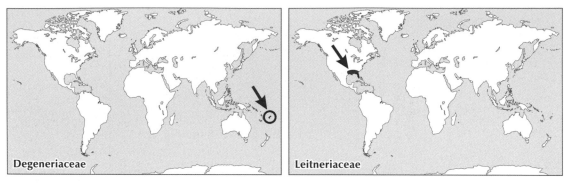

Figure 4.6 Two restricted and endemic plant families: the Degeneriaceae and Leitneriaceae.
Source: Adapted from Heywood (1978)

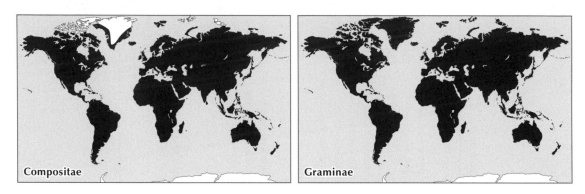

Figure 4.7 Two widespread plant families: the sunflower family (Compositae or Asteraceae) and the grass family (Poaceae or Graminae).
Source: Adapted from Heywood (1978)

broken (*disjunct*). Several factors cause broken distributions, including geological change, climatic change, evolution, and jump dispersal by natural and human agencies.

Evolutionary disjunctions

Evolutionary disjunctions occur under the following circumstances. A pair of sister species evolves on either side of an area occupied by a common ancestor. The common ancestor then becomes extinct. The extinction leaves a disjunct species pair. This mechanism may account for some amphitropical disjunct species, including the woody genera *Ficus*

and *Acacia* in Sonoran and Chihuahuan Deserts (North America) and the Chilean and Peruvian Deserts (South America) (Raven 1963).

Jump dispersal disjunctions

Jump dispersal is the rapid passage of individual organisms across large distances, often across inhospitable terrain, the jump taking less time than the individuals' life-spans (p. 37). Plants and animals that survive a long-distance jump and found a new colony lead to 'jump' disjunctions. The process is probably common, especially in plants. About 160 temperate or cool temperate

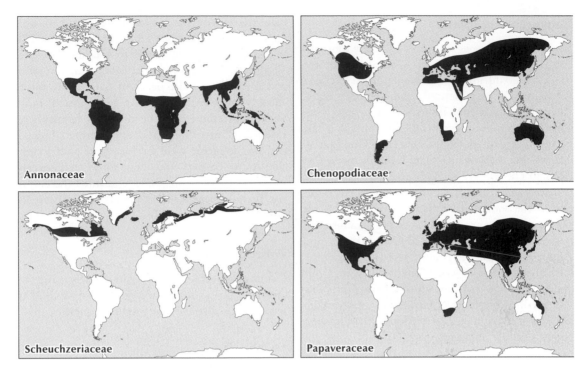

Figure 4.8 Zonal climatic distributions of four plant families: the pantropical sweetsop and soursop family (Annonaceae), the amphitropical sugarbeet, beetroot, and spinach family (Chenopodiaceae), the cold temperate arrowgrass family (Scheuchzeriaceae), and the north temperate poppy family (Papaveraceae).

Source: Adapted from Heywood (1978)

plant species or species groups have amphitropical distributions in the Americas (Raven 1963). Most of these arose from jump dispersal. Exceptions are members of the woodland genera *Osmorhiza* and *Sanicula* (Raven 1963). Tropical montane species of these genera almost bridge the gap in the disjunction. This suggests that the disjunctions are evolutionary in origin. The groups spread slowly along mountain chains. Later, members occupying the centre of the distribution became extinct, leaving the surviving members on either side of the tropics.

Humans carry species to all corners of the globe. In doing so, they create disjunct distributions. A prime example is the introduction of mammals to New Zealand. There are 54 such introduced species (C. M. King 1990), of which 20 came directly or indirectly from Britain and Europe, 14 from Australia, 10 from the Americas, 6 from Asia, 2 from Polynesia, and 2 from Africa. The package contained domestic animals for farming and household pets, and feral animals for sport or fur production. Farm animals included sheep, cattle, and horses. Domestic animals included cats and dogs. Sporting animals included pheasant, deer, wallabies, and rabbits. The Australian possum was introduced to start a fur industry. Captain James Cook liberated wild boar and goats on New Zealand. Many other species were introduced – European blackbirds, thrushes, sparrows, rooks, yellowhammers, chaffinches, budgerigars, hedgehogs, hares, weasels, stoats, ferrets, rats, and mice. Several species failed to establish themselves. These failed antipodean settlers include bandicoots, kangaroos, racoons, squirrels, bharals, gnus, camels, and zebras.

Geological disjunctions

Geological disjunctions are common in the southern continents, which formed a single landmass (Gondwana) during the **Triassic period** but have subsequently fragmented and drifted apart. Ancestral populations living on Gondwana were thus split and evolved independently. Their present-day distributions reflect their Gondwanan origins. Many flowering plant families reveal this history. The protea, banksia, and grevillea family (Proteaceae) is one of the most prominent families in the southern hemisphere (Figure 4.9). It provides numerous examples of past connections between South American, South African, and Australian floras. The genus *Gevuina*, for instance, has three species, of which one is native to Chile and the other two to Queensland and New Guinea. The break-up of **Pangaea** is also responsible for the disjunct distribution of the flightless running birds, or ratites (p. 308).

Climatic disjunctions

Climatic disjunctions result from a once widespread distribution being reduced and fragmented by climatic change. An example is the magnolia and tulip tree family (Magnoliaceae) (Figure 4.9). This family consists of about 12 genera and about 220 tree and shrub species native to Asia and America. The three American genera (*Magnolia*, *Talauma*, and *Liriodendron*) also occur in Asia. Fossil forms show that the family was once much more widely distributed in the northern hemisphere, extending into Greenland and Europe. Indeed, the Magnoliaceae were formerly part of an extensive Arcto-Tertiary deciduous forest that covered much northern hemisphere land until the end of the Tertiary period, when the decline into the Ice Age created cold climates.

Relict groups

Environmental changes, particularly climatic changes, and evolutionary processes that lead to a shrinking distribution produced these.

Climatic relicts

Climatic relicts are survivors of organisms that formerly had larger distributions. The alpine marmot (*Marmota marmota*), a large ground squirrel, lives on alpine meadows and steep rocky slopes in the Alps (Figure 4.10). It has also been

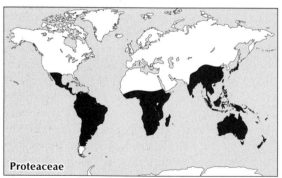
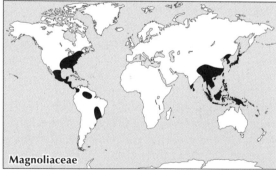

Figure 4.9 The protea, banksia, and grevillea family (Proteaceae) and the magnolia and tulip tree family (Magnoliaceae). The Proteaceae originated on Gondwana and have survived on all the southern continents. The Magnoliaceae once formed part of extensive Northern Hemisphere deciduous forests that retreated from high latitudes during the Quaternary ice age.

Source: Adapted from Heywood (1978)

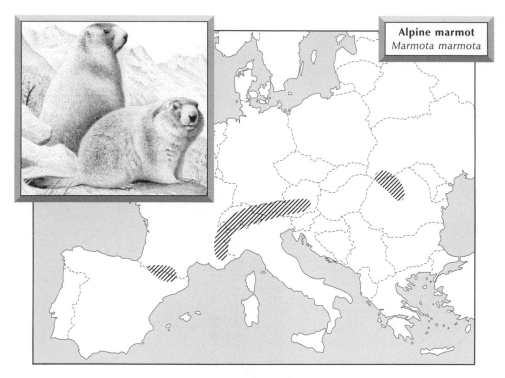

Figure 4.10 The alpine marmot (*M. marmota*) – a climatic relict.
Source: Adapted from Bjärvall and Ullström (1986)

introduced into the Pyrenees, the Carpathians, and the Black Forest. During the Ice Age, it lived on the plains area of central Europe. With Holocene warming, it became restricted to higher elevations and its present distribution, which lies between 1,000 m and 2,500 m, is a relict of a once much wider species distribution.

The Norwegian mugwort (*Artemisia norvegica*) is a small alpine plant. During the last ice age, it was widespread in northern and central Europe. Climatic warming during the Holocene epoch has left it stranded in Norway, the Ural Mountains, and three isolated sites in western Scotland.

Evolutionary relics

Evolutionary relicts are survivors of ancient groups of organism. The tuatara (*Sphenodon punctatus*) is the only native New Zealand reptile (Plate 4.2).

It is the sole surviving member of the reptilian order Rhyncocephalia, and is a 'living fossil'. It lives nowhere else in the world. Why has the tuatara survived on New Zealand but other reptiles (and marsupials and monotremes) have not? New Zealand in the Cretaceous period lay at latitude 60° to 70° S. The climate would have been colder, and the winter nights longer. Most reptiles and mammals could not have tolerated this climate. The tuatara has a low metabolic rate, remaining active at a temperature of 11°C, which is too cold to allow activity in any other reptile. It may have survived because it can endure cold conditions. Its long isolation on an island, free from competition with other reptiles and mammals, may also have helped.

Cycads belong to the family Cycadaceae, which comprises 9 genera and about 100 species, all of which are very rare and have highly restricted

Plate 4.2 Tuatara (*S. punctatus*).
Photograph by Pat Morris.

Plate 4.3 A cycad – *Zamia lindenii* – in Ecuador.
Photograph by Pat Morris.

distributions confined to the tropics and subtropics (Plate 4.3). Early members of the cycads were widely distributed during the Mesozoic era – specimens have been unearthed in Oregon, Greenland, Siberia, and Australia. They may have been popular items on dinosaurian menus. The reduction of cycad distribution since the Mesozoic may have partly resulted from competition with flowering plants (which reproduce more efficiently and grow faster). Climatic change may also have played a role. Over the last 65 million years, tropical climates slowly pulled back to the equatorial regions as the world climates underwent a cooling. Cycads are therefore in part climatic and in part evolutionary relicts. They commonly maintain a foothold in isolated regions – their seeds survive prolonged immersion in seawater and the group has colonized many Pacific islands.

RANGE REGULARITIES: AREOGRAPHY

Range size

The areas occupied by species vary enormously. In central and North America, the average area occupied by bear species (Ursidae) is 11.406 million km^2; for cat species (Felidae) it is 5.772 million km^2; for squirrel, chipmunk, marmot, and prairie dog species (Sciuridae) it is 0.972 million km^2; and for pocket gopher species (Geomyidae) it is 0.284 million km^2 (Rapoport 1982, 7). The range occupied by individual species spans 100 km^2 for such rodents as Desmarest's spiny pocket mouse (*Heteromys desmarestianus*) to 20.59 million km^2 for the wolf (*Canis lupus*).

Species range in central and North America is related to feeding habits (Table 4.5). Large carnivores tend to occupy the largest ranges. Large

Table 4.5 Mean geographical ranges of central and North American mammal species grouped by order

Order	Mean range area (millions of km²)
Carnivora (carnivores)	6.174
Artiodactyla (even-toed ungulates)	5.072
Lagomorpha (rabbits, hares, and pikas)	1.926
Chiroptera (bats)	1.487
Marsupialia (marsupials)	1.130
Insectivora (insectivores)	1.117
Xenarthra or Edentata (anteaters, sloths, and armadillos)	0.889
Rodentia (rodents)	0.764
Primates (primates)	0.249

Source: Adapted from Rapoport (1982)

herbivores come next, followed by smaller mammals. The larger range of carnivores occurs in African species, too. The mean species range for carnivores (Canidae, Felidae, and Hyaenidae) is 8.851 million km²; the mean species range for herbivores (Bovidae, Equidae, Rhinocerotidae, and Elephantidae) is 3.734 million km².

Geographical regularities in range size and shape

The *size* and *shape* of ranges are related. Figure 4.11a is a scattergraph of the greatest north–south and the east–west range dimensions for North America snake species (Brown 1995). In the graph, ranges equidistant in north–south and east–west directions will lie on the line of equality that slopes at 45 degrees upwards from the origin (Figure 4.11b). Ranges stretched out in a north–south direction will lie below the line of equality; ranges that stretch in an east–west direction lie above the line of equality. The pattern for North American snakes is plain enough (Figure 4.11a). Small ranges lie mainly above the line (these are stretched in a north–south direction); large ranges tend to lie below the line (these are stretched in an east–west direction).

There is a plausible reason for these patterns, which also occur in lizards, birds, and mammals (Brown 1995, 110–11). Local or regional environmental conditions limit species with small geographical ranges. The major mountain ranges, river valleys, and coastlines in North America run roughly north to south. The soils, climates, and vegetation associated with these north–south physical geographical features may determine the boundaries of small-range species. Large-range species are distributed over much of the continent. Local and regional environmental factors cannot therefore influence their ranges. Instead, they may be limited by large-scale climatic and vegetational patterns, which display a zonary arrangement, running east–west in wide latitudinal belts.

Range size tends to increase with increasing latitude. In other words, on average, geographical ranges are smaller in the tropics than they are near the poles. Moreover, species whose ranges are centred at increasingly higher latitudes (nearer the poles) tend to be distributed over an increasingly wider range of latitudes. This relationship is *Rapoport's rule* (G. C. Stevens 1989), named after its discoverer (see Rapoport 1982). The same pattern holds for altitudinal distributions: within the same latitude, the altitudinal range of a species increases with the midpoint elevation of the range (Stevens 1992). These biogeographical patterns are basic, holding for organisms from land mammals to **coniferous** trees. Five hypotheses may explain them – continental geometry, congenial environments, constraints on dispersal, climatic change, and tropical competition (Brown 1995, 112–14; see also Pither 2002) (Box 4.1).

Rapoport's rule does not apply everywhere. The range size of Australian mammals does not increase from the tropics towards the poles (F. D. M. Smith *et al.* 1994). Latitudinal and longitudinal variations in range size correlate with continental width. Moreover, the arid centre of Australia contains the fewest mammal species

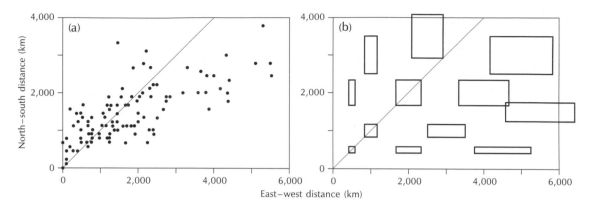

Figure 4.11 (a) A scattergraph of the greatest north–south versus the greatest east–west dimension of North American snake species ranges. The line represents ranges whose two dimensions are equal. (b) Some example shapes and sizes of ranges. For convenience, the ranges are shown as squares and rectangles, but they could be circles, ellipses, or any other shape. The diagonal line represents ranges with two equal dimensions. Above the line, ranges are stretched in a north–south direction; below the line, they are stretched in and east–west direction.

Source: Adapted from Brown (1995)

with the largest ranges, while the moist and mountainous east coast, and the monsoonal north, contain large numbers of species with small ranges.

Range change

Species ranges alter through dispersal and local extinctions. Acting in tandem, dispersal and extinction may lead to *range expansion* (through all or any of the dispersal processes), to *range contraction* (from local extinction), or to *range 'creep'* (through a mixture of spread and local extinction). It is far from easy to establish the processes involved in actual cases.

Some organisms – information is too scanty to say how many – have an actual geographical range that is smaller than their potential geographical range. In other words, some species do not occupy all places that their ecological tolerances would allow. Often, a species has simply failed to reach the 'missing bits' by dispersing. To an extent, the actual range of a species is a dynamic, statistical phenomenon that is constrained by the environment: in an unchanging habitat, the geographical range of a species can shift owing to the changing balance between local extinction and local invasion. Moreover, it may enlarge or contract owing to historical factors, as so plainly shown by the spread of many introduced species and chance colonizers in new, but environmentally friendly, regions.

SUMMARY

Species are not uniformly distributed over the land surface. Fauna and flora display regional differences. The largest regions of animals and plants are biogeographical regions, each bearing a distinctive fauna and flora. Some families and even some orders of animals are endemic to particular biogeographical regions. Others families are shared by two or more regions. A few families are cosmopolitan, being found in all biogeographical regions. Biological evolution and geological evolution have acted together to produce the

Box 4.1

ACCOUNTING FOR REGULARITIES IN RANGE SIZE

Continental geometry

Rapoport's rule and the tendency of small ranges to occur at low latitudes follow from the geometry of North America. The continent tapers from north to south so species become more tightly packed in lower latitudes. Against this idea, species ranges become smaller before the tapering becomes marked. Moreover, it does not account for the elevational version of Rapoport's rule.

Congenial environments

The abiotic environments in low latitudes are more favourable and less variable than those in high latitudes. Small-range species living in low latitudes can avoid extinction by surviving in 'sink' habitats, and recolonize favourable 'source' habitats and re-establish populations after local extinctions.

Constraints on dispersal

Barriers to dispersal are more severe in low latitudes, which may contribute to the evolution and persistence of narrowly endemic species. The logic of this idea is that a species living at high latitudes could pass over mountains in summer without experiencing harsher conditions than it experiences normally. In low latitudes, however, species used to living in tropical lowlands and trying to cross a mountain pass at an equivalent elevation would experience conditions never experienced before, no matter what the season. The same would be true of a tropical montane species trying to cross tropical lowlands in the winter.

Climatic change

Large-range species have adapted to the Pleistocene climatic swings at high latitudes.

Tropical competition

Range size increases and species diversity decreases in moving from the equator towards the poles. Interaction between species may be a stronger limiting factor in lower latitudes.

---

biogeographical regions, subregions, and the transition zones between them seen today. Species, genera, and families display three basic distributional patterns – large or small, widespread or restricted, and continuous or broken. Relict groups are remnants of erstwhile widespread groups that have suffered extinction over much of their former range, owing to climatic or evolutionary changes. Range size and shape display relatively consistent relationships with latitude and altitude.

ESSAY QUESTIONS

1 **Why do different regions carry distinct assemblages of animals and plants?**

2 **What conditions favour the survival of relict groups?**

3 **Why do species ranges tend to become bigger with increasing altitude and latitude?**

FURTHER READING

Cox, C. B. (2001) The biogeographic regions reconsidered. *Journal of Biogeography* 28, 511–23.
A very good paper.

Good, R. (1974) *The Geography of the Flowering Plants*, 4th edn. London: Longman.
Worth perusing.

Wallace, A. R. (1876) *The Geographical Distribution of Animals; With A Study of the Relations of Living and Extinct Faunas as Elucidating the Past Changes of the Earth's Surface*, 2 vols. London: Macmillan & Co.
All biogeographers should read this great work.

PART II ECOLOGICAL BIOGEOGRAPHY

5

HABITATS, ENVIRONMENTS, AND NICHES

Life is adapted to nearly all Earth surface environments. This chapter covers:

- places to live
- requirements of living
- constraints on living
- ways of living

A PLACE TO LIVE: HABITATS

Individuals, species, and populations, both marine and terrestrial, tend to live in particular places. These places are **habitats**. A specific set of environmental conditions – radiation and light, temperature, moisture, wind, fire frequency and intensity, gravity, **salinity**, currents, topography, soil, substrate, geomorphology, human disturbance, and so forth, characterizes each habitat.

Habitats come in all shapes and sizes, occupying the full sweep of geographical scales. They range from small (microhabitats), through medium (mesohabitats) and large (macrohabitats), to very large (megahabitats). *Microhabitats* are a few square centimetres to a few square metres in area (Table 5.1). They include leaves,

the soil, lake bottoms, sandy beaches, talus slopes, walls, riverbanks, and paths. *Mesohabitats* have areas up to about 10,000 km²; that is, a 100 × 100 kilometre square, which is about the size of Cheshire, England. Similar features of geomorphology and soils, a similar set of disturbance regimes, and the same regional climate influence each main mesohabitat. Deciduous woodland, caves, and streams are examples. *Macrohabitats* have areas up to about 1,000,000 km², which is about the size of Ireland. *Megahabitats* are regions more than 1,000,000 km² in extent. They include continents and the entire land surface of the Earth.

Landscape ecologists, who have an express interest in the geographical dimension of ecosystems, recognize three levels of 'habitat' – region,

Table 5.1 Habitat scales

Scale[a]	Approximate area (km²)	Terminology applied to landscape units at same scale[b]		
		Fenneman (1916)	Linton (1949)	Whittlesey (1954)
Microhabitat (small)	< 1	–	Site	–
Mesohabitat (medium)	1–10	–	–	–
	10–100	–	Stow	Locality
	100–1,000	District	Tract	District
	1,000–10,000	Section	Section	–
Macrohabitat (large)	10,000–100,000	Province	Province	Province
	100,000–1,000,000	Major division	Major division	Realm
Megahabitat (very large)	> 1,000,000	–	Continent	–

Notes: a These divisions follow Delcourt and Delcourt (1988). b The range of areas associated with these regional landscape units are meant as a rough-and-ready guide rather than precise limits

landscape, and landscape element. These correspond to large-scale, medium-scale, and small-scale habitats. Some landscape ecologists are relaxing their interpretation of a landscape to include smaller and larger scales – they have come to realize that a beetle's view or a bird's view of the landscape is very different from a human's view.

Landscape elements

Landscape elements are similar to microhabitats, but a little larger. They are fairly uniform pieces of land, no smaller than about 10 m, which form the building blocks of landscapes and regions. They are also called ecotopes, biotopes, geotopes, facies, sites, tesserae, landscape units, landscape cells, and landscape prisms. These terms are roughly equivalent to landscape element, but each has its own meaning (see Forman 1995; Huggett 1995).

Landscape elements are made of individual trees, shrubs, herbs, and small buildings. There are three basic kinds of landscape element – patches, corridors, and background matrixes:

- *Patches* are fairly uniform (homogeneous) areas that differ from their surroundings. Woods, fields, ponds, rock outcrops, and houses are all patches.

- *Corridors* are strips of land that differ from the land to either side. They may interconnect to form networks. Roads, hedgerows, and rivers are corridors.

- *Background matrixes* are the background ecosystems or land-use types in which patches and corridors are set. Examples are deciduous forest and areas of arable cultivation.

Landscape elements include the results of human toil – roads, railways, canals, houses, and so on. Such features dominate the landscape in many parts of the world and form a kind of 'designer mosaic'. Designed patches include urban areas, urban and suburban parks and gardens (greenspaces), fields, cleared land, and reservoirs. Designed corridors include hedgerows, roads and railways, canals, dykes, bridle paths, and footpaths. There is also a variety of undesigned patches – waste tips, derelict land, spoil heaps, and so on. Chapter 8 will give more detail on landscape elements and their ecological significance.

Landscapes and regions

Landscape elements combine to form *landscapes*. A landscape is a mosaic, an assortment of patches and corridors set in a matrix, no bigger than about 10,000 km². It is 'a heterogeneous land area composed of a cluster of interacting ecosystems that is repeated in similar form throughout' (Forman and Godron 1986, 11). By way of example, the recurring cluster of interacting ecosystems that feature in the landscape around the author's home, in the foothills of the Pennines, includes woodland, field, hedgerow, pond, brook, canal, roadside, path, quarry, mine tip, disused mining incline, disused railway, farm building, and residential plot.

Landscapes combine to form *regions*, more than about 10,000 km² in area. They are collections of landscapes sharing the same macroclimate. All Mediterranean landscapes share a seasonal climate characterized by mild, wet winters and hot, droughty summers.

THE BARE NECESSITIES: HABITAT REQUIREMENTS

It is probably true to say that no two species have exactly the same living requirements. There are two extreme cases – fussy species or habitat specialists and unfussy species or habitat generalists – and all grades of 'fussiness' between.

Habitat specialists and habitat generalists

Habitat specialists have very precise living requirements. In southern England, the red ant, *Myrmica sabuleti*, needs dry **heathland** with a warm south-facing aspect that contains more than 50 per cent grass species, and that has been disturbed within the previous five years (N. R. Webb and Thomas 1994). Other species are less pernickety and thrive over a wider range of environmental conditions. The three-toed woodpecker (*Picoides tridactylus*) lives in a broad swathe of cool temperate forest encircling the northern hemisphere. Races of the common jay (*Garrulus glandarius*) occupy a belt of oak and mixed deciduous woodland stretching from Britain to Japan.

Habitat generalists manage to eke out a living in a great array of environments. The human species (*Homo sapiens*) is the champion habitat generalist – the planet Earth is the human habitat. In the plant kingdom, the broad-leaved plantain (*Plantago major*), typically a species of grassland habitats, is found almost everywhere except Antarctica and the dry parts of northern Africa and the Middle East. In the British Isles, it seems indifferent to climate and soil conditions, growing in all grasslands on acid and alkaline soils alike. It also lives on paths, tracks, disturbed habitats (spoil heaps, demolition sites, arable land), pasture and meadows, road verges, riverbanks, mires, skeletal habitats, and as a weed in lawns and sports fields. In woodland, it lives only in relatively unshaded areas along rides. It does not live in aquatic habitats or tall herb communities.

Edge species and interior species

Interior species live in the core of a habitat and favour large patches, which have proportionally more core habitat than small patches. They actively avoid the habitat edges if they are able to meet their resource needs within their territories or home ranges. English woodland examples include the great spotted woodpecker (*Dendrocopos major*) and the nuthatch (*Sitta europaea*).

Edge species use a habitat edge and are more common in small patches. Two types of edge species are recognized, the first of which are intrinsically edge species, and the second of which are ecotonal species (McCollin 1998). Ecotonal species occur near the edge because the edge habitat suits them. They are not dependent on adjacent habitats for food, shelter, or anything else. Intrinsic edge species live near edges because the adjacent habitat provides resources. For instance, in highly fragmented agricultural landscapes, bird species living in woodland edges next to open country depend upon food resources offered by farmland. Examples are the rook (*Corvus frugilegus*) and the carrion crow (*C. corone corone*), which feed mainly on grain, earthworms and their eggs, and grassland insects, with the crow also taking small mammals and carrion; and the starling (*Sturnus vulgaris*), which feeds on leatherjackets and earthworms in the upper soil layers of pasture (McCollin 1998).

LIFE'S LIMITS: ECOLOGICAL TOLERANCE

Organisms live in virtually all environments, from the hottest to the coldest, the wettest to the driest, the most acidic to the most alkaline. Understandably, humans tend to think of their 'comfortable' environment as the norm. But moderate conditions are anathema to the **microorganisms** that love conditions fatal to other creatures. These are the **extremophiles** (Madigan and Marrs 1997). An example is high-pressure-loving microbes (barophiles) that flourish in deep-sea environments and are adapted to life at high pressures (Bartlett 1992). Many other organisms are adapted to conditions that, by white western human standards, are harsh, though not so extreme as the conditions favoured by the extremophiles. Examples are hot deserts and Arctic and alpine regions.

Limiting factors

A *limiting factor* is an environmental factor that slows down population growth. Justus von Liebig (1840), a German agricultural chemist, first suggested the term. He noticed that whichever **nutrient** happens to be in short supply limits the growth of a field crop. A field of wheat may have ample phosphorus to yield well, but if another nutrient – say nitrogen – should be lacking, then the yield lessens. No matter how much extra phosphorus is applied in fertilizer, the lack of nitrogen will limit wheat yield. Only by making good the nitrogen shortage could yields be improved. These observations led Liebig to establish a *'law of the minimum'*: the productivity, growth, and reproduction of organisms will be

constrained if one or more environmental factors lies below its limiting level.

Later, ecologists established a *'law of the maximum'*. This law applies where population growth is curtailed by an environmental factor exceeding an upper limiting level. In a wheat field, too much phosphorus is as harmful as too little – there is an upper limit to nutrient levels tolerated by plants.

Tolerance range

For every environmental factor, such as temperature or moisture, there are three 'zones': a lower limit, below which a species cannot live, an optimum range in which it thrives, and an upper limit, above which it cannot live (Figure 5.1). The upper and lower bounds define the tolerance range of a species for a particular environmental factor. The bounds vary from species to species. A species will prosper within its optimum range of tolerance; survive but show signs of physiological stress near its tolerance limits; and not survive outside its tolerance range (Shelford 1911). *Stress* is a widely used but troublesome idea in ecology. It may be defined as 'external constraints limiting the rates of resource acquisition, growth or reproduction of organisms' (Grime 1989).

Each species (or race) has a characteristic tolerance range (Figure 5.2). *Stenoecious* species have a wide tolerance; *euryoecious* species have a narrow tolerance. All species, regardless of their tolerance range, may be adapted to the low end (*oligotypic*), to the middle (*mesotypic*), or to the high end (*polytypic*) of an **environmental gradient**. Take the example of photosynthesis in plants. Plants adapted to cool temperatures (oligotherms) have photosynthetic optima at about 10°C and cease to photosynthesize above 25°C. Temperate-zone plants (mesotherms) have optima between 15°C and 30°C. Tropical plants (polytherms) may have optima as high as 40°C. Interestingly, these optima are not 'hard and fast'. Cold-adapted plants are able to shift their photosynthetic optima towards higher temperatures when they are grown under warmer conditions.

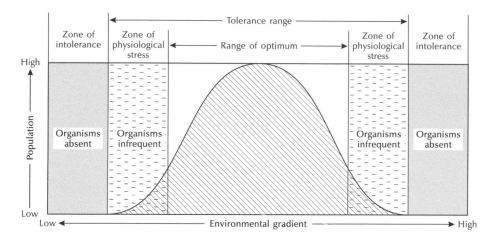

Figure 5.1
Tolerance range and limits.
Source:
Developed from Shelford (1911)

Ecological valency

Tolerance may be wide or narrow and the optimum may be at low, middle, or high positions along an environmental gradient. When combined, these contingencies produce six grades of *ecological valency* (Figure 5.2). The glacial flea (*Isotoma saltans*), a species of springtail, has a narrow temperature tolerance and likes it cold. It is an oligostenotherm. The midge *Liponeura cinerascens*, a grazing stream insect, has a narrow oxygen-level tolerance at the high end of the oxygen-level gradient. It is a polystenoxybiont. Other examples are shown on Figure 5.2.

ADAPTING TO CIRCUMSTANCES: NICHES AND LIFE-FORMS

Ways of living

Organisms have evolved to survive in the varied conditions found at the Earth's surface. They have come to occupy nearly all habitats and to fill multifarious roles within food chains.

Ecological niche

An organism's *ecological niche* (or simply *niche*) is its 'address' and 'profession'. Its address or home

is the habitat in which it lives, and is sometimes called the *habitat niche*. Its profession or occupation is its position in a food chain, and is sometimes called the *functional niche*. A skylark's (*Alauda arvensis*) address is open moorland (and, recently, arable farmland); its profession is insect-cum-seed eater. A merlin's (*Falco columbarius*) address is open country, especially moorland; its profession is a bird-eater, skylark and meadow pipit (*Anthus pratensis*) being its main prey. A grey squirrel's (*Sciurus carolinensis*) habitat niche is a deciduous woodland; its profession is a nut-eater (small herbivore). A grey wolf's (*Canis lupus*) habitat niche is cool temperate coniferous forest, and its profession is large-mammal-eater.

A distinction is drawn between the fundamental niche and the realized niche. The *fundamental* (or *virtual*) *niche* circumscribes where an organism would live under optimal physical conditions and with no competitors or predators. The *realized* (or *actual*) *niche* is always smaller, and defines the 'real-world' niche occupied by an organism constrained by biotic and abiotic limiting factors.

A niche reflects how an individual, species, or population interacts with and exploits its environment. It involves adaptation to environmental conditions. The **competitive exclusion principle** (p. 192) precludes two species occupying

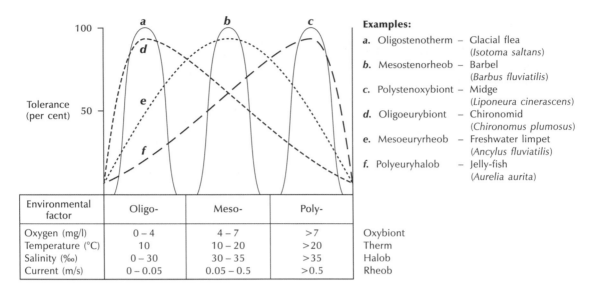

Examples:

a. Oligostenotherm – Glacial flea (*Isotoma saltans*)
b. Mesostenorheob – Barbel (*Barbus fluviatilis*)
c. Polystenoxybiont – Midge (*Liponeura cinerascens*)
d. Oligoeurybiont – Chironomid (*Chironomus plumosus*)
e. Mesoeuryrheob – Freshwater limpet (*Ancylus fluviatilis*)
f. Polyeuryhalob – Jelly-fish (*Aurelia aurita*)

Environmental factor	Oligo-	Meso-	Poly-	
Oxygen (mg/l)	0 – 4	4 – 7	>7	Oxybiont
Temperature (°C)	10	10 – 20	>20	Therm
Salinity (‰)	0 – 30	30 – 35	>35	Halob
Current (m/s)	0 – 0.05	0.05 – 0.5	>0.5	Rheob

Figure 5.2 Ecological valency, showing the amplitude and position of the optimum.
Source: Adapted from Illies (1974)

identical niches. However, a group of species, or **guild**, may exploit the same class of environmental resources in a similar way (R. B. Root 1967; Simberloff and Dayan 1991). In oak woodland, one guild of birds forages for arthropods from the foliage of oak trees; another catches insects in the air; another eats seeds. The foliage-gleaning guild in a California oak woodland includes members of four families: the plain titmouse (*Parus inornatus*, Paridae), the blue-grey gnat-catcher (*Polioptila caerulea*, Sylviidae), the warbling vireo and Hutton's vireo (*Vireo gilvus* and *V. huttoni*, Vireonidae), and the orange-crowned warbler (*Vermivora celata*, Parulidae) (R. B. Root 1967).

Ecological equivalents

Although only one species occupies each niche, different species may occupy the same or similar niches in different geographical regions. These species are **ecological equivalents** or **vicars**. A grassland ecosystem contains a niche for large herbivores living in herds. Bison and the prong-horn antelope occupy this niche in North America; antelopes, gazelles, zebra, and eland in Africa; wild horses and asses in Europe; the pampas deer and guanaco in South America; and kangaroos and wallabies in Australia. As this example shows, quite distinct species may become ecological equivalents through historical and geographical accidents.

Many bird guilds have ecological equivalents on different continents. The nectar-eating (nectivore) guild has representatives in North America, South America, and Africa. In Chile and California, the representatives are the hummingbirds (Trochilidae) and the African representatives are the sunbirds (Nectariniidae). One remarkable convergent feature between hummingbirds and sunbirds is the iridescent plumage.

Plant species of very different stock growing in different areas, when subjected to the same environmental pressures, have evolved the same life-form to fill the same ecological niche. The American cactus and the South African euphorbia, both living in arid regions, have adapted by evolving fleshy, succulent stems and by evolving spines instead of leaves to conserve precious moisture.

Life-forms

The structure and physiology of plants and, to a lesser extent, animals are often adapted for life in a particular habitat. These structural and physiological adaptations are reflected in life-form and often connected with particular ecozones.

The **life-form** of an organism is its shape or appearance, its structure, its habits, and its kind of life history. It includes overall form (such as herb, shrub, or tree in the case of plants), and the form of individual features (such as leaves). Importantly, the dominant types of plant in each ecozone tend to have a life-form finely tuned for survival under that climate.

Plant life-forms

A widely used classification of *plant life-forms*, based on the position of the shoot-apices (the tips of branches) where new buds appear, was designed by Christen Raunkiaer in 1903 (see Raunkiaer 1934). It distinguishes five main groups: therophytes, cryptophytes, hemicryptophytes, chamaephytes, and phanerophytes (Box 5.1).

A *biological spectrum* is the percentages of the different life-forms in a given region. The 'normal spectrum' is a kind of reference point; it is the percentages of different life-forms in the world flora. Each ecozone possesses a characteristic biological spectrum that differs from the 'normal spectrum'. Tropical forests contain a wide spectrum of life-forms, whereas in extreme climates, with either cold or dry seasons, the spectrum is smaller (Figure 5.4). As a rule of thumb, very predictable, stable climates, such as humid tropical climates, support a wider variety of plant life-forms than do regions with inconstant climates, such as arid, Mediterranean, and alpine climates. Alpine regions, for instance, lack trees, the dominant life-form being dwarf shrubs (chamaephytes). In the Grampian Mountains, Scotland, 27 per cent of the species are chamaephytes, a figure three times greater than the percentage of chamaephytes in

the world flora (Tansley 1939). Some life-forms appear to be constrained by climatic factors. Megaphanerophytes (where the regenerating parts stand over 30 m from the ground) are found only where the mean annual temperature of the warmest month is 10°C or more. Trees are confined to places where the mean summer temperature exceeds 10°C, both altitudinally and latitudinally. This uniform behaviour is somewhat surprising as different taxa are involved in different countries. Intriguingly, dwarf shrubs, whose life cycles are very similar to those of trees, always extend to higher altitudes and latitudes than trees do (Grace 1987).

Individual parts of plants also display remarkable adaptations to life in different ecozones. This is very true of leaves. In humid tropical lowlands, forest trees have evergreen leaves with no lobes. In regions of Mediterranean climate, plants have small, sclerophyllous evergreen leaves. In arid regions, stem succulents without leaves, such as **cacti**, and plants with entire leaf margins (especially among **evergreens**) have evolved. In cold wet climates, plants commonly possess notched or lobed leaf margins.

Animal life-forms

Animal life-forms, unlike those of plants, tend to match taxonomic categories rather than ecozones. Most mammals are adapted to basic habitats and may be classified accordingly. They may be adapted for life in water (*aquatic* or swimming mammals), underground (*fossorial* or burrowing mammals), on the ground (**cursorial** or running, and **saltatorial** or leaping mammals), in trees (*arboreal* or climbing mammals), and in the air (*aerial* or flying mammals) (Osburn *et al.* 1903). None of these habitats strongly relates to climate. That is not to say that animal species are not adapted to climate: there are many well-known cases of adaptation to marginal environments, including deserts (p. 99–101) (see Cloudsley-Thompson 1975b).

Box 5.1

PLANT LIFE-FORMS

Phanerophytes

Phanerophytes (from the Greek *phaneros*, meaning visible) are trees and large shrubs (Figure 5.3). They bear their buds on shoots that project into the air and are destined to last many years. The buds are exposed to the extremes of climate. The primary shoots, and in many cases the lateral shoots as well, are negatively geotropic (they stick up into the air). Weeping trees are an exception. Raunkiaer divided phanerophytes into 12 subtypes according to their bud covering (with bud-covering or without it), habit (deciduous or evergreen), and size (mega, meso, micro, and nano), and into three other subtypes – herbaceous phanerophytes, epiphytes, and stem succulents. A herbaceous example is the native cabbage (*Scaevola koenigii*). Phanerophytes are divided into four size classes: megaphanerophytes (> 30 m), mesophanerophytes (8–30 m), microphanerophytes (2–8 m), and nanophanerophytes (< 2 m).

Chamaephytes

Chamaephytes (from the Greek *khamai*, meaning on the ground) are small shrubs, creeping woody plants, and herbs. They bud from shoot-apices very close to the ground. The flowering shoots project freely into the air but live only during the favourable season. The persistent shoots bearing buds lie along the soil, rising no more than 20–30 cm above it. Suffructicose chamaephytes have erect aerial shoots that die back to the ground when the unfavourable season starts. They include species of the Labiatae, Caryophyllaceae, and Leguminosae. Passive chamaephytes have procumbent persistent shoots – they are long,

slender, comparatively flaccid, and heavy, and so lie along the ground. Examples are the greater stitchwort (*Stellaria holostea*) and the prostrate speedwell (*Veronica prostrata*). Active chamaephytes have procumbent persistent shoots that lie along the ground because they are transversely geotropic in light (take up a horizontal position in response to gravity). Examples are the heath speedwell (*V. officinalis*), the crowberry (*Empetrum nigrum*), and the twinflower (*Linnaea borealis*). Cushion plants are transitional to hemicryptophytes. They have very low shoots, very closely packed together. Examples are the hairy rock-cress (*Arabis hirusa*) and the house-leek (*Sempervivum tectorum*).

Hemicryptophytes

Hemicryptophytes (from the Greek *kryptos*, meaning hidden) are herbs growing rosettes or tussocks. They bud from shoot-apices located in the soil surface. They include protohemicryptophytes (from the base upwards, the aerial shoots have elongated internodes and bear foliage leaves) such as the vervain (*Verbena officinalis*), partial rosette plants such as the bugle (*Ajuga reptans*), and rosette plants such as the daisy (*Bellis perennis*).

Cryptophytes

Cryptophytes are tuberous and bulberous herbs. They are even more 'hidden' than hemicryptophytes – their buds are completely buried beneath the soil, thus affording them extra protection from freezing and drying. They include geophytes (with rhizomes, bulbs, stem tubers, and root tuber varieties) such as the purple crocus (*Crocus vernus*), helophytes or marsh plants

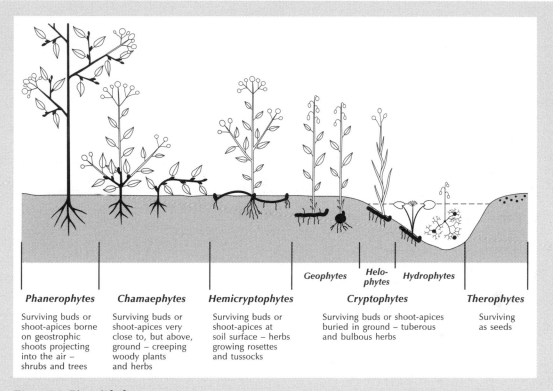

Figure 5.3 Plant life-forms.

Source: Adapted from Raunkiaer (1934)

such as the arrowhead (*Sagittaria sagittifolia*), and hydrophytes or water plants such as the rooted shining pondweed (*Potamogeton lucens*) and the free-swimming frogbit (*Hydrocharis morsus-ranae*).

Therophytes

Therophytes (from the Greek *theros*, meaning summer) or annuals are plants of the summer or favourable season and survive the adverse season as seeds. Examples are the cleavers (*Galium aparine*), the cornflower (*Centaurea cyanus*), and the wall hawk's-beard (*Crepis tectorum*).

Autoecological accounts

Detailed habitat requirements of individual species require careful and intensive study. A groundbreaking study comprised the *autoecological accounts* prepared for plants around Sheffield, England (Grime *et al*. 1988). The Natural

Environment Research Council's Unit of Comparative Plant Ecology (formerly the Nature Conservancy Grassland Research Unit) studied about 3,000 km^2 in three separate surveys. The region comprises two roughly equal portions: an 'upland' region, mainly above 200 m and with mean annual precipitation more than 850 mm,

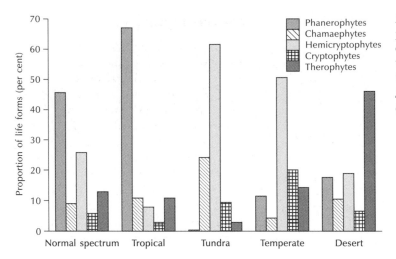

Figure 5.4 The proportion of plant life-forms in various ecozones. The 'normal' spectrum was constructed by selecting a thousand species at random.

Sources: From data in Raunkiaer (1934) and Dansereau (1957)

underlain by Carboniferous Limestone, Millstone Grit, and Lower Coal Measures; and a drier, 'lowland' region overlying Magnesian Limestone, Bunter Sandstone, and Keuper Marl.

Figure 5.5 is an example of the 'autoecological accounts' for the bluebell (*Hyacinthoides non-scripta*). The bluebell is a polycarpic perennial, rosette-forming geophyte, with a deeply buried bulb. It appears above ground in the spring, when it exploits the light phase before the development of a full summer canopy. It is restricted to sites where the light intensity does not fall below 10 per cent of the daylight between April and mid-June, in which period the flowers are produced. Shoots expand during the late winter and early spring. The seeds are gradually shed, mainly in July and August. The leaves are normally dead by July. There is then a period of aestivation (dormancy during the dry season). This ends in the autumn when a new set of roots forms. The plant cannot replace damaged leaves and is very vulnerable to grazing, cutting, or trampling. Its foliage contains toxic glycosides and, though sheep and cattle will eat it, rabbits will not. Its reproductive strategy is intermediate between a stress-tolerant ruderal and a competitor–stress-tolerator–ruderal (p. 177). It extends to 340 m around Sheffield, but is known to grow up to 660 m in the British Isles.

It is largely absent from skeletal habitats and steep slopes. The bluebell commonly occurs in woodland. In the Sheffield survey, it was recorded most frequently in broad-leaved plantations. It was also common in scrub and woodland overlying either acidic or limestone beds, but less frequent in coniferous plantations. It occurs in upland areas on waste ground and heaths, and occasionally in unproductive pastures, on spoil heaps, and on cliffs. In woodland habitats, it grows more frequently and is significantly more abundant on south-facing slopes. However, in unshaded habitats, it prefers north-facing slopes. It does not occur in wetlands. It can grow on a wide range of soils, but it most frequent and more abundant in the pH range 3.5–7.5. It is most frequent and abundant in habitats with much tree litter and little exposed soil, though it is widely distributed across all bare-soil classes.

Social factors

Home range

Individuals of a species, especially vertebrate species, may have a home range. A *home range* is the area traversed by an individual (or by a pair, or by a family group, or by a social group) in its

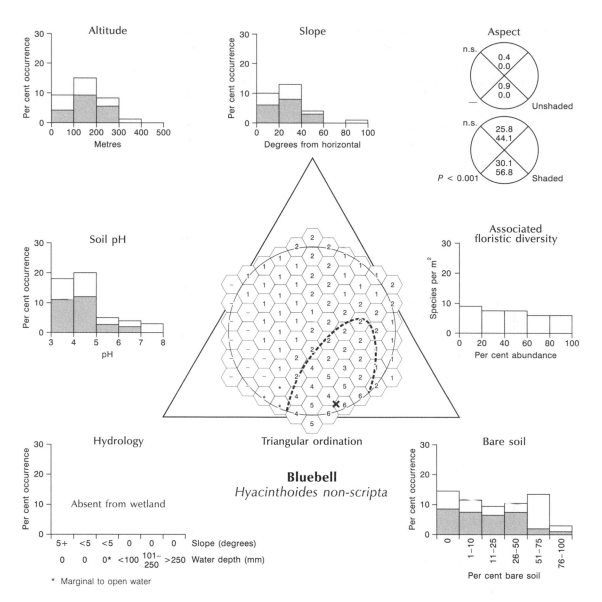

Figure 5.5 Autoecological accounts for the bluebell (*H. non-scripta*) in the Sheffield region, England. For the subject species, bluebell in the example, the blank bars show the percentage of simple occurrences over each of the environmental classes, and the shaded bars show the percentage of samples in which 20 per cent or more of the quadrat subsections contained rooted shoots. The 'triangular ordination' is explained in Figure 10.9. On the 'aspect' diagram, n.s. stands for 'not significant'.

Source: Adapted from Grime *et al.* (1988)

normal activities of gathering food, mating, and caring for young (Burt 1943). Home ranges may have irregular shapes and may partly overlap, although individuals of the same species often occupy separate home ranges. Land iguanas (*Conolophus pallidus*) on the island of Sante Fe, in the Galápagos Islands, have partly overlapping home ranges near cliffs and along plateaux edges (Figure 5.6a) (Christian *et al.* 1983). Red foxes (*Vulpes vulpes*) in the University of Wisconsin arboretum have overlapping ranges, too (Figure 5.6b) (Ables 1969).

The size of home ranges varies enormously (see Vaughan 1978, 306). In mammals, small home ranges are held by the female prairie vole (*Microtus ochrogaster*), mountain beaver (*Aplodontia rufa*), and the common shrew (*Sorex araneus*), at 0.014 ha, 0.75 ha, and 1.74 ha, respectively. The American badger (*Taxidea taxus*) has a medium-size home range of 8.5 km². A female grizzly bear (*Ursus arctos horribilis*) with three yearlings has a large home range of 203 km², and a pack of eight timber wolves (*Canis lupus*) has a very large home range of 1,400 km².

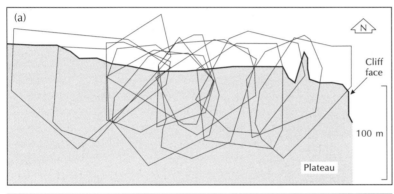

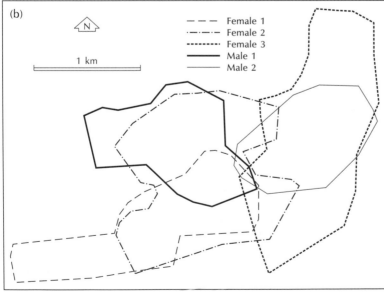

Figure 5.6 Home ranges. (a) Land iguanas (*C. pallidus*) on the island of Sante Fe, in the Galápagos Islands. (b) Red fox (*V. vulpes*) home ranges in the University of Wisconsin arboretum.

Sources: (a) Adapted from Christian *et al.* (1983). (b) Adapted from Ables (1969)

Territory

Some animals actively defend a part of their home range against members of the same species. This core area, which does not normally include the peripheral parts of the home range, is the **territory**. Species that divide geographical space in this way are territorial species. Animals may occupy territories permanently or temporarily. Breeding pairs of the tawny owl (*Strix aluco*) stay within their own territory during adulthood. Great tits (*Parus major*) establish territories only during the breeding season. The 'i'iwi (*Vestiaria coccinea*), a Hawaiian honeycreeper, exhibits territorial behaviour when food is scarce.

Habitat selection

As the conditions near the margins of ecological tolerance create stress, it follows that a species' ecological tolerance strongly influences its actual and potential geographical range. It is generally true that species with wide ecological tolerances are the most widely distributed. A species will occupy a habitat that meets its tolerance requirements, for it simply could not survive elsewhere. Nonetheless, even where a population is large and healthy, it does not necessarily occupy all favourable habitats within its geographical range, and there may be areas inside and outside its geographical range where it could live. In many cases, individuals 'choose' to live in particular habitats from those available and not to live in others, a process called **habitat selection**. Habitat selection appears to operate at different scales, with four levels being suggested (D. H. Johnson 1980; Fayt 1999; see also Pedlar *et al.* 1997):

1 *First-order selection* describes the geographical range of a given species.
2 *Second-order selection* describes the range of habitats within the home ranges of individuals.
3 *Third-order selection* describes the selection of habitats within an individual's home range.
4 *Fourth-order selection* represents the individual items of food.

Habitat selection is prevalent in birds. An early study was carried out in the Breckland of East Anglia, England (Lack 1933). The wheatear (*Oenanthe oenanthe*) lives on open heathland in Britain. It nests in old rabbit burrows. It does not occur in newly forested heathland lacking rabbits. Nesting-site selection thus excludes it from otherwise suitable habitat. The tree pipit (*Anthus trivialis*) and the meadow pipit (*A. pratensis*) are both ground nesters and feed on the same variety of organisms, but the tree pipit breeds only in areas with one or more tall trees. In consequence, in many treeless areas in Britain, the tree pipit does not live alongside the meadow pipit. David Lack found some tree pipits breeding in one treeless area close to a telegraph pole. The pipits used the pole merely as a perch on which to land at the end of their aerial song. Meadow pipits sing a similar song but land on the ground. This finding suggests that the tree pipit does not colonize heathland simply because it likes a perch from which to sing. The conclusion of the Breckland study was that the heathland and pine planation birds had a smaller distribution than they otherwise might because they selected habitat to live in. In short, they were choosy about where they lived.

SUMMARY

All living things live in particular places – habitats. Habitats range in size from a few cubic centimetres to the entire ecosphere. Species differ in their habitat requirements, the span going from habitat generalists, who live virtually anywhere, to habitat specialists, who are very choosy about their domicile. Species are constrained by limiting factors in their environment. Limiting factors include moisture, heat, and nutrient levels. Each species has a characteristic tolerance range and ecological valency. Life has to adapt to environmental conditions in the ecosphere. There are several 'ways of living', each of which corresponds to an ecological niche.

Ecological equivalents (or vicars) are species from different stock, and living in different parts of the world, that have adapted to the same environmental constraints. Adaptation to environmental conditions is also seen in life-forms. The overall habitat preferences of an individual species require detailed and intensive study. They may be summarized as autoecological accounts. Social factors affect how and where organisms live. Many species have home ranges and territories and select the habitat they wish to live in from the range of possible sites.

ESSAY QUESTIONS

1 **Compare and contrast habitat generalists and habitat specialists.**

2 **Why are vicars (ecological equivalents) so common?**

3 **How important is habitat selection in understanding the distribution of species?**

FURTHER READING

Forman, R. T. T. (1995) *Land Mosaics: the Ecology of Landscapes and Regions*. Cambridge: Cambridge University Press.
A weighty tome on landscape ecology with some useful sections for biogeographers.

Grime, J. P., Hodgson, J. G., and Hunt, R. (1988) *Comparative Plant Ecology: A Functional Approach to Common British Species*. London: Unwin Hyman.
Autoecological accounts in profusion.

CLIMATE AND LIFE

Climate is a master environmental factor imposing severe constraints on living things. This chapter covers:

- sunlight
- temperature
- moisture
- climatic zones

FLOWER POWER: RADIATION AND LIGHT

The Sun is the primary source of radiation for the Earth. It emits electromagnetic radiation across a broad spectrum, from very short wavelengths to long wavelengths (Box 6.1). The visible portion (sunlight) is the effective bit for photosynthesis. It is also significant in heating the environment. Long-wave (infrared) radiation emitted by the Earth is locally important around volcanoes, in **geothermal springs**, and in hydrothermal vents in the deep sea floor. Unusual organisms, including the thermophiles and hyperthermophiles that like it very hot, tap these internal sources of energy (p. 87).

Three aspects of **solar radiation** influence photosynthesis – the intensity, the quality, and the photoperiod or duration. The *intensity* of solar radiation is the amount that falls on a given area in a unit time. Calories per square centimetre per minute (cal/cm^2/min) were once popular units, but Watts per square metre (W/m^2) or kiloJoules per hectare (kJ/ha) are metric alternatives. The average annual solar radiation on a horizontal ground surface ranges from about 800 kJ/ha over subtropical deserts to less than 300 kJ/ha in polar regions. Equatorial regions receive less radiation than the subtropics because they are cloudier. A value of 700 kJ/ha is typical. **Heliophytes** are plants that grow best in conditions of high light intensity (full sunlight) and **sciophytes** are plants

---

Box 6.1

THE ELECTROMAGNETIC SPECTRUM EMITTED BY THE SUN

Electromagnetic radiation pours out of the Sun at the speed of light. Extreme ultraviolet radiation with wavelengths in the range 30 to 120 nanometres (nm) occupies the very short end of the spectrum. Ultraviolet light extends to wavelengths of 0.4 micrometres (µm). Visible light has wavelengths in the range 0.4 to 0.8 µm. This is the portion of the electromagnetic spectrum humans can see. Infrared radiation has wavelengths longer than 0.8 µm. It grades into radio frequencies with millimetre to metre wavelengths. The Sun emits most intensely near 0.5 µm, which is in the green band of the visible light. This fact might help to account for plants being green – they reflect the most intense band of sunlight.

---

adapted to conditions of low light intensity (shade).

The *quality* of solar radiation is its wavelength composition. This varies from place to place depending on the composition of the **atmosphere**, different components of which filter out different parts of the electromagnetic spectrum. In the tropics, about twice as much ultraviolet light reaches the ground above 2,500 m than at sea-level. Indeed, ultraviolet light is stronger in all mountains – hence incautious humans may unexpectedly suffer sunburn at ski resorts.

Photoperiod refers to seasonal variations in the length of day and night. This is immensely important ecologically because day-length, or more usually night-length, stimulates the timing of daily and seasonal rhythms (breeding, migration, flowering, and so on) in many organisms. *Short-day plants* flower when day-length is below a critical level. The cocklebur (*Xanthium strumarium*), a widespread weed in many parts of the world, flowers in spring when, as days become longer, a critical night-length is reached (Ray and Alexander 1966). *Long-day plants* flower when day-length is above a critical level. The strawberry tree (*Arbutus unedo*) flowers in the autumn as the night-length increases. In its Mediterranean home, this means that its flowers are ready for pollination when such long-tongued insects as bees are plentiful. *Day-neutral plants* flower after a period of vegetative growth, irrespective of the photoperiod.

In the high Arctic, plant growth is telescoped into a brief few months of warmth and light. *Positive* **heliotropism** (growing towards the Sun) is one way that plants can cope with limited light. It is common in Arctic and alpine flowers. The flowers of the Arctic avens (*Dryas integrifolia*) and the Arctic poppy (*Papaver radicatum*) track the Sun, turning at about 15° of arc per hour (Kevan 1975; see also Corbett *et al.* 1992). Their **corollas** reflect radiation onto their reproductive parts. The flowers of the alpine snow buttercup (*Ranunculus adoneus*) track the Sun's movement from early morning until mid-afternoon (Stanton and Galen 1989). Buttercup flowers aligned parallel to the Sun's rays reach mean internal temperatures several degrees Celsius above ambient air temperature. Internal flower temperature is significantly reduced as a flower's angle of deviation from the Sun increases beyond 45°.

Arctic and alpine animals and plants also have to cope with limited solar energy. Herbivores gear their behaviour to making the most of the short summer. Belding ground squirrels (*Spermophilus beldingi*), which live at high elevations in the western USA, are active for four or five summer months, and they must eat enough during that time to survive the winter on stored fat (Morhardt and Gates 1974). To do this, their body temperature fluctuates by 3–4°C (to a high of 40°C) so that valuable energy is not wasted in keeping body temperature constant. Should they need to cool down, they go into a burrow or else adopt a posture that lessens exposure to sunlight. A constant breeze cools them during the hottest part of the day.

SOME LIKE IT HOT: TEMPERATURE

Broadly speaking, average annual temperatures are highest at the equator and lowest at the poles. Temperatures also decrease with increasing elevation. The *average annual temperature range* is an important ecological factor. It is highest deep in high-latitude continental interiors and lowest over oceans, especially tropical oceans. In northeast Siberia, an annual temperature range of 60°C is not uncommon, whereas the range over equatorial oceans is less than about 3°C. Land lying adjacent to oceans, especially land on the western seaboard of continents, has an annual temperature range around the 11°C mark. These large differences in annual temperature range reflect differences in *continentality* (or *oceanicity*) – the winter temperatures of places near oceans will be less cold.

Many aspects of temperature affect organisms, including daily, monthly, and annual extreme and mean temperatures, and the level of temperature variability. Different aspects of temperature are relevant to different species and commonly vary with the time of year and the stage in an organism's life cycle. Temperature may be limiting at any stage of an organism's life cycle. It may affect survival, reproduction, and the development of seedlings and young animals. It may affect competition with other organisms and susceptibility to predation, parasitism, and disease when the limits of temperature tolerance are approached. Many flowering plants are especially sensitive to low temperatures between **germination** and seedling growth.

Microbes and temperature

Heat-loving microbes (**thermophiles**) reproduce or grow readily in temperatures over 45°C. **Hyperthermophiles**, such as *Sulfolobus acidocaldarius*, prefer temperatures above 80°C, and some thrive above 100°C. The most resistant hyperthermophile discovered to date is *Pyrolobus fumarii*. This microbe flourishes in the walls of 'smokers' in the deep-sea floor. It multiplies in temperatures up to 113°C. Below 94°C it finds it too cold and stops growing! Only in small areas that are intensely heated by volcanic activity do high temperatures prevent life. Cold-loving microbes (**psychrophiles**) are common in Antarctic sea ice. These communities include photosynthetic **algae** and **diatoms**, and a variety of **bacteria**. *Polaromonas vacuolata*, a bacterium, grows best at about 4°C, and stops reproducing above 12°C.

Animals and temperature

Upper and lower critical temperatures

In most animals, temperature is a critical limiting factor. Vital **metabolic processes** are geared to work optimally within a narrow temperature band. Cold-blooded animals (**poikilotherms**) warm up and cool down with environmental temperature (Figure 6.1a). They can assist the warming process a little by taking advantage of sunny spots or warm rocks. Most warm-blooded animals (**homeotherms**) maintain a constant body temperature amidst varying ambient conditions (Figure 6.1a). They simply regulate the production and dissipation of heat. The terms 'cold-blooded' and 'warm-blooded' are misleading because the body temperature of some 'cold-blooded' animals may rise above that of 'warm-blooded' animals.

Each homeothermic species has a characteristic *thermal neutral zone*, a band of temperature within which little energy is expended in heat regulation (Figure 6.1b) (Bartholomew 1968). Small

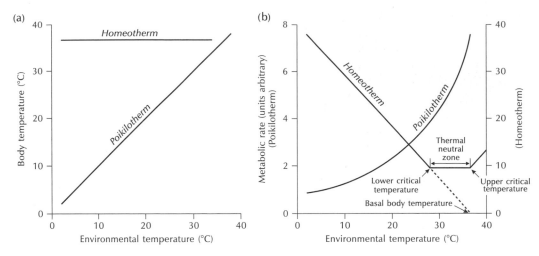

Figure 6.1 Temperature control in poikilotherms and homeotherms.
Source: Adapted from Bartholomew (1968)

adjustments are made by fluffing or compressing fur, by making local changes in the blood supply, or by changing position. The bottom end of the thermal neutral zone is bounded by a *lower critical temperature*. Below this temperature threshold, the body's central heating system comes on fully. The colder it gets, the more oxygen is needed to burn fuel for heat. Animals living in cold environments are well insulated – fur and blubber can reduce the lower critical temperature considerably. An Arctic fox (*Alopex lagopus*) clothed in its winter fur rests comfortably at an ambient temperature of –50°C without increasing its resting rate of metabolism (Irving 1966). Below the lower critical temperature, the peripheral circulation shuts down to conserve energy. An Eskimo dog may have a deep body temperature of 38°C, the carpal area of the forelimb at 14°C, and foot pads at 0°C (Irving 1966) (Figure 6.2). Hollow hair is also useful for keeping warm. It is found in the American pronghorn (*Antilocapra americana*), an even-toed ungulate, and enables it to stay in open and windswept places at temperatures far below 0°C. The polar bear (*Ursus maritimus*) combines hollow hair, a layer of blubber up to 11 cm thick, and black skin to produce a superb insulating machine. Each hair acts like a fibre-optic cable, conducting warming ultraviolet light to the heat-absorbing black skin. This heating mechanism is so efficient that polar bears are more likely to overheat than to chill down, which partly explains their ponderousness. Many animals also have behavioural patterns designed to minimize heat loss. Some roll into a ball, some seek shelter. Herds of deer or elk seek ridge tops or south-facing slopes.

Above the *upper critical temperature*, animals must lose heat to prevent their overheating. Animals living in hot environments can lose much heat. **Evaporation** helps heat loss, but has an unwanted side-effect – precious water is lost. Small animals can burrow to avoid high temperatures at the ground surface. In the Arizona desert, USA, most rodents burrow to a depth where hot or cold heat stress is not met with. Large size is an advantage in preventing overheating because the surface area is relatively greater than the body volume. Many desert mammals are adapted to high temperatures. Bruce's hyrax (*Heterohyrax brucei*) has an upper critical temperature of 41°C. Camels (*Camelus* spp.), oryx (*Oryx* spp.), common eland (*Taurotragus oryx*), and gazelle (*Gazella* spp.)

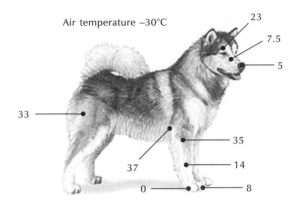

Figure 6.2 Temperatures at an Eskimo dog's extremities (°C) with an ambient air temperature of –30°C.

Source: Adapted from Irving (1966)

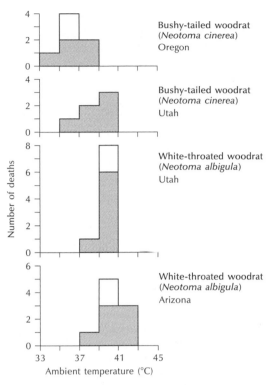

Figure 6.3 Lethal ambient temperatures for four populations of woodrats (*Neotoma*) living in the western USA. The numbers of deaths, shown by the shaded areas, are based on four-hour exposures.

Source: Adapted from Brown (1968)

let their body temperatures fluctuate considerably over a 24-hour period, falling to about 35°C towards dawn and rising to over 40°C during the late afternoon. In an ambient temperature of 45°C sustained for 12 hours under experimental conditions, an oryx's temperature rose above 45°C and stayed there for 8 hours without injuring the animal (Taylor 1969). It has a specialized **circulatory system** that helps it to survive such excessive overheating.

Animal distributions and temperature

Many mammal species are adapted to a limited range of environmental temperatures. Even closely related groups display significant differences in their ability to endure temperature extremes. The lethal ambient temperatures for four populations of woodrats (*Neotoma* spp.) in the western USA showed differences between species, and between populations of the same species living in different states (Figure 6.3).

Many bird species distributions are constrained by such environmental factors as food abundance, climate, habitat, and competition. Distributions of 148 species of North American land birds wintering in the conterminous USA and Canada, when compared with environmental factors,

revealed a consistent pattern (T. L. Root 1988a, b). Six environmental factors were used: average minimum January temperature; mean length of the frost-free period; potential vegetation; mean annual precipitation; average general humidity; and elevation. **Isolines** for average minimum January temperature, mean length of the frost-free period, and potential vegetation correlated with the northern range limits of about 60 per cent, 50 per cent, and 64 per cent, respectively, of the wintering bird species. Figure 6.4 shows the winter distribution and abundance of the eastern phoebe (*Sayornis phoebe*). The northern boundary is constrained by the –4°C isotherm of January minimum temperature. Just two environmental factors – potential vegetation and mean annual precipitation – coincided with eastern range boundaries, for about 63 per cent and 40 per cent of the species, respectively. On the western front, mean annual precipitation distribution coincided for 36 per cent of the species, potential vegetation 46 per cent, and elevation 40 per cent.

Why should the northern boundary of so many wintering bird species coincide with the average minimum January temperature? The answer to this poser appears to lie in metabolic rates (T. L. Root 1988b). At their northern boundary, the calculated mean metabolic rate in a sample of 14 out of 51 passerine (song birds and their allies)

species was 2.49 times greater than the basal metabolic rate (which would occur in the thermal neutral zone, see p. 87). This figure implies that the winter ranges of these 14 bird species are restricted to areas where the energy needed to compensate for a colder environment is not greater than around 2.5 times the basal metabolic rate. The estimated mean metabolic rate for 36 of the remaining 37 passerine species averaged about 2.5. This '2.5 rule' applies to birds whose body weight ranges from 5 g in wrens to 448 g in crows, whose diets range from seeds to insects, and whose northern limits range from Florida to Canada – a remarkable finding.

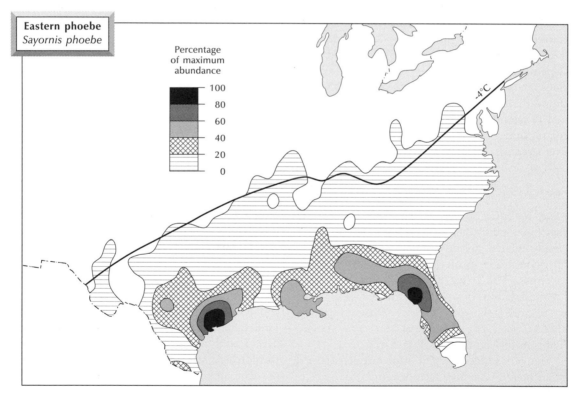

Figure 6.4 Winter distribution and abundance of the eastern phoebe (*S. phoebe*). The northern boundary is constrained by the –4°C average minimum January isotherm.

Source: Adapted from T. L. Root (1988a, b)

Plants and temperature

Temperature affects many processes in plants, including photosynthesis, respiration, growth, reproduction, and transpiration. Plants vary enormously in their ability to tolerate either heat or cold.

Cold tolerance

There are five broad categories of cold tolerance (Table 6.1). Temperatures lower than 10°C damage *chilling-sensitive* plants, which are mostly tropical. *Chilling-resistant* (*frost-sensitive*) plants can survive at temperatures below 10°C, but are damaged when ice forms within their tissues. *Frost-resistant* plants make physiological changes that enable them to survive temperatures as low as about −15°C. *Frost-tolerant* plants survive by withdrawing water from their cells, so preventing ice forming. The withdrawal of water also increases the concentration in sap and protoplasm, which acts as a kind of antifreeze, and lowers freezing point. Temperatures down to about −40°C can be tolerated in this way. **Lichens** can photosynthesize at −30°C, providing that they are not covered with snow. The reddish-coloured snow alga, *Chlamydomonas nivalis*, lives on ice and snowfields in the polar and nival zones, giving the landscape a pink tinge during the summer months. *Cold-tolerant* plants, which are mostly needle-leaved, can survive almost any subzero temperature.

Cold tolerance varies enormously at different seasons in some species. Willow twigs (*Salix* spp.) collected in winter can survive freezing temperatures below −150°C; a temperature of −5°C kills the same twigs in summer (Sakai 1970). Similarly, the red-osier dogwood (*Cornus stolonifera*), a hardy shrub from North America, could survive a laboratory test at −196°C by midwinter when grown in Minnesota (Weiser 1970). Nonetheless, dogwoods native to coastal regions with mild climates are often damaged by early autumn frosts. Temperatures of −5°C to −7°C kill plants growing on Mt Kurodake, Hokkaido Province, Japan, during the growing season. In winter, most of the same plants survive freezing to −30°C, and the willow ezo-mame-yanagi (*Salix pauciflora*), mosses, and lichens will withstand a temperature of −70°C (Sakai and Otsuka 1970). Acclimatization or cold hardening accounts for these differences. The coastal dogwoods did not acclimatize quickly enough. Timing is important in cold resistance, but absolute resistance can be altered. Many plants use the signal of short days in autumn as an early warning system. The short days trigger metabolic changes that stop the plant growing and produce resistance to cold. Many plant species, especially deciduous plants in temperate regions, need chilling during winter if they are to grow well the following summer. *Chilling requirements* are specific to species. They are often necessary for buds to break out of dormancy, a process called **vernalization**.

Table 6.1 Temperature tolerance in plants

Temperature sensitivity	Minimum temperature (°C)	Life-form
Chilling sensitive	>10	Broad-leaved evergreen
Chilling resistant (frost sensitive)	0 to 10	Broad-leaved evergreen
Frost resistant	−15 to 10	Broad-leaved evergreen
Frost tolerant	−40 to −15	Broad-leaved deciduous
Cold tolerant	< −40	Broad-leaved evergreen and deciduous; boreal needle-leaved

Source: After Woodward (1992)

Heat tolerance

Many plants require a certain amount of 'warmth' during the year. The total 'warmth' depends on the growing season length and the growing season temperature. These two factors are combined as *day-degree totals* (Woodward 1992). Day-degree totals are the product of the growing season length (the number of days for which the mean temperature is above a standard temperature, such as freezing point or 5°C), and the mean temperature for that period. The Iceland purslane (*Koenigia islandica*), a tundra annual, needs only 700 day-degrees to develop from a germinating seed to a mature plant producing seeds of its own. The small-leaved lime (*Tilia cordata*), a deciduous tree, needs 2,000 day-degrees to complete its reproductive development (Pigott 1981). Trees in tropical forests may need up to 10,000 day-degrees to complete their reproductive development.

Excessive heat is as detrimental to plants as excessive cold. Plants have evolved resistance to *heat stress*, though the changes are not so marked as resistance to cold stress (see Gates 1980, 1993, 69–72). Different parts of plants acquire differing degrees of heat resistance, but the pattern varies between species. In some species, the uppermost canopy leaves are often the most heat resistant; in other species, it is the middle canopy leaves, or the leaves at the base of the plant. Temperatures of about 44°C are usually injurious to evergreens and shrubs from cold-winter regions. Temperate-zone trees are damaged at 50–55°C, tropical trees at 45–55°C. Damaged incurred below about 50°C can normally be repaired by the plant; damaged incurred above that temperature is most often irreversible. Exposure time to excessive heat is a critical factor in plant survival, while exposure time to freezing temperatures is not.

Distributional limits in plants

Many distributional boundaries of plant species seem to result from extreme climatic events causing the failure of one stage of the life cycle (Grace 1987). The climatic events in question may occur rarely, say once or twice a century, so the chances of observing a failure are slim. Nonetheless, edges of plant distributions often coincide with isolines of climatic variables. The northern limit of madder (*Rubia peregrina*) in northern Europe sits on the 40°F (4.4°C) mean January isotherm (Salisbury 1926). Holly (*Ilex aquifolium*) is confined to areas where the mean annual temperature of the coldest month exceeds –0.5°C, and, like madder, seems unable to withstand low temperatures (Iversen 1944). Several frost-sensitive plant species, including the Irish heath (*Erica erigena*), St Dabeoc's heath (*Daboecia cantabrica*), large-flowered butterwort (*Pinguicula grandifolia*), and sharp rush (*Juncus acutus*), occur only in the extreme west of the British Isles where winter temperatures are highest. Other species, such as the twinflower (*Linnaea borealis*) and chickweed-wintergreen (*Trientalis europaea*), have a northern or northeastern distribution, possibly because they have a winter chilling requirement for germination that southerly latitudes cannot provide (Perring and Walters 1962). Low summer temperatures seem to restrict the distribution of such species as the stemless thistle (*Cirsium acaule*). Near to its northern limit, this plant is found mainly on south-facing slopes, for on north-facing slopes it fails to set seed (Pigott 1974). The distribution of grey hair-grass (*Corynephorus canescens*) is limited by the 15°C mean isotherm for July. This may be because its short life span (2 to 6 years) means that, to maintain a population, seed production and germination must continue unhampered (J. K. Marshall 1978). At the northern limit of grey hair-grass, summer temperatures are low, which delays flowering, and, by the time seeds are produced, shade temperatures are low enough to retard germination.

The small-leaved lime tree (*Tilia cordata*) ranges across much of Europe. The mean July 19°C isotherm marks its northern limit in England and Scandinavia (Pigott 1981; Pigott and Huntley 1981). The tree requires 2,000 growing day-degrees to produce seeds by sexual reproduction. But for the lime tree to reproduce, the flowers must develop and then the pollen must germinate and be transferred through a pollen tube to the ovary for fertilization. The pollen fails to germinate at temperatures at or below 15°C, germinates best in the range 17°C to 22°C, and germinates, but less successfully, up to about 35°C. A complicating factor is that the growth of the pollen tube depends on temperature. The growth rate is maximal around 20–25°C, diminishing fast at higher and lower temperatures. Indeed, the extension of the pollen tube becomes rapid above 19°C, which suggests why the northern limit is marked by the 19°C mean July isotherm.

Several models use known climatic constraints on plant physiology to predict plant species distribution. One study investigated the climatic response of boreal tree species in North America (Lenihan 1993). Several climatic predictor variables were used in a regression model. The variables were annual snowfall, day-degrees, absolute minimum temperature, annual soil-moisture deficit, and actual **evapotranspiration** summed over summer months. Predicted patterns of species' dominance probability closely matched observed patterns (Figure 6.5). The results suggested that the boreal tree species respond individually to different combinations of climatic constraints. Another study used a climatic model to predict the distribution of woody plant species in Florida, USA (Box *et al.* 1993). The State of Florida is small enough for variations in substrate to play a major role in determining what grows where. Nonetheless, the model predicted that climatic factors, particularly winter temperatures, exert a powerful influence, and in some cases a direct control, on species' distributions. Predicted distributions and observed distribution of the longleaf pine (*Pinus palustris*) and the Florida poison tree (*Metopium toxiferum*) are shown in Figure 6.6. The predictions for the longleaf pine are very good, except for a narrow strip near the central Atlantic coast. The match between

predicted and observed distributions is not so good for the Florida poison tree. The poison tree is a subtropical species and the model was less good at predicting the distribution of subtropical plants.

QUENCHING THIRST: MOISTURE

Protoplasm, the living matter of animal and plant cells, is about 90 per cent water – without adequate moisture there can be no life. Water affects land animals and plants in many ways. Air humidity is important in controlling loss of water through the skin, lungs, and leaves. All animals need some form of water in their food or as drink to run their excretory systems. **Vascular plants** have an internal plumbing system – parallel tubes of dead tissue called xylem – that transfers water from root tips to leaves. The entire system is full of water under stress (**capillary pressure**). If the *water stress* should fall too low, disaster may ensue – germination may fail, seedlings may not become established, and, should the fall occur during flowering, seed yields may be severely cut. An overlong drop in water stress kills plants, as anybody who has tried to grow bedding plants during a drought and hosepipe ban will know.

Bioclimates

On land, precipitation supplies water to ecosystems. Plants cannot use all the precipitation that falls. A substantial portion of the precipitation evaporates and returns to the atmosphere. For this reason, **available moisture** (roughly the precipitation less the evaporation) is a better guide than precipitation to the usable water in a terrestrial ecosystem. This point is readily understood with an example. A mean annual rainfall of 400 mm might support a forest in Canada, where evaporation is low, but in Tanzania, where evaporation is high, it might support a dry savannah.

Available moisture largely determines soil water levels, which in turn greatly influence plant

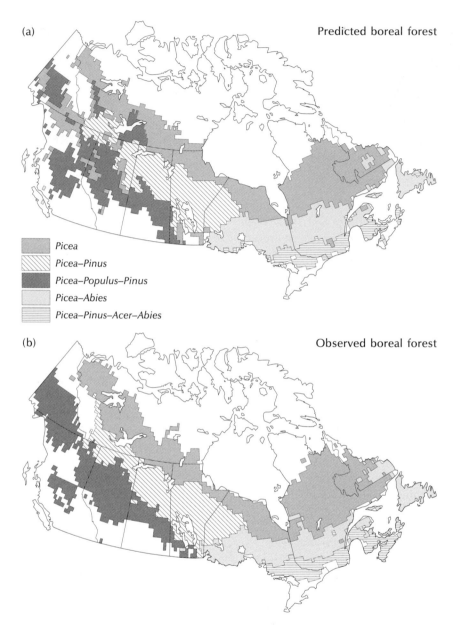

(a)   Predicted boreal forest

☐ Picea
☐ Picea–Pinus
■ Picea–Populus–Pinus
☐ Picea–Abies
☰ Picea–Pinus–Acer–Abies

(b)   Observed boreal forest

Figure 6.5 Boreal forest types in Canada. (a) Predicted forest types using a regression model. (b) Observed forest types.

Source: Adapted from Lenihan (1993)

Longleaf pine
Pinus palustris

Predicted

Observed

Florida poison tree
Metopium toxiferum

Predicted

Observed

Figure 6.6 Predicted and observed longleaf pine (*P. palustris*) and the Florida poison tree (*M. toxiferum*) distributions in Florida, USA.
Source: Adapted from Box *et al.* (1993)

growth. For a plant to use energy for growth, water must be available. Without water, the energy will merely heat and stress the plant. Similarly, for a plant to use water for growth, energy must be obtainable. Without an energy source, the water will run into the soil or run off unused. For these reasons, temperature (as a measure of energy) and moisture are *master limiting factors* that act in tandem. In tropical areas, temperatures are always high enough for plant growth and precipitation is the limiting factor. In cold environments, water is usually available for plant growth for most of the year – low temperatures are the limiting factor. This is true, too, of limiting factors on mountains where heat or water (or both) set lower altitudinal limits, and lack of heat sets upper altitudinal limits.

So important are precipitation and temperature that several researchers use them to characterize bioclimates. **Bioclimates** are the aspects of climate that seem most significant to living things. The most widely used bioclimatic classification is the 'climate diagram' devised by Heinrich Walter. This is the system of summarizing ecophysiological conditions that makes David Bellamy 'feel like a plant' (Bellamy 1976, 141)! *Climate diagrams* portray climate as a whole, including the seasonal round of precipitation and temperature (Figure 6.7). They show at a glance the annual pattern of rainfall; the wet and dry seasons characteristic of an area, as well as their intensity, since the evaporation rate is directly related to temperature; the occurrence or non-occurrence of a cold season; and the months in which early and late frost have been recorded. Additionally, they provide information on such factors as mean annual temperature, mean annual precipitation, the mean daily minimum temperature during the coldest month, the absolute minimum recorded temperature, the altitude of the station, and the number of years of record.

Knowing a species' bioclimatic requirement allows predictions to be made about its potential

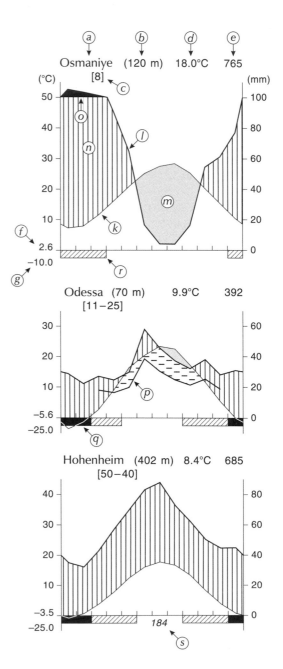

Figure 6.7 Examples and explanation of climate diagrams. The letters denote the following. *a* weather station. *b* Altitude (m above mean sea-level). *c* Number of years of observation. Where there are two figures, the first refers to temperature measurements and the second the precipitation measurements. *d* Mean annual temperature (°C). *e* Mean annual precipitation (mm). *f* Mean daily maximum temperature during the coldest month (°C). *g* Absolute minimum (lowest recorded) temperature (°C). *k* Curve of mean monthly temperature (1 scale graduation = 10°C). *l* Curve of mean monthly precipitation (1 scale graduation = 20 mm). *m* Relatively arid period or dry season (dotted). *n* Relatively humid period or wet season (vertical bars). *o* Mean monthly rainfall above 100 mm with the scale reduced by a factor of 0.1 (the black area in Osmaniye). *p* Curve for precipitation on a smaller scale (1 scale graduation = 30 mm). Above it, horizontal broken lines indicate the relatively dry period or dry season (shown for Odessa). *q* Months with a mean daily minimum temperature below 0°C (black boxes below zero line). *r* Months with an absolute minimum temperature below 0°C (diagonal lines). *s* Average duration of period with daily mean temperature above 0°C (shown as the number of days in standard type); alternatively, the average duration of the frost-free period (shown as the number of days in italic type, as for Honenheim). Mean daily maximum temperature during the warmest month (*h*), absolute maximum (highest recorded) temperature (*i*), and mean daily temperature fluctuation (*j*) are given only for tropical stations with a diurnal climate, and are not shown in the examples.

Source: Adapted from Walter and Lieth (1960–7)

fate under climatic change. Figure 6.8 shows changes in the bioclimatic envelopes of three British species using a worst-case scenario in a climate model (Berry *et al.* 2002; see also Berry *et al.* 2003). Notice that the bioclimatic envelope of the great burnet (*Sanguisorba officinalis*) is predicted to expand, that of the yellow-wort (*Blackstonia perfoliata*) to stay roughly the same, and that of the twinflower (*Linnaea borealis*) to contract. Other bioclimatic modelling studies predict the future distribution of tree species in Europe (Thuiller 2003) and identify the environmental limits for vegetation at biome and species scales in the fynbos biodiversity hotspot in South Africa, with a view to predicting the likely shrinkage under a warming climate (Midgley *et al.* 2002).

Wet environments

Plants are very sensitive to water levels. Hydrophytes are water plants and root in standing water. Helophytes are marsh plants. Mesophytes are plants that live in normally moist but not wet conditions. Xerophytes are plants that live in dry conditions.

Wetlands support **hydrophytes** and **helophytes**. The common water crowfoot (*Ranunculus aquatalis*) and the bog pondweed (*Potamogeton polygonifolius*) are hydrophytes; the greater bird's-foot trefoil (*Lotus uliginosus*) is a helophyte. These plants manage to survive by developing a system of air spaces in their roots, stems, or leaves. The air spaces provide buoyancy and improve internal ventilation. **Mesophytes** vary greatly in their ability to tolerate flooding. In the southern USA, bottomland hardwood forests occupy swamps and river floodplains. They contain a set of tree species that can survive in a flooded habitat. The water tupelo (*Nyssa aquatica*), which is found in bottomland forest in the southeastern USA, is well adapted to such wet conditions.

Flooding or high soil-moisture levels may cause seasonal changes in mammal distributions. The mole rats (*Cryptomys hottentotus*) in Zimbabwe focus their activity around the bases of termite mounds during the rainy seasons as they rise a metre or so above the surrounding grassland and produce relatively dry islands in a sea of water-logged terrain (Genelly 1965).

Many organisms are fully adapted to watery environments and always have been – the colonization of dry land is a geologically recent event. Some vertebrates have returned to an *aquatic* existence. Those returning to the water include crocodiles, turtles, extinct plesiosaurs and ichthyosaurs, seals, and whales. Some, including the otter, have adopted a *semi-aquatic* way of life.

Dry environments

Plants are very sensitive to **drought**, and **aridity** poses a problem of survival. Nonetheless, species of algae grow in the exceedingly dry Gobi desert. Higher plants survive in arid conditions by xerophytic adaptations – drylands support **xerophytes**. One means of survival is simply to avoid the drought as seeds (*pluviotherophytes*) or as below-ground storage organs (bulbs, tubers, or rhizomes). Other xerophytic adaptations enable plants to retain enough water to keep their protoplasts wet, so avoiding desiccation. Water is retained by several mechanisms. A very effective mechanism is water storage. *Succulents* are plants that store water in leaves, stems, or roots. The saguaro cactus (*Carnegiea gigantea*) and the barrel cactus (*Ferocactus wislizeni*) are examples. The barrel cactus stores so much water that Indians and other desert inhabitants have used it as an emergency water supply. Many succulents have a crassulacean acid metabolism (CAM) pathway for carbon dioxide assimilation. This kind of photosynthesis involves carbon dioxide being taken in at night with **stomata** wide open, and then being used during the day with stomata closed to protect against transpiration losses. Other xerophytes (sometimes regarded as *true xerophytes*) do not store water, but have evolved very effective ways of reducing water loss – leaves with thick **cuticles**, sunken and smaller stomata, leaves shed during dry periods, improved water uptake

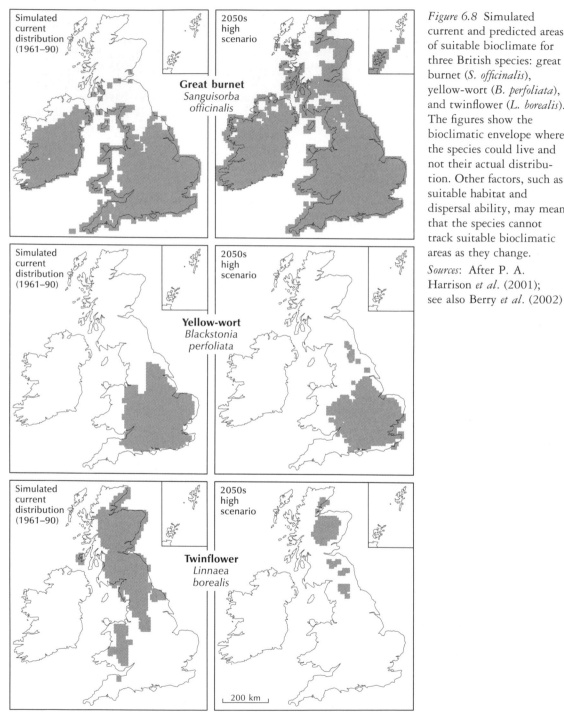

Figure 6.8 Simulated current and predicted areas of suitable bioclimate for three British species: great burnet (*S. officinalis*), yellow-wort (*B. perfoliata*), and twinflower (*L. borealis*). The figures show the bioclimatic envelope where the species could live and not their actual distribution. Other factors, such as suitable habitat and dispersal ability, may mean that the species cannot track suitable bioclimatic areas as they change.

Sources: After P. A. Harrison *et al*. (2001); see also Berry *et al*. (2002)

Great burnet
Sanguisorba officinalis

Yellow-wort
Blackstonia perfoliata

Twinflower
Linnaea borealis

Simulated current distribution (1961–90)

2050s high scenario

200 km

through wide spreading or deeply penetrating root systems, and improved water conduction. The big sagebrush (*Artemisia tridentata*) is an example of a true xerophyte. The creosote bush (*Larrea divaricata*) simply endures a drought by ceasing to grow when water is not available.

Plants vary enormously in their ability to withstand water shortages. Young plants suffer the worst. Drought resistance is measured by the specific survival time (the time between the point when the roots can no longer take up water and the onset of dessicative injury). The specific survival time is 1,000 hours for prickly-pear cactus, 50 hours for Scots pine, 16 hours for oak, 2 hours for beech, and 1 hour for forget-me-not (Larcher 1975, 172).

Desert-dwelling animals face a problem of water shortage, as well as high daytime temperatures. They have overcome these problems in several remarkable ways (Boxes 6.2 and 6.3).

Box 6.2

REPTILES IN DESERTS

Lizards are abundant in deserts in the daytime whereas mammals are not. The reason for this is not a reduction of evaporative water loss through the skin. Cutaneous water loss is about the same in mammals and reptiles. However, reptiles from dry habitats do have a lower skin permeability. Therefore, they lose less water through the skin than do reptiles from moist environments. The tropical, tree-living green iguana (*Iguana iguana*) loses about 4.8 mg/cm²/day through the skin and 3.4 mg/cm²/day through respiration; the desert-dwelling chuckwalla (*Sauromalus obesus*), which is active in daytime, loses about 1.3 mg/cm²/day through the skin and 1.1 mg/cm²/day through respiration. The difference between mammals and reptiles lies in three reptilian characteristics that predispose them for water conservation in arid environments: (1) low metabolic rates; (2) nitrogenous waste excretion as uric acid and its salts; and, in many taxa, (3) the presence of nasal salt glands (an alternative pathway of salt excretion to the kidneys). Low metabolic rates mean less frequent breathing, which means that less water is lost from the lungs. Uric acid is only slightly soluble in water and precipitates in urine to form a whitish, semi-solid mass. Water is left behind and may be reabsorbed into the blood and used to produce more urine. This recycling of water is useful to reptiles because their kidneys are unable to make urine with a higher osmotic pressure than that of their blood plasma. The potassium and sodium ions that do not precipitate are reabsorbed in the bladder. This costs energy, so why do it? The answer lies in the third water-conserving mechanism in reptiles – extra-renal salt excretion. In at least three reptilian groups – lizards, snakes, and turtles – there are some species that have salt glands. These glands make possible the selective transport of ions out of the body. They are most common in lizards, where they have been found in five families. In these families, a lateral nasal gland excretes salt. The secretions of the glands are emptied into the nasal passages and are expelled by sneezing and shaking of the head. Salt glands are very efficient at excreting. The total osmotic pressure of salt glands may be more than six times that of urine produced by the kidney. This explains the paradox of salt uptake from urine in bladders. As ions are actively reabsorbed, water follows passively and the animal recovers both water and ions from the urine. The ions can then be excreted through the salt gland at much higher concentrations. There is thus a proportional reduction in the amount of water needed to dispose of the salt.

Box 6.3

MAMMALS IN DESERTS

Rodents are the dominant small mammals in arid environments. Population densities may be higher in deserts than in temperate regions. As with reptiles, several features of rodent biology predispose them to desert living. Many rodents are nocturnal and live in burrows. Although nighttime activity might be thought to avoid the heat stress of the day, heat stress can also occur at night when deserts can be cold. In rodents (and birds and lizards), there is a countercurrent water-recycler in the nasal passages that is important in the energy balance of these organisms. While breathing in, air passes over the large surface area of the nasal passages and is warmed and moistened. The surface of the nasal passages cools by evaporation in the process. When warm, saturated air from the lungs is breathed out, it condenses in the cool nasal passages. Overall, this process saves water and energy. Indeed, the energetic savings are so great that it is unlikely that a homeotherm could survive without this system – ethmo-turbinal bones in the fossil *Cynognathus* are persuasive evidence that mammal-like reptiles (therapsids) were homeotherms. However, this countercurrent exchange of heat and moisture is not an adaptation to desert life; it is an inevitable outcome of the anatomy and physiology of the nasal passages.

Water is also lost in faeces and urine. Rodents generally can produce fairly dry faeces and concentrated urine. Kangaroo rats, sand rats, and jerboas can produce urine concentrations double to quadruple the urine concentration in humans. The spinifex hopping mouse or dargawarra (*Notomys alexis*), which lives throughout most of the central and western Australian arid zone, is a hot contender for 'world champion urine concentrator'; its urine concentration is six times higher than in humans (Plate 6.1). Low evaporative water loss through nocturnal habits, concentrated urine, and fairly dry faeces mean that many desert rodents are independent of water – they can get all the water they need from air-dried seeds. Part of this water comes from the seeds and part comes from the oxidation of food (metabolic water). Interestingly, the water content of some desert plants varies with the relative humidity of the air. In parts of semi-arid Africa, *Disperma* leaves have a water content of 1 per cent by day, but at night, when the relative humidity increases, their water content rises to 30 per cent. The leaves are forage for the oryx (*Oryx gazella*). By feeding at night, the oryx takes in 5 litres of water, on which it survives through several water-conserving mechanisms (Taylor 1969). The banner-tailed kangaroo rat (*Dipodomys spectabilis*) from Arizona, which is not as its name

Plate 6.1 Spinifex hopping mouse (*N. alexis*) – world champion urine concentrator.

Photograph by Pat Morris.

implies a pouched mammal, stores several kilo-grams of plant material in its burrow where it is exposed to a relatively high relative humidity of the burrow atmosphere (Schmidt-Nielsen and Schmidt-Nielsen 1953). Hoarding food in a burrow provides not only a hedge against food shortages, but also an enhanced source of water. In Texas, burrows of the plains pocket gopher (*Geomys bursarius*) have relative humidities of 86–95 per cent (Kennerly 1964). In sealed burrows, the humidities can be up to 95 per cent while the soil of the burrow floor contains only 1 per cent water. Nonetheless, although high temperatures and low humidities are avoided in burrows, other stresses do occur – carbon dioxide concentrations may be 10–60 times greater than in the normal air.

Snow

This is a significant ecological factor in polar and some boreal environments. A *snow cover* that persists through the winter is a severe hardship to large mammals. To most North American artiodactyls, including deer, elk, bighorn sheep, and moose, even moderate snow imposes a burden by covering some food and making it difficult to find. In mountainous areas, deer and elk avoid deep snow by abandoning summer ranges and moving to lower elevations. South-facing slopes and windswept ridges, where snow is shallower or on occasions absent, are preferred at these times. In areas of relatively level terrain, deer and moose respond to deep snow by restricting their activities to a small area called 'yards' where they establish trails through the snow. Prolonged winters and deep snow take a severe toll on deer and elk populations.

For small mammals, snow is a blessing. It forms an insulating blanket, a sort of crystalline duvet, under which is a ground-surface microenviron-ment where activity, including breeding in some species, continues throughout the winter. To these small mammals, which include shrews (*Sorex*), pocket gophers (*Thomomys*), voles (*Microtus*, *Clethrionomys*, *Phenacomys*), and lemmings (*Lemmus*, *Dicrostonys*), the most stressful times are autumn, when intense cold descends but snowfall has not yet moderated temperatures at the ground surface, and in the spring, when rapid melting of a deep snowpack often results in local flooding. Another advantage of a deep snowpack is that green vege-tation may be available beneath it, and several species make tunnels to gain access to food.

Even in summer, snow may be important to some mammals. Alpine or northern snowfields commonly last through much of the summer on north-facing slopes and provide a cool microcli-mate unfavoured by insects. Caribou (*Rangifer tarandus*) and bighorn sheep (*Ovis* spp.) sometimes congregate at these places to seek relief from pesky warble flies.

THE BIG PICTURE: CLIMATIC ZONES

Terrestrial ecozones

On land, characteristic animal and plant com-munities are associated with nine basic climatic types, variously called **zonobiomes** (Walter 1985), ecozones (Schultz 1995), and ecoregions (Bailey 1995, 1996) (Figure 6.9):

1 *Polar and subpolar zone.* This zone includes the Arctic and Antarctic regions. It is associated with tundra vegetation. The Arctic tundra regions have low rainfall evenly distributed throughout the year. Summers are short, wet, and cool. Winters are long and cold. Antarctica is an icy desert, although summer warming around the fringes is causing it to bloom.
2 *Boreal zone.* This is the cold-temperate belt supporting coniferous forest (taiga). It usually

has cool, wet summers and very cold winters lasting at least six months. It is only found in the northern hemisphere where it forms a broad swathe around the pole – it is a circumpolar zone.

3 *Humid mid-latitude zone*. This zone is the temperate or nemoral zone. In continental interiors it has a short, cold winter and a warm, or even hot, summer. Oceanic regions, such as the British Isles, have warmer winters and cooler, wetter summers. This zone supports broad-leaved deciduous forests.

4 *Arid mid-latitude zone*. This is the cold-temperate (continental) belt. The difference between summer and winter temperatures is marked and rainfall is low. Regions with a dry summer but only a slight drought support temperate grasslands. Regions with a clearly defined drought period and a short wet season support cold desert and semi-desert vegetation.

5 *Tropical and subtropical arid zone*. This is a hot desert climate that supports thorn and scrub savannahs and hot deserts and semi-deserts.

6 *Mediterranean subtropical zone*. This is a belt lying between roughly 35° and 45° latitude in both hemispheres with winter rains and summer drought. It supports sclerophyllous (thick-leaved), woody vegetation adapted to drought and sensitive to prolonged frost.

7 *Seasonal tropical zone*. This zone extends from roughly 25° to 30° north and south. There is a marked seasonal temperature difference. Heavy rain in the warmer summer period alternates with extreme drought in the cooler winter period. The annual rainfall and the drought period increase with distance from the equator. The vegetation is tropical grassland or savannah.

8 *Humid subtropical zone*. This zone has almost no cold winter season, and short wet summers. It is the warm temperate climate in Walter's zonobiome classification. Vegetation is subtropical broad-leaved evergreen forest.

9 *Humid tropical zone*. This torrid zone has rain all year and supports evergreen tropical rain forest. The climate is said to be diurnal because it varies more by day and night than it does through the seasons.

Marine ecozones

The marine biosphere also consists of 'climatic' zones, which are also called ecozones. The main surface-water marine ecozones are the polar zone, the temperate zone, and the tropical zone (Bailey 1996, 161):

1 *Polar zone*. Ice covers the polar seas in winter. Polar seas are greenish, cold, and have a low salinity.

2 *Temperate zone*. Temperate seas are very mixed in character. They include regions of high salinity in the subtropics.

3 *Tropical zone*. Tropical seas are generally blue, warm, and have a high salinity.

Biomes

Each ecozone supports several characteristic communities of animals and plants known as **biomes** (Clements and Shelford 1939). The deciduous forest biome in temperate western Europe is an example. It consists largely of woodland with areas of heath and moorland. A plant community at the biome scale – all the plants associated with the deciduous woodland biome, for example – is a **plant formation**. An equivalent animal community has no special name; it is simply an animal community. Smaller communities within biomes are usually based on plant distribution. They are called *plant associations*. In England, associations within the deciduous forest biome include beech forest, lowland oak forest, and ash forest. Between biomes are transitional belts where the climate changes from one type to the next. These are called **ecotones**.

Zonobiomes

All the biomes around the world found in a particular ecozone constitute a zonobiome.

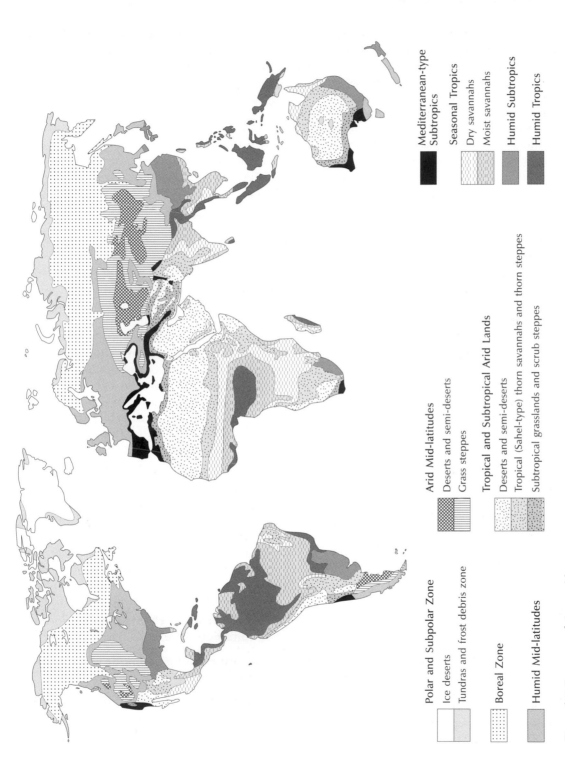

Mediterranean-type
Subtropics

Seasonal Tropics
Dry savannahs
Moist savannahs
Humid Subtropics
Humid Tropics

Arid Mid-latitudes
Deserts and semi-deserts
Grass steppes

Tropical and Subtropical Arid Lands
Deserts and semi-deserts
Tropical (Sahel-type) thorn savannahs and thorn steppes
Subtropical grasslands and scrub steppes

Polar and Subpolar Zone
Ice deserts
Tundras and frost debris zone

Boreal Zone

Humid Mid-latitudes

Figure 6.9 Ecozones of the world.
Source: Adapted from Schultz (1995)

A plant community at the same large scale is a *formation-type* or *zonal plant formation*. The broad-leaved temperate forests of western Europe, North America, eastern Asia, southern Chile, southeast Australia and Tasmania, and most of New Zealand comprise the humid temperate zonobiome. Between the zonobiomes are transitional belts where the climate changes from one type to the next. These are called *zonoecotones*.

Freshwater communities (lakes, rivers, marshes, and swamps) are part of continental zonobiomes. They may be subdivided in various ways. Lakes, for instance, may be well mixed (*polymictic* or *oligomictic*) or permanently layered (*meromictic*). They may be wanting in nutrients and biota (*oligotrophic*) or rich in nutrients and algae (*eutrophic*). A thermocline (where the temperature profile changes most rapidly) separates a surface-water layer mixed by wind (*epilimnon*), from a more sluggish, deep-water layer (*hypolimnon*). And, as depositional environments, lakes are divided into a **littoral** (near-shore) zone, and a **profundal** (basinal) zone.

Marine ecozones, and the deep-water regions, consist of biomes (equivalent to terrestrial zonobiomes). The chief *marine biomes* are the intertidal (**estuarine**, littoral marine, algal bed, coral reef) biome, the open sea (pelagic) biome, the upwelling zone biome, the **benthic** biome, and the hydrothermal vent biome.

Orobiomes

Mountain areas possess their own biomes called **orobiomes**. The basic environmental zones seen on ascending a mountain are submontane (colline, lowland), montane, subalpine, alpine, and nival (p. 125). On south-facing slopes in the Swiss Alps around Cortina, the submontane belt lies below about 1,000 m. It consists of oak forests and fields. The montane belt ranges from about 1,000 m to 3,000 m. The bulk of it is Norway spruce (*Picea abies*) forest, with scattered beech (*Fagus sylvatica*) trees at lower elevations. Mountain pines (*Pinus montana*) with scattered Swiss stone pines

(*P. cembra*) grow near the treeline. The subalpine belt lies between about 3,000 m and 3,500 m. It contains diminutive forests of tiny willow (*Salix* spp.) trees, only a few centimetres tall when mature, within an alpine grassland. The alpine belt, which extends up to about 4,000 m, is a meadow of patchy grass and a profusion of alpine flowers – poppies, gentians, saxifrages, and many more. The mountaintops above about 4,000 m lie within the nival zone and are covered with permanent snow and ice.

SUMMARY

Limiting climatic factors are radiation and light, various measures of temperature (e.g. annual mean, annual range, occurrence of frost), various measures of the water balance (e.g. annual precipitation, effective precipitation, drought period, snow cover), windiness, humidity, and many others. Of these climatic factors, temperature and water are master limiting factors and constrain the distribution of many species. Bioclimates, which are summarized in climate diagrams, characterize the climatic factors that strongly affect living things. Animals and plants have adapted to dry and wet environments. Ecozones are large climatic regions sharing the same kind of climate. The Mediterranean ecozone is an example. Ecozones are equivalent to zonobiomes, which in turn are composed of similar biomes.

ESSAY QUESTIONS

1 **Explain how vertebrates have adapted to conditions of extreme aridity.**

2 **To what extent do climatic factors limit species distributions?**

3 **How useful is the idea of bioclimates?**

FURTHER READING

Bailey, R. G. (1996) *Ecosystem Geography*, with a foreword by Jack Ward Thomas, Chief, USDA Forest Service. New York: Springer.

A gentle introduction to the world's ecoregions.

Huggett, R. J. (1995) *Geoecology: An Evolutionary Approach*. London: Routledge.

A survey of all environmental factors. Climate is covered in Chapters 4 and 5.

Schultz, J. (1995) *The Ecozones of the World: The Ecological Divisions of the Geosphere*. Hamburg: Springer.

A detailed look at the world's ecozones.

Stoutjesdijk, P. and Barkman, J. J. (1992) *Microclimate, Vegetation and Fauna*. Uppsala, Sweden: Opulus Press.

An unusual book dealing with microclimate and life.

1

SUBSTRATE AND LIFE

The material that animals and plants live in or on influences the distribution of species and communities. This chapter covers:

- plant-life and substrate
- animal-life and substrate

ROCKY FOUNDATIONS: PLANTS AND SUBSTRATE

Substrate lovers

Some plants specialize in living on bare rocks, others in living in soils rich in certain chemicals.

Bare rock specialists

Rock plants (**petrophytes**) grow on bare rock surfaces. Some algae and lichens attach themselves to the surface; these are **exolithophytes**. Some lichens penetrate tiny cracks in the rock with their rhizoids; these are **rhizolithophytes**. Saxicolous species live on rocky terrain, in or on cliffs, rocks, and talus. Some saxicolous species are **chomophytes** that favour small ledges where **detritus** and humus have collected. Others are *crevice plants*

or **chasmophytes** that prefer small crevices in the rock surface where some humus has formed. In the Peak District of Derbyshire, England, the wallflower (*Cheiranthus cheiri*) is a common and colourful chomophyte and maidenhair spleenwort (*Asplenium trichomanes*) is a common chasmophyte (P. Anderson and Shimwell 1981, 142).

Vascular plants on talus at high altitudes tend to cluster around stones. This could be because the areas of rock accumulation are relatively more stable than areas of thin, fine-grained talus. Another possibility is that soil moisture is more readily available between and below stones where trapped fine-grained material holds water. Differences in temperature might also affect plant distribution. Francisco L. Pérez (1987, 1989, 1991) conducted a revealing study of soil moisture and temperature influences on the distribution of tall frailejón (*Coespeletia timotensis*), a giant caulescent (stemmed) Andean rosette plant, on sandy and blocky talus slopes in the Páramo de Piedras Blancas, Venezuela (Plate 7.1). Rosette density and cover increased down the talus slope in parallel with increasing particle size and substrate stability (Figure 7.1). Rosette density was not so much associated with slope position per se, as with the increasing proportion of the talus surface occupied by large rocks downslope. The rosette plants virtually all grew in areas of blocky talus and in areas downslope from isolated boulders embedded in finer sandy material. The roots of the plants always grew upslope and beneath stones. Water content of the surface soil was always 10 to 20 times greater under blocky talus and beneath boulders than in contiguous areas of bare sandy talus. The amount of water available for plant growth was also higher beneath stones, even 20 cm into the soil. Soil texture was similar (sand to sandy loam) on both talus types. The extra water found under the stones could result from any or all of three processes. First, while rain is falling, water flows over the stones and accumulates in the sandy soil matrix between and under the stones. Second, after the rain has stopped falling, the stone layer preventing water rising to the surface by capillary action reduces

Plate 7.1 Two rosettes of tall frailejón (*C. timotensis*) located directly downslope from several boulders embedded in a mobile talus slope. The largest plant is about 200 cm tall. Páramo de Piedras Blancas, Venezuela. Photograph by Francisco L. Pérez.

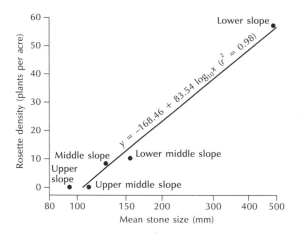

Figure 7.1 Rosette density of tall frailejón (*C. timotensis*) versus average stone size (length of longest axis in mm) on sandy and blocky talus slopes in the Páramo de Piedras Blancas, Venezuela. The regression line is significant at *p* > 0.001.

Source: Adapted from Pérez (1989)

evaporative loss. Third, at sunset, falling temperatures promote condensation in the hollow spaces between the stones. The rosettes favour blocky talus areas because water is available under stones, even through the dry season. The bare sandy talus areas are more difficult to colonize because they dry out during the dry season.

Substrate specialists

Some plants and microorganisms love or hate particular elements or compounds in their substrate. Six groups of substrate specialists are common:

1 **Calcicoles** (or **calciphiles**) are plants that favour such calcium-rich rocks as chalk and limestone (Figure 7.2). Calcicolous species often grow only on soil formed in chalk or limestone. An example from England, Wales, and Scotland is the meadow oat-grass (*Helictotrichon pratense*), the distribution of which picks out the areas of chalk and limestone and the calcium-rich schists of the Scottish Highlands (Figure

7.3a). Other examples are traveller's joy (*Clematis vitalba*), the spindle tree (*Euonymus europaeus*), and the common rock-rose (*Helianthemum nummularium*). Some plants are capable of living on the most forbidding of carbonate surfaces (Box 7.1).

2 **Calcifuges** (or **calciphobes**) avoid calcium-rich soils, preferring instead acidic rocks deficient in calcium. An example is the wavy hair-grass (*Deschampsia flexuosa*) (Figure 7.3b). However, many calcifuges are seldom entirely restricted to exposures of acidic rocks. In the limestone Pennine dales, the wavy hair-grass can be found growing alongside meadow oatgrass. Acid-loving microbes (**acidophiles**) prosper in environments with a pH below 5. *Sulfolobus acidocaldarius*, as well as being a hyperthermophile (p. 87), is also and acidophile.

3 **Neutrophiles** are acidity 'middle-of-the-roaders'. They tend to grow in the range pH 6–8. In the Pennine dales, strongly growing, highly competitive grasses that make heavy demands on water and nutrient stores are the most common neutrophiles.

4 **Alkaliphiles** (or **alkalophiles**), also termed **basophiles** and **basiphiles**, prefer alkaline conditions, with acidity in the range pH 8–11. The alkali-loving microbe *Natronobacterium gregoryi*, which lives in soda lakes, is an example.

5 **Halophiles** are organisms that live in areas of high salt concentration. Salt-loving microbes live in intensely saline environments. They survive by producing large amounts of internal solutes that prevent rapid dehydration in a salty medium. An example is *Halobacterium salinarium*.

6 **Nitrophiles** and **phosphatophiles** are found in agricultural landscapes, which often have raised levels of nitrogen, phosphorus, and other nutrients at forest edges. Some of the most typical edge species in Europe are examples – elder (*Sambucus nigra*) in the small-shrub layer and common nettle (*Urtica dioica*) and cleavers or goosegrass (*Galium aparine*) in the herb layer.

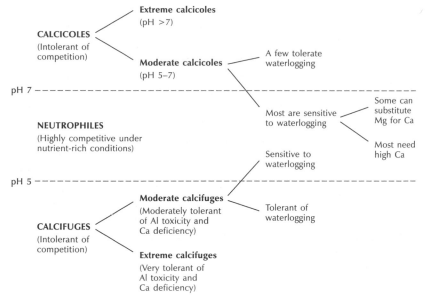

CALCICOLES
(Intolerant of
competition)

Extreme calcicoles
(pH >7)

Moderate calcicoles
(pH 5–7)

A few tolerate
waterlogging

pH 7 –

NEUTROPHILES
(Highly competitive under
nutrient-rich conditions)

Most are sensitive
to waterlogging

Some can
substitute
Mg for Ca

Most need
high Ca

Sensitive to
waterlogging

pH 5 –

Moderate calcifuges
(Moderately tolerant
of Al toxicity and
Ca deficiency)

Tolerant of
waterlogging

CALCIFUGES
(Intolerant of
competition)

Extreme calcifuges
(Very tolerant of
Al toxicity and
Ca deficiency)

Figure 7.2 A classification
of calcicoles and
calcifuges. The neutral
zone between pH 5 and 7
may be occupied by
highly demanding and
strongly competitive
species that exclude the
moderate calcicoles and
calcifuges.

Source: Adapted from
Etherington (1982, 270)

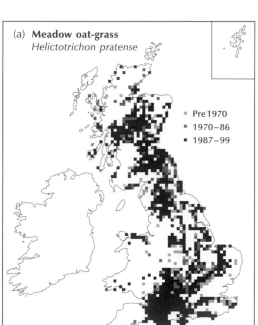

(a) **Meadow oat-grass**
Helictotrichon pratense

Pre 1970
1970–86
1987–99

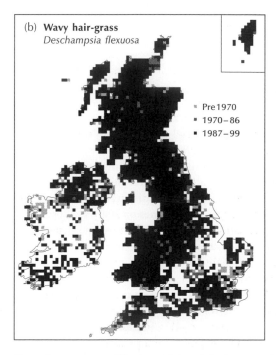

(b) **Wavy hair-grass**
Deschampsia flexuosa

Pre 1970
1970–86
1987–99

Figure 7.3 Calcioles and calcifuges. (a) Distribution of meadow oat-grass (*H. pratense*), a calcicolous species, in
the British Isles. (b) Distribution of wavy hair-grass (*D. flexuosa*), a calcifuge species, in the British Isles.
Source: After Preston *et al.* (2002)

Box 7.1

EXTREME SUBSTRATE ADAPTATIONS IN MESQUITE TREES

Perhaps the most extreme adaptation to a harsh environment is seen in two species of mesquite trees – tamarugo (*Prosopis tamarugo*) and Argentine mesquite (*P. alba*) – that grow in the Pampa del Tamagural, a closed basin, or salar, in the rainless region of the Atacama Desert, Chile. These plants manage to survive on concrete-like carbonate surfaces (Ehleringer *et al.* 1992). Their leaves abscise and accumulate to depths of 45 cm. Because there is virtually no surface water, the leaves do not decompose and

nitrogen is not incorporated back into the soil for recycling by plants. The thick, crystalline pan of carbonate salts prevents roots from growing into the litter. To survive, the trees have roots that fix nitrogen in moist subsurface layers, and extract moisture and nutrients from groundwater at depths of 6–8 m or more through a taproot and a mesh of fine roots lying between 50 and 200 cm below the salt crust. A unique feature of this ecosystem is the lack of nitrogen cycling.

A host of basiphiles, calciphiles, and other substrate specialists thrive on Scottish serpentine outcrops at Shetland, Rhum, Coyles of Muick, and Meikle Kilrannoch (Spence (1970) (Table 7.1).

Plant communities and substrate

Within the world's zonal biomes are areas of **intrazonal** and **azonal soils** that, in some cases, support distinctive vegetation. These non-zonal vegetation communities are sometimes styled **pedobiomes** (Walter and Breckle 1985). Several different pedobiomes are distinguished on the basis of soil type: lithobiomes on stony soil, psammobiomes on sandy soil, halobiomes on salty soil, helobiomes in marshes, hydrobiomes on waterlogged soil, peinobiomes on nutrient-poor soils, and amphibiomes on soils that are flooded only part of the time (e.g. riverbanks and mangroves). Pedobiomes commonly form a mosaic of small areas and exist in all zonobiomes. There are instances where pedobiomes are extensive: the Sudd marshes on the White Nile in south-central Sudan, which cover some 32,000 km² in the wet season; glaciofluvial sandy plains; and the nutrient poor soils of the Campos Cerrados in Brazil.

Stony soil biomes (lithobiomes)

Lithobiomes are associated with the *talus slopes* common in alpine, Arctic, and desert regions. Talus forms by the accumulation of loose rock debris of varying sizes. Plants appear to have difficulty in colonizing talus. Where colonization has taken place, plants are commonly associated with specific talus zones or substrate types. In the Jura Mountains, central Europe, Roman Bach (1950) found that talus slopes formed of limestone fragments are graded: the small fragments accumulate beneath rock outcrops, the source of the talus, while the biggest (blocks with diameters of about 50 cm) lie at the foot of the talus slope. This gradation of particle size creates a *lithosequence* of parent materials, soils, and vegetation. On the upper slope, rendzina soils evolve. They comprise a deep and gravel-rich sandy loam with a granular structure topped by 60 to 100 cm of mull humus. Their pH ranges from 6.5 to 7.8. These productive soils support a forest of mountain maple (*Acer* sp.), with shrubs of ash (*Sorbus* spp.) and hazel (*Corylus* sp.), and a herb layer predominated by ferns and members of the Cruciferae. Towards the foot of the talus slope, blocky raw carbonate soils evolve. There is no fine soil

Table 7.1 Substrate specialists on serpentine outcrops in Scotland

Specialism	Species
Basiphiles	Kidney vetch (*Anthyllis vulneraria*), Arctic sandwort (*Arenaria norvegica* subsp. *norvegica*), green spleenwort (*Asplenium viride*), black spleenwort (*A. adiantum-nigrum*), alpine mouse-ear (*Cerastium alpinum*), Shetland mouse-ear (*C. nigrescens*), lady's bedstraw (*Galium verum*), three-flowered rush (*Juncus triglumis*), crested hair-grass (*Koeleria macrantha*), meadow oat-grass (*Helichtotrichon pratense*), alpine catchfly (*Lychnis alpina*), alpine bistort (*Persicaria viviparum*), stone bramble (*Rubus saxatilis*), moonwort (*Botrychium lunaria*), frog orchid (*Coeloglossum viride*), black bog-rush (*Schoenus nigricans*), glaucous sedge (*Carex flacca*), knotted pearlwort (*Sagina nodosa*), mossy saxifrage (*Saxifraga hypnoides*)
Calciphiles	Brittle bladder-fern (*Cystopteris fragilis*), hoary whitlow grass (*Draba incana*), autumn gentian (*Gentianella amarella*), a large liverwort – *Plagiochila asplenoides*
Magnesium- or calcium-rich specialists	A subspecies of common scurvygrass (*Cochlearia officianalis* subsp. *scotica*), Pyrenean scurvygrass (*C. pyrenaica*), sea campion (*Silene vulgaris* subsp. *maritima*), thrift (*Armeria maritime*), sea plantain (*Plantago maritime*)

material, and a layer of mor humus, some 30 cm thick, lies directly on the limestone boulders. Some organic matter accumulates between the boulders and feeds roots. Spruce (*Picea abies*) forest and *Hylocomium* mosses grow in this geomorphologically active landscape. The spruce does not grow far up the talus slope because it cannot endure the frequent salvos of rolling boulders and the motion of the soil.

Rock-loving biomes occur on *inselbergs* (Porembski *et al.* 2000). Inselbergs are often rocky outcrops of varying size. They provide several habitats that support a range of vegetation types (Table 7.2). It is common for the diversity of vascular plant species to increase with increasing inselberg size. In the Ivory Coast, Africa, a sample of nearly 100 inselbergs, ranging in size from 200 m² to 7 km² (excluding forested sites), showed this species–area pattern (Porembski *et al.* 2000). The larger inselbergs tend to house more plant species because they contain more types of habitat, including mat communities and ephemeral flush vegetation. Moreover, large inselbergs have large populations of plants, so reducing the risk of local extinctions. Inselberg size affects the relative abundance of plant life-forms, as well as **species richness**. Figure 7.4 depicts the life-form spectra of three inselbergs in the savannah zone. Therophytes dominate the smallest outcrop (500 m²), followed by hemicryptophytes and cryptophytes. Chamaephytes and phanerophytes are absent. The importance of therophytes declines on the medium (15,000 m²) and large (200,000 m²) outcrops, while the percentage of all other life-forms, including chamaephytes and phanerophytes, rises. The disturbance regime on inselbergs may explain this pattern of life-form occurrence. Unpredictable climatic fluctuations (such as the amount and distribution of rainfall) are higher on small rock outcrops and encourage a very high percentage of annuals, which tend to be pioneer species adapted to short-term disturbance. Larger inselbergs have more stable growth conditions and favour perennial plants. In support of this idea, the mat-forming chamaephyte *Afrotrilepis pilosa* (a sedge) seldom occurs on outcrops smaller than 1 ha, while trees such as *Hildegardia barteri* and *Hymenodictyon floribundum* need outcrops of about 5 ha (Plate 7.2). Proximity to other outcrops is a complicating factor: *Afrotrilepis pilosa* does occur on outcrops smaller than 1 ha that lie within 300 m of a larger inselberg.

Table 7.2 Habitats and associated vegetation on inselbergs

Primary habitat type	Secondary habitat type	Vegetation
Rock surfaces	Rock surfaces	Cryptogamic biofilm of cyanobacterial lichens (e.g. *Peltula* spp.) or nitrogen-fixing cyanobacteria (e.g. *Stigonema* spp. and *Scytonema* spp.)
	Boulders	Chlorophytic lichens (on lichen inselbergs); lichens on basal and overhanging parts (on cyanobacterial inselbergs)
	Drainage channels	Cyanobacteria and, to a lesser degree, cyanobacterial lichens; mosses and vascular plants are rare
	Wet flush	Thick (up to 1 cm when wet) cyanobacterial crust with colonizing vascular plants; some moss patches
	Level rock faces	Epilithic vascular plants – succulents or xerophytic species; some epiphytes
Rock crevices	Horizontal and vertical crevices	With thin soil (<2 cm), short-lived vascular plants; rhizatomous ferns and mosses
	Clefts	Thicker soils than in crevices support perennial shrubs and trees (mostly deciduous)
	Boulder bases	Sparse cover of mosses, shade-tolerant ferns, and small trees
Depressions	Seasonal rock pools (weathering pits and grinding holes)	Highly specialized vascular plants (including some poikilohydric species) in some pools, usually short-lived herbs rooted in the sandy substrate and sometimes (arriving by chance) free-floating water plants (e.g. *Lemna* spp.), often with epilithic and endolithic cyanobacteria and certain lichens on rock-pool walls
	Permanently water-filled rock ponds	Rare habitats with occasional vascular plants of the type that prefer wet ground
	Rock debris	Mosses, liverworts, and cyanobacteria forming dense swards where wet; therophytes and some poikilohydric vascular plants
	Soil-filled depressions	Cryptogams, therophytes, and some geophytes; occasional small shrubs and trees. Trees and shrubs may dominate where the soil is deep
Base of steeply inclined slopes		Ephemeral flush: species-rich and meadow-like during rainy season; sparse and dried-out plant remains during dry season. Carnivorous plants on the poor soils (e.g. *Drosera indica*). Grasses and sedges thrive
Smooth rocky slopes		Mat of vascular poikilohydric plants that cling directly to the rock surface. Mats normally formed of monocotyledons such as *Afrolepsis* (in west Africa), sometimes with orchids. In some places, cushion-like mats of poikilohydric fern genus *Selaginella* occur. Dicot mats are rare

Source: Based on discussion in Porembski *et al.* (2000)

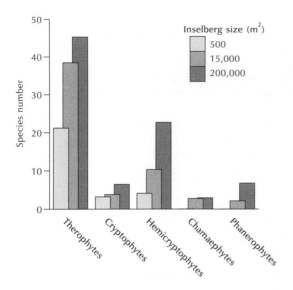

Figure 7.4 Life-form spectra versus inselberg size in the Ivory Coast, Africa.

Source: Adapted from Porembski *et al.* (2000a)

Sandy soil biomes

On sandy acidic soils found along the coastal lowlands and offshore islands of southeast Queensland, Australia, between the New South Wales border and Rockhampton, grows wallum. Wallum is coastal heathland vegetation named after the aboriginal name for *Banksia aemula*, commonly called wallum banksia. Wallum vegetation is low, dense, shrubby, and very diverse. As well as wallum banksia, other banksias, wattles (acacias), she-oaks (allocasuarinas), gum trees (eucalypts), tea trees (leptospermums), boronias, sedges, and reeds thrive. Where the water table reaches the surface, wet wallum forms, a sedgeland dominated by paperbarks (*Melaleuca* spp.) with an understorey including wallum bottlebrush (*Callistemon pachyphyllus*), swamp banksia (*Banksia robur*), and wild may (*Leptospermum liversidgei*). Dry wallum forms on higher ground where scattered trees often have a shrubby understorey of wallum species.

Nutrient-poor soil biomes

Quartzite is sandstone that has been converted into solid quartz rock. Lacking the pores of sandstone, it is not easily weathered and tends to produce nutrient-poor soils. In Shropshire, England, the Stiperstones range is primarily a quartzite ridge with such rugged tors as the Devil's Chair and Cranberry Rock that gives rise to acidic and extremely nutrient-poor soils supporting species-poor heathland vegetation of heather (*Calluna vulgaris*) and whinberry (*Vaccinium myrtillus*). The slopes lie mainly on Mytton Flags, which are generally acidic in reaction, with some slight base enrichment where springs percolate through to the surface producing flushes of wetland vegetation. Therefore, the typical Stiperstones semi-natural vegetation is heathland and acidic grassland.

Serpentinites, which are ultramafic rocks, support eye-catching vegetation communities (Brooks 1987). Being deficient in aluminium, clay formation is slow in soils formed from serpentinites, which are consequently typically highly erodible, shallow, and stock few nutrients. These peculiar features affect vegetation growth. Outcrops of serpentine sustain small islands of brush and bare ground in a sea of forest and grassland. Native floras with many endemic species populate these islands. In a locality some 5.5 km north of Geyserville, California (Jenny 1980, 248), rocky outcrops of schist support coast live oak trees (*Quercus agrifolia*), while soils derived from schists on the extensive slopes, known as the Raynor Series, carry native bunch grass (*Stipa* sp.) and wild oats (*Avena* sp.). Adjacent soils derived from serpentine, called the Montara Series, support a forest of digger or grey pine (*Pinus sabiniana*). The junction between schist and serpentine is sharp – no more than a metre wide. Grass in the oak–savannah grows to a height of 40–110 cm, then, a mere stride away, plummets to 5–15 cm in the digger-pine forest.

The effect of serpentine on vegetation is clearly seen in Cuba. In a transect across Monte Libano,

(a)

(b)

Plate 7.2 (a) *A. pilosa* and *H. floribundum* on Mount Niangbo in the Sudanian region of the Ivory Coast. Both species are typical elements of the West African inselberg vegetation. The Cyperaceae *A. pilosa* is a so-called resurrection plant that may lose most of its water content during periods of drought. It may survive in the desiccated state for months or even years. Among angiosperms this is absolutely outstanding and is a characteristic adaptation of a number of Cyperaceae, Poaceae, and Velloziaceae that are common on inselbergs. Altogether, there are only about 300 species of vascular plants that are desiccation-tolerant. Most of them occur on inselbergs. (b) *A. pilosa* growing on an inselberg near Bouaké, which lies in the transition zone between dry and wet forest and Ivory Coast. Photographs by Stefan Porembksi.

a serpentine overlies limestone (Borhidi 1991) (Figure 7.5). The vegetation on the limestone changes from royal palm–rain tree (*Roystonea–Samanea*) grassland in footslopes, to semi-deciduous forest and mogote forest on backslopes. The serpentine, which underlies most of the upper slopes and summits, supports pine forest and, where bands of limestone occur, pine forest with agave. More sheltered sites on serpentine support sclerophyllous montane forest, while similar sites on limestone support submontane rain forest.

Within the British Isles, there are 19 outcrops of ultramafic rock. Of these outcrops, 11 occur in Scotland and bear vegetation characteristic of serpentine. The remaining outcrops – 7 in Scotland and 1 at the Lizard, Cornwall – bear vegetation uncharacteristic of serpentine. The Scottish island of Unst has characteristic serpentine vegetation. The vegetation mainly forms a closed canopy. There are also areas of scree where the plant association is Arctic sandwort (*Arenaria norvegica* subsp. *norvegica*) and northern rock-cress (*Arabis petraea*). Other species, including sea

campion (*Silene vulgaris* subsp. *maritima*) and Shetland mouse-ear (*Cerastium nigrescens*), occur on these barren areas. Shetland mouse-ear is the only serpentine-endemic plant confined to the British Isles, and indeed to Unst, where it grows on two adjacent hills. The uncharacteristic serpentine vegetation of the 50 km² of serpentine rocks of the Lizard Peninsula comprises four types of 'heath', each with dominant species:

1 Rock heath with heather (*Calluna vulgaris*) and sheep's fescue (*Festuca ovina*) in shallow soil pockets among serpentine rocks;
2 Mixed heath with Cornish heath (*Erica vagans*) and gorse (*Ulex europaeus*) on well-drained brown earths;
3 Tall heath with Cornish heath, purple moor-grass (*Molinia caerulea*), and black bog-rush (*Schoenus nigricans*) in shallow valleys and depressions on seasonally waterlogged soils;
4 Short heath with bristle bent (*Agrostis curtisii*), heather, bell heather (*Erica cinerea*), cross-leaved heath (*E. tetralix*), purple moor-grass,

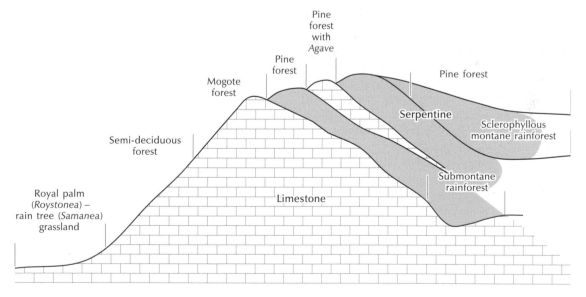

Figure 7.5 Vegetation transect across Monte Libano, Cuba.
Source: Adapted from Borhidi (1991)

and gorse, on level but raised portions of the plateau. None of the these species is endemic to the serpentine areas of the Lizard Peninsula, but the rare Cornish heath is almost restricted to serpentine outcrops, as is a possible sub-species of spring sandwort (*Minuartia verna*) – *M. verna* subsp. *gerardi*.

Another remarkable nutrient-poor soil biome occurs in the western Great Basin desert of the USA, where '*tree islands*' grow in a sea of sage-brush vegetation. The islands take the form of about 140 small stands of Sierra Nevada conifers (mainly *Pinus ponderosa* and *Pinus jeffreyi*), from one to several hectares in area, lying up to 60 km east of the eastern margins of the Sierra Nevada montane forest. They are restricted to outcrops of hydrothermally (hot water) altered andesitic bedrock, from which base cations have been leached on exposure, that produce localized patches of azonal soil (Billings 1950). The soils derived from andesitic bedrock in the Great Basin are primarily Xerollic Haplargids, typical of desert brown soils, whereas the soils derived from the altered bedrock form shallow Lithic Entisols, light yellow in colour, acid in reaction, and low in base cations and phosphorus (DeLucia and Schlesinger 1990). Principal component analysis of 18 soil parameters indicated that soils formed in altered rock from different sites have much in common, but soils formed in unaltered rock differ according to vegetation cover – forest, piñon pine–juniper woodland, or sagebrush (Figure 7.6) (Schlesinger *et al.* 1989). The first principal component relates to acidity, alkalinity, and calcium content; the second to carbon and nitrogen content. Taken together, these two axes account for 48 per cent of the variance in the data.

Soil nutrient levels may play a decisive role in influencing the vegetation within a region. In England, Scotland, and Wales, vegetation data for 13,614 individual samples in both 1978 and 1990 were grouped into 100 vegetation classes using TWINSPAN (Bunce *et al.* 1999). The 100 classes were then analysed using a statistical ordination technique to measure the degree of similarity between them. The ordination axes extracted the greatest amount of variation from the data. The first ordination axis is a gradient of soil nutrients levels. It separates the vegetation of arable fields, which is highly disturbed and found on nutrient-rich soils, from various types of grass-land, and then moorland and heath and bog, which occur on nutrient-poor peats and podzols (Figure 7.7). Notice in Figure 7.7 that the second ordination axis separates short-lived, herbaceous species tolerant of disturbance (crops and weeds) from long-lived, large plants less tolerant of fre-quent disturbance (woods). This second axis cap-tures an environmental gradient of disturbance and shade. A third axis, not shown in the figure, is a moisture gradient that probably differentiates the classes that lie near to each other.

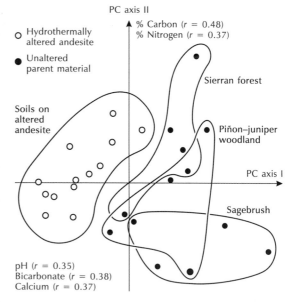

Figure 7.6 Principal component analysis of soil data from vegetation in and around 'tree islands' of the western Great Basin Desert, USA. The first 2 axes account for 48 per cent of the variance and are mostly strongly correlated with the variables indicated.

Source: Adapted from DeLucia and Schlesinger (1990)

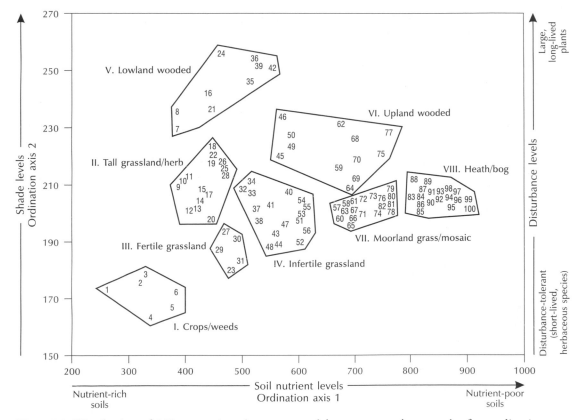

Figure 7.7 Distribution of 100 vegetation classes, grouped by aggregate class, on the first ordination axes, showing the importance of substrate.

Source: Adapted from Bunce *et al*. (1999)

Calcium-rich soil biomes

Limestone areas support vegetation communities that are tolerant of lime-rich soils, although local variations in soil acidity and in soil moisture often produce variant types (Box 7.2).

Wet soil biomes

Wetlands are terrain in which the soil is saturated with water, and soil saturation is the chief determinant of soil development and communities. They occur from tundra to the tropics and on every continent except Antarctica. *Marshes*, *swamps*, *bogs*, and *fens* are all wetlands. Regional and local differences result from variations in soils,

topography, climate, hydrology, water chemistry, vegetation, and other factors, including human disturbance.

Blanket bog is an interesting example of a wetland. It needs a wet and cool climate to let bog mosses (*Sphagnum* spp.) grow. Suitable sites occur in western Scotland, western Ireland, and the northern English hills. Under acidic and water-logged conditions, the decomposition of bog moss and other plants is slow and peat forms. The peat builds up on flat ground, spreads over large tracts of gently sloping ground, and fills in hollows, so creating a widespread mantle or blanket of peat, hence its name. Blanket bogs attract a range of acid-tolerant plants, including heather (*Calluna*

Box 7.2

LIMESTONE GRASSLANDS OF THE DERBYSHIRE AND STAFFORDSHIRE DALES, ENGLAND

The Carboniferous Limestone of the southern Pennines supports a diverse community of calcareous grasslands (Anderson and Shimwell 1981, 114–28). Two groups of communities dominate the steeper dale slopes overlying calcareous soils. The first is a short-turf grassland rich in grasses and herbs; the second is a tall grassland vegetation of grasses and herbs.

Short-turf grassland

The species-rich short-turf grassland often has more than 45 species of grass, herb, and sedge per square metre and is the 'typical' limestone grassland. The chief species are normally five neutrophiles – sheep's fescue, ribwort plantain (*Plantago lanceolata*), common bird's-foot-trefoil (*Lotus corniculatus*), and harebell (*Campanula rotundifolia*); and five moderate calcicoles – crested hair-grass, meadow oat-grass, glaucous sedge, spring sedge (*Carex caryophyllea*), and fairy flax (*Linum catharticum*). A host of moderate and exclusive regional calcicoles grow in these communities, including quaking-grass (*Briza media*), limestone bedstraw (*Galium sterneri*), and Nottingham catchfly (*Silene nutans*). In addition, several dales have extra uncommon species. For instance, Flag Dale has patches of tor-grass (*Brachypodium pinnatum*) and upright brome (*Bromus erectus*). These grassland communities occupy grazed sites with high soil pH, mainly on south-facing slopes where drought conditions sometimes occur.

Four variants of this 'typical' limestone grassland occur: a dry or xeric variant, a damp variant, a grassland-scrub transition variant, and a species-poorer variant on slopes less than 15 degrees. The dry variant favours areas of shallow soils around rock outcrops, screes, and anthills. It is usually dominated by sheep's fescue, while wild thyme (*Thymus praecox*) and black medic (*Medicago lupulina*) are more important and such small annual species as hairy rock-cress (*Arabis hirsuta*), thyme-leaved sandwort (*Arenaria serpyllifolia*), and barren strawberry (*Potentilla sterilis*) occur. In the damp variant, which lies mainly at higher altitudes and on north- and west-facing slopes, flea sedge (*Carex pulicaris*) becomes important, and carnation sedge (*Carex panicea*), grass-of-parnassus (*Parnassia palustris*), and devil's-bit scabious (*Succisa pratensis*) occur. The grassland–scrub transition variant boasts a colourful summertime community of bloody cranesbill (*Geranium sanguineum*), Nottingham catchfly, marjoram (*Origanum vulgare*), wild basil (*Clinopodium vulgare*), and rough hawkbit (*Leontodon hispidus*). Sheep's fescue (*Festuca ovina*) and glaucous sedge (*Carex flacca*) dominate the species-poor community on the shallower slopes, and the presence of such species as dog's mercury (*Mercularils perennis*), wood sage (*Teucrium scorodonia*), and false brome (*Brachypodium sylvaticum*) suggest a conversion from hazel scrub over the last century or so.

Tall grassland

The tall grass–herb vegetation is also rich in species. The chief plants are neutrophilous grasses and herbs that may grow up to 80 cm. The co-dominant species are false oat-grass, red fescue, yellow oat-grass (*Trisetum flavescens*), and, in some stands, smooth meadow grass (*Poa pratensis*) or rough meadow-grass (*P. trivialis*). The most luxuriant stands grow on damp rendzinas with a slight acidic reaction, on north- and west-facing slopes. Jacob's-ladder (*Polemonium caerulum*), a nationally rare but locally frequent plant, favours these damper tall-herb habitats. Along with the globeflower (*Trollius europaeus*), it is a boreal plant normally found in light, open woodlands. It may have survived in the tall-herb communities from the herb layer of moist ash woodland and scrub that were invaded by coarse grasses after clearing. The presence of tufted hair-grass (*Deschampsia cespitosa*) and other woodland herbs, also associated with the herb layer of ash woodlands, supports this view.

Other grassland types

The more acidic brown earths of the gently sloping upper dale sides support a variety of neutral grasslands. Common bent (*Agrostis tenuis*) and sweet vernal-grass (*Anthoxanthum odoratum*), which occur infrequently of the steeper slopes, come to dominate with red fescue. Yellow oat-grass and Yorkshire fog (*Holcus lanatus*) may be abundant. Downy oat-grass (*Helictotrichon pubescens*) tends to replace its less robust relative, meadow oat-grass. Herb species are less common and mosses rare. These grasslands are similar to the unimproved, neutral meadow grasslands of the limestone plateaux.

Wavy hair-grass dominates areas with deep, non-calcareous soils with an acid reaction (pH 5.0 or less). In Coombs Dale, wavy hair-grass forms a dense carpet in acidic grassland. Wild thyme (*Thymus praecox*) is the only moderate calcicole present, heath bedstraw (*Galium saxitale*) replaces limestone bedstraw (*G. sterneri*), and tormentil (*Potentilla erecta*) occurs. These grasslands, along with the transitional types to the neutral grasslands, are the main habitat for the mountain pansy (*Viola lutea*).

In some dales, where there are areas of acidic plateau downwash or where loessal material fills limestone fissures, irregular patches of bracken (*Pteridium aquilinium*) and matt grass (*Nardus stricta*) run down from the plateau edge.

The damp alluvial soils of the valley bottoms, which are enriched in calcium washed from the dale slopes, support a grazed, tussocky grassland dominated by tufted hair-grass (*Deschampsia cespitosa*) and deficient in other grass species and calcicolous or tall herbs. Common daisy (*Bellis perennis*), common mouse-ear (*Cerasatium fontanum*), lesser celandine (*Ranunculus ficaria*), field wood-rush (*Luzula campestris*), and meadow saxifrage (*Saxifraga granulata*) occupy moderately rich, lightly grazed soils in these areas, forming a community similar to that found on the limestone plateaux.

vulgaris), bearberry (*Arctostaphylos uva-ursi*), dwarf birch (*Betula nana*), bog orchid (*Hammarbya paludosa*), bog ashpodel (*Narthecium ossifragum*), purple moor-grass (*Molinia caerulea*), broad-leaved cotton-grass (*Eriophorum latifolium*), deergrass (*Trichophorum cespitosum*), and insectivorous plants such as round-leaved sundew (*Drosera rotundifolia*) and butterwort (*Pinguicula vulgaris*).

Saltwater biomes

Plants living in areas of *salt water*, either at coastal sites or inland, are characteristically **halophytes**. A common structural adaptation of halophytes is a succulence of the plant body. It is common to find plants zoned along a gradient of saltiness. Figure 7.8 shows the zonation of land plants at the German coast relative to mean highwater mark. Glasswort (*Salicornia europaea*) penetrates down to about 30 cm below mean highwater mark. Sea aster (*Aster tripolium*) and annual sea-bite (*Suaeda maritima*) hardly ever go below it. Saltmarsh grass (*Puccinella maritima*) and sea arrow-grass (*Triglochin maritimum*) occupy higher levels. A similar zonal pattern, involving different species, is found on calcareous sand dunes, where salt concentrations are highest in the areas between the dunes and may be high enough to

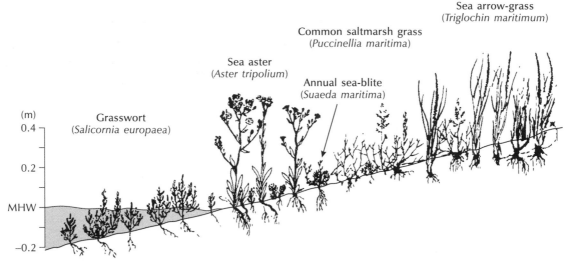

Figure 7.8 Zonation of land plants at the German coast.
Source: Adapted from Ellenberg (1986)

prevent any plants from growing. *Halocnemum strobilaceum* is very tolerant of salty soils and grows at the base of dunes or on muddy salt flats.

Multifactor biomes

The distribution of biomes usually reflects the combined action of several environmental factors, of which substrate is often an important one. To illustrate this point, consider the distribution of central European broadleaved communities (Figure 7.9). Figure 7.9a shows the approximate areas occupied by the alliances and suballiances according to moisture status and acidity, which is largely controlled by substrate. Figure 7.9b shows the tree species forming the woodlands under different moisture and acidity conditions.

LIFE ON THE ROCKS: ANIMALS AND SUBSTRATE

Soil and substrate affect some animals. For instance, the type and texture of soil or substrate

is critical to two kinds of mammal: those that seek diurnal refuge in burrows, and those that have modes of locomotion suited to rough surfaces.

Rock-living animals

Saxicolous species grow in, or live among, rocks. Some woodrats (*Neotoma*) build their homes exclusively in cliffs or steep rocky outcrops. The dwarf shrew (*Sorex nanus*) seems confined to rocky areas in alpine and subalpine environments. Even some saltatorial (jumping) species are adapted to life on rocks. The Australian rock wallabies (*Petrogale* and *Petrodorcas*) leap adroitly among rocks. Traction-increasing granular patterns on the soles of their hind feet aid this accomplishment. Rocky Mountain pikas (*Ochotona princeps*) in the southern Rocky Mountains, USA, normally live on talus or extensive piles of gravel (Hafner 1994). Those living near Bodie, a ghost town in the Sierra Nevada, utilize tailings of abandoned gold mines (A. T. Smith 1974, 1980). The yellow-bellied marmot (*Marmota flaviventris*) is

(a)

(b)

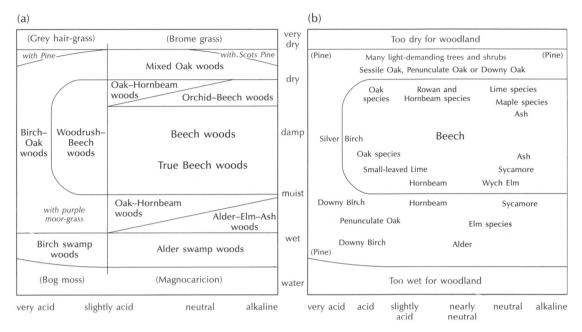

Figure 7.9 Distribution of central European broadleaved communities. (a) Approximate areas occupied by the woodland alliances and suballiances according to moisture status and acidity. (b) Tree species forming the woodlands under different moisture and acidity conditions. The species in brackets occur only in some parts of central Europe.

Source: Adapted from Ellenberg (1986)

another saxicolous species, and commonly occurs with the Rocky Mountain pika. The entire life style of African rock hyraxes (*Heterohyrax*, *Procavia*) centres on their occupancy of rock piles and cliffs. Most of their food consists of plants growing among, or very close by, rocks. The scent of urine and faeces on the rocks bonds their social system. The rocks provide useful vantage points to keep an eye out for predators, hiding places, and an economical means of conserving energy. Terry A. Vaughan (1978, 431) reported observing a colony of Bruce's hyrax or bush hyrax (*Heterohyrax brucei*) on the Yatta Plateau, southern Kenya, in July 1973. On first emerging from their nocturnal retreats, deep in rock crevices, they avoid touching the cool rock surfaces with their undersides, turn broadside to catch the first rays of the Sun, and bask. As the temperature soars during the morning, the bush hyraxes move

to dappled shade beneath the sparse foliage of trees or bushes. When the ambient temperature tops 30°C, they move to deep shade where they lie sprawled on the cool rock and remain there during the hot afternoon. Before dark, they move to the open and lie full length on warm rock.

The South American flat-headed bat (*Molossops mattogrossensis*) has a dorsoventrally flattened head and body that enables it to enter narrow cracks on the exfoliating granite of inselbergs, and tubercular projections on its forearms to help it develop traction against the rock surface when it is wedged in the crack. Remarkably, in rocky areas of the arid Namib desert of Africa, and in Ethiopia, Kenya, and the Sudan, bats of the same family have also adapted to live in the cracks of exfoliating granite. An example is the flat-headed free-tail bat (*Platymops setiger*) from southeastern Sudan, southern Ethiopia, and Kenya, which shows similar adaptations, including the tubercles on the forearms, although some individuals bear warty protuberances instead of tubercles. Similarly, the South African flat-headed bat (*Sauromys petrophilus*), which is found in Namibia, Zimbabwe, Botswana, South Africa, the Tete district of western Mozambique, and possibly Ghana, is similar to the flat-headed free-tail bat except that it lacks wart-like granulations on its forearms.

Some mammals are adapted to living on inselbergs (Mares and Seine 2000). An inselberg in Namibia supports various rock specialists: rock mice (*Petromyscus* sp.), rock elephant shrews (*Elephantulus myurus*), dassie rats (*Petromus typicus*), a mid-sized rock hyrax (*Procavia capensis*), and large-bodied klipspringers (*Oreotragus oreotragus*). Other species frequent inselbergs within their range. For example, leopards (*Panthera pardus*) and black mamba snakes (*Dendroaspis polylepis*) hunt on inselbergs and shelter in their rocks. Hawks, owls, gemsboks (*Oryx gazella*), spotted hyenas (*Crocuta crocuta*), and other animals use inselbergs as places of food and shelter.

Burrowing animals

Burrowing species, which tend to be small, commonly live on a particular kind of soil. For instance, many desert rodents display marked preferences for certain substrates. In most deserts, no single species of rodent lives on all substrates; and some species occupy only one substrate. Heteromyid rodents (pocket mice, kangaroo rats, and kangaroo mice), for instance, are weak diggers and prefer sandy soil. Four species of pocket mice (*Perognathus*) occur in Nevada, USA (Hall 1946). Their preferences for soil types are largely complementary. The little pocket mouse (*Perognathus longimembris*) lives on moderately firm soils of slightly sloping valley margins. The long-tailed pocket mouse (*P. formosus*) is restricted to slopes where stones and cobbles are scattered and partly embedded in the ground. The desert pocket mouse (*P. penicullatus*) is associated with the fine, silty soil of the bottomland. The Great Basin

pocket mouse (*P. parvus*), a substrate generalist, can survive on a variety of soil types.

Ord's kangaroo rat (*Dipodomys ordii*) lives on loose, sandy soils throughout its range, and rarely occupies clay or gravel. It sometimes occurs with Merriam's kangaroo rat (*D. merriami*), but Merriam's kangaroo rat generally favours harder soils. Ord's kangaroo rats seem to prefer to dig into inclines where available, such as banks along roads and fencerows, in areas of sandy soils. The burrows have several entrances, but the rats usually use one at given time. In dry sand, footprints and tail tracks lead to the entrance currently used.

Animal communities and substrate

Different assemblages of rodents occur along catenae in deserts, owing to variations in substrate (Vaughan 1978, 276). A typical catena comprises rocky outcrops with boulders, gravelly lower slopes, and a sandy desert floor (Figure 7.10). The preferences of different species for these three substrates reflect digging ability, style of locomotion, and foraging technique. Merriam's kangaroo rat (*D. merriami*) avoids danger by fast and erratic hops. Although it will live on several kinds of terrain, it prefers open ground where it may exercise its mastery of hopping to the full. In contrast, the cactus mouse (*Peromyscus eremicus*) cannot run fast, but is good at climbing and scrambling. For this reason, it lives in rocky terrain, or sometimes in areas with cactus or brush, where safe havens are within clambering distance.

To some animals, substrate may mean vegetation rather than soil or rock. In submontane northern England, spiders are a major invertebrate predator. Their community structure appears to be determined by the architecture of the vegetation. Analysis of spider species trapped in pits from the dominant vegetation types on three submontane summit plateau areas in the northern Pennines revealed that vegetation density, which is reflected in vegetation community types (upland grassland, blanket bog, and so on), is the major determinant of spider distribution (Figure 7.11).

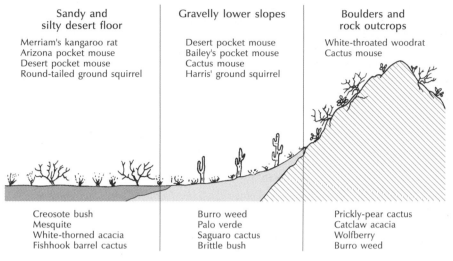

Sandy and silty desert floor	Gravelly lower slopes	Boulders and rock outcrops
Merriam's kangaroo rat Arizona pocket mouse Desert pocket mouse Round-tailed ground squirrel	Desert pocket mouse Bailey's pocket mouse Cactus mouse Harris' ground squirrel	White-throated woodrat Cactus mouse

Figure 7.10 Communities of mammals and plants living on contrasting substrates in North American deserts.

Source: Adapted from Vaughan (1978)

Creosote bush Mesquite White-thorned acacia Fishhook barrel cactus	Burro weed Palo verde Saguaro cactus Brittle bush	Prickly-pear cactus Catclaw acacia Wolfberry Burro weed

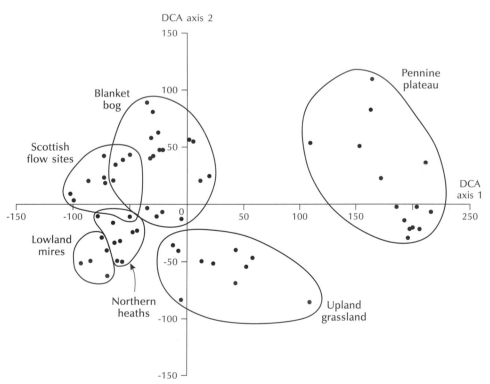

Figure 7.11 Ordination diagram of spider fauna from north Pennines sites, and flow sites in Sutherland, Scotland. Vegetation type (captured on DCA Axis 1, which is mainly a vegetation density axis) largely determines the spider assemblages.

Source: After Downie *et al.* (1995)

SUMMARY

Substrate wields a strong influence over species and communities, especially at a local level. Some plants love living on bare rock surfaces and others specialize in living on certain soils – calcium-rich soils, calcium-deficient soils, acid soils, neutral soils, saline soils. Substrates can favour the formation of distinct plant communities: stony soils and talus, sandy soils, nutrient-poor soils formed on quartzite and serpentine, limestone soils, water-logged soils (marsh, swamp, bog, and fen), and saline soils all tend to support characteristic biomes. Such animals as rock hyraxes and rock kangaroos live on rocks. Burrowing animals are greatly affected by substrate. Animal communities are to an extent structured according to the nature of the substrate on which they live.

FURTHER READING

Brooks, R. R. (1987) *Serpentine and Its Vegetation: A Multidisciplinary Approach*. London and Sydney: Croom Helm.

A wealth of information for dipping into.

Huggett, R. J. (1995) *Geoecology: An Evolutionary Approach*. London: Routledge.

A survey of all environmental factors. Chapter 6 covers substrate.

Porembski, S. and Barthlott, W. (eds) (2000) *Inselbergs: Biotic Diversity of Isolated Rock Outcrops in Tropical and Temperate Regions* (Ecological Studies, vol. 146). Berlin: Springer.

An excellent collection of essays with colour illustrations.

ESSAY QUESTIONS

1 **How do plants manage to live in rocky places?**

2 **How important is substrate in explaining the distribution of animal and plant communities?**

3 **How do animals manage to live in rocky places?**

8

TOPOGRAPHY AND LIFE

Many topographic factors influence living things. The most influential factors are altitude, slope direction, slope gradient, and landscape pattern. This chapter covers:

- elevation and life
- aspect and life
- slopes and life
- life in landscapes

Many topographic factors affect animals and plants. The most powerful factors are altitude, aspects, slope inclination, and the pattern of landscape patches, corridors, and matrices.

MOUNTAIN CLIMBING: ALTITUDE AND LIFE

Altitude exerts a strong influence on animals and plants. It affects communities and individual species.

Communities and altitude

Altitudinal floral zones

The plant communities girdling the Earth as broad zonal belts are paralleled in the plant communities encircling mountains – orobiomes (p. 104). Individual animal and plant species often occur within a particular elevational band, largely owing to climatic limits of tolerance. However, altitudinal ranges are influenced by a host of environmental factors, and not just by climatic ones. For instance, treelines are not always purely the result of climatic constraints on tree growth, but involve pedological and biotic interactions as well.

Carl Linnaeus, the eighteenth-century Swedish botanist, was aware that, on ascending mountainsides, a traveller would pass though different zones of life. Alexander von Humboldt, the German philosopher and geographer, recognized that a mountain was a miniature version of a hemisphere, carrying altitudinal replicates of tropical, temperate, boreal, and frigid floral zones. Clinton Hart Merriam (1894) came to a similar conclusion and estimated that each mile of altitude was equivalent to about 800 miles of latitude. Today, altitudinal vegetation zones are well established, the basic zones in ascending order being submontane, montane, upper montane, subalpine, alpine, subnival, and nival (Figure 8.1). However, the causes of this zonary arrangement are not wholly resolved and are the subject of continuing research (e.g. discussion in Huggett and Cheesman 2002).

Altitudinal faunal zones

Animal communities are zoned according to altitude, but not so neatly as are plant communities. Animals tend to associate themselves with particular types of plant community, so it is not surprising that their communities show some accordance with altitudinal vegetation zones.

Species and altitude

Plant species and altitude

Studies of individual species' distributions with increasing elevation tend to show continuous variation, with species' distributions overlapping. David W. Shimwell (1971) studied an altitudinal transect up the mountain Corserine in the Rhinns of Kells Range, southwest Scotland (Figure 8.2).

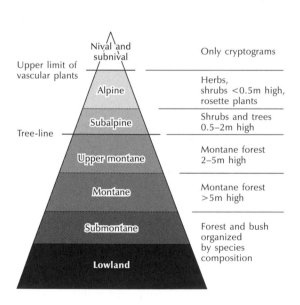

Figure 8.1 Altitudinal vegetation zones. The physiognomic character of the vegetation refers to tropical and warm-temperate oceanic islands, but would be typical of many other mountains.

Source: Adapted from Leuschner (1998)

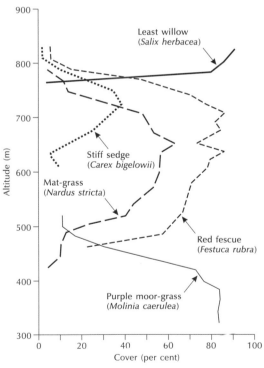

Figure 8.2 An altitudinal transect up the mountain Corserine, Rhinns of Kells Range, southwest Scotland.

Source: Adapted from Shimwell (1971)

Purple moor-grass (*Molinia caerulea*) was the dominant species up to 460 m but did not grow above 520 m. Above 490 m, red fescue (*Festuca rubra*) and mat-grass (*Nardus stricta*) were co-dominant up to 780 m. Stiff sedge (*Carex bigelowii*) ranged from 600 to 820 m but did not attain dominant status. Above 780 m, dwarf willow (*Salix herbacea*) rapidly became dominant.

Similarly, George Forrest, a plant collector who made seven plant and seed expeditions to Yunnan in western China between 1904 and 1932, found that rhododendron species have distinct altitudinal ranges (Figure 8.3). *Rhododendron griersonianum* and *R. diaprepes* occupy the cool-temperate zone, *R. forrestii* and *R. roxieanum* the subalpine and lower alpine zone, and *R. impeditum* and *R. keleticum* the alpine zone. *R. keleticum* is a semi-

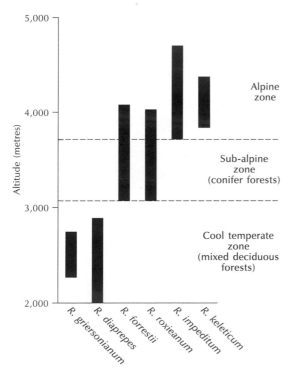

Figure 8.3 Altitudinal range of rhododendrons in Yunnan, China.

Source: Adapted from University of Liverpool Botanic Gardens (undated)

prostrate shrub that grows on alpine meadows and cliffs and screes, and *R. impeditum* is a dwarf rhododendron.

Animal species and altitude

Like their floral counterparts, animals tend to occupy particular altitudinal ranges. Box 8.1 discusses the altitudinal ranges of bird species on an elevational transect on Mauna Loa, Hawaii.

Animal species richness varies with altitude. Studies reveal three general patterns: a fall in species richness with increasing elevation (e.g. G. C. Stevens 1992); a mid-elevation peak in species richness for a wide variety of taxa (e.g. Rahbek 1995); and a peak species richness at high elevations for some taxa (e.g. Sanders *et al.* 2003). The peak at high elevations applies to ants in Spring Mountains, eastern Mojave desert, southern Nevada, USA (Sanders *et al.* 2003). The explanation for this pattern may lie in lower temperatures and higher precipitation encouraging higher levels of primary production than in the more arid lower elevations, which causes lower levels of physiological stress on the ants. Interestingly, in the southern African Nama Karoo, plant species richness was higher on mesas (table mountains) than on the surrounding plains in some study sites (Burke *et al.* 2003).

GETTING ORIENTATED: ASPECT AND LIFE

Plants and aspect

Aspect (compass direction) or *slope exposure* affects local climate (topoclimate) and microclimate light, windiness, and soil conditions. These environmental factors in turn influence vegetation. North-facing and south-facing slopes in extratropical regions commonly display the starkest contrasts. In the northern hemisphere, south-facing slopes tend to be warmer, and so more prone to drought, than north-facing slopes. The difference may be greater than imagined. In

Box 8.1

BIRDS SPECIES AND ELEVATION ON MAUNA LOA, HAWAI'I

The flora and fauna along an elevational transect on Mauna Loa, running 35 km from 3,660 m near the summit to 1,190 m near the summit of Kilauea, were studied in detail as part of the International Biological Programme (Mueller-Dombois *et al.* 1981). Figure 8.4 summarizes the results. The vegetation forms seven zones. Zone I, the alpine zone, and Zone II, the subalpine zone, extend through areas of lava-rock outcrop that contain only small amounts of fine soil material. They are occupied primarily by shrub species, though the lehua (*Metrosideros collina* subspecies *polymorpha*), a tree species, appears in Zone II. Zones III and IV, mountain parkland and savannah, are underlain by lava covered by a blanket of volcanic ash that thickens and becomes more continuous downslope. In these zones, shrub-species diversity decreases, tree-species diversity increases, and herbaceous life-forms, especially grasses, become more abundant. In Zone V, lehua dry forest, most of the shrubs, except those confined to high altitudes – 'ohelo (*Vaccinium peleanum*) and pukiawe (*Styphelia douglasii*) – reappear while the other tree species disappear. In Zones VI and VII (open and closed lehua rain forest), ferns become prevalent, shrubs associated with rock outcrops decrease in Zone VII, and lehua is dominant only on deeper soil in the rain forest.

Twenty-two bird species were recorded on the Mauna Loa transect. Figure 8.4 shows the elevational distribution of nine of them. Four species groups were recognized. Group 1 has one member – the Hawaiian thrush or 'oma'o (*Phaeornis obscurus obscurus*). This is the only bird living beyond the treeline in the sparse alpine scrub. It also lives in the closed rain forest. Its distribution is therefore rather odd, occupying high and low elevations but not the intervening areas. Its absence from the middling elevations

appears to result from competition with another frugivore, the exotic leiothrix (*Leiothrix lutea*). Group 2 comprises a spatially heterogeneous set of open area birds. The Hawaiian goose or nene (*Branta sandwichensis*) occurs from the treeline throughout the subalpine forest and mountain parkland zones. The golden plover or kolea (*Pluvialis dominica*), an indigenous bird, occurs mainly in the mountain parkland and open lehua dry-forest. It is absent from the savannah zone, probably because it dislikes the tall grass growing there. The Hawaiian owl or pueo (*Asio flammeus sandwichensis*) is confined to the mountain parkland and savannah zones. Group 3 comprises three endemic species ranging from the lower subalpine scrub to the rain forest. The Hawaiian hawk or 'io (*Buteo solitarius*) is found throughout the range but the distribution of both the Hawai'i 'elepaio (*Chasiempis sandwichensis sandwichensis*) and 'i'iwi (*Vestiaria coccinea*) is interrupted by the lehua dry-forest zone. The differences in distribution are attributable to differences in general behaviour and feeding habits: the 'elepaio and 'i'iwi favour colonies of koa (*Acacia koa*) trees or other forest groves with closed canopies, neither of which grow in the lehua dry-forest zone. Group 4 consists of the two most abundant native honey-creepers – the Hawai'i 'amakihi (*Loxops virens virens*) and the 'apapane (*Himatione sanguinea*). The wide distribution of these species indicates broad tolerance of temperature and rainfall regimes. The slight fall in 'apapane density in the lehua dry-forest suggests that it prefers closed tree canopies.

The elevational distribution of these birds on Mauna Loa suggests three things. First, distinct patterns of bird distribution are related to the distribution of major vegetation types. Second, the pattern of bird distribution is correlated to

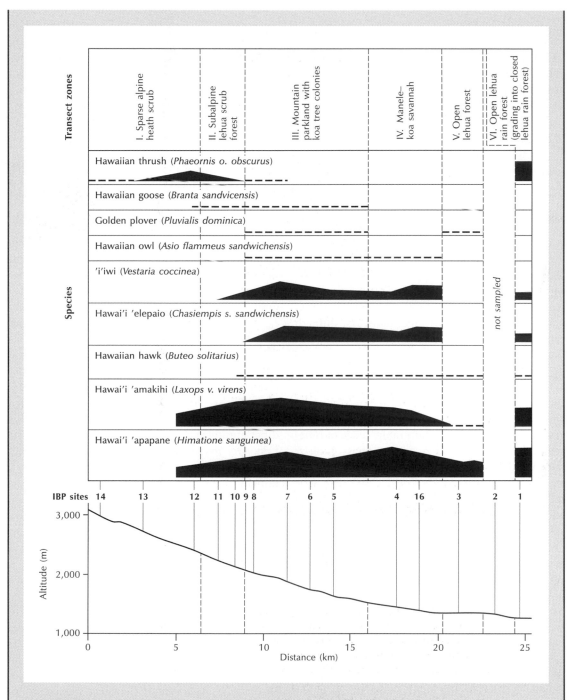

Figure 8.4 Distribution of native bird species along the Mauna Loa altitudinal transect.
Source: Adapted from Mueller-Dombois *et al.* (1981)

environmental factors, including vegetation – witness the upper limits of the 'apapane and Hawai'i 'amakihi that end suddenly at the treeline. Above the treeline, no individual 'apapane or Hawai'i 'amakihi are seen, but just below the treeline they live at high densities. The 'i'iwi and Hawaii 'elepaio exhibit distinct density changes where savannah vegetation changes to open lehua dry forest. Third, vegetation structure exerts such a profound influence upon bird species distributions that the effect of competition between species is of little importance. This contrasts with a similar altitudinal gradient in the Peruvian Andes where competition plays a significant role (Terborgh and Weske 1975). The difference between the two transects may reflect the contrasting settings – Mauna Loa is an island setting with low avian diversity; the Peruvian Andes is a continental setting with high avian diversity.

a Derbyshire dale, England, the summer mean temperature was 3°C higher on a south-facing slope than on a north-facing slope (Rorison *et al.* 1986), a difference equivalent to a latitudinal shift of hundreds of kilometres! Aspect affects slopes at all scales, from mountainsides, through sand dunes, to the sides of anthills.

North-facing and south-facing contrasts

Exposure-related vegetation differences can be spectacular. The Lägern is an 800-m high, east–west running mountain range made of limestone. It lies near Baden, in the Swiss Jura. A narrow mountain ridge, some 50 cm wide, separates the north-facing and south-facing slopes. The microclimate on the north-facing side is cold and supports beech (*Fagus sylvatica*) forest with subalpine elements such as alpine penny-cress (*Thlaspi alpestre*) and green spleenwort (*Asplenium viride*). The warm south-facing slope supports warmth-loving oak (*Quercus pubescens*) woodland with such southern elements as pale-flowered orchid (*Orchis pallens*), wonder violet (*Viola mirabilis*), and bastard balm (*Melittis melissophylum*). This dramatic botanical contrast on either side of the ridge is equivalent to an altitudinal difference of 1 km and a latitudinal difference of 1,000 km. On coastal sand dunes in the Netherlands, north-facing slopes support dense stands of boreal crowberry (*Empetrum nigrum*),

while south-facing slopes support open cover of grey hair-grass (*Corynephorus canescens*) and lichens. More subtle differences in aspect-related topoclimates are commonly recorded. In southeastern Ohio, microclimatic differences between northeast-facing and southwest-facing valley sides lead to a mixed-oak association on southwest-facing slopes and a mixed mesophytic plant association on the moister northeast-facing slopes (Finney *et al.* 1962). On Cushetunk Mountain, New Jersey, life-form spectra on north-facing slopes, south-facing slopes, and the entire ridge differ significantly (Cantlon 1953). In Crete, the Leka Ori massif supports calcareous woodland. Mediterranean cypress (*Cupressus sempervirens*), growing in association with Cretan maple (*Acer sempervirens*) and Kermes oak (*Quercus coccifera*), dominates the eastern slopes; Cretan zelkova (*Zelkova abelicea*), an endangered species, is present on slopes with a northerly aspect (Vogiatzakis *et al.* 2003).

Slope exposure tends to produce distinctive patterns in plant species distributions. In the northern hemisphere, plants with a southern distribution tend to grow on south-facing slopes at the northern edge of their range, while plants with a northern distribution tend to grow on north-facing slopes towards the southern end of their range. As a rule of thumb, a plant close to the edge of its distribution range tends to find itself on a slope that looks towards the centre of the range. For example, the wild strawberry (*Fragaria vesca*) is restricted to south-facing slopes in northern Norway, prefers level areas and gentle slopes in temperate Europe, and occurs mainly on north-facing slopes in the Mediterranean lowlands (Stoutjesdijk and Barkman 1992, 77).

Aspect affects the flowering phenology and seed development of plants. Near Innsbruck, Austria, two alpine sedges growing at 2,200–2,300 m showed phenological and developmental differences according to topoclimate (Wagner and Reichegger 1997). The two sedges are the curved sedge (*Carex curvula*) and the cushion sedge (*C. firma*), which are key species of grassland communities in the upper alpine belt of the European Alps. Curved sedge was found growing on Mt Patscherkofel (2,247 m), an isolated mountain 6 km south of the Inn valley, while cushion sedge was found on Mt Hafelekar

(2,334) in the northern mountain range 4 km north of the Inn valley. Curved sedge was studied at three sites and cushion sedge at two sites (Table 8.1). Both sedge species exhibited striking differences in their flowering phenology between the sites. The timing of their reproductive development was largely influenced by the time of snowmelt. On early-thawing sites, the prefloration period (the time between snowmelt to the onset of flowering) was about 20 days for curved sedge (south-facing and west-facing sites) and 28 days for cushion sedge (west-facing site). On late-thawing sites, the prefloration period was 14 days for curved sedge (north-facing site) and 8 days for cushion sedge (north-facing site). These differences in anthesis (the date of flowering) influenced the length of the postfloration period (the time between flowering and the formation of mature fruits). Curved sedge on the west-facing

Table 8.1 Study sites of sedges in the Austrian Alps

Aspect and description	Altitude (m)	Slope (°)	Distribution of sedges
Mt Patscherkofel			
South-facing slope on a windswept ridge with shallow snow in winter	2,240–2,245	20–25	Tussocks of curved sedge sparsely distributed among rock outcrops and scree
North-facing slope in a wind-protected hollow, 50 m (horizontally) from the ridge site. In spring, snow lasts 2–3 weeks longer than on the ridge	2,225–2,200	28–30	Tussocks of curved sedge scattered among the vegetation
West-facing slope with moderate wind exposure and snow cover during winter	2,185–2,200	20	Curved sedge tussocks near their lower limit compete with alpine grasses and ericaceous dwarf shrubs
Mt Hafelekar			
West-facing slope exposed to Sun and wind with little or no snow cover in winter and ground frost lasting far into spring	2,310–2,320	18–22	Cushions of cushion sedge
North-facing slope with snow drifts of 2–3 m in winter and spring	2,310–2,320	30–32	Cushions of cushion sedge distributed along a gradient from the margin of the snowpack down to where snow remains longest, which is their survival limit

Source: Based on information in Wagner and Reichegger (1997)

Mt Patscherkofel site bloomed at the earliest date and took 60 days to produce mature fruits, while the later-flowering plants on the north-facing site needed 50 days to complete seed development. The equivalent figures for cushion sedge on Mt Hafelekar were 70 days (west-facing site) and 55 days (north-facing site). This compensation for shorter growing seasons by curbing the time of prefloration is common in many arctic and alpine plants (see p. 86).

Leeward and windward slopes

Aspect determines exposure to prevailing winds. Leeward slopes, especially on large hills and mountains, normally lie within a **rain shadow**. Rain-shadow effects on vegetation are pronounced in the Basin and Range Province of the USA: the climate of the Great Basin and mountains are influenced by the Sierra Nevada, and the climates of the prairies and plains are semi-arid owing to the presence of the Rocky Mountains. In the Cascades, the eastern, leeward slopes are drier than the western, windward slopes. Consequently, the vegetation changes from western and mountain hemlock (*Tsuga heterophylla* and *T. mertensiana*) and Pacific silver fir and subalpine fir (*Abies amabilis* and *A. lasiocarpa*), to western larch (*Larix occidentalis*) and ponderosa pine (*Pinus ponderosa*), and finally to sagebrush (*Artemisia* spp.) desert (Billings 1990).

Animals and aspect

Slope exposure influences the geographical ranges of some animal species. In the mountainous regions of the western USA, valleys tend to lie on an east–west axis. Accordingly, south-facing slopes are drier and warmer than adjacent north-facing slopes. These microclimatic differences strongly influence the distribution of animals and plants. For instance, in the steep-sided mountains of southern California, where the climax vegetation is chaparral, the biotas on adjacent north-facing and south-facing slopes are altogether different. Some species of small mammal, such as

the San Diego pocket mouse (*Chaetodipus fallax*), are confined to south-facing slopes, and others, such as the dusky-footed woodrat (*Neotoma fuscipes*), are restricted to north-facing slopes (Vaughan 1954).

Detailed studies of Alpine marmots (*Marmota marmota*) in the French Alps revealed a significant role for slope exposure in habitat preference and growth rates (Allainé *et al.* 1994, 1998). Two high-altitude sites in the Vanoise National Park were investigated – the Réserve de la Grande Sassière (La Sassière) and the Vallon de la Lenta (La Lenta). The altitudinal range at La Sassière is 2,300–2,800 m, and for La Lenta it is 2,100–2,400 m. Slopes are gentler and plant cover greater at La Lenta, and aspect is east–west, as opposed to north–south in La Sassière. Disturbance at La Sassière is chiefly occasioned by hikers harassing the marmots, but at La Lenta it is mainly the result of agricultural practices, especially hay harvesting. One study (Allainé *et al.* 1994) found that marmot settlement, which occurred in 59 of the 88 quadrats measured at La Sassière and 26 out of the 35 quadrats measured at La Lenta, was significantly affected by slope (only at La Sassière), exposure to the Sun (at both sites), and plant cover (at both sites). At La Sassière, the following percentage of quadrats had marmot settlements: 48 per cent of north-facing quadrats, 71 per cent of valley quadrats, and 79 per cent of south-facing quadrats. The equivalent figures for La Lenta were 63.6 per cent of west-facing quadrats, 38 per cent of valley quadrats, and 100 per cent of east-facing quadrats. Taking all environmental factors into account, it was concluded that alpine marmots prefer sites with a southern or eastern aspect (where snow melts relatively early), intermediate slopes (15–45°), moderate plant cover (25–75 per cent), and a low level of human disturbance.

Later work at La Sassière looked at the effects of several environmental factors on the post-weaning growing pattern of wild juvenile Alpine marmots (Allainé *et al.* 1998). Factors considered were the year of birth, exposure to the Sun (aspect) in the home range (classified as south-facing, valley, or north-facing), litter size, and sex of young. In brief, the results showed that juveniles born on southern exposures emerged heavier, grew faster, and benefited from a longer time to gain weight than juveniles born on northern exposures.

UPS AND DOWNS: INCLINATION AND LIFE

Slopes of all inclinations affect microclimates and topoclimates to greater or lesser degrees. Slope gradient, in combination with slope exposure, often explains certain features of animal and plant distributions, as some of the examples in the previous section revealed. Two other significant aspects of slope are (1) slope steepness and (2) linked sequences of slopes – toposequences or catenas.

Steep slopes

Steep slopes affect animals directly by offering a distinct habitat and indirectly through local climatic effects. Under the right conditions, steep slopes are associated with the growth of thermals, uprising columns of air that can assist flying animals. Thermals support soaring birds such as vultures, gulls, and eagles. In India, vertical walls in towns absorb early-morning solar radiation, warm up, and produce heated air that starts to rise. Vultures begin soaring from towns an hour earlier than they do in the open country (Cone 1962).

Steep slopes influence animals by their physical presence and by their contribution to topographic heterogeneity in an area. Several mammals are adapted to living on steep slopes in mountainous terrain. In Europe, the chamois (*Rupicapra rupicapra*) and the alpine ibex (*Capra ibex ibex*) are well adapted to life in precipitous and rocky terrain. The chamois is found in the Alps, Pyrenees, and Carpathians. The alpine ibex,

an adroit jumper adapted to cliffs, lives in several European and north African mountains. Some large birds of prey nest on high ledges in mountainous terrain. In Europe, the golden eagle (*Aquila chrysaetos*) nests high on rocky crags or a cliff face, though it will sometimes nest in a tall, exposed tree.

Cliffs increase topographic heterogeneity of landscapes and affect physical and biological processes. In grasslands, cliffs create an array of vegetative characteristics that are only observed on cliff sites (Ward and Anderson 1988). Although heterogeneity may occur in hilly terrain without cliffs, cliffs provide benefits for some wildlife species that hilly terrain alone cannot provide. For instance, they provide secure nesting sites, and afford protection against predation and extreme environmental conditions.

Toposequences

Slope commonly varies in a regular sequence over undulating terrain – from summit, down hillside, to valley bottom – to form a geomorphic **catena** or **toposequence**. The geomorphic toposequence is mirrored in *soil* and *vegetation toposequences*. Figure 8.5 shows a soil and vegetation toposequence on Hodnet Heath, Shropshire, England. Humus–iron podzols form on dry summit slopes and support either dry heather (*Calluna vulgaris*) heath or silver birch (*Betula pendula*) wood with bracken (*Pteridium aquilinium*) and wavy hair-grass (*Deschampsia fleuosa*). Gley podzols form on the wetter mid-slopes and support damp cross-leaved heath (*Erica tetralix*) and heather, with purple moor-grass (*Molinia caerulea*) and common cotton-grass (*Eriophorum angustifolium*) becoming more important towards the bottom end. Peaty gley soils and peat soils form in the wet lower slopes and support purple moor-grass, common cotton-grass, and bog mosses (*Sphagnum* spp.), with tussocks of hare's-tail cotton-grass (*Eriophorum vaginatum*) in bogs and rushes (*Juncus* spp.) in acid pools of water.

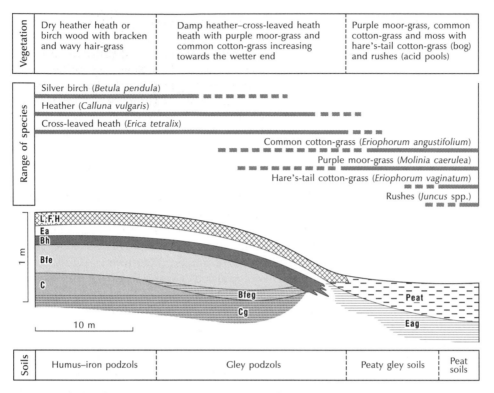

Figure 8.5 Soil and vegetation toposequences on Hodnet Heath, Shropshire, England. The letters L, F, H, Ea, etc. denote soil horizons.

Source: Adapted from Burnham and Mackney (1968)

Combined topographic effects

Altitude, latitude, aspect, slope, and other land-form attributes wield individual influences upon animals and plants. However, they act in concert and it is instructive to investigate multivariate effects in the field at local and regional scales.

On the small scale, aspect and topographical position were good predictors of funnel ant (*Aphaenogaster longiceps*) nests in a 20-ha eucalypt-forest site at Wog Wog, southeastern New South Wales (Nicholls and McKenzie 1994). At a larger scale, a study of myrtle beech (*Nothofagus cunninghamii*) in the cool-temperate rain forest of south-eastern Australia demonstrated the multivariate subtlety of plant–environment interactions (Lindenmayer *et al.* 2000). Topographic, climatic,

disturbance, and community factors all influenced the distribution of myrtle beech. Five variables were discovered to be significant in explaining the presence of myrtle beech: the age of the overstorey eucalypt stands, the dominant species of overstorey eucalypt tree, topographic position, slope gradient, and the estimated amount of rainfall in the warmest quarter of the year. Myrtle beech was more likely to occur in old-growth eucalypt stands, in gullies rather than mountainsides, and in places with high rainfall in the warmest quarter of the year. It was present on flatter areas dominated by alpine ash (*Eucalyptus delegatensis*) and shining gum (*E. nitens*) and on steeper slopes dominated by Australian mountain ash (*E. regnans*).

On a larger scale, elevation and altitudinal range partly explained the distribution of 61 willow (*Salix*) species in Europe, though latitude and July mean temperature have somewhat more explanatory power (Myklestad and Birks 1993).

LIVING ROOM: LIFE IN LANDSCAPES

Landscape ecology identifies landscape patterns and structures. Patterns are spatial arrangements of land units. In landscape ecology, the units are called patches, corridors, and matrixes (p. 72). Most of the characteristics of landscape patterns are described by landscape metrics, which are usually derived from digital elevation models and GIS databases (see Huggett and Cheesman 2002). Landform and landscape elements may combine to form landscape networks, mosaics, and regions. Landscape structures are identified as sets of inter-related landscape elements (systems) that are thought to function as a whole. Examples are drainage networks, transport systems, urban settlements, and agricultural systems.

Landscape elements

The current mainstay of landscape ecology, the *patch–corridor–matrix model*, has had extraordinary success in explaining many features of species patterns and dynamics. The three landscape elements are themselves made of individual plants (trees, shrubs, herbs), small buildings, roads, fences, small water bodies, and the like (see p. 72). They include natural and human-made landscape components, so the patch–corridor–matrix model integrates the biological and physical aspects of landscapes:

- *Patches* are fairly uniform (homogeneous) areas that differ from their surroundings – woods, fields, parks, ponds, rock outcrops, houses, gardens, and so forth. A wealth of quantitative patch metrics has been devised (e.g. Bogaert *et al.* 2000a, b, 2001a; Bogaert 2001), and is applied to nature conservation problems and

nature reserve design (e.g. Bogaert *et al.* 2001b, c).

- *Corridors* are strips of land that differ from the land to either side, and are inextricably linked with patches. They comprise trough corridors (e.g. roads and roadsides, powerlines for electricity transmission, gas lines, oil pipelines, railways, dykes, and trails); wooded strip corridors (e.g. hedgerows and fencerows); and stream and river (riparian) corridors. Greenways are hybrid corridors comprising parks, trails, waterways, scenic roads, and bike paths. Much research points to a central role played by corridors in landscape ecology and management. For instance, riparian corridors are vital to water and landscape planning and to the restoration of aquatic systems (Naiman and Décamps 1997; Décamps 2001).

- *Matrixes* are the background ecosystems or land-use types in which patches and corridors are set. The matrix is simply the dominant ecosystem or land-use – forest, grassland, heathland, arable, residential, greenhouses, or whatever – in an area. Its identification is problematic when two, three, or more ecosystems and land-uses co-dominate in a landscape. In these situations, using area, connectivity, or other criteria may single out the matrix.

The patch–corridor–matrix model represents at root a heterogeneous mosaic of 'islands of nature' connected by corridors and surrounded by 'hostile' environment (Turkry and Bogaert 2000, 62). It is not the only model of landscape ecology. A *gradient model* may be more appropriate where differences in habitats and species along an environmental gradient make a landscape heterogeneous but not patchy. The *variegation model* deals with species for which the landscape forms a continuum of migration and that display no qualitative or functional difference between patch and matrix (McIntyre and Barrett 1992).

The biogeographical and ecological effects of landscape elements are multifarious. Some of the main effects are explained on the following pages.

Patches

Patch size affects the occurrence of animal and plant species. Larger patches tend to house more species than smaller patches. Three explanations arise: (1) a small patch contains a small 'sample' of the original habitat and is less likely therefore to contain all the species found in a larger patch; (2) habitat diversity tends to be lower in small patches, so affecting the number of species that a small patch can support; (3) species in small patches tend to have lower populations than those in larger patches, with the result that fewer species can maintain viable populations in small patches. Species with large home ranges are more sensitive to this process than species with small home ranges.

Interestingly, latitude and longitude affect species–patch size relationships. Bird species in small woods within agricultural landscapes of the Netherlands, the UK, Denmark, and Norway displayed species–area relationships (Hinsley *et al.* 1998). Species richness across all woods declined with increasing latitude, as did the proportion of resident species; the proportion of migrant species rose. In addition, the slopes and intercepts of the species–area curves (p. 240) declined with increasing latitude. So, not only were there fewer species available to colonize woods at higher latitudes, but the gain of species richness for a unit increase in wood area was smaller than at lower latitudes.

Large and small patch restrictions

Some species are restricted to large patches while other species are restricted to small patches. Species restricted to large patches are far more common than species restricted to small patches, partly owing to differing 'minimum area requirements'. Some species require a larger area of continuous habitat in which to live than others, the smallest area necessary being the called the 'minimum area'. In a study of bird species in 16 small and isolated woodlands in northern Humberside, England, 4 species out of 49 recorded had minimum area requirements (McCollin 1993). These species were the great tit (*Parus major*) and spotted flycatcher (*Muscicapa striata*), with minimum area requirements of about 1 ha, and the jay (*Garrulus glandarius*) and turtle dove (*Streptopelia turtur*), with minimum area requirements of about 4 ha. Chaffinch (*Fringilla coelebs*), blue tit (*Parus caeruleus*), robin (*Erithacus rubecula*), wren (*Troglodytes troglodytes*), and blackbird (*Turdus merula*) occurred in all 16 woodlands, which suggested that their minimum area requirements were less than 0.73 ha, the area of the smallest wood.

Species restricted to small patches are uncommon, but they do exist. Grey squirrels (*Sciurus carolinensis*) in New Jersey, USA, are common in small woodland patches, but scarce or absent in large patches, probably because small patches do not house the owl predators that large patches do. In New Zealand, large (up to 8 cm diameter) carnivorous kauri snails (*Paryphanta busbyi*) are more likely to survive in small patches because in large patches they are prey to wild pigs.

Patch shape and edge effects

Smaller patches usually have a higher ratio of perimeter to area than larger patches. For this reason, smaller patches are more vulnerable to edge effects than are larger patches (Figure 8.6). Patently, large patches contain more 'interior' than 'edge'. Indeed, there is a minimum patch size below which there is no 'interior', the whole patch being an 'edge'. Even 10 years ago, research revealed that many surviving woodland fragments in England and Wales lay in the range 1 to 5 ha (44.0 per cent by number, 9.6 per cent by area) and therefore were predominantly 'edge' habitat (Spencer and Kirby 1992). Two aspects of edge effects are particularly interesting: edges and interior species, and edges as ecological traps.

Edge species and interior species. Studies commonly reveal the prevalence of edge species

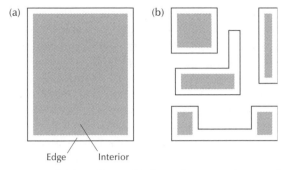

Figure 8.6 The effects of habitat fragment shape and size on edge effects. Large fragments with a low ratio of perimeter to area (a) contain proportionally less edge than small fragments with a high ratio of perimeter to area (b). Some small fragments may be all edge, as may narrow linkages between fragments.

in small patches and interior species in large patches. Interior species live in the core of a habitat. They actively avoid the habitat edges if they are able to meet their resource needs within their territories or home ranges. English woodland examples include the great spotted woodpecker (*Dendrocopos major*) and nuthatch (*Sitta europaea*). Edge species use a habitat edge. Two types of edge species are recognized, the first of which are intrinsically edge species, and the second of which are ecotonal species (McCollin 1998). *Ecotonal species* occur near the edge because the edge habitat suits them. They are not dependent on adjacent habitats for food, shelter, or anything else. *Intrinsic edge species* live near edges because the adjacent habitat provides resources. For instance, in highly fragmented agricultural landscapes, bird species living in woodland edges next to open country depend upon food resources offered by farmland. Examples are the rook (*Corvus frugilegus*) and carrion crow (*C. corone corone*), which feed mainly on grain, earthworms and their eggs, and grassland insects, with the crow also taking small mammals and carrion; and the starling (*Sturnus vulgaris*), which feeds on leatherjackets and earthworms in the upper soil layers of pasture (McCollin 1998).

Edges as 'ecological traps'. Species diversity in edge habitats tends to be high, and frequently includes a number of exotic species. Edge habitats also typically support a higher diversity of herbivores and predators than adjacent habitats: the 'rich pickings' of food supplies attract the herbivores, and the abundance of herbivores lures the predators. For instance, many game birds commonly live at higher densities in edges than in patch interiors and provide a banquet for predators. Hawks, cats, canines, and other predators often centre their foraging in edge habitats. However, predation rates vary with edge type. Three types of edge – forest–farm edge, forest–river edge, and levee–swamp edge – in a bottomland forest lying along the Roanoke River in North Carolina, USA, had different levels of predation during the 1996 breeding season (Saracco and Collazo 1999). Predation rates of northern bobwhite (*Colinus virginianus*) eggs and clay eggs were significantly higher along forest–farm edges than along the other two edges, where the predation rates were roughly the same. Taken with higher avian predator abundance on forest–farm edges, the pattern of egg predation indicated that avian predators exerted more predation pressure along these edges, a finding consonant with other studies where agricultural encroachment into forested landscapes may have a deleterious effect upon breeding birds.

Corridors

Roads

Roadsides, and even roads themselves, provide habitats that are different to surrounding habitats. They act as wildlife refuges for some species. In particular, they often support 'edge' species from surrounding forest. They are also potential avenues of movement for various groups of animals and plants through what otherwise might be unpleasant terrain. Grassy roadsides provide routes for grassland species across forested and intensive agricultural regions. The length of the routes can be enormous: the Interstate Highway System in the USA has created about 70,000 km of potential movement corridors. Movement may take place on the road surface and on the vehicles that move along it, in the open space above the road, and in the roadside habitat. In general, animals make little use of the open road and roadside, both of which harbour predators and other hidden dangers (such as cars and lorries). Some frogs and snakes are dispersed by 'hitching a ride', though this is not a common process, a documented case being the establishment of the spotted grass frog (*Limnodynastes tasmaniensis*) at Kununurra, in northwestern Australia, some 1,800 km from its southeastern Australia homeland (Martin and Tyler 1978). Such predators as the red fox (*Vulpes vulpes*), wolf (*Canis latrans*), dingo (*C. familiaris dingo*), cheetah (*Acinonyx jubatus*), and lion (*Panthera leo*) use tracks and lightly used roads, especially at night. They use them as clear pathways, uncluttered with vegetation, along which to move and hunt. Animals do not generally use metalled road surfaces as conduits, though moose (*Alces canadensis*) sometimes amble along them at night (and charge at cars). Bats use open spaces above roads as flight paths and foraging spaces.

Roads act as barriers or filters to animals that would cross them. Almost every animal, from spiders and beetles to kangaroos and deer, is a potential road-crosser. In wetlands, roads pose barriers to the free movement of aquatic animals, and may divide and isolate populations. The chief components of roads that deter would-be crossers are: the bare road surface itself; the altered roadside habitat; and the noise, movements, emission, and lights that are part of road traffic (Bennett 1991). On wide roads with much traffic, all three elements conspire to create a daunting and formidable barrier to wildlife. The Florida panther (*Puma concolor coryi*), which once ranged through much of the southeastern USA, lives in southwest Florida in national and state parks and on nearby private lands. With a population of about 70 adults, the Florida panther is among the most critically endangered animals in the world. Fragmentation of the panther's habitat by roads is partly a cause for its reduction that in 1992 was so severe that biologists gave the subspecies between 24 and 60 years until extinction.

In cases where a species will not cross a road, the population splits into two, the interbreeding rates on either side of the corridor change, and two subpopulations evolve. Genetic differentiation in the two subpopulations may ensue. In Britain, roads separating common frog (*Rana temporaria*) populations have led to genetically differentiated subpopulations (Reh 1989).

Trails

Some plants use trails as conduits. The invasion of cheatgrass (*Bromus tectorum*) over huge, dry areas of northwestern North America took place largely along cattle trails and railroad corridors. Many terrestrial mammals make trails within their home range, which they use for foraging and so forth. Humans are inveterate trail-makers. They use their trails for movement – on foot, on horseback, on a motorcycle, or other kind of vehicle.

Powerlines

Electricity transmission lines, gas lines, oil lines, and dykes tend to be fairly straight with sharp boundaries, and to have a fairly constant width over which disturbance or maintenance is evenly distributed. They favour edge and generalist species. In a forested Tennessee landscape in the USA, almost all the birds in powerline corridors were edge species (Anderson *et al.* 1977). Just two forest interior species – the scarlet tanager (*Piranga olivacea*) and the red-eyed vireo (*Vireo olivaceus*) – ranged into narrow powerline corridors.

Powerlines act as strong filters, largely because they make a loud 'buzzing hum' that deters would-be crossers, especially in wet weather. To some crossers they are a hazard. In south-central Nebraska, USA, spring migrating sandhill cranes (*Grus canadensis*) may crash into powerlines. A study marked alternating spans of powerline with 30 cm-diameter yellow aviation balls (Morkill and Anderson 1991). The number of cranes flying over marked and unmarked powerline spans did not differ, but cranes reacted more often to marked than unmarked spans, mainly by gaining altitude or by changing the direction of flight. More dead cranes were found under unmarked than marked powerline spans.

Hedgerows and other wooded strips

Hedgerows and windbreaks are line corridors dominated by edge species living at high densities. Forest interior species are normally present, albeit in low numbers. Width is a key factor in determining the species richness and abundance of many species living in hedgerows and windbreaks, though vertical structure also exerts a major influence of bird species diversity. Hedgerows support a large variety of game birds. If a hedgerow is associated with a wall, fence, ditch, or soil bank, then even greater species richness is encouraged. A ditch attracts amphibians and reptiles, while the sunny side of a soil bank encourages drought-tolerant species. Some of the species harboured by a hedgerow or windbreak may frequent adjacent fields where they eat crops. Conversely, many birds that live in fields use hedgerows for perching or for foraging. The high density of animals in hedgerows draws in predators from the surroundings.

A variety of wildlife uses hedgerows and other wooded strips as conduits. The high number of road-kill victims found where a woodland strip is broken by a road attests to this fact. Not all species will cross hedgerow gaps. The dormouse (*Muscardinus avellanarius*) is an arboreal (tree) habitat specialist that is averse to crossing even narrow gaps in hedgerows, though they are prepared to move across grass fields (Bright 1998). Plants may also move along wooded strips, though the evidence is patchy. A recent study of remnant and regenerated hedgerows in Tompkins County, in the Finger Lakes region of central New York State, USA, indicated that forest herbs colonize hedgerows from source areas (Corbit *et al.* 1999).

Stream and river corridors

Stream or riparian corridors play a starring role in many ecosystems, exerting a big influence over animals and plants. Many species rely on stream corridors for food and water, for shelter, for travel and rest, and for reproduction. The varied habitats and excellent food base (water plants, herbs and shrubs bearing berries and seeds, leafy foliage) foster high species diversity. A reason for this richness of wildlife and wildlife habitats is the fact that riparian ecosystems receive water, nutrients, and energy from upstream ecosystems. These upstream additions allow greater species richness and help to maintain a roughly constant supply of resources (L. D. Harris 1984, 142). Permanent lakes and rivers support many amphibians and aquatic birds and mammals. They also contain fish and other aquatic organisms that form the seat of several food webs.

The chief habitats in stream corridors are riverbank and floodplain. Changing water levels, changing soil moisture levels, and erosion and deposition of sediment characterize riverbank habitats. Floodplains are often rich in wetland habitats, different types being dictated by the frequency and the degree of inundation – marsh or bog, shrub swamp or thicket, forested swamp, vegetation of rarely flooded levees, ridges, and hummocks. All these floodplain habitats share periodic flooding, poor soil drainage, occasional surface deposition of sediment, and nutrient-rich soils. In dry regions, such as Arizona, riparian corridors sometimes form 'linear oases' (thin green lines), which are supported primarily by groundwater, and contain rare species. In tropical grasslands, biodiverse 'gallery forest' meanders across the plains, providing water, food, and shade for many species in a grassland matrix.

Many mammals use river corridors as conduits. Mountain lion (*Felix concolor*), bobcat (*Lynx rufus*), grizzly bear (*Ursus arctos horribilis*), and black bear (*Ursus americanus*) are known to move many kilometres along river corridors. If river meanders are present, people and animals (e.g. egrets, kingfishers, and river otters) may move directly between meanders, rather than follow the circuitous channel. Hawks and related birds of prey commonly migrate along windward edges of ridges, where updrafts of air facilitate gliding. Some exotic (non-native) plants use river corridors to spread (p. 44).

Networks

All networks – veins, arteries, roads, rivers – may be reduced to simple topological forms called graphs, that is, arrays of points (nodes) that are connected or not connected to one another by lines (edges). Many natural and human-made networks are linear flow systems, which include corridor networks, tree networks, and circuit networks. Network analysis therefore provides a method for probing biological, physical, and human-made nets. To be sure, several quantitative properties of networks – density and frequency, order, degree of integration, and accessibility – may be computed for natural networks and human-made networks.

Many properties of corridor and circuit networks affect the distribution of abundance of animals and plants. Location within a circuit is sometimes a significant property and determines the species present. In Brittany, France, the carabid beetle fauna differs in various parts of the hedgerow network. Some species live near the periphery, some nearer the centre, and some in wide corridors within the network (Burel and Baudry 1990). Similarly, while single corridors tend to favour edge species, networks that involve the proximity of two corridors (at right angles or an acute angle) may be able to support some patch species, even in the absence of a patch. This appears to be the case for kangaroos using roadside strips in Western Australia (Arnold *et al.* 1991).

Some studies hint at relationships between species distribution and stream networks. Some relationships are broad, as in seasonally migrating mammals using river networks to go from lowlands to uplands. Examples in the Grand Tetons of North America are elk (*Cervus canadensis*), mule deer (*Odocoileus hemionus*), and moose (*Alces americana*), which migrate to subalpine meadows in early summer and back in autumn to winter feeding areas in the plains. Another example is the increased abundance of some animals near streams. Raccoons (*Procyon lotor*) and opossums (*Didelphis virginiana*) are predators of forest songbird eggs and nestlings. In Missouri, USA, abundances of raccoons and opossums were related to, among other things, stream density (Dijak and Thompson 2000).

Other relationships are more specific and relate species distribution to stream order. Plants in first-order and second-order streams are usually different from plants in high-order streams (e.g. mosses and ferns in headwaters). Because terrain elevation changes with stream order, floodplain vegetation also changes. In Europe, riverine hardwood forests of big river plains contain similar species and genera that often recur in different geographical regions. Typical species are common hawthorn (*Crataegus monogyna*) or black hawthorn (*C. nigra*), ash (*Fraxinus excelsior*) or narrow-leaved ash (*F. angustifolia*), fly honeysuckle (*Lonicera xylosteum*) or perfoliate honeysuckle (*L. caprifolium*), pedunculate oak (*Quercus robur*) or holm oak (*Q. ilex*), hoary willow (*Salix eleagnos*) or common osier (*S. viminalis*) (Schnitzler 1994). Similarly, a sequence of terrestrial mammal species occupying similar functional niches occurs along different-sized streams according to the river-continuum concept (Vannote *et al.* 1980). In the western Cascades of North America, a series of carnivorous, amphibious mammals with similar ecological roles eat prey of different sizes, live in streams of different orders, and occur at different elevations in a drainage basin (Figure 8.7). The northern water shrew (*Sorex palustris*) and marsh shrew (*S. bendirei*) live in the headwater

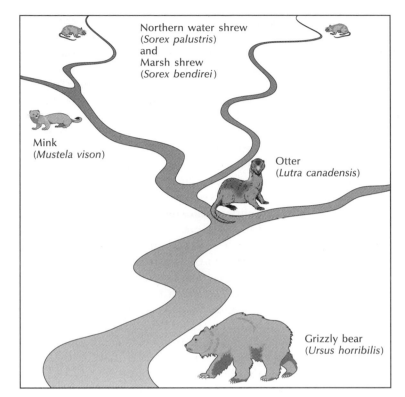

Figure 8.7 The association of different-sized carnivorous mammals with stream order and typical food-particle size according to the river-continuum concept.

Source: Adapted from Huggett and Cheesman (2002, 215)

and lower order streams, the mink (*Mustela vison*) lives in slightly higher-order streams, the otter (*Lutra canadensis*) lives in middle-order streams, and the grizzly bear (*U. arctos horribilis*) lives in higher-order streams.

Mosaics

Landscape elements (patches, corridors, and matrixes) combine to form landscape mosaics, within which there is a range of *landscape structures*. These structures are distinct spatial clusters of ecosystems or land-uses or both. Although patches, corridors, and matrixes combine in sundry ways to create landscape mosaics, six fundamental types of landscape have been identified: large-patch landscapes, small-patch landscapes, dendritic landscapes, rectilinear landscapes, chequerboard landscapes, and inter-digitated landscapes (Figure 8.8).

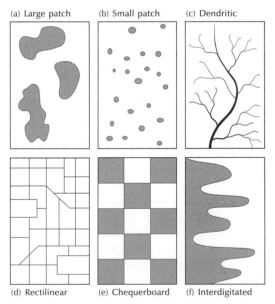

Figure 8.8 Six basic landscape patterns.

Sources: Adapted from Forman (1990; 1995, 309)

The arrangement of landscape elements in landscape mosaics affects the ease with which animals and plants may move through a landscape. In its turn, the ease of movement affects population dynamics. *Landscape resistance* describes the ease of movement. It is a mosaic property determined by the structural landscape properties, including the degree of connectedness, that impede movement. There is no standard way of measuring landscape resistance, though several landscape factors contribute to it, including high fragmentation of patches, gaps in corridors, a lack of corridors between patches, and the presence of main roads. In southern Holland, landscape resistance affected the movement of woodland birds, including the nuthatch (*Sitta europaea*) (Harms and Opdam 1990). In this study, built areas, glasshouses, and busy roads increased landscape resistance, while wooded vegetation decreased it.

Landscape resistance is specific to each species: the resistance offered by a landscape to one species, or group of species, may not be the same as the resistance offered by the same landscape to another species. So, if the spatial structure of a landscape should change, species will fare differently. For instance, species with small home ranges will respond differently to species with large home ranges. Indeed, many wide-ranging species (caribou, tigers, black bears, and vultures) are sensitive to the arrangement of regional landscapes, and commonly use two or more landscapes that need to be close together (Forman 1995, 25). In addition, habitat generalists may find it easier to move through heterogeneous landscapes created by habitat fragmentation than may habitat specialists (e.g. Mabry *et al.* 2003).

Landscape structure and composition, and particularly the connections between corridors and the matrix, helped to explain seasonal fluctuations in wood mouse (*Apodemus sylvaticus*) populations living in cultivated landscapes (Ouin *et al.* 2000). A study conducted in the polders of the Mont Saint-Michel Bay, western France, used GIS to look into the seasonal dispersal of wood mice from hedges to crops at field and landscapes scales.

Ninety per cent of the area was under intensive agriculture, with wheat (*Triticum aestivum*), maize (*Zea mays*), peas (*Pisum sativum*), and carrots (*Daucus carotta*) as the main crops. The semi-natural habitats (hedges and grassy linear habitats) were distributed over a dyke network. The results suggested that the summertime drop in hedgerow populations of wood mice resulted from a movement into the crops. Hedgerows serve as a source of wood mice in spring, and the rate at which they colonize fields in summer depends on the quality of the crops, as reflected in the crop cover and seed availability in fields and in landscapes.

Finally, *landscape context* is sometimes a significant factor in explaining species richness. This was demonstrated in a study of butterfly species richness in calcareous grassland fragments near the city of Göttingen in Germany (Krauss *et al.* 2003). Figure 8.9a shows that the richness of butterfly species in the grassland fragments increases with increasing habitat area, for specialists more than for generalists. However, in Figure 8.9b, butterfly species richness is unrelated to an index of habitat isolation. Figure 8.9c indicates a connection with landscape context in that generalist butterfly richness increases with landscape diversity.

SUMMARY

Elevation, aspect, and slope inclination can affected the distribution of species and communities at local and regional scales. Floral, and to a lesser extent faunal, zones encircle mountains. Animal and plant species tend to have definite altitudinal ranges. Slope aspect may exert a strong influence of animals and plants, especially on north-facing and south-facing slopes. Windward and leeward slopes also tend to bear different types of vegetation. The microclimatic influences of aspect on animals can be subtle. Some animals use steep slopes and cliffs as a habitats. Slope sequences from hilltop to valley produce vegetation catenas or toposequences. In most situations,

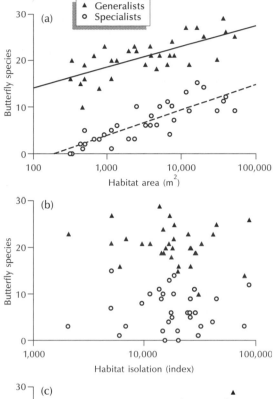

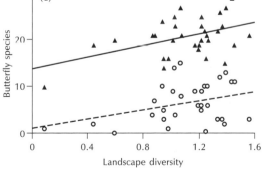

all topographic factors act in concert. Landscapes are built of three elements – patches, corridors, and matrixes. Patch size, shape, and edginess have a large influence on species richness. Corridors include roads, trails, powerlines, hedgerows, fencerows, and streams. They play enormously important ecological roles. Some corridors interconnect to create networks. Landscape elements combine to form landscape mosaics. Landscape resistance is an important property of landscape mosaics, determining the ease with which a species can move through a landscape.

ESSAY QUESTIONS

1 Why do altitudinal treelines form?

2 How important is aspect in influencing animal and plant species distributions?

3 How does habitat fragmentation affect species diversity?

FURTHER READING

Bennett, A. F. (1999) *Linkages in the Landscape: The Role of Corridors and Connectivity in Wildlife Conservation*. Gland, Switzerland: IUCN (International Union for the Conservation of Nature).

An excellent book on many aspects of landscape ecology.

Forman, R. T. T. (1995) *Land Mosaics: The Ecology of Landscapes and Regions*. Cambridge: Cambridge University Press.

A mine of information on landscape ecology.

Huggett, R. J. and Cheesman, J. E. (2002) *Topography and the Environments*. Harlow: Prentice Hall.

An intermediate-level book devoted to the topographical factor.

Figure 8.9 Relationships between the number of specialist and generalist butterfly species in grassland fragments and three landscape measures. (a) Butterfly species richness versus habitat area. (b) Butterfly species richness versus an index of habitat isolation. (c) Butterfly species richness versus landscape diversity.
Source: Adapted from Krauss *et al.* (2003)

9

DISTURBANCE

Physical agencies (fire, flood, wind) and biological agencies (pests, pathogens, grazers, humans) disturb ecosystems. This chapter covers:

- the nature of disturbance
- physical disturbance by wind, water, and fire
- biological disturbance by pathogens, grazers, and people

WHAT IS DISTURBANCE?

Disturbing agents

Disturbance is any event that disrupts an ecosystem. Disturbing agencies may be physical or biological. Strong wind, fire, flood, landslides, and lightning cause *physical disturbance*. Pests, pathogens, and the activities of animals, plants, and humans cause *biological* or *biotic disturbance*. The effects of these disturbing agencies on ecosystems can be dramatic. Pathogens, for example, are forceful disrupters of ecosystems – witness the efficacy of Dutch elm disease in England and the chestnut blight in the Appalachian region, eastern USA (p. 151).

Physical disturbing agencies are external to an ecosystem while biological disturbing agencies are internal. It is useful to think of disturbance as a continuum between purely external agencies and purely internal agencies. Classical disturbance lies at the external agency end of the continuum. It is the outcome of physical processes acting at a particular time, creating sharp patch boundaries, and increasing resource availability. Internal community cycles, such as the **hummock–hollow cycle** in peatland, lie at the internal end of the disturbance continuum and are driven by internal community dynamics.

Random and point disturbance

Some disturbances act essentially *randomly* within a landscape to produce disturbance patches. Strong winds commonly behave in this way. The patches produced by random disturbance can be extensive: the **biomantles** formed by burrowing animals are a case in point (D. L. Johnson 1989, 1990), as are the patches of eroded soil created by grizzly bears (*Ursus arctos horribilis*) excavating dens, digging for food, and trampling well-established trails (Butler 1992; see also Butler 1995). Second, some disturbances, such as fire, pests, and pathogens, tend to start at a *point* within a landscape and then spread to other parts. In both cases, the disturbances operate in a heterogeneous manner because some sites within landscapes will be more susceptible to disturbing agencies than other sites.

The agencies of disturbance often work in tandem. In forested landscapes of the southeastern

USA, individual pine trees are disturbed by lightning strikes. Once struck, pine trees are susceptible to colonization by bark beetles whose populations can expand to epidemic proportions and create forest patches in which gap-phase succession is initiated; thus the bark beetle appears to magnify the original disturbance by lightning (Rykiel *et al.* 1988).

Scales of disturbance

Disturbance affects all scales of landscape (Figure 9.1). At *small scales* (defined as 1 to 500 years and a square metre to a square kilometre), wildfire, wind damage, clear-cut, flood, and earthquake are the dominant causes of disturbance events.

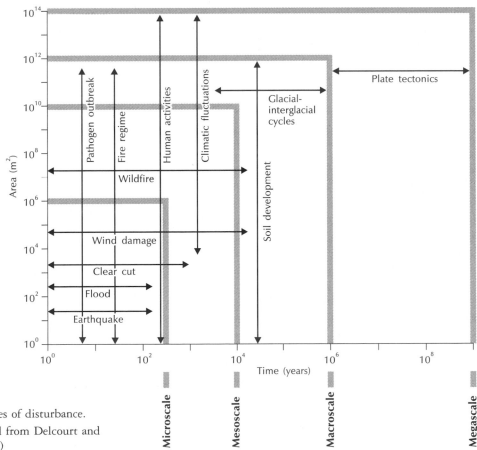

Figure 9.1 Scales of disturbance.
Source: Adapted from Delcourt and Delcourt (1988)

Vegetational units at this scale range from individual plants and forest stands, and landscapes range from sample plots to first-order drainage basins. Local disturbances lead to patch dynamics within individual vegetation patches.

Disturbance events over *medium scales* (defined as half a millennium to 10 millennia and 1 km^2 to 10,000 km^2), encompass interglacial stages and landscapes ranging from second-order drainage basins to mountain ranges up to 1° × 1° (latitude × longitude). On the lower end of this scale, a prevailing disturbance regime, such as pathogen outbreaks and frequent fires, influences patch dynamics over a landscape mosaic. In the upper range of the scale, the prevailing disturbance regimes may themselves change so causing changes within patches and between patches that in turn alter the landscape mosaic. Disturbances at large scales (defined as 10 millennia to 1 million years, and 10,000 km^2 to 1 million km^2), span one to several glacial-interglacial cycles and affect landscapes ranging in size from physiographic provinces to subcontinents. At this scale, changes in prevailing disturbance regime are effected by regional and global environmental changes.

Disturbances at the *largest scale*, defined as 1 million years to 4,600 million years, and areas more than 1,000,000 km^2 (continents, hemispheres, the globe), are driven by plate tectonics which alters global climates and influences biotic evolution. It would be useful to add bombardment and volcanic activity to this scheme. Bombardment disrupts ecosystems within seconds, disturbing small or very large areas depending on the size of the impactor. Astronomical theory suggests that the bombardment regime may be periodic. Volcanic activity, like bombardment, acts very quickly and may disturb the entire ecosphere, though local disturbances are far more common. Repeated disturbance during a bout of volcanism is, arguably, more likely to cause severe stress in the ecosphere, than is a single volcanic explosion.

This chapter will examine disturbance by wind, fire, and grazing. Many other forms of disturbance could be discussed, but lack of space precludes comprehensive coverage. However, a study of disturbance regimes will be mentioned as it shows how multivariate analysis and dynamic systems models are being used to study the effect of environmental change on ecosystems.

EARTH, AIR, WATER, FIRE: PHYSICAL DISTURBANCE

Wind

Tree-throw by strong winds, and to lesser extent other factors, may have a considerable impact on communities (Plate 9.1). In many forests, disturbance by uprooting is the primary means by which species richness is maintained (Schaetzl *et al.* 1989a, b). A fallen tree creates a gap in the forest that seems vital to community and vegetation dynamics and successional pathways: it provides niches with much sunlight for pioneer species, encourages the release of suppressed, shade-tolerant saplings, and aids the recruitment of new individuals. Even tropical forests, once thought immune to physical disturbance, have been shown to contain seemingly haphazard patterns of tree forms, age classes, and species. The tropical rain forests of the Far East may be analysed as 'gap phases' (Whitmore 1975). **Gaps** are openings made in the forests by various disturbances, and the phases are the stages of tree growth in the gaps, from seedling to maturity and death. Small gaps, about 0.04 ha, are opened up by the fall of individual large trees, with crowns 15–18 m in diameter. Lightning strikes open up gaps with an area of about 0.6 ha. Local storms cause larger gaps (up to 80 ha), while typhoons and tornadoes destroy even larger areas. These gaps are an integral part of the tropical forest system.

In 1938, a hurricane travelled across New England in the USA and destroyed whole stands of trees. Some of the trees had their trunks broken, but most were uprooted. On the ground, the fallen trees created pit-and-mound topography

Plate 9.1 A wind-thrown tree on the Kingston Lacy estate, Dorset, England, six months after the 1987 hurricane. Photograph by Shelley S. Huggett.

that influenced the distribution of understorey plants through microclimatological and microtopographic effects. The distribution of tree seedlings was related to microtopography through soil morphology, nutrition, and moisture content at pit-and-mound sites. In Harvard Forest, Petersham, Massachusetts, USA, 62 pit-and-mound pairs were found (Stephens 1956). Investigation revealed that they were formed in four episodes: 1938, 1815 (during a hurricane on 20 September that year), some time in the first half of the seventeenth century (possibly 1635 when a hurricane was recorded in the Plymouth Colony), and the second half of the fifteenth century. Even in areas with no pit-and-mound microtopography, soil horizons invariably showed traces of overturned horizons, a sure sign of much older wind throws. In the forests of eastern North America generally, disturbance has caused curious stands of trees, in which all individuals are the same age, in remnants of presettlement forests. These have a patchy distribution owing to 'the varying paths of storms and to the fickle behaviour of the winds in local areas' (H. M. Raup 1981, 43).

Fire

Fire is relatively common in many terrestrial environments. It influences the structure and function of some plant species and many communities (e.g. Whelan 1995). The effects of fire are at once beneficial and detrimental. There are three main *detrimental effects*. First, many organisms are lost from the community. Second, after a severe fire, the ground is left vulnerable to soil erosion. Third, minerals are lost in smoke and through **volatilization**, which releases large quantities of nitrogen and sulphur. *Beneficial effects* are that dead litter is burnt to ash, which boosts mineral recycling, and that **nitrogen-fixing legumes** often appear in the aftermath of a moderate surface (as opposed to crown) fire.

Adapting to fire

In areas where fires are frequent, many plants are tolerant of fire, and some require a fire to prompt seed release and germination. In the Swartboskloof catchment, near Stellenbosch, Cape Province, South Africa, mountain fynbos (Mediterranean-type) plants possess a wide range of regeneration strategies and fire-survival mechanisms (van Wilgen *et al.* 1992). Most species can sprout again after a fire, and are resilient to a range of fire regimes. Few species are reliant solely on seeds for regeneration. Those that are obligate reseeders regenerate from seeds stored in the soil, and just a few, such as oleander-leaved protea (*Protea neriifolia*), maintain seed stores in the canopy. Most of the fynbos plants flower within a year of a fire. Some species, including rosy watsonia (*Watsonia borbonica*) and fire lily (*Cyrtanthus ventricosus*), are stimulated into flowering by fires, though the trigger appears to act indirectly through changes in the environment (such as altered soil temperature regimes), and not directly through heat damage to leaves or **apical buds**.

It has long been recognized that fire influences the structure and function of some plant species and many communities (e.g. Ahlgren and Ahlgren 1960). In some cases, the effects of fire appear to outweigh climatic effects. The elevational distribution of three pine species – border piñon (*Pinus discolor*), chichahua pine (*P. leiophylla*), and Apache pine (*P. engelmannii*) – in the Chiricahua Mountains, Arizona, USA, are not, as would be expected, determined by light levels, shade tolerance, and drought resistance, but chiefly by fire (Barton 1993). Similarly, fire greatly influences the structure and composition of certain tree stands along an elevational gradient in the central Himalayas (Tewari *et al.* 1989).

Fires and changing climates

Climatic factors undoubtedly influence fires in natural vegetation. Warmer and drier climates tend to favour forest fires, though the link between climate and wildfires is uncertain (Christensen *et al.* 1989). Analysis of historical records of fires and climate in Yellowstone National Park, USA, from 1895–1989, revealed that 36 per cent of the temporal variance in annual burn area was explained by surface climate variables that induce droughty conditions: increasing temperatures in the fire season; decreasing precipitation in the antecedent season; and a trend to drought conditions in the antecedent season (Balling *et al.* 1992). The wildfires disturb biological systems and greatly enhance the sediment transport through the landscape. Analogues of the debris flow deposits and sedimentation events in streams following the 1988 fires in the park (Christensen *et al.* 1989) have been discovered in older alluvial fan deposits in the northeastern section, allowing past sedimentation events related to fires to be studied (Meyer *et al.* 1992). It was found that alluvial fans aggrade (build up) during frequent sedimentation events related to fires, and that periods of fire outbreaks appear to be associated with times of drought or high climatic variability. During wetter periods, sediment is removed from alluvial fan storage and transported down streams, resulting in floodplain aggradation. In short, the dominant fluvial activity in the area is modulated by climate, with fire, through its disturbing action on vegetation, acting as a catalyst for sediment transport during droughts.

Over the last century, there has been a trend to a set of climatic conditions that favour the outbreak of fires in Yellowstone National Park (Balling *et al.* 1992). Whether this is a greenhouse signal is not clear. Numerical climate models predict increasing aridity in the area associated with higher rainfall but even higher evapotranspiration. The historical records for the area show that aridity appears to relate to increased temperature, which is generally consistent with climate model predictions, and a decrease in precipitation in the antecedent season, which is not consistent with the climate model predictions. Whatever its cause, the trend towards increasing aridity enhances the likelihood of fires, as is clear from the historical records.

Global warming is predicted to influence the frequency, intensity, duration, and timing of fires. Modelling results predict great variations in future fire patterns in northern North America (Flannigan *et al.* 2000; Dale *et al.* 2001). The Seasonal Severity Rating of fire hazard (SSR), a measure of fire weather severity and a rough guide to area burned, increases over large parts of North America in two climate model scenarios – the Canadian General Circulation Model (GCM) and the Hadley GCM (Figure 9.2). The Hadley scenario predicts wetter conditions and small decreases in SSR for the northern Great Plains, with increases of less than 1 per cent for much of the continent. The warmer and drier Canadian GCM predicts a 30 per cent increase in SSR for the southeastern USA and Alaska, with around 10 per cent increases elsewhere. Other studies also bring out the importance of fire in vegetation change under warmer climatic regimes (e.g. Weber and Flannigan 1997; Rupp *et al.* 2000a, b).

Fire models

Mathematical models have helped in understanding the nature of fires and their impact. A model of the daily spread of a fire front in Glacier National Park used six environmental gradients (elevation, topographic moisture, time since last burn, primary succession, drainage, and alpine wind–snow exposure), and four categories of disturbance (intensity of last burn, slide disturbances, hydric disturbances, and influences of heavy winter grazing by ungulates) (Kessell 1976, 1979). The procedure was to use gradient analysis

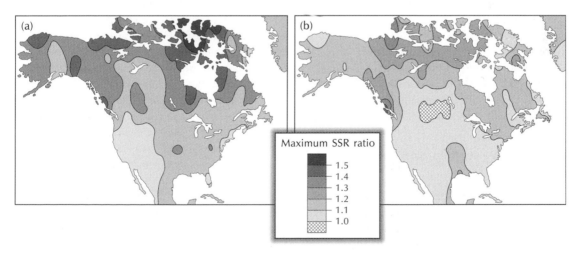

Figure 9.2 Predicted changes in the mean seasonal severity rating of fire hazard (SSR) between 1998 and 2060 under two climate scenarios. Isolines of 1.0 mean no change, ratios greater than 1.0 mean an increase in SSR, and ratios lower than 1.0 mean a decrease in SSR. (a) Canadian GCM. (b) Hadley GCM.
Source: Adapted from Flanningan *et al.* (2000)

to estimate the vegetation and fuels present throughout a large area, run the fire model in each spatial cell, and then simulate post-fire succession deterministically according to habitat types or to life history traits of the species.

Other models consider the spatial patterns of fire and species dispersal, but do not incorporate the actual spread of fire.

The BRIND model combined a forest growth model with several phenomena, including fire, that are important in Australia (Shugart and Noble 1981). The model was built to explore the long-term dynamics of *Eucalyptus*-dominated forests in the Brindabella Range, near Canberra. Altitudinal differences in community composition were predicted: at 850 m, mixed stands of narrow-leaved peppermint (*Eucalyptus robertsonii*), candlebark (*E. rubida*), and manna gum (*E. viminalis*) occurred; at 1,050 m, narrow-leaved peppermint and brown barrel (*E. fastigata*) became dominant; at 1,300 m, pure stands of alpine ash (*E. delegatensis*) or, in some simulation runs, stands dominated by either mountain gum (*E. dalrympleana*) or snow gum (*E. pauciflora*), arose; and at 1,500 m, on mountain peaks, stands dominated by snow gum appeared, either as pure stands or in mixtures with silver banksia (*Banksia marginata*) and silver wattle (*Acacia dealbata*). These different communities resulted from altitudinal variations in degree-days (ranging from about 2,400 at 850 m to 1,200 at 1,500 m) and from wildfire probabilities. They closely matched observed tree communities in the Brindabella Range.

Disturbance regimes and forest dynamics

Global warming should favour a rise in the rate of forest disturbance owing to an increase in meteorological conditions likely to cause forest fires (drought, wind, and natural ignition sources), convective winds and thunderstorms, coastal flooding, and hurricanes. Simulations suggest that changes in forest composition associated with global warming would depend upon the distur-

bance regime (Overpeck *et al*. 1990). Two sets of simulations were run. In the first set ('step function' experiments), simulated forest was grown from bare ground under present-day climate for 800 years. This enabled the natural variability of the simulated forest to be characterized. At year 800, a single climatic variable was changed in a single step to a new mean value, which perturbed the forest. The simulation was then continued for a further 400 years. In each perturbation experiment, the probability of a catastrophic disturbance was changed from 0.00 to 0.01 at year 800. This is a realistic frequency of about one plot-destroying fire every 115 years when a 20-year regeneration period (during which no further catastrophe takes place) of the trees in a plot is assumed. In each of the step-function simulations, three types of climatic change (perturbation) were modelled: a 1°C increase in temperature; a 2°C increase in temperature; and a 15 per cent decrease in precipitation. In the second set of simulation runs ('transient' experiments), forest growth was, as in the step-function experiments, started from bare ground and allowed to run for 800 years under present climatic conditions. Then, from the years 800 to 900, the mean climate, both temperature and precipitation, was changed linearly year by year to simulate a twofold increase in the level of atmospheric carbon dioxide, until the year 1600 when mean climate was again held constant. As in the step-function experiments, the probability of forest disturbance was changed from 0.00 to 0.01 at year 800. In all simulation runs, a relatively drought-resistant soil was assumed, and the results were averaged from 40 random plots into a single time series for each model run.

The model was calibrated for selected sites in the mixed coniferous–hardwood forest of Wisconsin, USA, and the southern **boreal forest** of Quebec, Canada. Selected summary results are presented in Figure 9.3. An increase in forest disturbance will probably create a climatically induced vegetation change that is equal to, or greater than, the same climatically induced change of vegetation without forest disturbance. In many cases, this enhanced change resulting from increased disturbance is created by rapid rises in the abundances of species associated with the early stages of forest succession, which appear because of the increased frequency of forest disturbance. In some cases, as in Figures 9.3a, d, and e, a step-function change of climate by itself does not promote a significant change in forest **biomass**, but the same change working hand-in-hand with increased forest disturbance does have a thoroughgoing effect on forest composition and biomass. Interestingly, the altered regimes of forest disturbance, as well as causing a change in the composition of the forests, also boost the rate at which forests respond to climatic change. For instance, in the transient climate change experiments, where forest disturbance is absent through the entire duration of the simulation period, vegetation change lags behind climatic change by about 50 to 100 years, and simulated vegetation takes at least 200 to 250 years to attain a new equilibrial state. In the simulation runs where forest disturbance occurred from year 800 onward, the vegetational change stays hard on the heels of climatic change and takes less than 180 years to reach a new equilibrium composition after the climatic perturbation at year 800.

DISEASES, PESTS, AND HUMANS: BIOTIC DISTURBANCE

Pathogens

The plight of the American chestnut (*Castanea dentata*) vividly shows the disturbing effects of some pathogens. Around 1900, the American chestnut comprised 25 per cent of the native eastern hardwood forest in the USA. It was a valuable forest species – its hardwood being used for furniture, its tall straight timbers for telegraph poles, its tannin for leather tanning, and its nuts for food. By 1950, most trees were dead or dying because of chestnut blight, a parasitic disease caused by a fungal parasite – the sac fungus,

Cryphonectria (Endothia) parasitica. The fungus enters the host tree through a bark wound made by woodpeckers or bark-boring insects. It grows into the bark and outer sapwood forming a spreading, oozing sore that eventually encircles or 'girdles' the truck or branch. Once girdled, the tree dies because nutrient supplies to and from the roots are stopped. The time from initial attack to death is 2–10 years.

The sac fungus is native to Asia. In 1913, it was found on its natural host in Asia – the Chinese chestnut (*C. mollissima*) – to which it does no serious harm. It was introduced into New York City by accident, probably in 1904, on imported Asian nursery plants. The first infected trees were found in the Bronx Zoo. The American chestnut is so susceptible to *Cryphonectria* that it was removed from almost its entire range within 40 years. All that remained were a few individuals lucky enough to have escaped the fungal spores, and adventitious shoots spouting from surviving root systems.

In the wake of the chestnut blight, several oak species, beech, hickories, and red maple have become co-dominants: the oak–chestnut forest has turned into oak or oak–hickory forests. This is evident in Watershed 41, western North Carolina (Figure 9.4). Here, the disease first attacked in the late 1920s. The survey carried out in 1934 shows the original forest composition, as the disease had not killed many trees by that time. The American chestnut is plainly the dominant species, occupying over twice the basal area of various hickory species. By 1954, after the forest had borne the brunt of the fungal attack, the tree species composition had altered dramatically. The chestnut had all but disappeared, and hickories (*Carya* spp.), the chestnut oak (*Quercus prinus*), the black oak (*Q. velutina*), and the yellow poplar (*Liriodendron tulipifera*) had become co-dominants.

Despite the ravages of the chestnut blight, replanting with backcrossed trees derived from blight resistant hybrid chestnuts has led to hopes for a successful replanting programme. Mesic sites (not too dry, not too wet), especially shallow

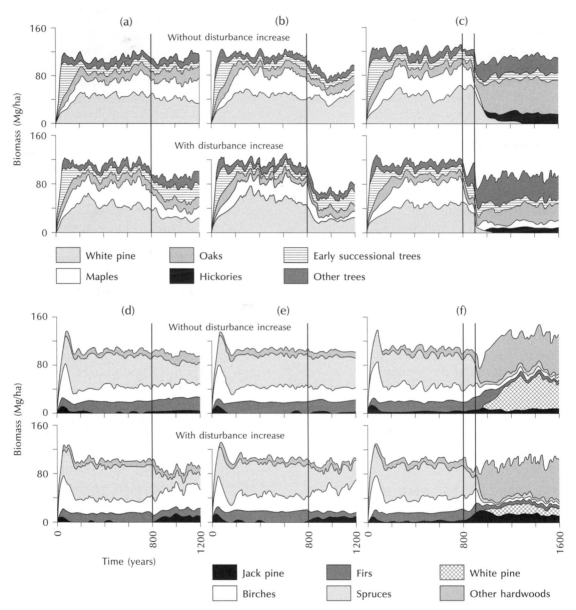

Figure 9.3 Simulated changes in species composition of forests at two sites investigated in eastern North America. (a) to (c) is a site in Wisconsin, and (d) to (f) is a site in southern Quebec. At both sites, experiments were run with an increase in disturbance at year 800 (top of figure for each site) and without an increase in disturbance at year 800 (bottom figure for each site). Additionally, three climatic change scenarios were simulated: a 1°C temperature increase at year 800 (left-hand figures, a and d); a 15 per cent decrease in precipitation (middle figures, b and e); and a transient change in which mean monthly precipitation and temperature were changed linearly from year 800 to year 900 and thereafter held constant (right-hand figures, c and f).

Source: Adapted from Overpeck *et al.* (1990)

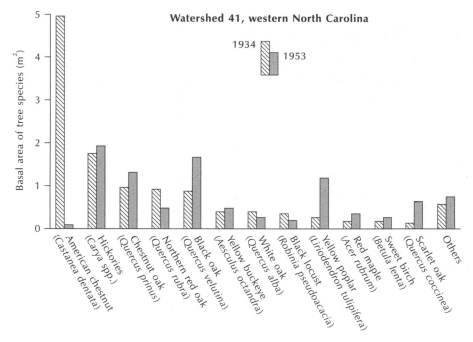

Figure 9.4 The effect of the chestnut blight in Watershed 41, western North Carolina, USA.
Source: From data in Nelson (1955)

coves, with surviving chestnuts may have the greatest potential for chestnut growth and **biocontrol** of blight, if effective hypovirulence, blight resistance, and forest management measures are developed. In addition, surviving root systems, which occur throughout the chestnut's natural range, give hope that a genetically engineered, less virulent strain of sac fungus might be released and help to restore a valuable forest species (Choi and Nuss 1992).

Grazing

At least since the time of Charles Darwin, it has been known that grazing by herbivores alters the structure and composition of plant communities. Controversy surrounds the role of herbivory in ecosystems. Proponents of the food-limitation hypothesis claim that herbivorous animals regulate the abundance of plant populations and may

control the functioning of entire ecosystems, and that the abundance of most herbivore populations is limited by food. Opposing beliefs are that herbivores are normally scarce relative to their food supply owing to the depredations of natural enemies and the weather, or that herbivory is, in the main, beneficial to plants. These issues are not settled, but grazing is without doubt an important element in nearly all ecosystems, and may greatly influence community stability and productivity.

The overall effects of grazing have been ascertained by compiling data from a range of environments (Milchunas and Lauenroth 1993). A worldwide set of data was compiled at 236 sites where species composition, aboveground net primary production, root biomass, and soil nutrients in grazed versus protected ungrazed plots had been measured. The data set was subjected to multivariate regression analysis. It was found that changes in species composition with grazing depended primarily on aboveground net primary production and the evolutionary history of the grazing site.

A complication in establishing the effects of grazing is that it often acts in concert with other disturbing agencies. This is true on Cumberland Island, a barrier island lying off the coast of Georgia, USA, where Monica G. Turner and Susan P. Stratton (1987) studied the propagation and behaviour of disturbances caused by grazing and fire (Box 9.1).

Box 9.1

DISTURBANCE ON CUMBERLAND ISLAND, GEORGIA, USA

Cumberland Island is composed of a core of stable Pleistocene sediments, rarely subject to storm over-wash, surrounded by a periphery of more dynamic Holocene beach, dune, and marsh sediments that are readily refashioned by storm events. Vegetation on the island is arranged in longitudinal belts that are chiefly the result of physical influences such as tidal energy, salt spray, and topography (Figure 9.5). The upland forests tend to be patchy. During the twentieth century, horses (*Equus caballus*), cattle (*Bos bovine*), and hogs (*Sus scrofa*) roamed free on the island, until in 1974 the cattle were removed and the hog population reduced through trapping. Some 180 horses still lived there in the 1980s, and the native white-tailed deer (*Odocoileus virginianus*) population had grown large following release from hunting and removal of predators. Intensive grazing by these animals depleted the forest understorey and clipped interdune and high marsh vegetation to the height of a mown lawn. Besides grazing disturbance, large fires on the island have been sparked by lightning, especially during dry summers. Approximately 1,300 hectares were burnt since 1900, much of it repeatedly on a 20–30 year rotation. Fire size varies with community type: in scrub, fires are large, hot, and often burn into other vegetation types; in mature pine stands and freshwater marshes, fires are less intense, although they may still burn into fringing areas.

Taken together, fire and grazing are important determinants of the landscape mosaic on Cumberland Island, and, although they behave very differently, they both 'consume' biomass as they proceed and, owing to the influence of landscape heterogeneity, induce a similar response in the landscape. Feral horses mainly graze the salt marshes and grassland, which are both highly disturbed and resilient, but also influence the maritime forest that cannot alone support the horse herd. The salt marshes and grassland are energy sources for a perturbation that may change the forest community. Deer graze the interdune meadows, open grassy fields, and upper edges of the salt marshes, and will browse most of the broad-leaved woody species on the island. Disturbance arises when the deer population grows and it moves from its preferred habitats to other landscape patches. As browsing by deer is selective, it may change forest composition. Fires will only start in communities with adequate fuel supplies. On Cumberland Island, they generally begin in the scrub or freshwater marsh where most of the available plant material is consumed. The scrub, and the freshwater marshes, are highly resilient and replace themselves quickly after a burn. The oak forests are fire resistant, but when burned recover slowly. Scrub patches within the landscape create a focus for fire that can move into the forest and convert it to scrub.

In conclusion, fire and grazing disturbances on Cumberland Island seem to be driven by resilient patches within the landscape, yet influence resistant patches from which little energy is obtained. It should be pointed out that fire and grazing are only part of the disturbance regime on the island – feral pigs, human visitors, and storms play a part.

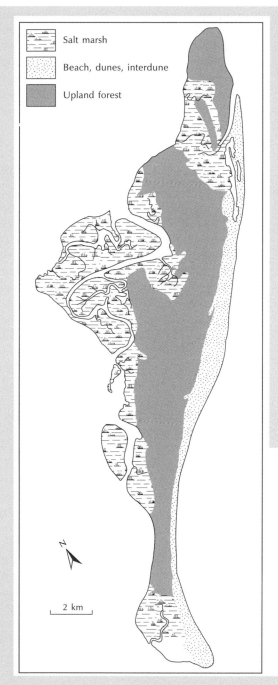

Salt marsh

Beach, dunes, interdune

Upland forest

2 km

Figure 9.5 Vegetation on Cumberland Island, Georgia, USA.
Source: Adapted from Turner and Stratton (1987)

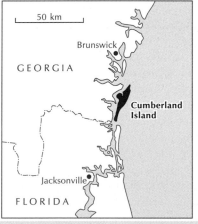

50 km

Brunswick

GEORGIA

Cumberland
Island

Jacksonville

FLORIDA

The effects of grazing on ecosystems are complex. Dynamic systems models of plant–herbivore dynamics underscore this fact. An example is a model of plant–herbivore interactions in which plant growth is constrained by a carrying capacity (Crawley 1983, 249):

$$\frac{dP}{dt} = \frac{aP(K - P)}{K} - bHP$$

$$\frac{dH}{dt} = cHP - dH$$

In these equations, P is the plant population, H is the herbivore population, K is the carrying capacity of plants, and a, b, c, and d are parameters: a is the intrinsic growth rate of plants in the absence of herbivores; b is the depression of the plant population per encounter with a herbivore, a measure of herbivore searching efficiency; c is the increase in the herbivore population per encounter with plants, a measure of herbivore growth efficiency; and d is the decline in herbivores (death rate) in the absence of plants. The parameters and the carrying capacity affect system stability and steady-state populations numbers. In brief, increasing the intrinsic growth rate of plants, a, increases system stability and herbivore steady-state density, but has no effect on plant abundance (Figure 9.6a). Increasing herbivore-searching efficiency, b, reduces, somewhat surprisingly, herbivore steady-state density, but has no effect on system stability or on steady-state plant abundance (Figure 9.6b). Increasing herbivore growth efficiency, c, reduces steady-state plant abundance and so reduces system stability, but increases steady-state herbivore numbers (Figure 9.6c). Increasing the carrying capacity of the environment for plants, K, increases herbivore steady-state density and reduces system stability, but, paradoxically, has no effect on steady-state plant abundance (Figure 9.6d).

Clearly, the dynamics displayed by the plant–herbivore model, even in such a relatively simple case, are complex. Adding more realistic features, such as more complex functional responses of herbivores, plant compensation, and impaired plant regrowth, further complicates the system's dynamics. Importantly, some features of the dynamics, such as the paradox of enrichment (whereby an increased plant carrying capacity has no effect on steady-state plant abundance), are counterintuitive.

Introduced predators on islands

The Indian mongoose (*Herpestes auropunctatus*) is one of the most potent predators on diurnal ground-foraging lizards. It has been introduced to various islands worldwide in the hope of controlling rats and other vertebrate pests. Its success in doing so has been mixed. Its success in reducing and causing the extinction of native bird and reptile populations is spectacular (Case and Bolger 1991). This is demonstrated by a diurnal lizard census on Pacific islands with and without mongooses (Figure 9.7). Islands without mongooses have nearly a 100 times greater diurnal lizard abundance than islands with mongooses.

Other potent introduced island predators are rats and domestic cats and dogs. Cats and the tree rat (*Rattus rattus*) are capable of climbing trees and affect species that the mongoose is less likely to capture. New Zealand lizards became far less common after the mid-nineteenth century, owing to cover reduction through forest clearance and predation by cats. Today, islands around New Zealand without mammalian predators house extraordinarily high numbers of lizards, with one lizard per 3 m^2 being recorded. The same pattern is found elsewhere. High numbers of diurnal lizards are found on rat-free Cousin Island in the Seychelles and on San Pedro Martir in the Sea of Cortez, Mexico. Some introduced reptiles have also caused extinctions. The introduced brown tree snake (*Boiga irregularis*) on Guam, the largest of the Mariana Islands, has eliminated 13 of 22 native bird species, 2 of 3 native bat species, and 4 of 10 native lizard species.

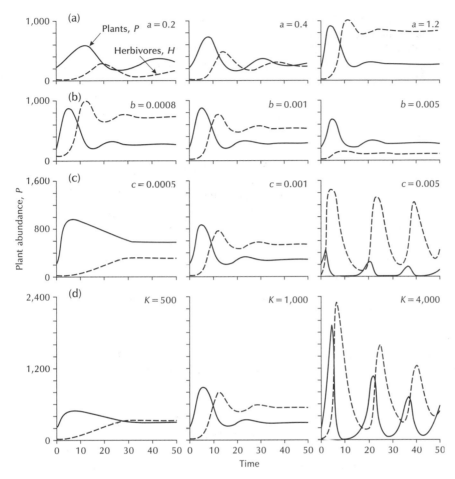

Figure 9.6 Dynamics of a plant–herbivore system with resource-limited plants (growth constrained by an environmental carrying capacity). (a) The effect of increasing growth rate of plants, *a* (*b* = 0.001, *c* = 0.001, *d* = 0.3, *K* = 1,000) is to increase stability and herbivore equilibrium density but has no effect on equilibrium plant abundance. (b) The effect of increasing herbivore searching efficiency, *b* (*a* = 0.5, *c* = 0.001, *d* = 0.3, *K* = 1,000) is to *reduce* herbivore equilibrium density but leaves the stability and equilibrium plant abundance unaltered. (c) The effect of increasing herbivore growth efficiency, *c* (*a* = 0.5, *b* = 0.001, *d* = 0.3, *K* = 1,000) is to reduce equilibrium plant abundance and so reduce stability but increase equilibrium herbivore numbers. (d) The effect of increasing plant carrying capacity, *K* (*a* = 0.5, *b* = 0.001, *c* = 0.001, *d* = 0.3) is to increase herbivore equilibrium density but to leave plant equilibrium unchanged and to reduce stability. This is the paradox of enrichment.

Source: Adapted from Crawley (1983)

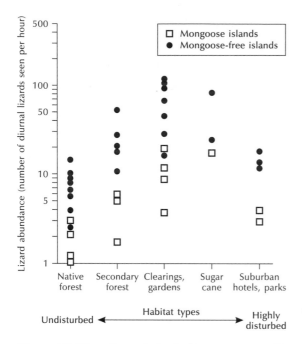

Figure 9.7 The effect of the Indian mongoose (*H. auropunctatus*) on Pacific island lizards. The censuses were conducted during sunny days from 1984 to 1988. Each point represents the average of 2–4 censuses.

Source: Adapted from Case and Bolger (1991)

Introduced predators produce a distinctive biogeographical pattern, namely a reciprocal co-occurrence pattern. This means that many native amphibians and reptiles occur on islands without predators, but are absent from islands with predators. An example is the tuatara (*Sphenodon punctatus*) and its predator, the Polynesian rat (*Rattus exulans*), in New Zealand. On islands where the rat is present, the tuatara is either absent or not breeding. The largest surviving frog, the Hamilton frog (*Leiopelma hamiltoni*), occurs only on rat-free islands. On Viti Levu and Vanua Levu, the two largest islands of Fiji, the combination of cats and mongooses has proved devastating. Two ground-foraging emos skinks – the black skink (*Emoia nigra*) and *E. trossula* – are locally extinct, and the two largest skinks in Fiji have not been seen on these islands for over a century, although they do survive on mongoose-free islands.

SUMMARY

All ecosystems are to some degree disturbed by wind, fire, floods, grazing, humans, and so forth. Some communities are geared to function under severe disturbances from, for example, hurricanes and fires. Plants growing in areas where burning is common have adapted to fire in various ways. Global warming is likely to alter physical disturbance regimes, which will have knock-on effects on vegetation. Disturbance by organisms includes diseases, grazing, and predation. The effects of these biotic agents can be substantial, involving for instance the near annihilation of the American chestnut and the decimation of native birds and reptiles on islands with introduced predators.

ESSAY QUESTIONS

1 **How important is disturbance in shaping the form and function of communities?**

2 **How do plants adapt to fire?**

3 **How potent are biotic agents of disturbance?**

FURTHER READING

Bradshaw, G. A. and Marquet, P. A. (eds) (2003) *How Landscapes Change: Human Disturbance and Ecosystem Fragmentation in the Americas* (Ecological Studies, vol. 162). New York and London: Springer.

Instructive, up-to-date essays.

Frelich, L. E. (2002) *Forest Dynamics and Disturbance Regimes: Studies from Temperate Evergreen–Deciduous Forests*. Cambridge: Cambridge University Press.

Good as a case study on the effects of disturbance on forests.

Rundel, P. W., Montenegro, G., and Jaksic, F. M. (eds) (1998) *Landscape Disturbance and Biodiversity in Mediterranean-type Ecosystems* (Ecological Studies, vol. 136). Berlin: Springer. Examples from Mediterranean-type ecosystems.

M. G. Turner (ed.) (1987) *Landscape Heterogeneity and Disturbance* (Ecological Studies, vol. 64), pp. 85–101. New York: Springer. A good set of case studies.

Whelan, R. J. (1995) *The Ecology of Fire*. Cambridge: Cambridge University Press. An excellent introduction to the topic.

POPULATIONS

Most species exist as groups of interbreeding individuals – populations. This chapter covers:

- the form and function of single populations
- ways in which populations survive
- human exploitation and control of populations

BIRTH, SEX, AND DEATH: THE DEMOGRAPHY OF SINGLE POPULATIONS

Populations

Organisms are all to some extent sociable beings, interacting with one another to survive. Their interactions create populations and communities. *Populations* are loose collections of individuals belonging the same species. Red deer (*Cervus elaphas*) in Britain constitute a population. All of them could interbreed, should the opportunity arise. In practice, most populations, including the red deer population in Britain, exist as sets of *local populations* or *demes*. A local red deer population lives in the grounds of Lyme Park, Cheshire. Its members form a tightly linked, interbreeding

group and display features typical of many populations (Box 10.1).

Population growth

A population changes due to natality, immigration, mortality, and emigration. *Natality* covers a variety of ways of 'coming into the world' – live birth, hatching, germinating, and fission. In mammals, natality is the same thing as the birth rate. *Mortality* covers the only way of 'leaving the world' – dying. Death may occur through old age (senescence). It is more likely to result from disease, starvation, or predation. Two aspects of reproduction are distinguished. *Fertility* measures the actual number of new arrivals in a population. In humans, the fertility rate averages about one birth every eight years per female of

Box 10.1

THE RED DEER POPULATION IN LYME PARK, CHESHIRE, ENGLAND

Lyme Park lies on the western flanks of the Pennines. Its 535 ha consist of park grassland, moorland, and woodland. Red deer (*Cervus elaphas*) have lived in Lyme Park for over 400 years. They are a remnant of a larger population once present in Macclesfield Forest, which used to extend many kilometres along the west side of the Pennines. Up to 1946, while the park was privately owned, the population numbered 170 to 242. Some deer were shot for sport and some for the dinner table. After ownership was passed to the National Trust and Stockport Metropolitan Borough Council, the population rose to over 500. Culling was introduced in 1975 and the population is now maintained at 250 to 300.

A study of the red deer population was carried out from 1975 to 1983 (Goldspink 1987). Hinds dominated the population. There were more than three times more hinds than stags. Hinds were recorded in three main areas – the park, the moorland, and Cluse Hay (which is a subdivision of the moorland) (Figure 10.1). All tagged hinds remained close to their areas of capture. Hinds tended to form large groups of 30 or more throughout the study area. Most calves were born in the first two weeks of June. Maximum densities of hinds and calves per km^2 varied from 30 in the park to 77 on the moorland. Three bachelor herds of stags were present. Moor stags spent much of the summer on a small area (30 ha) of ground above Cluse Hay. They dispersed more during winter, probably to avoid exposure and windchill. The Knott group, which lives in the park, consisted of prime stags, more than six years old, during the winter. Younger stags were recorded on Cage Hill. Stags in the park area were more tolerant of disturbance than those living on the moorland. The rut started in mid-September and went on well into November. The median date of observed

copulations was 16 October. The apparent gestation period (to the median date of calving) was about 237 days. Stags from the Knott rutted on Cluse Hay; stags from Cage Hill preferred to rut on the moor or the park. Moor stags appeared to play but a small part in the rut, despite their numerical abundance. Dominant stags, which commonly had 40-hind harems for 2–3 weeks during the rut, were usually replaced within 2 years.

Growth rates of stags and hinds were low on the moorland. Calf-to-hind ratios were generally low and within the range 0.20–0.34; maxima of

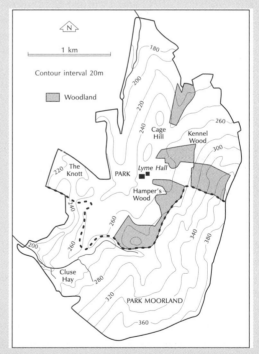

Figure 10.1 The main areas occupied by the red deer (*C. elaphas*) population, Lyme Park, Cheshire, England.

Source: Adapted from Goldspink (1987)

0.55–0.60 were recorded in woodland sites. Conditions on the moorland were severe during the winter due to the poor quality of herbage, mainly purple moor-grass (*Molinia caerulea*), and a lack of shelter. Annual calf mortality varied from 10 to 30 per cent. It was particularly high during the cold winters of 1978–9 and 1981–2. Population density and winter food were probably principal factors limiting the performance of deer on the moor. Growth rates of stags and hinds in the park were high, but calf-to-hind ratios were low at 0.15–0.26. Culling has reduced levels of natural mortality but further improvements in performance are unlikely to be achieved without a reduction in sheep stocks, some improvement in habitat, and the provision of shelter.

childbearing age. **Fecundity** measures the potential number of new arrivals. In humans, the fecundity rate is about one birth per 9–11 months per female of childbearing age. Not many women fulfil that particular potential!

Most classic population studies focused on natality and mortality, and assumed that immigration and emigration balance one another or are too small to contribute greatly to population change. Obviously, *emigration* and *immigration* are important components of long-term biogeographical change. They are also important in more recent population studies that emphasize small-scale comings and goings within a species range. This idea, in the guise of metapopulation theory, will be discussed later.

Exponential growth

Populations grow *exponentially* when the population increase per unit time is proportional to population size and is unconstrained. This is readily appreciated by a simple example. Take two individuals. After a year they have produced four offspring. Reproducing at the same rate, those four offspring will themselves have produced eight offspring after a further year. The growth of the population follows a geometric progression – 2, 4, 8, 16, 32, 64, and so on. This exponential growth rapidly produces huge numbers. A female common housefly (*Musca domestica*) lays about 120 eggs at a time (Leland Ossian Howard, cited in Kormondy 1996, 194).

Around half of these develop into females, each of which potentially lays 120 eggs. The number of offspring in the second generation would therefore be 7,200. There are seven generations per year. If all individuals of the reproducing generation should die off after producing, the first fertile female will be a great, great, great, great, great grandmother to over 5.5 trillion flies.

The following equation describes growth in populations with overlapping generations and prolonged or continuous breeding seasons:

$$\frac{dN}{dt} = rN$$

Notice that the growth rate, dN/dt, is directly proportional to population size, N, and to the *intrinsic rate of natural increase* (per capita rate of population growth), r. The intrinsic rate of natural increase, sometimes called the *Malthusian parameter*, is defined as the specific birth rate, b, minus the specific death rate, d. In symbols, $r = b - d$. (Specific rates are rates per individual per unit time.) Exponential growth, then, simply depends upon how many more births there are than deaths, and on how large the population is. No other factors restrict population growth. Unhindered growth of this kind is the **biotic potential** of a population (R. N. Chapman 1928). It is described by the solution to the population growth equation:

$$N = N_0 e^{rt}$$

This equation shows that the starting population, N_0, increases exponentially (e is the base of the natural logarithms) at a rate determined by the intrinsic rate of natural increase, r. It describes a J-shaped curve (Figure 10.2). The *maximum intrinsic rate of natural increase* varies enormously between species. In units of per capita per day, it is about 60 for the bacterium *Escherichia coli*; 1.24 for the protozoan *Paramecium aurelia*; 0.015 for the Norway rat (*Rattus norvegicus*); 0.009 for the domestic dog (*Canis domesticus*); and 0.0003 for humans. For the red deer population in Lyme Park, from 1969 to 1975, it is 0.07696.

The *doubling time* is the time taken for a population to double its size. It may be calculated from values of r. The doubling times for the species mentioned above, assuming geometric increase prevails, are: *E. coli* – 0.01 days; *P. aurelia* – 0.56 days; *R. norvegicus* – 46 days; *C. domesticus* – 77 days; and humans – 6.3 years. For Lyme Park red deer, it is 9.0 years.

Logistic growth

The English house sparrow (*Passer domesticus*) was introduced to the USA around 1899. Concern was expressed that, in a decade, a single pair could lead to 275,716,983,698 descendants, and by 1916 to 1920 there would be about 575 birds per 40 ha (Kormondy 1996, 194). In the event, there were only 18 to 26 birds per 40 ha, less than 5 per cent of the expected density. Patently, something had prevented the unfettered geometric increase in the immigrant house sparrow population.

A growing population cannot increase indefinitely at a geometric rate. There is a ceiling or **carrying capacity**, a limit to the number of individuals that a given habitat can support. An environmental resistance curtails the biotic potential of a population (R. N. Chapman 1928). After a period of rapid growth, the house sparrow population reached its population ceiling, partly because native hawks and owls took to the new English dish on the menu.

The letter K usually denotes the population ceiling or carrying capacity. It alters the population growth equation in this way:

$$\frac{dN}{dt} = rN(1 - N/K)$$

This equation for growth rate is called the *logistic equation*. It describes a sigmoid or S-shaped curve (with the top of the 'S' pushed to the right) where population grows fast initially but tapers off towards the carrying capacity (Figure 10.3). When the population size is small, compared with the carrying capacity, the environmental resistance to growth is minimal. The population responds by virtually unbounded exponential growth. This is the bottom part of the 'S'. When the population size is large and approaching the carrying capacity, environmental resistance is great. The population barely grows. This is the distorted top part of the 'S'. If the population should exceed the carrying capacity, negative growth occurs; in other words, the population falls until it drops below the carrying capacity.

Population growth (or decline) rates are affected by *density-dependent factors* and by *density-independent factors*. The effects of density-independent factors

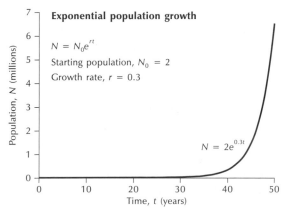

Figure 10.2 Exponential growth of a hypothetical population. Notice that, with a bit of artistic licence, the population growth describes a J-shaped curve.

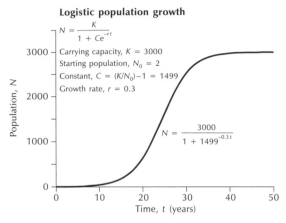

Logistic population growth

$$N = \frac{K}{1 + Ce^{-rt}}$$

Carrying capacity, $K = 3000$
Starting population, $N_0 = 2$
Constant, $C = (K/N_0) - 1 = 1499$
Growth rate, $r = 0.3$

$$N = \frac{3000}{1 + 1499^{-0.3t}}$$

Figure 10.3 Logistic growth of a hypothetical population. Notice that the population growth curve is a stretched S-shape.

do not vary with population density and they will affect the same proportion of organisms at any density. Weather, pests, pathogens, and humans commonly affect populations in that way. The effects of density-independent factors do vary with population density, so that the portion of organisms affected increases or decreases with population density. Birth rates and death rates both depend on population density: birth rates tend to decrease with increasing density, while death rates tend to increase with increasing density.

Population irruptions

Irruptions are common features of many populations. They involve a population surging to a high density and then sharply declining. This boom-and-bust pattern occurs in some introduced ungulates. It happened to the Himalayan tahr (*Hemitragus jemlahicus*) in New Zealand (Caughley 1970). The tahr is a goat-like ungulate from Asia. Introduced into New Zealand in 1904, it spread through a large part of the southern Alps. After its introduction, the population steadily rose. The birth rate fell slightly but the death rate increased, largely due to greater juvenile

mortality. After a period when the population was large, a decline set in. This decline was caused by reduced adult fecundity and further juvenile losses. Changes in the tahr population may have resulted from changes in food supply. As the tahr grazed, they altered the character of the vegetation that they fed on. In particular, snow tussocks (*Chionochloa* spp.), which are evergreen perennial grasses, were an important food source late in winter. They cannot abide even moderate grazing pressure and disappeared from land occupied by the tahr. With no snow tussocks to eat, the tahr browsed on shrubs in winter and managed to kill some of these.

Another irruption occurred on Southampton Island, Northwest Territories, Canada (D. C. Heard and Ouellet 1994). By 1953, caribou (*Rangifer tarandus groenlandicus*) had been hunted to extinction on Southampton Island. In 1967, 48 caribou were captured on neighbouring Coat's Island and released on Southampton Island. The population of caribou older than one grew from 38 in 1967 to 13,700 in 1991. The annual growth rate during this period was 27.6 per cent and did not decrease with increasing population density. On Southampton Island, caribou did not suffer high winter mortality in some years, but they did on Coat's Island. Caribou density was higher on Coat's, which suggests that adverse weather has a minimal effect when animal density is low.

Irruptions also occur because of unusually high immigration. In the winter of 1986–7, many more rough-legged buzzards (*Buteo lagopus*) moved into Baden-Württemberg, southwest Germany, than in any other winter during at least the preceding century (Dobler *et al.* 1991). From 14 January to 7 April, rough-legged buzzards were observed daily. The highest numbers observed were 109 individuals on 1 February and 110 on 8 February. The buzzard influx was caused by high snow cover and cold spells in the eastern parts of central Europe. Depressions over southern Europe may have blocked the route south with cloud and snowfall.

Population crashes

Some populations *crash*. In early 1989, two-thirds of the Soay sheep (*Ovis aries*) population on St Kilda, in the Outer Hebrides, Scotland, died within a 12-week period. The cause of the crash was investigated by post-mortem examination and laboratory experiments (Gulland 1992). Post-mortem examination showed emaciated carcasses with a large number of nematode (*Ostertagia circumcincta*) parasites, and that the probable cause of death was protein-energy malnutrition. However, well-nourished Soay sheep artificially infected with the parasite in the laboratory showed no clinical signs or mortality, even when their parasite burdens were the same as those recorded in the dead St Kilda sheep. Thus, parasites probably contributed to mortality only in malnourished hosts, exacerbating the effects of food shortage.

Marine iguanas (*Amblyrhynchus cristatus*) living on Sante Fe, in the Galápagos Islands, suffered a 60–70 per cent mortality due to starvation during the 1982–3 **El Niño**–Southern Oscillation event (Laurie and Brown 1990a, b). Adult males suffered a higher mortality than adult females while food was short, but size explained most of the mortality differences between the sexes. Almost no females bred after the event. In the next year, the frequency of reproduction doubled, the age of first breeding decreased, and mean clutch size increased from two to three. These demographic changes are examples of density-dependent adjustments of population parameters that help to compensate for the crash.

Population crashes may have repercussions within communities. In Africa, savannahs are maintained as grassland partly because of browsing pressures by large and medium herbivores. In Lake Manyara National Park, northern Tanzania, bush encroached on the grassland between 1985 and 1991, shrub cover increasing by around 20 per cent (Prins and Van der Jeugd 1993). Since 1987, poaching has caused a steep decline in the African elephant (*Loxodonta africana*) population

in the Park. However, shrub establishment preceded the elephant population decline, and is not attributable to a reduction in browsing. In two areas of the Park, shrub establishment coincided with **anthrax** epidemics that decimated the impala (*Aepyceros melampus*) population. In the northern section of the Park the epidemic was in 1984; in the southern section it was in 1977. An even-aged stand of umbrella thorn (*Acacia tortilis*) was established within the grassland in 1961, which date coincided with another anthrax outbreak among impala. Similarly, another even-aged stand of umbrella thorn was established at the end of the 1880s, probably following a rinderpest pandemic. The evidence suggests that umbrella-thorn seedling establishment is a rare event, largely owing to high browsing pressures by such ungulates as impala. Punctuated disturbances by epidemics, which cause population crashes in ungulate populations, create narrow windows for seedling establishment. This process may explain the occurrence of even-aged umbrella thorn stands.

Chaotic growth

Irregular fluctuations observed in natural populations, including explosions and crashes, were traditionally attributed to external random influences such as climate and disease. In the 1970s, Robert M. May recognized that these wild fluctuations could result from *intrinsic non-linearities* in population dynamics (May 1976). For instance, a relatively simple population growth equation of the form

$$N^t_{+1} = N^t e[r(1 - N^t/K)],$$

which is applicable to populations that suffer epidemics at high densities, possesses an amazing range of dynamic behaviour – stable points, stable cycles, and chaos – depending on the growth rate, *r* (Figure 10.4).

Much of the work on *chaotic population dynamics* is theoretical. Some studies have shown that chaos

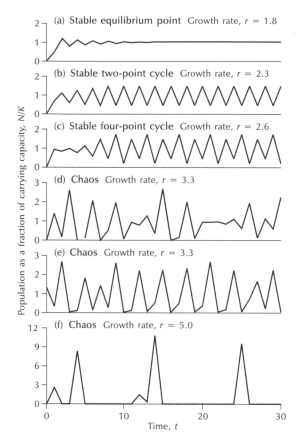

Figure 10.4 Internally-driven population dynamics: stable points, stable cycles, and chaos. NB: The growth rates, *r*, in (d) and (e) are the same, but the ratios of the initial population to the carrying capacity, N_0/k, are different: in (d), $N_0/k = 0.075$; in (e), $N_0/k = 1.5$.
Source: Adapted from May (1981)

does occur in natural populations. The ticklegrass or hairgrass (*Agrostis scabra*) displays oscillations and chaos in experimental plots (Tilman and Wedin 1991). Monocultures of ticklegrass were planted at two different densities on 10 different soil mixtures. Populations growing in unproductive soils, low in nitrogen, maintained relatively stable aboveground biomass. Populations growing on richer soil showed biomass oscillations. Populations growing on the richest soils displayed a 6,000-fold crash (Figure 10.5). No other species

growing in the garden where the experiment was conducted crashed in 1988, which suggests that environmental agencies were not responsible. The dynamics for the crashing populations were shown to be chaotic. The chaotic regime resulted from growth inhibition by litter accumulation. The litter causes a one-year delay between growth and the inhibition of future growth. Litter production is greater in more productive plots. The magnitude of the time-delayed inhibitory effect therefore increases with productivity. This may lead to an oscillation between a year of low litter and high living biomass, and a year of high litter and low living biomass. The implications of this study are profound – litter feedback may cause population oscillations and possibly chaos in productive habitats. However, litter accumulates where it falls, and litter-driven chaotic dynamics occurs at small spatial scales – it may avoid detection if looked for at medium and large scales.

Since the late 1980s, part of theoretical ecology has focused on the spatio-temporal dynamics generated by simple ecological models. To a large extent, the results obtained have changed views of complexity (Bascompte and Sole 1995). Specifically, simple rules are able to produce complex spatio-temporal patterns. Consequently, some of the complexity underlying nature does not necessarily have complex causes – it may have simple causes. The emerging framework has far-reaching implications in ecology and evolution. It is improving the comprehension of such topics as the scale problem, the response to habitat fragmentation, the relationships between chaos and extinction, and how higher diversity levels are supported in nature.

Age and sex structure

The aggregate population size, *N*, obscures the fact that individuals differ in age and in ability to reproduce. The population age structure is the proportion of individuals in each age class. A stable age distribution with a zero growth rate

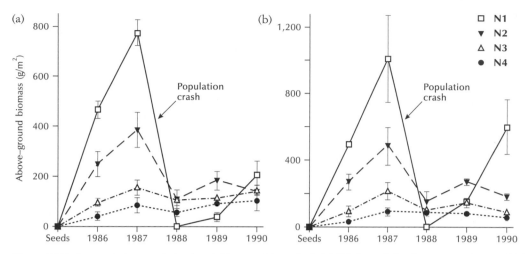

Figure 10.5 A population crash in the ticklegrass (*A. scabra*). (a) High seed density monocultures. (b) Low seed density monocultures. The soil mixtures are divided into four groups, N1–N4. Group N1 is poorest in nitrogen, and Group N4 is the richest in nitrogen.

Source: Adapted from Tilman and Wedin (1991)

has about the same number of individuals in each age class, except for fewer in post-reproductive classes. In contrast, a growing population has a pyramid-shaped structure, with large numbers of pre-reproductive individuals.

Life tables

A **life table**, like an actuarial table, shows the number of individuals expected in each age class based upon fecundity and survival rates. Survival rates, s, are related to death rates, d, in the following way: $s = 1 - d$. Life-table statistics are normally computed for females. This is because fecundity for males is very difficult to determine where mating occurs promiscuously. A life table for a hypothetical population of females is shown in Table 10.1. Column 1 lists the age classes. Column 2 gives the survivorship expressed as the fraction surviving at age x; this is the probability than an average newborn will survive to that age. Column 3 gives the fecundity; this is the number of offspring produced by an average individual of age x during that period. The sum of fecundity terms over all ages, or the total number of

offspring that would be produced by an average organism with no mortality, is called the *gross reproductive rate*. It is 9.8 in the example. Column 4 shows the number of expected offspring by age class. The figures are simply the product of survivorship and fecundity. The sum of expected offspring gives the *net reproductive rate*, R. This is the expected number of female offspring to which a newborn female will give birth in her lifetime. It is 7.322 in the example. Column 5 shows the product of age and expected offspring. The total of these products divided by the net reproductive rate defines the *mean generation time*, T. This is the average age at which females produce offspring. It is 5.26714 years in the example. The earlier young are born, the earlier they, in turn, will have offspring and the more rapid population growth will be. The calculations show that the population will increase by a factor of 7.322 times every 5.26714 years. Plainly, growth rates expressed on a generation basis would be difficult to compare. To avoid this problem, growth rates are normally converted to an annual figure and designated by the Greek letter lambda (λ). The annual growth rate of the hypothetical population is 1.459.

Table 10.1 Life table of a hypothetical population

Column 1 Age, x (years)	Column 2 Survivorship, l_x	Column 3 Fecundity, b_x	Column 4 Expected offspring, $l_x b_x$	Column 5 Product of age and expected offspring, $x l_x b_x$
0	1	0	0	0
1	0.99	0.2	0.2	0.2
2	0.98	0.4	0.392	0.784
3	0.96	0.8	0.768	2.304
4	0.92	1.2	1.104	4.416
5	0.86	1.6	1.376	6.880
6	0.78	2.2	1.716	10.296
7	0.68	1.4	0.952	6.664
8	0.56	0.8	0.448	3.584
9	0.42	0.6	0.252	2.268
10	0.26	0.4	0.104	1.040
11	0.06	0.2	0.012	0.132
12	0	0	0	0
		9.8 This is the gross reproductive rate, GRR	**7.322** This is the net reproductive rate, R	**38.566** This is the total weighted age

Notes: Mean generation time, $T = 38.566/7.322 = 5.26714$ years. Annual growth rate, λ, is given by $\lambda = R^{1/T}$. It is $7.322^{1/5.26714} = 1.459$

Survivorship curves

A *survivorship curve* shows the fraction of a **cohort** of newborn (or newly hatched) alive individuals in subsequent years. Natural populations have a great range of survivorship curves. Three basic types are recognized (Pearl 1928) (Figure 10.6):

1 *Type I* survivorship curves are 'rectangular' or convex on semi-logarithmic plots. They show low mortality initially that lasts for more than half the life-span, after which time mortality increases steeply. This survivorship pattern is common amongst mammals, including the Dall mountain sheep (*Ovis dalli dalli*) and humans. It is also displayed by such reptiles as the desert night lizard (*Xantusia vigilis*).

2 *Type II* survivorship curves are 'diagonal' or straight on semi-logarithmic plots. They show a reasonably constant mortality with age. This survivorship pattern is common in most birds, including the American robin (*Turdus migratorius migratorius*), and reptiles.

3 *Type III* survivorship curves are concave on semi-logarithmic plots. They show extremely high juvenile morality and relatively low mortality thereafter. This survivorship pattern is common in many fish, marine invertebrates, most insects, and plants. It is also characteristic of the British robin (*Erithacus rubecula melophilus*).

These three basic survivorship curves do not exhaust all the possibilities – many intermediate types occur.

Cohort-survival models

Models of population disaggregated by age and sex are called *cohort-survival models* (Leslie 1945, 1948). A cohort-survival model starts with the age structure of a female population at a given

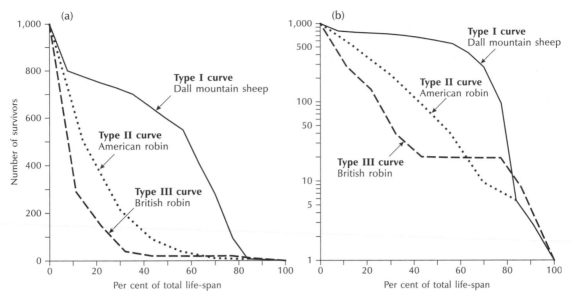

Figure 10.6 Survivorship curves of the Dall mountain sheep (*O. dalli dalli*), the American robin (*T. migratorius migratorius*), and the British robin (*E. rubecula melophilus*). (a) Arithmetic plot. (b) Semi-logarithmic plot.

time, and with the birth rates and survival rates per individual in each age group. From this information it predicts the age structure of the female population in future years.

Cohort-survival models are used to study population change. One application considered the effects of overexploitation on the blue whale (*Balaenoptera musculus*) population (Usher 1972). It used data on the blue whale population collected in the 1930s, before the near extinction of the species in the early 1940s. Age-specific birth rates and survival rates for each of seven age groups were calculated (Table 10.2). The birth terms (fecundity terms in this study) show that female blue whales do not breed in the first four years of life. At full breeding, each cow produces one calf every two years (since the sex ratio is 1:1, this calf has a 50 per cent chance of being female). The survival terms, all but one of which is 0.87, are based on the best available estimate of natural mortality. The survival rate of the 12-plus age class is 0.8. This is because not all

members of the oldest age-group die within a two-year period; indeed, some live to be 40. The survival rate of 0.8 per two years gives a life expectancy of 7.9 years for old whales.

The parameters in the table may be arranged as a growth matrix, from which the growth characteristics of the whale population may be extracted. The dominant latent root, λ_1, of the growth matrix is 1.0986. This shows that the population is capable of growing. The intrinsic rate of natural increase of the blue whale population is given by the natural logarithm of the dominant latent root; it is ln 1.0986 = 0.094. The dominant latent root may be used to estimate the harvest of whales that can be taken without causing a decline in the population. The population size increases from N to $\lambda_1 N$ over a two-year period. The harvest that may be taken, H, expressed as a percentage of the total population, is

$$H = 100 \{(\lambda_1 - \lambda)/\lambda_1\}$$

Table 10.2 Parameters for a cohort-survival model of the blue whale (*Balaenoptera musculus*) population

Parameters (per cow per 2 years)	Age groups						
	0–1	2–3	4–5	6–7	8–9	10–11	12+
Birth rate	0	0	0.19	0.44	0.50	0.50	0.45
Survival rate	0.87	0.87	0.87	0.87	0.87	0.87	0.80

Source: Adapted from Usher (1972)

This is about 4.5 per cent of the total population per year. If the harvest rate were exceeded, the population would decrease, unless homeostatic mechanisms should come into play and alter fecundity and survival rates. Evidence suggests that, when under pressure, the blue whale population shows a slight increase in the pregnancy rate and that individuals reach maturity earlier in life, and that they may generally grow more rapidly. The big question is: can these responses offset the effects of exploitation? For a trustworthy answer to this question, long-term studies involving data collection and modelling are needed.

The wandering albatross (*Diomedia exulans*) is a splendid bird, with a wingspan of over three metres. It breeds on southern cool-temperate and sub-Antarctic islands, and forages exclusively in the southern hemisphere. Wandering albatross populations around the world have declined since the 1960s. The major factor contributing to this decline is accidental deaths associated with longline fishing, which has increased rapidly since the 1960s. Entanglement and collisions with netsonde monitor cables on trawlers may also contribute to albatross mortality. Longline operations involve baited hooks attached to weighted lines being thrown overboard. Seabirds lured to the bait swallow the hooks and die by drowning when they are dragged beneath the surface as the weighted line sinks, or die when the lines are hauled. Some 44,000 albatrosses, including about 10,000 wandering albatrosses, are killed in this manner each year (Brothers 1991). An age-structured model of a wandering albatross population,

based on demographic data from the population at Possession Island, Crozets, simulated population trends over time (Moloney *et al.* 1994). The aim was to investigate the potential impacts of new longline fisheries, such as that for the Patagonian toothfish (*Dissostichus eleginoides*) in Antarctica. The model consisted of 12 year age-classes, comprising 4 age categories: chicks (prior to fledging, 0–1 years), juveniles (before the first return to the breeding colony, 1–4 years), immatures (after returning to the colony but before the first breeding, 5–10 years), and breeding adults (more than 10 years). The model kept track of the population in each age class, N_i. The populations change owing to eggs being laid (fecundity) and survival from one age class to the next. Fecundity and survival terms are summarized in Table 10.3, together with starting numbers for a stable population. The simulations predicted a decreasing population. The rate of decrease is constant at 2.29 per cent per year and is associated with a stable age structure of 14 per cent chicks, 24 per cent juveniles, 15 per cent immatures, and 47 per cent adults. Further simulations were carried out to test the sensitivity of the wandering albatross population to altered survival rates. On the assumption that longline fishing operations affect juveniles more than adults, there is a time lag of 5–10 years before further decreases in population numbers are affected in the breeding populations. In addition, because wandering albatrosses are long-lived, population growth rates take about 30–50 years to stabilize after a perturbation, such as the introduction of a new longline fishery.

Table 10.3 Demographic parameters for a wandering albatross (*Diomedia exulans*) population

Age class (years)	Age category	Survival (per year)	Fecundity (per year)	Starting population (for a stable age structure)
0	Eggs and chicks	0.640	0	313
1	Juveniles	0.715	0	201
2	Juveniles	0.715	0	147
3	Juveniles	0.715	0	107
4	Juveniles	0.715	0	78
5	Immatures	0.980	0	57
6	Immatures	0.980	0	58
7	Immatures	0.980	0	58
8	Immatures	0.980	0	59
9	Immatures	0.980	0	59
10	Immatures	0.980	0	60
11	Adults	0.922	0.294	1,066

Source: Adapted from Moloney *et al.* (1994)

Metapopulations

There are four chief types of **metapopulation**, normally styled 'broadly defined', 'narrowly defined', 'extinction and colonization', and 'mainland and island' (Figure 10.7). The broad and narrow categories are more concisely named loose and tight metapopulations, respectively.

Loose metapopulations

A *loose metapopulation* is simply a set of subpopulations of the same species. Rates of mating, competition, and other interactions are much higher within the subpopulations than they are between the subpopulations. The subpopulations may arise through an accident of geography – the habitat patches in which the subpopulations live lie farther apart than the normal dispersal distance of the species. The conspecific subpopulations that comprise such a metapopulation may or may not be interconnected. By this definition, most species with large and discontinuous distributions form metapopulations.

Where habitat patches are so close together that most individuals visit many patches in their lifetime, the subpopulations behave as a single

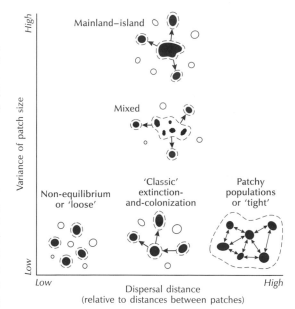

Figure 10.7 Types of metapopulations classed according to the variance of patch size and the dispersal distance.

population – all individuals effectively live together and interact. Where habitat patches are so scattered that dispersal between them almost never happens, the subpopulations behave effectively as separate populations. This situation applies to mammal populations living on desert mountaintops in the American southwest (Brown 1971).

Tight metapopulations

A *tight metapopulation* is a set of conspecific subpopulations living in a mosaic of habitat patches, with a significant exchange of individuals between the patches. Migration or dispersal among the subpopulations stabilizes local population fluctuations. This **rescue effect** helps to prevent local extinctions through the recolonization of new habitats or habitats left vacant by local extinctions (or both). In theory, a tight metapopulation structure occurs where the distance between habitat patches is shorter than the species is physically capable of travelling, but longer than the distance most individuals move within their lifetimes. An example is the European nuthatch (*Sitta europaea*) population. The nuthatch inhabits mature deciduous and mixed forest and has a fragmented distribution in the agricultural landscapes of western Europe (Verboom *et al.* 1991). It shows all the characteristics of a tight metapopulation: the distribution is dynamic in space and time; the extinction rate depends on patch size and habitat quality; and the patch colonization rate depends on the density of surrounding patches occupied by nuthatches.

Extinction-and-colonization metapopulations

In the narrowest possible sense, and as originally defined (Levins 1970), a metapopulation consists of subpopulations characterized by frequent turnover. In these '*extinction-and-colonization*' metapopulations (Gutiérrez and Harrison 1996), all subpopulations are equally susceptible to local extinction. Species persistence therefore depends on there being enough subpopulations, habitat patches, and dispersal to guarantee an adequate rate of recolonization. This metapopulation structure seems to apply to the pool frog (*Rana lessonae*) on the Baltic coast of east-central Sweden (Sjögren 1991). Frog immigration mitigates against large population fluctuations and **inbreeding**, and 'rescues' local populations from extinction. Extinction-and-colonization metapopulation structure also applies to an endangered butterfly, the Glanville fritillary (*Melitaea cinxia*) in a fragmented landscape (Hanski *et al.* 1995, 1996b). This butterfly became extinct on the Finnish mainland in the late 1970s. It is now restricted to the Åland islands. It has two larval host plants on the islands – ribwort plantain (*Plantago lanceolata*) and spiked speedwell (*Veronica spicata*) – both of which grow on dry meadows. The metapopulation was studied on a network of 1,502 discrete habitat patches (dry meadows), comprising the entire distribution of this butterfly species in Finland (Figure 10.8). Survey of the easily detected larval groups revealed a local population in 536 patches. The butterfly metapopulation satisfied the necessary conditions for a species to persist in a balance between stochastic local extinctions and recolonizations.

Mainland–island metapopulations

In some cases, there may be a mixture of small subpopulations prone to extinction and a large persistent population. The viability of the '*mainland–island*' *metapopulations* (Gutiérrez and Harrison 1996) is less sensitive to landscape structure than in other types of metapopulation. An example is the natterjack toad (*Bufo calamita*) metapopulation at four neighbouring breeding sites in the northern Rhineland, Germany (Sinsch 1992). Over 90 per cent of all reproductive males showed a lifelong fidelity to the site of first breeding; females did not prefer certain breeding sites. Owing to the female-biased exchange of individuals among neighbouring sites, the genetic distance between local populations

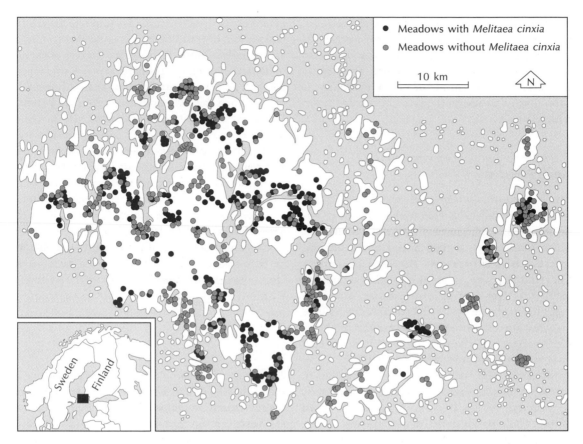

Figure 10.8 The Glanville fritillary (*M. cinxia*) population in Åland, Finland, in 1993. Shaded circles are empty meadows; solid circles are occupied meadows.

Source: Adapted from Hanski *et al.* (1996b)

was generally low but increased with geographical distance. This spatial pattern is consistent with the structure of a mainland–island metapopulation. Up to three mass immigrations of males per breeding period, replacing previously reproductive individuals, suggested the existence of temporal populations successively reproducing at the same locality. Genetic distances were considerably greater between temporal populations than between local ones, indicating partial reproductive isolation. An exchange of reproductive individuals between the temporal populations at each site was not detected, but gene flow due to the recruitment of first-breeders originating from

offspring other than their own seemed probable. Thus, natterjack metapopulations consist of interacting local and temporal populations. Three out of four local populations had low reproductive success, as well as the latest temporal population. The persistence of these populations depended entirely on the recruitment of juveniles from the only self-sustaining local population. This 'rescue-effect' impeded local extinction.

Metapopulations and conservation

Metapopulation theory is now widely applied in conservation biology. It is especially important where *habitat fragmentation* is the central concern. The 'classic' example is the northern spotted owl in the USA, for which bird the metapopulation approach to management has been thoroughly developed. Other examples abound, including the euro and Leadbeater's possum, both from Australia.

The spotted owl in northwestern North America

The northern spotted owl (*Strix occidentalis caurina*) is closely associated with mature and old-growth coniferous forests in the Pacific northwest. These forests once formed a continuous habitat, but they have shrunk over the last half-century. They now survive as small remnants occupying just 20 per cent of their former extent. This fragmentation of the spotted owl habitat reduces dispersal success, which may jeopardize the long-term survival of the species. Northern spotted owls require territories of around 1,000 ha. They avoid open areas, where they are vulnerable to predation by the larger great-horned owl (*Bubo virginianus*), and will cross only small cleared strips. Their survival therefore depends on the connectivity of the forest habitat (Ripple *et al.* 1991). The preservation of a viable owl population is required by the Endangered Species Act and by laws governing the US Forest Service, which owns 70 per cent of the owl's remaining habitat. The US government commissioned scientists to discover how much forest, and in what form, was required to ensure the owl's survival. The result was a spatial model of the owl population (Lamberson *et al.* 1992; see also S. Harrison 1994; Gutiérrez and Harrison 1996; Noon and Franklin 2002).

The model let a single spatial element represent the territory of a pair of owls. The dynamics of each pair were modelled using rates of pair-formation, fecundity, and survival measured in the field. Dispersal was estimated from juvenile owls tagged with radio transmitters. Owl demography was linked to forest fragmentation by juvenile dispersal success. To join the breeding population, a newly fledged owl needed to find a vacant and suitable territory. Its chances of doing so depended upon the proportion of forested landscape. The results predicted a sharp threshold, below which the owl population could not persist. With less than 20 per cent forest, the juvenile recruitment to the breeding population was too low to balance mortality of adult territory holders, and the population collapsed. In addition, when juveniles were required to find mates as well as empty territories, a second survival threshold emerged, below which extinction soon ensued. The recommendation was that 7.7 million acres (about 3 million ha) of **old-growth forest** should be preserved in a system of patches, each patch large enough for more than 20 owl territories within 12 miles (about 20 km) of the nearest patch.

The euro

The common wallaroo or euro (*Macropus robustus*), called the biggada by aborigines, is a large, rather shaggy-haired kangaroo. Its range includes most of Australia, except the extreme south and the western side of the Cape York Peninsula, Queensland. Its habitat usually features steep escarpments, rocky hills, or stony rises; and, in areas where extreme heat is experienced for long periods, caves, overhanging rocks, and ledges for shelter.

Members of a euro population living in a fragmented landscape occupied a very large (1,196 ha) patch of remnant vegetation (G. W. Arnold *et al.* 1993). They were sedentary and did not use any other vegetation remnants. However, some young males from this remnant dispersed up to 18 km to other remnants. In areas that had a large remnant (more than 100 ha), individuals were also sedentary. Some individuals included in their home ranges smaller remnants within 700 m. To access these, they usually moved along connecting corridors between remnants. Occasionally, a few individuals made excursions of several kilometres outside their normal home ranges, either overnight or over several months. In areas where all remnants were small (less than 30 ha), individuals lived alone or in small groups, moving frequently between several remnants. Overall, the euros in the 1680-km^2 study area appeared to be separated into a number of metapopulations, some of which had very small numbers of animals. Within the metapopulations, movements between populations appear to be dependent on the availability of 'stepping stones' and corridors. In two populations that had low densities of euros, the numbers of juveniles per adult female were significantly lower than in systems with higher densities. Long-term survival of these two populations is questionable.

Leadbeater's possum

Leadbeater's possum (*Gymnobelideus leadbeateri*) lives in the cool, misty mountain forests of the central highlands of Victoria, Australia (Plate 10.1). These forests are timber-producing areas. Forest fragmentation is threatening the survival of the possum, which is an endangered species.

The viability of possum metapopulations in the fragmented old-growth forests was assessed using a population model (Lindenmayer and Lacy 1995). Computer simulations with subpopulations of 20 or fewer possums were characterized by very rapid rates of extinction, most metapopulations typically failing to persist for longer than 50 years. Increases in either the migration rate or the number of small subpopulations worsened the metapopulations' demographic instability, at least when subpopulations contained fewer than 20 individuals and when migration rates were kept within realistic values for Leadbeater's possums dispersing between disjunct habitat patches. The worsening demographic situation was reflected in lower rates of population growth and depressed probabilities of metapopulation persistence. These effects appeared to be associated with substantial impacts of chance demographic events on very small subpopulations,

together with possum dispersal into either empty patches or functionally extinct (single-sex) subpopulations. Increased migration rates and the addition of subpopulations containing 40 possums produced higher rates of population growth, lower probabilities of extinction, and longer persistence times. Extinctions in these scenarios were also more likely to be reversed through recolonization by dispersing individuals.

A common problem with fragmented populations is that genetic information within the gene

Plate 10.1 Tall mountain ash forest in the O'Shannassy catchment, Victoria, Australia, typical of mature forest occupied by Leadbeater's possum (*G. leadbeateri*). Photograph by Andrew Bennett.

pool becomes less well mixed. At the highest rates of migration, subpopulations of 40 possums were thoroughly mixed and behaved genetically as a single larger population. However, over a century, the gene-pool mixing would decline and lead to a significant loss in genetic variability. This loss would occur even with highest migration rates among five 40-animal subpopulations. While demographic stability might occur in metapopulations of 200 individuals, considerably more individuals than this might be required to avoid a significant decline in genetic variability over a century.

A QUESTION OF SURVIVAL: POPULATION STRATEGIES

Animals and plants possess primary ecological strategies. These strategies are recurrent patterns of specialization for life in particular habitats or niches. They involve the fundamental activities of an organism, including resource capture, growth, and reproduction. There are two main schemes for population strategies. The first involves two basic strategies – opportunism and competitiveness; the second involves three basic strategies – competitiveness, stress toleration, and ruderalism.

Opportunists and competitors

Two basic population strategies depend on how long a habitat is favourable. A crucial factor is a population's generation time compared with the 'life-span' of a stable habitat. One extreme case is where the generation time is about the same as a stable habitat's 'life-span'. This would apply, for example, to fruit flies (*Drosophila*) living in a tropical forest. A fruit fly lives on ripe fruits, which plainly are temporary habitats. If a fruit fly should attain adulthood, it simply migrates to another ripe fruit. Under these circumstances, there is nothing to be lost by exceeding the habitat's carrying capacity because the resources left for future generations will not be jeopardized. For this rea-son, short-lived habitats tend to be occupied by exploiters or opportunists with boom-and-bust population dynamics. Such opportunistic species are called *r-strategists*, after the intrinsic growth rate, *r*, in population growth equations.

Another extreme case occurs where the generation time is tiny compared with a stable habitat's 'life-span'. This would apply, for instance, to an orang-utan (*Pongo pygmaeus*), to whom an entire forest is a stable and permanent habitat. Under these circumstances, it pays species to look to the future. Long-lived habitats with a fairly stable carrying capacity would be degraded if population density should overshoot the carrying capacity. This would reduce the habitat's carrying capacity for future generations. Species in these habitats are adapted to harvesting food in a crowded environment. They are highly competitive and they tend to have low population growth rates. They are called *K-strategists*, after the carrying capacity parameter, *K*, in population growth equations.

The *r*-strategists

The *r*-strategists continually colonize temporary habitats. Their strategy is essentially opportunistic. They have evolved high population growth rates, produced by a high fecundity and a short generation time. Migration is a major component of their population dynamics, and may occur as often as once per generation. The habitats they colonize are commonly free from rivals. This favours a lack of competitive abilities and small size. They defend themselves against predators by synchronizing generations (hence satiating predators) and by being mobile, which enables them to play a game of hide and seek. The greatest *r*-strategists are insects and bacteria. The African armyworm (*Spodoptera exempta*) is a good example. This migrant moth is a crop pest in eastern Africa. It lays up to 600 eggs 'at a sitting', has a generation time of a little over three weeks, and large population outbreaks occur on the young growth of grasses, including maize (*Zea mays*). Some birds are *r*-strategists. The quail (*Coturnix coturnix coturnix*) is a case in point (Puigcerver *et al*. 1992).

The *K*-strategists

K-strategists occupy stable habitats. They maintain populations at stable levels, and are intensely competitive. They are often selected for large size. Many vertebrates are *K*-strategists. They are characteristically large, long-lived, very competitive, have low birth and death rates, and invest much time and effort in raising offspring. Their low growth rate helps to avoid the habitat degradation that would follow an overshooting of the carrying capacity. Animal and plant *K*-strategists alike tend to invest much energy in defence mechanisms. Animal *K*-strategists often have small litters or clutches and parents are very careful with their young. An extreme example is the wandering albatross (*Diomedia exulans*). This bird breeds in alternate years (when successful), lays just one egg, and is immature for 9–11 years, the longest period of immaturity for any bird (see p. 170).

Not all vertebrates are *K*-strategists. The nomadic Australian zebra finch (*Taeniopygia castanotis*) is *r*-selected, and the blue tit (*Parus caerulus*), with the largest recorded clutch size for a passerine bird, is an opportunist. If their populations should fall dramatically, then the birth rate will often increase to bring the population bouncing back to its stable level. This happens by increasing the litter or clutch size.

Closely related species may display different population strategies. Two trefoil species (*Trifolium*) in Europe – the pale trefoil (*T. pallescens*) and Thal's clover (*T. thalii*) – prefer distinct habitats and reproduce only by seeds (Hilligardt 1993). Pale trefoil grows on alpine screes, often in morainic areas; Thal's clover is predominantly found in subalpine and alpine pastures. Thal's clover, a plant of more competitor-influenced habitats, is a *K*-strategist; pale trefoil has characteristics of an invasive, pioneer *r*-strategist. It has less developed vegetative structures, higher seed production, higher seed-bank reserves, smaller seeds, and higher rates of seedling mortality.

Competitors, stress-tolerators, and ruderals

Some populations are finely adjusted to habitats that vary considerably in time. Where temporal habitat variations are pronounced, populations may experience adverse conditions or stress. A three-strategy model, which adds a stress-toleration strategy to competitive and opportunistic strategies, has been applied to plants (Grime 1977; Grime *et al*. 1988).

External environmental factors that affect plants fall into two chief categories – stress and disturbance. Stress involves all factors that restrict photosynthesis (mainly shortages of light, water, or nutrients; and suboptimal temperatures). Disturbance involves partial or total destruction of plant biomass arising from herbivores, pathogens, and humans (trampling, mowing, harvesting, and ploughing), and from wind-damage, frosting, droughting, soil erosion, and fire. Combining high and low stress with high and low disturbance

Table 10.4 Environmental contingencies and ecological strategies in plants

Disturbance intensity	Stress intensity	
	Low	High
Low	Competitors	Stress-tolerators
High	Ruderals	(No practicable strategy)

Source: Adapted from Grime (1977)

yields four possible contingencies to which vegetation adapts (Table 10.4). Notice that there is no practicable strategy in plants for living in a stressful and highly disturbed environment. This is because continuous and severe stress in highly disturbed habitats prevents a sufficiently swift recovery or re-establishment of the vegetation. (Some humans – teachers come to mind – manage to survive under those conditions.) The remaining three environmental contingencies appear to have produced three basic ecological strategies in plants – *competitors*, *stress-tolerators*, and *ruderals*. Competitors correspond to *K*-strategists, and ruderals to *r*-strategists. Stress-tolerators are an additional group whose strategy is a response to the temporal variability and adversity of the physical environment. In addition, there are four intermediate ecological strategies – competitive ruderal, stress-tolerant competitor, stress-tolerant ruderal, and competitor–stress-tolerator–ruderal strategy (or C–S–R). Strategies of different herbaceous plants may be found using a dichotomous key (Table 10.5).

The primary and intermediate ecological strategies may be displayed on a triangular diagram (Figure 10.9). The plants associated with each strategy share the same adaptive characteristics. The following examples are all British plants:

1 *Competitors* exploit low stress and low disturbance environments. The rosebay willow-herb (*Chamerion angustifolium*) occurs in a wide variety of undisturbed habitats. It is notably common in fertile, derelict environments (urban clearance sites, cinders, building rubble, mining and quarry waste, and other spoil heaps). It is abundant in woodland glades, wood margins, scrub, and young plantations; it is widespread in open habitats such as walls, cliffs and rock outcrops, waysides and waste ground, and riverbanks; and seedlings are sometimes found in wetland (but not submerged areas), arable fields, and pastures.

2 *Stress-tolerators* exploit high stress, low disturbance, low competition environments. The heath grass (*Danthonia decumbens*) favours mountain grassland. It is also found in pastures and heathland, and occasionally in road verges, grassy paths, scree slopes, rock outcrops, and wetland.

3 *Ruderals* exploit high disturbance, low stress, and low competition environments. The shepherd's purse (*Capsella bursa-pastoris*) lives on disturbed, fertile ground. It is especially abundant as a weed on arable land and in gardens, and occurs in a range of disturbed artificial habitats such as demolition sites, paths, spoil heaps, and manure heaps.

4 *Competitive ruderals* exploit low stress, medium competition, and medium disturbance environments. The policeman's helmet (*Impatiens glandulifera*) is abundant on riverbanks, shaded marshland, and disturbed, lightly shaded areas in woodland and scrub.

5 *Stress-tolerant competitors* exploit medium competition, medium stress, and low disturbance environments. The male fern (*Dryopteris filix-mas*) is restricted to two very different habitats – woodland habitats and skeletal habitats (mainly cliffs and walls, but also quarry spoil, rock outcrops and screes, riverbanks, and some cinder tips).

6 *Stress-tolerant ruderals* exploit medium disturbance, medium stress, and low competition environments. The carline thistle (*Carlina vulgaris*) occurs in areas of discontinuous turf associated with calcareous pastures, rock outcrops, lead-mine heaps, and derelict wasteland. It is also frequent on sand dunes.

7 *Competitor–stress-tolerator–ruderal* (*C–S–R*) **strategists** occupy a middle-of-the-road position (medium competition, medium stress, medium disturbance) that enables them to live in a wide range of conditions. The Yorkshire fog (*Holcus lanatus*) is recorded as seedlings in every type of habitat. It is most abundant in meadows and pastures, and is prominent in spoil heaps, waste ground, grass verges, paths, and in hedgerows. It is less common on riverbanks, arable land, marsh ground, and in skeletal habitats such as

Table 10.5 A key for identifying ecological strategies of herbaceous plants

Levels of dichotomous key					Strategy
1	2	3	4	5	
Annual	→ Potentially fast-growing	→ Flowering precocious			Ruderal
		→ Flowering delayed			Competitive–ruderal
	→ Small and slow-growing				Stress-tolerant–ruderal
Biennial	→ Potentially large and fast-growing				Competitive–ruderal
	→ Small and slow-growing				Stress-tolerant–ruderal
Perennial	→ Vernal geophyte				Stress-tolerant–ruderal
	→ Not vernal geophyte	→ Rapid leaf turnover	→ Rapid proliferation and fragmentation of shoots		Competitive–ruderal
			→ Shoots not fragmenting rapidly	→ Shoots tall and laterally extensive	Competitor
				→ Shoots short or creeping	Competitive–stress-tolerant–ruderal (C–S–R)
		→ Slow leaf turnover	→ Shoots tall, or laterally extensive, or both		Stress-tolerant competitor
			→ Shoots short and not laterally extensive		Stress-tolerator

Source: Adapted from a diagram in Grime et al. (1988)

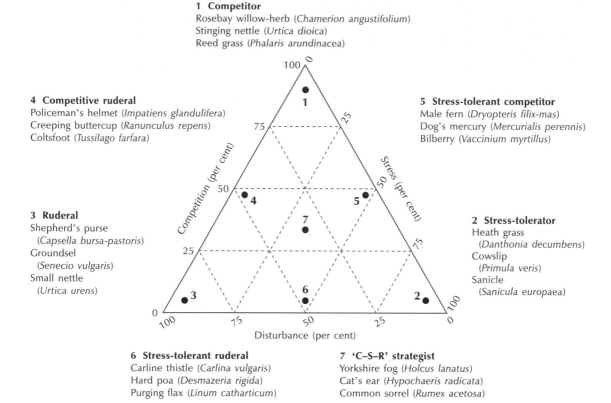

1 Competitor
Rosebay willow-herb (*Chamerion angustifolium*)
Stinging nettle (*Urtica dioica*)
Reed grass (*Phalaris arundinacea*)

4 Competitive ruderal
Policeman's helmet (*Impatiens glandulifera*)
Creeping buttercup (*Ranunculus repens*)
Coltsfoot (*Tussilago farfara*)

5 Stress-tolerant competitor
Male fern (*Dryopteris filix-mas*)
Dog's mercury (*Mercurialis perennis*)
Bilberry (*Vaccinium myrtillus*)

3 Ruderal
Shepherd's purse
   (*Capsella bursa-pastoris*)
Groundsel
   (*Senecio vulgaris*)
Small nettle
   (*Urtica urens*)

2 Stress-tolerator
Heath grass
   (*Danthonia decumbens*)
Cowslip
   (*Primula veris*)
Sanicle
   (*Sanicula europaea*)

6 Stress-tolerant ruderal
Carline thistle (*Carlina vulgaris*)
Hard poa (*Desmazeria rigida*)
Purging flax (*Linum catharticum*)

7 'C–S–R' strategist
Yorkshire fog (*Holcus lanatus*)
Cat's ear (*Hypochaeris radicata*)
Common sorrel (*Rumex acetosa*)

Figure 10.9 Ecological strategies in plants. The triangular diagram is used to define various mixes of three basic strategies – competitor, stress tolerator, and ruderal. Pure competitors lie at the top corner of the triangle, pure stress-tolerators at the bottom right corner, and pure ruderals at the bottom left corner. Intermediate strategies are shown. Example species are all British.

Source: Adapted from Grime *et al.* (1988)

walls and rock outcrops. It occurs in low frequency in scrub and woodland clearings, and is found also in grassy habitats near the sea and on mountains.

Different life-forms tend to have a narrow range of ecological strategies. Trees and shrubs centre on a point between the competitor–stress-tolerator–ruderal and competitive–ruderal positions. Herbs have a wide range of strategies, though competitor–stress-tolerator–ruderal is common. **Bryophytes** (mosses and liverworts) are

mainly ruderals and stress-tolerator–ruderals. Lichens are mainly stress-tolerator–ruderals and stress-tolerators.

PLAYING WITH NUMBERS: POPULATION EXPLOITATION AND CONTROL

Exploiting populations

The *commercial exploitation* of marine mammals, fish, big game, and birds has caused the extinction of several species and the near-extinction of several others. The cases of the passenger pigeon, great auk, northern fur seal, and the American bison illustrate this point.

Passenger pigeon

These graceful birds once flourished in North America. On 1 September 1914, the last passenger pigeon (*Ectopistes migratorius*), called 'Martha', died in the Cincinnati Zoological Garden. The rapid extinction of such a flourishing population resulted from three unfortunate circumstances. First, the pigeons were very tasty when roasted or stewed. Second, they migrated in huge and dense flocks, perhaps containing two billion birds, that darkened the sky as they passed overhead. Third, they gathered in vast numbers, perhaps several hundred million birds, to roost and to nest. A colony in Michigan was estimated to be 45 km long and an average of around 5.5 km wide. Professional hunters – pigeoners – formed large groups to trap and kill the pigeons. Nesting trees were felled to collect the squabs (young, unfledged pigeons). The clearing of forest for farmland and the expansion of the railroad made huge tracts of the pigeon's range accessible and the persecution intensified. By about 1850,

several thousand people were employed in catching and marketing passenger pigeons. Every means imaginable was used to obtain the tasty bird. Special firearms, cannons, and forerunners to the machine gun were built. In 1855 alone, one New York handler had a daily turnover of 18,000 pigeons. In 1869, 7,500,000 birds were captured at one spot. In 1879, a billion birds were captured in the state of Michigan. The harvesting rate disrupted breeding and the population size began to plummet. Even by 1870, large breeding assemblies were confined to the Great Lakes states, and the northern end of the pigeon's former range. The last nest was found in 1894. The last bird was seen in the wild in 1899.

Auks

The 'penguin of the Arctic', the great auk (*P. impennis*), was an early casualty of uncontrolled harvesting. This flightless bird was the largest of the Alcidae, which family includes guillemots, murres, and puffins. It used to nest around the north Atlantic Ocean, in northeastern North America, Greenland, Iceland, and the north-western British Isles (Figure 10.10). The last pair died in Iceland in 1844. A few museum specimens survive.

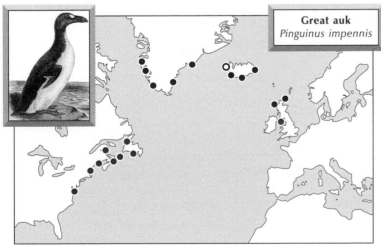

Great auk
Pinguinus impennis

Figure 10.10 The great auk's (*P. impennis*) former distribution. Black dots are former nesting grounds. The white circle marks the site where the last two birds died in 1844.

Sources: Map adapted from Ziswiler (1967); picture from Saunders (1889)

The razorbill (*Alca torda*) is the largest living auk. For many years, this species was killed for its feathers, and by fishermen for use as bait. This exploitation caused the species to decline. The razorbill is now protected and is staging a comeback, so it may avoid the fate of its close relative, the great auk.

Fur seals

Millions of northern fur seals (*Callorhinus ursinus*) used to breed on the Pribilof Islands, in the Bering Sea, Alaska, USA. Between 1908 and 1910, Japanese seal hunters slaughtered four million of them. The Fur Seal Treaty of 1911 was signed by Japan, Russia, Canada, and the USA. This was the first international agreement to protect marine resources. The signatories agreed rates of harvesting for pelts. The happy result was that the northern fur seal population bounced back. New colonies occasionally appear. Northern fur seal pups were first observed on Bogoslof Island, in the southeast Bering Sea, in 1980

(Loughlin and Miller 1989). By 1988 the population had grown to more than 400 individuals, including 80 pups, 159 adult females, 22 territorial males, and 188 juvenile males. Some animals originated from rookeries of the Commander Island; others were probably from the Pribilof Islands.

American bison

Humans have hunted terrestrial, as well as marine, mammal species to extinction or near extinction. A spectacular, if shameful, example concerns the American bison or 'buffalo' (*Bison bison*) (Figure 10.11). In 1700, the buffalo population on the Great Plains was around 60 million. Native Indians had hunted this population sustainably for thousands of years. A new era of buffalo hunting started in the 1860s, when the Union Pacific Railway was opened. At first, special hunting parties were hired to supply the railway workers with meat. William 'Buffalo Bill' Cody rose to fame at this time for having killed

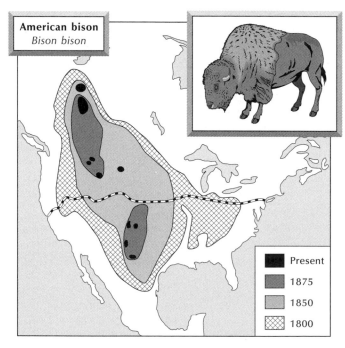

Figure 10.11 The shrinking range of the American bison (*B. bison*).

Source: Adapted from Ziswiler (1967)

250 buffalo in a single day! The tracks of the Union Pacific cut right across the heart of the buffalo country. Once the railway was operating, the railway company encouraged buffalo shoot-outs from the trains. This perpetrated a senseless and bloodthirsty slaughter. Tens of thousands of animals were shot and left to rot alongside the tracks. By the 1890s, only a few hundred buffalo were still alive in the USA, and a few hundred more of the wood buffalo variety in Canada. In 1902, the population was less than a hundred. Happily, buffalo breed well under animal husbandry. Thanks to William T. Hornaday of the New York Zoological Society, the American Bison Society (founded in 1905), and the US and Canadian governments, the species was saved. There was a vigorous programme of captive breeding and releasing of animals onto protected range, such as that in Yellowstone Park. Today, the bison has recovered to a size of tens of thousands and has been introduced into Alaska.

Controlling populations

Much conservation effort is currently expended on *reintroducing* species to at least some of their former domain, and keeping successfully introduced species under control. These two aspects of population management are exemplified by beaver reintroductions in Sweden and by managing feral goats in New Zealand and feral cats on Marion Island.

European beavers

The European beaver (*Castor fiber*) is Europe's largest rodent. It lives beside shores of water-courses and lakes, in excavated burrows or constructed lodges. The beaver was heavily hunted in Europe during the nineteenth century, by the end of which about 1,200 animals survived in eight isolated populations (Figure 10.12). During the twentieth century, protection, natural spread, and reintroductions produced a powerful comeback that continues today (Halley and Rosell 2002). The fate of the European beaver in Sweden typifies the story in many countries (Bjärvall and Ullström 1986, 77). Overexploited during the nineteenth century, protection came too late – the last animals had been shot two years earlier. In 1922, reintroductions were begun using animals captured in southern Norway. The population grew to about 400 animals in 1939 and 2,000 in 1961. During the 1960s, the population tripled and increased to 40,000 animals by 1980. Given such a large increase, some hunting was permitted from 1973. The hunting appears to have been justified. The reintroduced beaver population in Sweden has behaved as introduced ungulate populations – it 'exploded' and then started to decline (Hartman 1994). In two study areas, the population growth-rate turned negative after 34 and 25 years and at densities of 0.25 and 0.20 colonies per km^2, respectively. Management policy for an irruptive species should probably allow hunting during the irruptive phase. Hunting keeps the irruptive population in check and so maintains food resources and avoids uncontrolled population decline.

If range expansion and reintroductions continue at current rates, *C. fiber* will be common in much of Europe. It may even gain a foothold in the British Isles, where a reintroduction is planned for western Scotland and another has happened in Kent. The Kent site is fenced, so is technically a captive population rather than a reintroduction. However, the fence is there as much to protect the beavers as it is to prevent their spreading – the habitat in southeast England is highly fragmented and beavers have, for example, no road sense (Halley, personal communication 2003).

New Zealand goats

Feral goats (*Capra hircus*) live in about 11 per cent of New Zealand, mostly on land reserved for conservation of the indigenous biota. They were introduced from Europe before 1850. They were recognized as a threat to native vegetation from

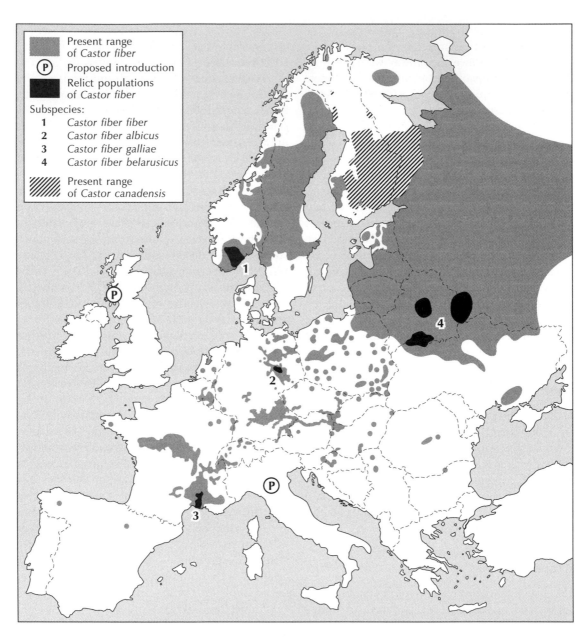

Figure 10.12 Beavers in Europe.

Source: After Halley and Rosell (personal communication 2003)

the 1890s onward and soon attained the rank of a pest. The effect of their browsing is most noticeable on islands, where subtropical forest has in some cases been converted to rank grassland (Rudge 1990). Uncontrolled goat densities are usually less than 1 per ha, but have reached 10 per ha in one area (Parkes 1993). The total population is at least 300,000. Goats are highly selective in what they eat, but switch their diets as the more palatable food species are eliminated from their habitat.

Unharvested feral goat populations in New Zealand form stably dynamic *K*-strategist relationships with their resources. Their impacts are thus chronic, but the relationship can be manipulated to favour the resource if management can be applied 'chronically', that is, as sustained control. The Department of Conservation has attempted to organize its control actions in the highest priority areas either to eradicate goats where possible or to sustain control where eradication is not presently possible. In 1992–3, 6,160 hunter-days of ground-control effort and 230 hours of helicopter hunting effort were expended against feral goats, and 10.2 km of goat fencing was constructed (though goats are notoriously hard to fence in). If was hoped that, if the then present effort was maintained through to the first decade of the twenty-first century, about half the feral goat populations then on the conservation estate would be eradicated or controlled. However, goats are a valuable farming resource, and this may assure the survival of feral populations in hobby farms and hill country farms bordering forests (Rudge 1990).

Marion Island cats

Marion Island is the largest of the Prince Edward Islands and lies in the southern Indian Ocean, about midway between Antarctica and Madagascar. It is a volcanic island with an area of 29,000 ha that supports a tundra biota. Domestic cats (*Felis catus*) were introduced to Marion Island as pets in 1949. By 1975, the population was about 2,139 and had risen to 3,405 in 1977. As on other sub-Antarctic islands where cats were introduced, the feral cats played havoc with the local bird populations. They probably caused the local extinction by 1965 of the common diving petrel (*Pelecanoides urinatrix*), and reduced the populations and breeding successes of winter-breeding great-winged petrels (*Pterodroma macroptera*) and grey petrels (*Procellaria cinerea*). Control of the cat population was begun in 1977. The viral disease feline panleucopaenia was chosen as the most efficient and cost-effective method. The population fell from 3,405 cats in 1977 to 615 cats in 1982. However, the decline slowed down and evidence suggested that **biological control** was becoming ineffective. A full-scale hunting effort was launched in the southern spring of 1986 (Bloomer and Bester 1992). The aim was to eradicate the cat population from the island. In 14,725 hours of hunting, 872 cats were shot dead and 80 were trapped. The number of cats sighted per hour of night hunting decreased, which was a strong indication that the population had declined. But, by the end of the third season, it became evident that hunting alone was no longer removing enough animals to sustain the population decline, and *trapping* was incorporated into the eradication programme. Trapping with baited walk-in cage traps has a drawback: other species, such as sub-Antarctic skuas (*Catharacta antarctica*), also become ensnared. However, a small loss of non-target species has to be offset against the long-term benefits of the cat-trapping programme to the bird population.

SUMMARY

Populations are interbreeding groups of individuals (demes). Unconstrained population growth describes a rising exponential curve; when constrained, it describes a logistic curve, or may behave chaotically. Population irruptions and crashes are common. Age and sex are important components of population growth. They are summarized in life tables and survivorship curves, and they are studied using cohort-survival models. Metapopulations are defined according to the geographical location of individuals within a population – whether they form one large interacting group, or whether they live in several relatively isolated subpopulations, for example. Recent work on the conservation of threatened populations draws heavily on metapopulation theory. Populations adopt various strategies to maximize their chances of survival. Opportunists (*r*-strategists) and competitors (*K*-strategists) are two basic groups. Stress-tolerators form another strategic option. Wild populations suffer heavily at the hands of humans. Some species were hunted to extinction. Others were hunted to near extinction but managed to recover. Population control is often necessary to prevent damage to agricultural resources and **environmental degradation**. Control involves culling or eradicating introductions, such as goats and cats. And it may involve re-establishing a species over vacated parts of its former range.

ESSAY QUESTIONS

1 Why do some populations irrupt and crash?

2 What are the advantages and disadvantages of the different population survival strategies?

3 Why do some populations need controlling?

FURTHER READING

Hanski, I. and Gilpin, M. E. (1997) *Metapopulation Biology: Ecology, Genetics, and Evolution*. New York: Academic Press.
Advanced but worth reading.

Kormondy, E. J. (1996) *Concepts of Ecology*, 4th edn. Upper Saddle River, NJ: Prentice Hall.
Relevant sections worth reading.

Kruuk, H. (1995) *Wild Otters: Predation and Populations*. Oxford: Oxford University Press.
A good case study.

Perry, J. H., Woiwod, I. P., Smith, R. H., and Morse, D. (1997) *Chaos in Real Data: Analysis on Non-linear Dynamics from Short Ecological Time Series*. London: Chapman & Hall.
If mathematics is not your strong point, forget it.

Taylor, V. J. and Dunstone, N. (eds) (1996) *The Exploitation of Mammal Populations*. London: Chapman & Hall.
Examples of exploitation.

Tuljapurkar, S. and Caswell, H. (1997) *Structured-Population Models*. New York: Chapman & Hall.
Somewhat demanding.

INTERACTING POPULATIONS

Populations interact with one another. This chapter covers:

- cooperating populations
- competing populations
- plant-eating populations
- flesh-eating populations
- biological population control

Two populations of the same species, living in the same area, may or may not affect each other. Interactions between pairs of populations take several forms, depending on whether the influences are beneficial or detrimental to the parties concerned (Table 11.1). In mutually beneficial relationships, both populations gain; in mutually detrimental relationships, both lose. Other permutations are possible. One population may gain and the other lose, one may gain and the other be unaffected; one may lose and the other be unaffected.

LIVING TOGETHER: COOPERATION

Four types of interaction between populations or species favour peaceful coexistence – neutralism, protocooperation, mutualism, and commensalism. **Neutralism** involves no interaction between two populations. It is probably very rare, because there are likely to be indirect interactions between all populations in a given ecosystem. The other forms of interaction are common.

Protocooperation

Protocooperation involves both populations benefiting one another in some way from their interaction. However, neither population's

Table 11.1   Interactions between population pairs

Kind of interaction	Species		Comments
	A	B	
Neutralism	No effect	No effect	The populations do not affect one another
Protocooperation	Gains	Gains	Both populations gain from their interaction, but the interaction is not obligatory
Mutualism	Gains	Gains	Both populations gain from their interaction, and the interaction is obligatory (they cannot survive without it)
Commensalism	Gains	No effect	Population A (the commensal) gains; population B (the host) is unaffected
Competition	Loses	Loses	The populations inhibit one another
Amensalism	Loses	No effect	Population A is inhibited; population B is unaffected
Predation	Gains	Loses	Population A (the predator) kills and eats members of population B (the prey)

survival depends on the interaction (the interaction is not obligatory). Some small birds ride on the backs of water buffalo – the birds obtain food and in doing so rid the buffalo of insect pests. Some birds pick between the teeth of crocodiles – the birds get a meal and the crocodile gets a free dental hygiene session.

Animal–animal protocooperation

A remarkable example of protocooperation is the partnership between an African bird, the greater honey-guide (*Indicator indicator*), and a mammal, the honey badger or ratel (*Mellivora capensis*). The honey-guide seeks out a beehive, then flitters around making a chattering cry to attract the attention of a honey badger, or any other interested mammal, including humans. The honey badger rips open the hive and feeds upon honey and bee larvae. After the honey badger has had its fill, the bird feeds. Its digestive system can even cope with the nest-chamber wax, which is broken down by **symbiotic** wax-digesting bacteria living in its intestines. In captivity, individuals have been kept alive for up to 32 days feeding exclusively on wax! The honey-guide is adept at finding beehives but is incapable of opening them; the honey badger is good at opening beehives but is not good at finding them. The protocooperation clearly profits both species.

Plant–plant protocooperation

Protocooperation seems to occur among plants. Two species of bog moss, Magellan's sphagnum (*Sphagnum magellanicum*) and papillose sphagnum (*S. papillosum*), were grown under laboratory conditions (Li *et al.* 1992). Under dry conditions, *S. magellanicum* outcompeted *S. papillosum* for water, largely because it is better designed for transporting and storing water. *S. magellanicum* is a drought-avoider and prefers peat hummocks, while *S. papillosum* is a drought-tolerator and prefers lower sites on a hummock. However, *S. magellanicum* benefits *S. papillosum* growing at higher hummock positions by encouraging lateral flow of water. Both species grow better when mixed, suggesting a degree of protocooperation.

Animal–plant protocooperation

Plants often form cooperative partnerships with animals. The most common examples are associations between plants and pollinators, and between plants and seed dispersers. Plants are rooted to the place that they grow so they must employ a go-between to carry their pollen to other plants and their seeds to other sites. Colourful flowers with nectar and brightly coloured fruit have presumably evolved to attract animals as potential delivery agents. Some herbivores eat fruits, the seeds in which may pass through the herbivore intestines unharmed and grow into new plants from the droppings. Mammals and birds carry seeds in their fur and feathers. The sheep carries a garden-centreful of plant seeds around in its fleece (p. 38).

Birds, bees, and butterflies are the commonest pollinators. Some mammals are pollinators, including humans, nectar-feeding bats, and the slender-nosed honey possum or noolbender (*Tarsipes rostratus*). The noolbender is a diminutive Australian marsupial that feeds on nectar, pollen, and, to some extent, small insects that live in flowers.

An unusual example of plant–animal cooperation is the alliance between some species of acacia and epiphytic ants. Species of acacia that normally support ant colonies are highly palatable to herbivorous insects; those that do not normally house an ant fauna to ward off herbivores are less palatable. Acacias guarded by ant gangs produce nectaries and swollen thorns that attract and benefit the ants. The plants put matter and energy into attracting ants that will protect their leaves, rather than into chemical warfare. This antiherbivore ploy is broad-based, since the ants fiercely attack a wide range of herbivores.

Mimics

Müllerian mimicry, named after Fritz Müller, a nineteenth-century German zoologist, occurs when two or more populations of distasteful or dangerous species come to look similar. An example is bees and wasps, which are normally banded with yellow and black stripes. In Trinidad, the unpalatable butterflies *Hirsutis megara* (from the family Ithomiidae) and *Lycorea ceres* (from the family Danaidae) resemble one another (Figure 11.1). The poisonous passionflower butterflies (*Heliconius*) are confined to the Americas. There are some 40 species, most of which are restricted to rain forest. In different parts of South America, the various species tend to look much alike and form Müllerian mimicry rings. It is not too difficult to see the advantage of such biological photocopying. Tenderfoot predators presumably learn to avoid unsavoury prey by trial and error, killing and eating at random. If distasteful species varied widely in appearance, then the predators would be forced to kill many of each before learning those to avoid. However, when the unsavoury prey is coloured in the same way, then the predator quickly learns to avoid one basic pattern. Its sample snacks are therefore spread out over many prey species. By resembling one another, unsavoury prey populations keep their losses by predation to a minimum.

Batesian mimicry, named after the nineteenth century English naturalist Henry Walter Bates, occurs when a palatable animal comes to look like an unpalatable one. The harmless dronefly

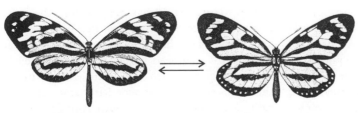

Hirsutis megara   Lycorea ceres

Figure 11.1 The unpalatable Trinidadian butterflies *H. megara* (from the family Ithomiidae) and *L. ceres* (from the family Danaidae) resemble one another – they are Müllerian mimics.
Source: After Brower (1969)

(*Eristalis tenax*) has warning coloration much like a honeybee's (*Apis mellifera*). The advantage in such mimicry is avoidance by would-be predators mistaking it for a bee with a sting in its tail. The dronefly and other mimics are thus 'sheep in wolf's clothing', bluffing their way through life on the strength of masquerade. Batesian mimicry is quite common in the animal kingdom. Many species of harmless snakes mimic poisonous snakes – harmless and poisonous coral snakes in central America are so similar that only an expert can tell them apart. Batesian mimicry is disadvantageous to the poisonous 'model' because predators will eat harmless mimics and so take longer to learn to avoid the model's warning colours. So, technically speaking, it is a brand of predation.

Mutualism

In a mutualistic relationship, both populations benefit from interacting. However, they are dependent on their interaction to survive. **Mutualism** is far less common than protocooperation, of which it is a more extreme form. It is the evolutionary equivalent of 'putting all your eggs in one basket' – the reliance of one species on another is complete. Certain Australian termites, for example, cannot produce enzymes to digest the cellulose in wood. They exploit wood as a food source by harbouring a population of protozoans (*Myxotricha paradoxa*) capable of making cellulose-digesting enzymes in their guts. The protozoans are **intestinal endosymbionts**. Neither the termites nor the protozoans would survive without the other. One generation of termites passes on the protozoans to the next by exchanging intestinal contents.

Lichens are thought to depend on a mutualistic alliance between a fungus and an alga – the fungus provides the shell, the alga the photosynthetic powerhouse. However, the algae in some lichens can be grown without their fungal host, so the relationship may in some cases be protocooperation.

Mutualism is also known in marine environments. Two encrusting sponges (*Suberites rubrus* and *S. luridus*) protect the queen scallop (*Chlamys opercularis*) from predation by a starfish, the common starfish (*Asterias rubens*) (Pond 1992). They probably do so by making it difficult for the common starfish to grip the scallop with their tube-like feet, and by excluding other organisms settling on the scallop, so hindering its mobility. The sponge benefits from the association by being afforded protection from the sea lemon (*Archidoris pseudoargus*), a shell-less nudibranch-gastropod predator, and more generally by being carried to favourable locations.

Commensalism

Commensalism occurs when one population (the *commensal*) benefits from an interaction and the other population (the *host*) is unharmed. Bromeliads and orchids grow as epiphytes on trees without doing any harm. The house mouse (*Mus musculus domesticus*) has lived commensally with humans since the first permanent settlements were established in the Fertile Crescent (a sweep of well-watered land in Mesopotamia, Syria, and Palestine in which early Near Eastern civilizations flourished) (Boursot *et al.* 1993). The house mouse benefits from the food and shelter inadvertently provided by humans, while the humans suffer no adverse effects. Four other examples are as follows.

Cattle egrets and cattle

Cattle egrets (*Bubulcus ibis*), which are native to Africa and Asia, follow cattle grazing in the Sun, pouncing on crickets, grasshoppers, flies, beetles, lizards, and frogs flushed from the grass as the cattle approach (Plate 11.1). The cattle are unaffected by the birds, while the birds appear to benefit – they feed faster and more efficiently when associated with the cattle. Interestingly, the cattle egret has expanded its range, successfully colonizing Guyana, South America, in the 1930s. It filled a niche created by extensive forest

Plate 11.1 Cattle egrets (*B. ibis*). Photograph by Pat Morris.

clearance. Without large herbivores to stir up its meal, it has taken to feeding in the company of domestic livestock, which also disturbs insects and other prey living in grass.

Midge and mosquito larvae in pitcher-plant pools

In Newfoundland, Canada, the carnivorous common pitcher plant (*Sarracenia purpurea*) has water-filled leaves holding decaying invertebrate carcasses (S. B. Heard 1994). Larvae of the pitcher-plant midge (*Metriocnemus knabi*) and the pitcher-plant mosquito (*Wyeomyia smithii*) feed on the carcasses. The two insect populations are limited by the carcass supply, but they use the carcasses at different stages of decay and live commensally. The midges feed by chewing upon solid materials, while the mosquitoes filter-feed on tiny bits that break off the decaying matter and on bacteria. The mosquitoes' eating of particles has no effect on the midges, but midges feeding on solid material creates a supply of organic fragments and, by increasing the surface area of decaying matter, bacteria.

Mynas and king-crows

In India, two bird species, the common myna (*Acridotherus tristis*) and the jungle myna (*A.*

fuscus), forage in pure and mixed flocks on fallow land (Veena and Lokesha 1993). Another bird species, the king-crow (*Dicrurus macrocercus*), a drongo, feeds on insects disturbed by the foraging mynas. King-crows tend to consort with larger myna flocks, comprising 20 or more individuals. The king-crows and mynas eat different foods, so they do not compete. The king-crows benefit from the association, whereas the mynas are unaffected – a case of commensalism.

River otters and beavers

In the boreal forest of northeast Alberta, Canada, a partial commensalism has evolved between river otters (*Lutra canadensis*) and beavers (*Castor canadensis*) (Reid *et al*. 1988; Plate 11.2). In winter, otters dig passages through the beaver dams to guarantee underwater access to adjacent water bodies, which are frozen over. The passages lead to a reduction in pond water-level, which may increase the otters' access to air under the ice and concentrate the fishes upon which the otters feed. Beavers repair the passage during times of open water, so otters benefit from the creation and maintenance of fish habitats by the beavers, who themselves are unaffected by the otters' activities.

(a)

(b)

Plate 11.2 (a) Canadian beaver (*C. canadensis*). (b) Beaver dam and lodge, Rocky Mountains National Park, Colorado, USA.

Photographs by Pat Morris.

STAYING APART: COMPETITION

While some species cooperate with one another, others compete for food or territory.

Competing for resources

Competition is a form of population interaction in which both populations suffer. It occurs when two or more populations compete for a common resource, chiefly food or territory, that is limited. It may engage members of the same species (**intraspecific competition**) or members of two or more different species (**interspecific competition**). The *limiting resource* prevents one or both the species (or populations) from growing.

Competitive exclusion

The outcome of competing for a limiting resource is not obvious and depends upon a variety of circumstances. It would be tempting to think either that both species may coexist, though at reduced density, or else that one species may displace the other. However, coexistence is exceedingly unlikely. As a very strict rule, no two species can coexist on the same limiting resource. If they should try to do so, one species will outcompete the other and cause its extinction. *Gause's principle*, named after Georgii Frantsevich Gause, encapsulates this finding. Also called the *competitive exclusion principle* (Hardin 1960), Gause's principle was first established in mathematical models by Alfred James Lotka (1925) and Vito Volterra (1928), and later in laboratory experiments by Gause (1934).

Gause used a culture medium to grow two *Paramecium* species – *P. aurelia* and *P. caudatum*. Each population showed a logistic growth curve when cultured separately. When grown together, they displayed a logistic growth curve for the first six to eight days, but the *P. caudatum* population slowly fell and the *P. aurelia* population slowly rose, though it never attained quite so high a level as when it was cultured separately. Many other laboratory experiments using various competing populations give comparable results. Competitors used include protozoa, yeasts, hydras, *Daphnia*, grain beetles, fruit flies, and duckweed (*Lemna*).

Scramble competition

Competition between populations occurs in two ways – *scramble competition* and *contest competition*. First, scramble, exploitation, or resource competition occurs when two or more populations use the same resources (food or territory), and when those resources are in short supply. This competition for resources is indirect. One population affects the other by reducing the amount of food or territory available to the others.

A classic example of scramble competition concerns the native red squirrel (*Sciurus vulgaris*) and the introduced grey squirrel (*S. carolinensis*) in England, Wales, and Scotland (Figure 11.2; Plate 11.3). The red squirrel favours coniferous forest, but also lives in deciduous woodland, particularly beech woods. The grey squirrel, a native of eastern North America, is adapted to life in broad-leaved forests, but is adaptable and will colonize coniferous forests. In Britain, the red squirrel's range progressively shrunk during the twentieth century as the grey squirrel, introduced from America in the late nineteenth and early twentieth centuries, has extended its range. A dramatic decline in red squirrel range occurred from 1920 to 1925, largely due to disease following great abundance (Shorten 1954). The grey squirrel rapidly advanced into the areas relinquished by the red squirrel, and into areas around their introduction sites. Up to around 1960, the red squirrel still had a few strongholds in the southern parts of England, especially East Anglia and the New Forest. The grey squirrel has advanced into these areas and replaced the red squirrel in much of East Anglia (H. R. Arnold 1993). Red squirrels still thrive on the Isle of Wight and Brownsea Island in Poole Harbour, Dorset.

The superior adaptability to the herbivore niche at canopy level in deciduous woodland may account for the success of the grey squirrel (C. B. Cox and Moore 1993, 61). However, a recent study highlights the complexity of squirrel competition (Kenward and Holm 1993). Conservative demographic traits combined with feeding competition could explain the red squirrel's replacement by grey squirrels. In oak–hazel woods, oak (*Quercus* spp.) and acorn abundance largely determine grey squirrel foraging, density, and productivity. In contrast, red squirrels forage where hazels (*Corylus avellana*) are abundant, and low hazelnut abundance seems to account for their relatively low density and breeding success. Red squirrels fail to exploit good acorn crops, although acorns are more abundant than hazels. However, in Scots pine (*Pinus sylvestris*) forests, their density and their breeding success is as high as grey squirrels' in deciduous woods. Captive grey squirrels thrive on a diet of acorns, but red squirrels have a comparative digestive efficiency of only 59 per cent, seemingly because they are much less able than grey squirrels to neutralize acorn polyphenols. A model with simple competition for the autumn hazel crop, which grey squirrels eat before the acorn crop, shows that red squirrels are unlikely to persist with grey squirrels in woods with more than 14 per cent oak canopy. Oak trees in most British deciduous woods give grey squirrels a food refuge that red squirrels fail to exploit. Thus, red squirrel replacement by grey squirrels may arise solely from feeding competition, which the postwar decline in coppiced hazel made worse.

Another example is the scramble competition between the mountain hare (*Lepus timidus*) and the brown hare (*L. europaeus*) in Europe (Thulin 2003). The mountain hare is an arctic and subarctic species that lives in Ireland and the Scottish Highlands, at high altitudes in the Alps, in two isolated forests in Poland, and throughout most of Norway, Sweden, and Finland. Its range has gradually shrunk over the last 10,000 years, with declines in Scandinavia and Russia recorded during the twentieth century. The brown hare is an open landscape specialist that occupies most of the lowlands in northern and central Europe, except northern Scandinavia and the Iberian Peninsula. Competition between the two species seems likely, as the brown hare will occupy the preferred habitat of the mountain if the mountain hare is absent, and vice versa. Moreover, the mountain hare lives at high altitudes and in deep forests where the brown hare is present, and evidence suggests that, in direct contact with brown hares, mountain hares disappear from the

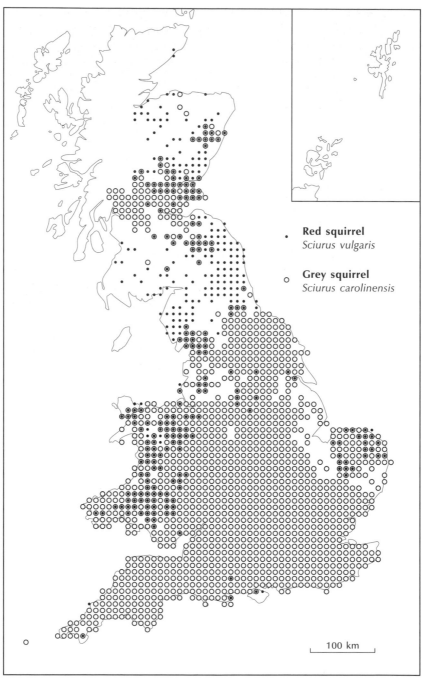

Red squirrel
Sciurus vulgaris

Grey squirrel
Sciurus carolinensis

100 km

Figure 11.2 The distributions of the native red squirrel (*S. vulgaris*) and the introduced grey squirrel (*S. carolinensis*) in England, Wales, and Scotland.
Source: Adapted from H. R. Arnold (1993)

(a)

Plate 11.3 Competing squirrels in the UK (a) Red squirrel (*S. vulgaris*). (b) Grey squirrel (*S. carolinensis*).
Photographs by Pat Morris.

(b)

optimal brown hare habitats and that brown hares have the potential to expand into mountain hare territory. Other factors may help to account for the mountain hare's decline. Hybridization between the two species leads to the loss of mountain hares, as, unlike brown hare males, mountain hare males do not appear to guard their mates. It is also possible that certain pathogens affect mountain hares more than they do brown hares, and indeed brown hares may be sub-clinical vectors of the European Brown Hare Syndrome (EBHS), which may cause instant death in mountain hares.

Contest competition

Contest or interference competition occurs through direct interaction between individuals. The direct interaction may be subtle, as when plants produce toxins or reduce light available to other plants, or overtly aggressive, as when animal competitors 'lock horns'.

Direct competition occurs in four duckweed species (*Lemna minor*, *L. natans*, *L. gibba*, and *L. polyrhiza*) (Harper 1961). Lesser duckweed (*L. minor*) grows fastest in uncrowded conditions. Under crowded conditions, when all species are grown together, lesser duckweed grows the slowest. Nutrient levels have no effects on the growth rates, so competition for light causes the changes: the duckweed species interfere with one another.

Several animal and plant species interfere with competitors by chemical means. The chemicals released have an inhibiting, or allelopathic, impact on rivals. The nodding thistle (*Carduus nutans*) is **allelopathic**, releasing soluble inhibitors that discourage the growth of pasture grasses and legumes, at two phases of its development – at the early bolting phase when larger rosette leaves are decomposing and releasing soluble inhibitors, and at the phase when bolting plants are dying (Wardle *et al.* 1993). Moreover, thistle-tissue additions to the soil seem to stimulate nodding thistle seedlings. The thistle plants may weaken pasture and, at the same time, encourage recruitment of their own kind. Crowberry (*Empetrum hermaphroditum*) inhibits the growth of Scots pine (*P. sylvestris*) and aspen (*Populus tremula*) (Zackrisson and Nilsson 1992). It releases water-soluble **phytotoxic** substances from secretory glands on the leaf surface. These toxins interfere with Scots pine and aspen seed germination on the forest floor.

Aggressive competition is not so common as is popularly believed. As a rule, organisms avoid potentially dangerous encounters. But aggression does occur. A classic example is two acorn barnacles – Poli's stellate barnacle (*Chthamalus stellatus*) and the common barnacle (*Balanus balanoides*) –

that grow on intertidal-zone rocks around Scotland. A study on the shore at Millport (Figure 11.3) showed that, when free-swimming larvae of *Chthamalus* decide to settle, they could attach themselves to rocks down to mean tide level (Connell 1961). Where *Balanus* is present, they attach only down to mean high neap tide level. During neap tides, the range between low water mark and high water mark is at its narrowest. Young *Balanus* grow much faster than *Chthamalus* larvae – they simply smother them or even prise them off the rocks. This behaviour is direct, aggressive action to secure the available space. When *Balanus* is removed, *Chthamalus* colonized the intertidal zone down to mean tide level. However, without *Chthamalus, Balanus* is unable to establish a population above mean high neap tide level, seemingly because of adverse weather (warm and calm conditions) promoting desiccation.

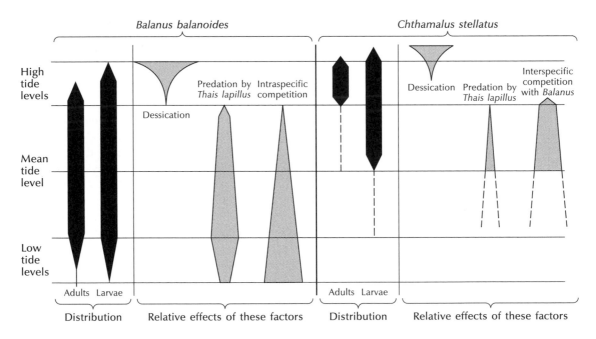

Figure 11.3 The vertical distribution of larvae and adult acorn barnacles on intertidal-zone rocks, Millport, Scotland. Two competing species are shown: *B. balanoides* and *C. stellatus. Thais lapillus* is the predatory dog whelk.

Source: Adapted from Connell (1961)

Mechanisms of coexistence

Species coexistence is the rule in nature. It depends upon avoiding competition and is achieved through several mechanisms: the environment is normally diverse and contains hiding places and food patches of various sizes; many species use a range of resources rather than one; animals have varying food preferences and commonly switch diets. The result is that competition is often diffuse: a species competes with many other species for a variety of resources, and each resource represents a small portion of the total resource requirements.

Resource partitioning

Species avoid niche overlap by partitioning their available resources according to size and form, chemical composition, and seasonal availability. Five warbler species (genus *Dendroica*) live in spruce forests in Maine, USA. They each feed in a different part of the trees, use somewhat different foraging techniques to find insects among the branches and leaves, and have slightly different nesting dates (Figure 11.4) (MacArthur 1958). The differences in feeding zones are large enough to account for coexistence of the blackburnian warbler (*D. fusca*), black-throated green

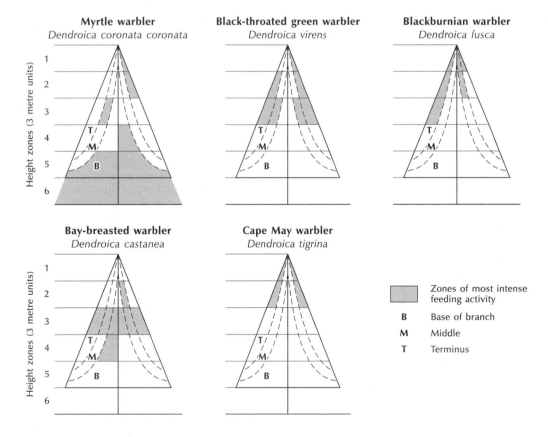

Figure 11.4 Warblers in spruce forests, Maine, USA.
Source: Adapted from MacArthur (1958)

warbler (*D. virens*), and bay-breasted warbler (*D. castanea*). The Cape May warbler (*D. tigrina*) is different. Its survival depends upon outbreaks of forest insects to provide a glut of food. The myrtle warbler (*D. coronata coronata*) is not so common as the other species and is less specialized.

Some species partition resources solely according to prey size. Among such predatory species as hawks, larger species normally eat larger prey than smaller species. In North America, the goshawk (*Accipiter gentilis*) eats prey weighing up to 1,500 g (Storer 1966). The smaller Cooper's hawk (*A. cooperi*) takes prey almost as large, but only very occasionally. It mainly eats prey in the 10–200 g range. The smaler-still sharp-shinned hawk (*A. striatus*) will eat prey up to 100 g, and sometimes a little more, but most of its prey is in the 10–50 g range. In these species, the goshawk is 1.3 times longer than Cooper's hawk, which is itself 1.3 times longer than the sharp-shinned hawk.

Character displacement

Competition between two species may be avoided by evolutionary changes in both. Two predators may eat the same size prey and therefore, all other factors being constant, they are competitors. Under these circumstances, it would not be uncommon for one species gradually to tackle slightly large prey and the other species to tackle slightly smaller prey. The two species would thus diverge by evolutionary changes. For example, in Israel, small members of the cat family show *character displacement* in the diameter of the upper canine teeth (Dayan *et al*. 1990). In each cat species, the ratio of the male upper canine is constant within and between each felid species. This means that each sex of each cat species is a separate morphospecies. The size differences in the canines probably relate to the spacing of the vertebrae in prey species, which is related to prey size. In consequence, competition between sexes and between species is driving resource partitioning in prey size. In central Brazil, maned wolves (*Chrysocyon brachyurus*), crab-eating foxes

(*Cerdocyon thous*), and hoary foxes (*Lycalopex vetulus*) live sympatrically (Juarez and Marinho-Filho 2002). The hoary foxes are frugivore–insectivores and show little overlap of food items with maned wolves and crab-eating foxes. The more generalist feeding maned wolves and crab-eating foxes do have overlapping diets, but maned wolves tend to eat larger prey than the crab-eating foxes.

Evolutionary divergence can only be inferred from the phenomenon of character displacement when a species' appearance or behaviour differs when a competitor is present from when it is absent. An example is afforded by the beak size of ground finches (Geospizinae) living on the Galápagos Islands (Figure 11.5) (Lack 1947). Abingdon and Bindloe Islands have three species of *Geospiza* that partition seed resources according to size – the large ground finch (*G. magnirostris*), the medium ground finch (*G. fortis*), and the small ground finch (*G. fuliginosa*). *G. magnirostris* has a large beak adapted to husk large seeds that smaller finches would fail to break open. *G. fuliginosa* has a small beak that can husk smaller seeds more efficiently than the larger species, *G. fortis*. *G. fortis* feeds on intermediate-sized seeds. Now, *G. magnirostris* does not reside on Charles or Chatham Islands. On these islands, *G. fortis* tends to have a heavier beak, on average, than on Abingdon or Bindloe Islands. On Daphne Island, where *G. fortis* lives without *G. fuliginosa*, its beak is intermediate in size between the two species on Charles and Chatham Islands. On Crossman Islands, *G. fuliginosa* lives without the other two finches, and its beak is intermediate in size there. The habitats on all the islands are very similar, and the explanation for this divergence in beak size is that competition has caused character displacement.

Spatial competition

All organisms, and particularly terrestrial plants and other sessile species, interact mainly with their neighbours. Neighbourhoods may differ considerably in composition, largely because of

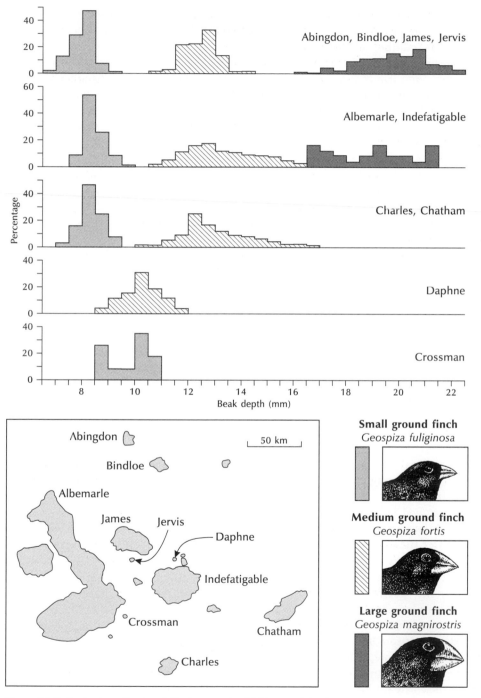

Figure 11.5 Beak size of ground finches on the Galápagos Islands.
Source: Adapted from Lack (1947)

colonization limitation (propagules have not yet arrived at some sites). In areas of apparent uniformity, there is a great number of 'botanical ghettos'. This spatial variety enables more species to coexist than in a uniform environment. Traditional competition theory holds that there can be no more consumer species than there are limiting resources. The *spatial competition hypothesis* suggests that neighbourhood competition, coupled with random dispersal among sites, allows an unlimited number of species to survive on a single resource (Tilman 1994). Such rich coexistence occurs because species with sufficiently high dispersal rates persist in sites that are unoccupied by superior competitors.

The spatial competition hypothesis appears to explain the coexistence of numerous grassland plant species in the Cedar Creek Natural History Area, Minnesota, USA. The grassland plants compete strongly below ground for nitrogen, which is the only limiting resource. The best competitor is the little bluestem (*Schizachyrium scoparium*), a grass. According to classical competition theory, as there is just one limiting resource (soil nitrogen), this species should outcompete all others and take over. In garden plots it does just that. But in the natural grassland it shares the environment with over 100 other species. The answer to this enigma may lie in the relative energy allocation for root growing and for reproduction. Species investing in root growth are good competitors where nitrogen levels are limiting, but they are also poor dispersers. Species investing in reproduction, such as the ticklegrass (*Agrostis scabra*) and the quackgrass (*Agropyron repens*), though not the best nitrogen competitors, are good dispersers. So, the 'good-dispersers-but-poor-competitors' invade abandoned fields immediately, and are not dominated by the 'slow-dispersers-but-good-competitors' until 30 or 40 years later. It is possible, therefore, that their poor colonizing abilities prevent the superior competitors from occupying the entire landscape, and that this provides sites in which numerous species of inferior competitors can persist.

VEGETARIAN WILDLIFE: HERBIVORY

Three forms of population interaction benefit one species but are detrimental to the other. *Predation* occurs when a predatory population (the winner) exploits a prey population (the loser). Normally, a predator has a seriously harmful effect on its prey – it kills it and eats it, but not necessarily in that order! This is classic carnivory. Plants lose leaves to herbivores. Herbivory is thus a form of predation. *Parasitism* is a relatively mild form of predation, though it may still lead to death. Parasitism is like predation but the host (a member of the population adversely affected), rather than being consumed immediately, is exploited over a period. *Amensalism* occurs when one population is adversely affected by an interaction but the other species is unaffected.

Plant eaters

Types of herbivore

Herbivores are animals that eat live plants. *Herbivory* is a special kind of predation where plants are the prey. All parts of a plant – flowers, fruits, seeds, sap, leaves, buds, galls, herbaceous stems, bark, wood, and roots – are eaten by something, but no one herbivore eats them all. There are several specialized herbivore niches – grazers, browsers, grass, grain, or seed feeders (graminivores), grain and seed eaters (granivores), leaf eaters (folivores), fruit and berry eaters (frugivores), nut eaters (nucivores), nectar eaters (nectarivores), pollen eaters (pollenivores), and root eaters. Detritivores eat dead plant material and will be discussed in the next chapter.

A vast army of invertebrate herbivores munches, rasps, sucks, filters, chews, and mines its way through the phytosphere. A smaller but potent force of vertebrate herbivores grazes and browses its way through the phytosphere (Table 11.2). Some mammalian herbivore niches are unusual. The giant panda (*Ailuropoda melanoleuca*)

Table 11.2 Herbivorous tetrapods

Order	Name	Herbivores within Order	Example	Example of plant tissue eaten
Amphibians				
Anura	Frogs, toads	Few	Common frog (*Rana temporaria*) tadpoles	Water weed
Reptiles				
Chelonia	Tortoises and turtles	Some	Giant tortoise (*Testudo gigantea*)	Grasses, sedges
Squamata	Snakes and lizards	Few	Marine iguana (*Amblyrhynchus cristatus*)	Seaweeds
Birds				
Anseriformes	Ducks and geese	Many	Southern screamer (*Chauna torquata*)	Aquatic plants, grasses, and seeds
Falconiformes	Eagles and hawks	Few	Palm-nut vulture (*Gypohierax angolensis*)	Palm nuts, fish, molluscs, crabs
Galliformes	Game birds	Most	Red grouse (*Lagopus lagopus*)	Heather shoots, insects
Columbiformes	Pigeons	All	Wood pigeon (*Columba palumbus*)	Legume leaves, seeds
Psittaciformes	Parrots	Most	Blue-and-yellow macaw (*Ara ararauna*)	Fruits, kernels
Apodiformes	Hummingbirds and swifts	Some	Sword-billed hummingbird (*Ensifera ensifera*)	Nectar, insects
Passeriformes	Perching birds	Many	Plantcutter (*Phytotoma rutila*)	Fruit, leaves, shoots, buds, seeds
Mammals				
Marsupialia	Kangaroos, etc.	Many	Honey possum or noolbender (*Tarsipes spencerae*)	Nectar, pollen, some insects
Chiroptera	Bats	Few	Long-tongued bat (*Glossophaga soricina*)	Fruit, nectar, some insects
Dermoptera	Flying lemurs	All	Flying lemurs (*Cynocephalus* spp.)	Leaves, buds, flowers, fruit
Edentata (Xenarthra)	Sloths and armadillos	Some	Three-toed sloths (*Bradypus* spp.)	Leaves, fruit
Primates	Apes, monkeys, and lemurs	Most	Human (*Homo sapiens*)	Everything except wood
Rodentia	Voles, mice, squirrels, etc.	Most	Alpine marmot (*Marmota marmota*)	Grass, plants, roots
Lagomorpha	Rabbits, hares, and pikas	All	Mountain hare (*Lepus timidus*)	Fine twigs, sprigs of bilberry and heather
Carnivora	Dogs, cats, bears, etc.	Few	Giant panda (*Ailuropoda melanoleuca*)	Bamboo
Hyracoidea	Hyraxes (conies)	All	Rock hyraxes (*Heterohyrax* spp.)	Grass
Proboscidea	Elephants	All	African elephant (*Loxodonta africana*)	Grass, leaves, twigs
Sirenia	Sea cows	All	Manatee (*Trichechus manatus*)	Aquatic plants
Perissodactyla	Odd-toed ungulates[a]	All	Brazilian tapir (*Tapirus terrestris*)	Succulent vegetation, fruit
Artiodactyla	Even-toed ungulates[b]	All	Goats (*Capra* spp.)	Everything

Source: Partly adapted from Crawley (1983)
Notes: a Horses and zebras, tapirs, rhinoceroses. b Pigs, peccaries, hippopotamuses, camels, deer, cows, sheep, goats, antelope

is a purely herbivorous member of the Carnivora. The palm-nut vulture (*Gypohierax angolensis*) is the only herbivorous member of the Falconiformes. It eats nuts of the oil palm (*Elaeis guinensis*) and raphia palm (*Raphia ruffia*), as well as fish, molluscs, and crabs. Even invertebrate niches can be surprising. An adult tiger beetle (*Cicindela repanda*), a common and widespread water-edge species, was recently seen feeding on fallen sassafras fruits lying on a Maryland beach, USA (Hill and Knisley 1992). This was the first report of **frugivory** by a tiger beetle, which may be an opportunistic frugivore, especially in the autumn just after emergence, when fruits would provide a valuable energy resource before **overwintering**.

Browsers and **grazers** are chiefly mammals, but a few large reptiles and several bird species occupy this niche, too. Some of the geese and goose-like waterfowl crop grass. The South American hoatzin (*Opisthocomus hoazin*), a highly aberrant cuckoo, mainly eats the tough rubbery leaves, flowers, and fruits of the tall, cane-like water-arum (*Montrichardia*) and the white mangrove (*Avicenna*). The curious New Zealand kakapo (*Strigops habroptilus*) is a ground-dwelling parrot. It extracts juices from leaves, twigs, and young shoots by chewing on them and detaching them from the plant, or it fills its large crop with browse and retires to a roost where it chews up the plant material, swallows the juices, and 'spits out' the fibre as dry balls. Members of the grouse family (Tetraonidae) browse on buds, leaves, and twigs of willows and other plants of low nutritive value. Their digestive systems house symbiotic bacteria that digest the vegetable materials and are comparable to the digestive system of ruminants. A few birds eat roots and tubers. Pheasants dig into the soil with their beaks, while others scratch with their feet. Some cranes, geese, and ducks are adapted for obtaining the roots, rhizomes, and bulbous parts of aquatic plants.

Grain, seed, nectar, fruit, and nut eating are specialities of many birds. Sometimes, the relationship between a fruit-producing plant and its 'predator' is close (see Box 11.1). Many mammals eat fruit, but very few are specialized fruit eaters. An exception is the fruit-eating bats, which include nearly all members of the Pteropodidae. These animals have reduced cheek teeth as an adaptation for their fluid diet.

Box 11.1

THE HAWTHORN AND THE AMERICAN ROBIN: A FRUIT–FRUGIVORE SYSTEM

European hawthorn (*Crataegus monogyna*) grows in western Oregon, USA. Only one frugivore, the American robin (*Turdus migratorius*), forages on the fruits, making this an unusually straightforward fruit–frugivore system (Sallabanks 1992). The fruits carry seeds that are dispersed by the American robin, who bears them away from the parent bush. Dispersal efficiency is low. An average 21 per cent of seeds are dispersed each year. Most of the fruits simply fall to the ground. Robins dropped 20 per cent of the fruits that they picked. They defecated or regurgitated 40 per cent of the swallowed fruits (seeds) beneath parent bushes. Bushes with more fruit were visited more frequently, had a greater dispersal success (more seeds were dispersed), and had seeds dispersed more efficiently (a greater proportion of seeds was dispersed, and the seeds were more successful in propagating). The optimal fruiting strategy for the European hawthorn is, therefore, to grow as big as possible as quickly as possible by delaying fruiting until later in life.

Defence mechanisms in plants

As prey species, plants have a distinct disadvantage over the animal herbivore counterparts – they cannot move and escape from predators; they have to stand firm and cope with herbivore attack as best they may. The world is green, so obviously plants can survive the ravages of plant-eating animals. They do so in two main ways. First, some herbivore populations evolve self-regulating mechanisms that prevent their destroying their food supply; or other mechanisms, particularly predation, may hold herbivore numbers in check. Second, all that is green may not be edible – plants have evolved a battery of defences against herbivores.

Plant defences are formidable. Some discourage herbivores by structural adaptations such as thorns and spikes. Some engage in chemical warfare, employing by-products of primary metabolic pathways. Many of these chemical by-products, including terpenoids, steroids, acetogenins (juglone in walnut trees), phenylpropanes (in cinnamon and cloves), and alkaloids (nicotine, morphine, caffeine) are distasteful or poisonous and are very effective deterrents (see Larcher 1995, 21).

Herbivores have developed ways of sidestepping plant defence systems. Some have evolved enzymes to detoxify plant chemicals. Others time their life cycles to avoid the noxious chemicals in the plants. Two examples will illustrate these points – cardiac glycosides in milkweed and tannins in oaks.

Poisonous milkweed

The bloodflower (*Asclepias curassavica*), a milkweed, is abundant in Costa Rica and other parts of central America. It contains secondary plant substances called cardiac glycosides (or cardenolides) that affect the vertebrate heartbeat and are poisonous to mammals and birds (Brower 1969). Cattle will not eat the milkweed, even though it grows abundantly in grass. They are wise not to do so, for it causes sickness and occasionally death. However, certain insects, including the larvae of danaid butterflies (Danainae), eat it without any deleterious effects. The danaid butterflies are distasteful to insect-eating birds and serve as models in several mimicry complexes (p. 189). The danaids have evolved biochemical mechanisms for eating the milkweed and storing the poison in their tissues. Thus, they acquire protection from predators from the plants that they eat.

Unpalatable oak-leaves

The common oak (*Quercus robur*) in western Europe is attacked by the larvae of over 200 species of butterflies and moths (Feeny 1970). It protects itself against this massive assault by using tannins for chemical and structural defences. The tannins lock proteins in complexes that insects cannot digest and utilize, and they also make leaves tough and unpalatable. The herbivorous attackers partly get round these defences by concentrating feeding in early spring, when the leaves are young and, because they contain less tannin, less tough. They also alter their life cycles in summer and autumn – many late-feeding insects overwinter as larvae and complete their development on the soft and tasty spring leaves. Some insect species do feed on oak leaves in summer and autumn. These tend to grow very slowly, which may be an adaptation to a low-nitrogen diet when proteins are locked away in older leaves.

Herbivore and plant interactions

Herbivores interact with plants in two chief ways. First, in *non-interactive herbivore–plant systems*, herbivores do not affect vegetation growth, even when there are very large numbers of them. Second, in *interactive herbivore–plant systems*, herbivores affect vegetation growth. Two examples will illustrate these different systems – seed predation and grazing.

Seed predation

Herbivores that feed upon plant seeds (seed preda-tors) do not normally influence plant growth. Birds eating plant seeds have no effect on plant growth, though they could influence the long-term survival of a plant population by eating a large portion of the annual seed production. On the other hand, plants do influence herbivore pop-ulations – seed predator populations will be low in times of seed shortages. British finches feed either on herb seeds or on tree seeds. The herb-seed feeders have stable populations, whereas the tree-seed feeders have fluctuating populations (Newton 1972). Herbs are usually consistent in the annual number of seeds produced, but tree-seed production is variable and, in some cases, irregular. The North American piñon pine (*Pinus edulis*) illustrates variable tree-seed production. The piñon pine yields a heavy cone crop at irreg-ular intervals. The crop satiates its invertebrate seed and cone predators. It also allows efficient for-aging by piñon jays (*Gymnorhinus cyanocephalus*), which disperse the seeds and store them in sites suitable for germination and seedling growth (Ligon 1978). Many other trees are mast fruiters. In mast fruiting, all trees of the same species within a large area produce a big seed crop in one year. They then wait a long time before produc-ing anything other than a few seeds. The advan-tage of such synchronous seeding is that seed predators are satiated and are unlikely to eat the entire crop before germination has begun.

Grazing

In interactive herbivore–plant systems, the herbi-vores affect the plants, as well as the plants affecting the herbivores. As a rule, plants have more impact on herbivore populations than herbi-vores have on plant populations – plant abund-ance, quality, and distribution greatly affect herbivore numbers. Nonetheless, herbivory can affect plant populations. This is the case where vegetation is grazed. Sheep, for example, are 'woolly lawnmowers' that maintain pasture. The effects of grazing on vegetation are plainest to see where a fence separates grazed and ungrazed areas. The grazed area consists mainly of grasses, while shrubs and tall herbs dominate the ungrazed area. Given enough time, the ungrazed area would revert to woodland. Domestic live-stock sometimes has such a large influence on plant communities that the herbivore may be identified from the vegetation it produces (Crawley 1983, 295). In the UK, dock (*Rumex obtusifolius*) is abundant in horse pastures; silver-weed (*Potentilla anserina*) is common where geese graze. Spiny shrubs and aromatic, sticky herbs flourish where goats graze in Mediterranean **maquis**.

Patterns of herbivore–plant interaction are complex and generalization is difficult. Population models, similar to those used to study predator–prey relationships, suggest some basic possibilities (see p. 156).

HUNTER AND HUNTED: CARNIVORY

Flesh eaters

Carnivores are animals that eat animals. *Carnivory* is predation where animals are the prey. It includes predators eating herbivores as well as top predators eating predators. A predator benefits from interaction with a prey – it gets a meal; the prey does not benefit from interaction with a predator – it normally loses its life. Rare excep-tions are lizards that lose their tails to predators but otherwise survive attacks. Predation is plain to see and easy to study. It is evident when one animal eats another, often in a gory spectacle.

Predators come in a variety of shapes and sizes. Highlights from a predator catalogue might be the 'small but deadly' – the least weasel (*Mustela nivalis*) is the smallest carnivore ever – and the 'large and equally deadly' – the African lion (*Panthera leo*). Some fossil carnivores were awesome in size and possibly in speed. The most famous examples, even before their appearance in Jurassic Park, are the dinosaurs *Tyrannosaurus rex* and *Velociraptor*. But more recent avian predators were just as terrifying. The phorusrhacoids were a group of large, flightless, flesh-eating birds (L. G. Marshall 1994). They lived from 62 million to about 2.5 million years ago in South America and became the dominant carnivores on that continent. They ranged in height from 1 to 3 m, and were able to kill and eat animals the size of small horses.

Carnivorous specialization

Predators eat all animals, except those at the very top of food chains. The zoological menu is vast, but predators tend to specialize. Carnivore niches include general flesh eaters (faunavores), fish eaters (piscivores), blood feeders (haematophages or sanguivores), egg eaters (ovavores), insect eaters (insectivores), ant eaters (myrmecovores), and coral eaters (corallivores). Omnivores are unspecialized carnivores-cum-herbivores. **Scavengers** feed on dead animal remains. Coprophages feed on dung or faeces.

A few carnivorous niches are highly specialized and rather unusual. There are few *sanguivores* (blood eaters); the most notorious (apart from Count Dracula) are the New World vampire bats. The sharp-beaked ground finch (*Geospiza difficilis*), one of the Galápagos finches, obtains some of its nourishment by biting the bases of growing feathers on boobies (*Sula* spp.) and eating the blood that oozes from the wound. *Coprophagy* (dung eating) is more widespread than might be supposed. Several birds eat animal dung. The ivory gull (*Pagophila eburnea*) feeds on the dung dumped on ice by polar bears, walruses, and seals. Puffins and petrels eat whale dung floating on the water. Certain vultures and kites feed solely on human and canine faeces around native villages. *Myrmecophagy* (ant eating) is found in several species in Australia, South America, Africa, and Eurasia that are adapted to feed upon ants and

termites (Figure 11.6). Teeth, jaws, and skulls possess extreme adaptations for this diet. The teeth reduce to flat, crushing surfaces, as in the aardvark (*Orycteropus afer*), or they are missing, as in the giant or great anteater (*Myrmecophaga tridactyla*). Tongues are long, mobile, and worm-like, with uncommon protrusibility ('stick-out-ability'). Enlarged salivary glands produce a thick, sticky secretion that coats the tongue. Ant-eating mammals also tend to have elongated snouts and powerful forelimbs for digging and ripping open nests.

Some plants are carnivorous. Indeed, about 600 species of carnivorous plants grow worldwide and the carnivorous habit appears to have evolved independently in at least six angiosperm subclasses (Ellison *et al*. 2003). Plants have evolved four methods of trapping their prey: adhesive traps, snapping traps, pitcher plants, and suction traps. Sundews, such as the British great sundew (*Drosera anglica*), have *adhesive traps*. Plants with active traps that snap shut when trigger hairs are touched – *snapping traps* – include the Venus flytrap (*Dionaea muscipula*), native to the Carolinas in the USA. *Pitcher plants* occur worldwide. Those of the genus *Heliamphora* (Sarraceniaceae) in the Guayana Highlands of Venezuela have carnivorous traits (Jaffe *et al*. 1992). *H. tatei* is a true carnivore. It attracts prey through special visual and chemical signals, traps and kills prey, digests prey through secreted enzymes, contains commensal organisms, and has wax scales that aid prey capture and nutrient absorption. Other species of *Heliamphora* possess all these traits except that they seem neither to secrete proteolytic enzymes nor to produce wax scales. The pattern of the carnivorous syndrome among *Heliamphora* species suggests that the carnivorous habit has evolved in nutrient-poor habitats to improve the absorption of nutrients by capturing mainly ants. Only *H. tatei* further evolved mechanisms for rapid assimilation of organic nutrients and for capturing a variety of flying insects. The carnivorous traits are lost in low light conditions, indicating that nutrient

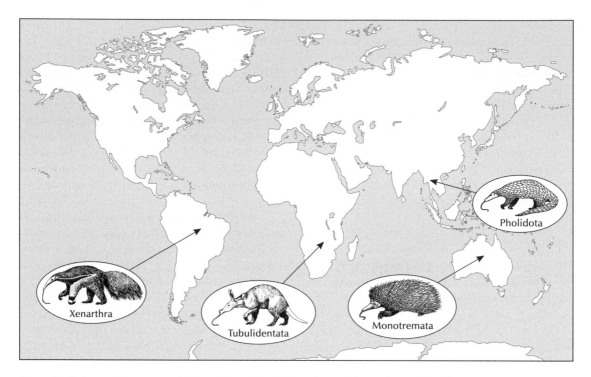

Figure 11.6 Ant-eating mammals from four orders: Xenarthra – all members of the family Myrmecophagidae, which includes the giant or great anteater (*M. tridactyla*); Tubulidentata – the aardvark (*O. afer*); Pholidota – eight species of pangolin (*Manis*); and Monotremata – the Australian and New Guinean spiny anteaters or echidnas, *Tachyglossus aculeatus* and *Zaglossus bruijni*. The numbat (*Myrmecobius fasciatus*), which is not illustrated, is an Australian marsupial anteater.

Source: Animal drawings adapted from Rodríguez de la Fuente (1975)

supply is limiting only under fast growth. Finally, the bladderworts (*Utricularia*), which have a worldwide distribution, bear *suction traps*. The greater bladderwort (*U. vulgaris*) in Britain is an example. Bladderworts have water-filled bladders comprising a thin-walled (two cells thick) sac with an inwards-opening door. Small animals, such as crustaceans, touching the sensitive hair near the door, cause the bladder to 'fire', producing an inrush of water as the door opens that carries the animals inside.

Prey switching

The switching of prey follows the principle of optimal foraging. Predators select the prey that provides the largest net energy gain, considering the energy expended in capturing and consuming prey. This usually means the prey that happened to be most abundant at the time. It also means that predators normally take very young, very old, or physically weakened prey. Some carnivores show local feeding specializations within their geographical range. In Mediterranean environments, the European badger (*Meles meles*), a carnivore species with morphological, physiological, and behavioural traits proper to a feeding generalist, prefers to eat the European rabbit (*Oryctolagus cuniculus*) (Martin *et al.* 1995). It will eat other prey according to their availability when the rabbit kittens are not abundant. The badger is a poor hunter, and its ability to catch relatively immobile baby rabbits may explain the curious specialization.

Prey selection

Where several carnivores compete for the same range of prey species, coexistence is possible if each carnivore selects a different prey. In the tropical forests of Nagarahole, southern India, there is a wide range of large mammalian prey species, mainly ungulates and primates, which are eaten by the tiger (*Panthera tigris*), the leopard (*P. pardus*), and the dhole (*Cuon alpinus*) (Karanth and Sunquist 1995). The three predators show significant selectivity among prey species. Tigers prefer gaur (*Bos gaurus*), whereas wild pig (*Sus scrofa*) is underrepresented in leopard diet, and the langur or leaf monkey (*Presbytis entellus*) is underrepresented in dhole diet. Tigers selected prey weighing more than 176 kg, whereas leopard and dhole focused on prey in the 30–175 kg size class. Average weights of principal prey killed by tiger, leopard, and dhole were, respectively, 91.5 kg, 37.6 kg, and 43.4 kg. Tiger predation was biased towards adult males in chital or axis deer (*Axis axis*), sambar (*Cervus unicolor*), and wild pig, and towards young gaur. Dholes selectively preyed on adult male chital, whereas leopards did not. If there is choice, large carnivores selectively kill large prey, and non-selective predation patterns reported from other tropical forests may be the result of scarcity of large prey. Because availability of prey in the appropriate size classes is not a limiting resource, selective predation may facilitate large carnivore coexistence in Nagarahole.

Carnivore communities

Carnivore communities normally consist of predators with overlapping tastes. For this reason, they possess complex feeding relationships. A dry tropical forest in the 2,575-km² Huai Kha Khaeng Wildlife Sanctuary, Thailand, contained 21 carnivore species from 5 families (Rabinowitz and Walker 1991). The carnivores fed on at least 34 mammal species, as well as birds, lizards, snakes, crabs, fish, insects, and fruits. Nearly half the prey identified in large carnivore faeces, which was mainly deposited by Asiatic leopard (*Panthera pardus*) and clouded leopard (*Neofelis nebulosa*), was barking deer (*Muntiacus muntjak*), with sambar deer (*Cervus unicolor*), macaques (*Macaca* spp.), wild boar (*Sus scrofa*), crestless Himalayan porcupine (*Hystrix hodgsoni*), and hog badger (*Arctonyx collaris*) being important secondary prey items. Murid rodents, especially the yellow rajah-rat (*Maxomys surifer*) and the bay bamboo-rat (*Cannomys badius*), accounted for a third of identified food items in small carnivore faeces. Non-mammal prey accounted for just over a fifth, and fruit seeds for an eighth, of all food items found in small carnivore faeces.

Predators and prey

Commonly, predator and prey populations evolve together. Predators better able to find, capture, and eat prey are more likely to survive. Natural selection tends to favour those hunter's traits within predator populations. In turn, predation pushes the prey population to evolve in a direction that favours individuals good at hiding and escaping. Over thousands of generations, evolutionary forces acting upon predator and prey populations lead to elaborate and sophisticated adaptations. These features are seen in the social hunting behaviour of lions and wolves, the folding fangs and venom-injecting apparatus of viperine snakes, and spiders and their webs.

Predator–prey cycles

Several pairs of predators and prey appear to display cyclical variations in population density. The pairs include: sparrow–hawk (Europe), muskrat–mink (central North America), hare–lynx (boreal North America), mule deer–mountain lion (Rocky Mountains), white-tailed deer–wolf (Ontario), moose–wolf (Isle Royale, Michigan), caribou–wolf (Alaska), and white sheep–wolf (Alaska).

The rationale behind *predator–prey cycles* is deceptively straightforward. Consider a population of weasels, the predator, and a population of voles, the prey. If the vole population should be large, then, with a glut of food scampering around, the weasel population will flourish and increase. As the weasel population grows, so the vole population will shrink. Once vole numbers have fallen low enough, the prey shortage will cause a reduction in weasel numbers. The vole population, enjoying the dearth of predators, will then rise again. And so the cycle continues.

Although the rationale behind predator–prey cycles is plausible, sustained oscillations in predator and prey systems are not always the result of population interactions. Canadian lynx (*Lynx canadensis*) populations show a 9–10-year peak in density (Figure 11.7). Lynx pelt records kept by Hudson's Bay Company in Canada reveal this cycle. The Canadian lynx feeds mainly upon the snowshoe hare (*Lepus canadensis*), which also follows a 10-year cycle. Field investigations suggested that the cycle in snowshoe hare density is correlated with the availability of food (the terminal twigs of shrubs and trees). The lynx population, therefore, follows swings in snowshoe hare numbers, and does not drive them.

Oscillations in predator–prey systems were modelled over 70 years ago (Lotka 1925; Volterra 1926, 1931). Refined versions of these models have produced three important findings. First, stable population cycles occur only when the prey's carrying capacity (in the presence of predators) is large. This finding goes against intuition. It is called the 'paradox of enrichment' and states that an increase in carrying capacity decreases stability. Second, the population cycles are only stable when the prey population grows faster than the predator population, or when the predators are relatively inefficient at catching their prey. Third, some predator–prey 'cycles' display the features of chaotic dynamics.

Chaos in Finnish weasel and vole populations

Long-term studies of predator (weasel) and prey (vole) dynamics in Finland have revealed population cycles of 3–5 years (Figure 11.8). A predator–prey model with seasonal effects included (by allowing weasels to breed only when the vole density exceeds a threshold) predicted population changes that closely resembled the observed changes in weasel and vole populations (Hanski *et al.* 1993). The study suggested that the cycles in vole populations are chaotic and driven by the weasel predation. Further work on boreal voles and weasels supports these findings and shows the importance of multispecies predator–prey assemblages with field-vole-type rodents as the **keystone species** (Hanski and Korpimäki 1995; Hanski and Henttonen 1996).

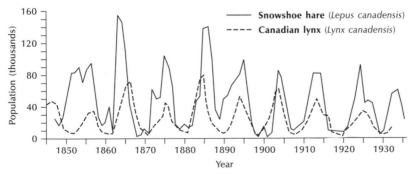

Figure 11.7 Cycles in the population density of the Canadian lynx (*L. canadensis*) and the snowshoe hare (*L. canadensis*), as seen in the number of pelts received by the Hudson's Bay Company.
Source: Adapted from MacLulich (1937)

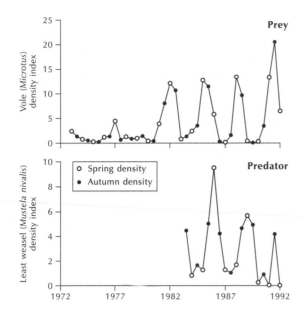

Figure 11.8 Observed cycles in predators and prey, Alajoki, western Finland. The prey is the vole (*Microtus* spp.) and the predator is the least weasel (*Mustela nivalis*).

Source: Adapted from Hanski *et al.* (1993)

Geographical effects

Laboratory experiments

Gause (1934) studied a simple predator–prey system in the laboratory (Figure 11.9). He used two microscopic protozoans, *Paramecium caudatum* and *Didinium nasutum*. *Didinium* prey voraciously on *Paramecium*. In one experiment, *Paramecium* and *Didinium* were placed in a test-tube containing a culture of bacteria supported in a clear **homogeneous** oat medium. The bacteria are food for the *Paramecium*. The outcome was that *Didinium* ate all the *Paramecium* and then died of starvation. This result occurred no matter how large the culture vessel was, and no matter how few *Didinium* were introduced. In a second experiment, a little sediment was added to the test-tube. This made the medium **heterogeneous** and afforded some refuges for the prey. The outcome of this experi-

ment was the extinction of the *Didinium* and the growth of the prey population to its carrying capacity. In a third experiment, the prey and predator populations were 'topped up' every three days by adding one *Paramecium* and one *Didinium* to the test-tube. In other words, immigration was included in the experiment. The outcome was two complete cycles of predator and prey populations. Gause concluded that, without some form of external interference such as immigration, predator–prey systems are self-annihilating.

Later, laboratory experiments using an orange → herbivorous mite → carnivorous mite food chain showed that coexistence was possible (Huffaker 1958). When the environment was homogeneous (oranges placed close together and spaced evenly), the predatory mite (*Typhlodromus occidentalis*) overate its phytophagous prey (*Eotetranychus sexmaculatus*) and died out, as Gause's *Didinium* had done. Increasing the distance between oranges simply lengthened the time to extinction. But when the environment was made heterogeneous, the system stabilized. Several different 'environments' were constructed. The basic 'landscape' was 40 oranges placed on rectangular trays like egg cartons, with some of the oranges partly covered with paraffin or paper to limit the available feeding area. Complications were introduced by using varying numbers of rubber balls as 'substitute' oranges. In some cases, entire new trays of oranges were added and artificial barriers of Vaseline constructed that the mites could not cross. Stability occurred in a 252-orange landscape with complex Vaseline barriers. The prey were able to colonize oranges in a hop, skip, and jump fashion, and to keep one step ahead of the predator, which exterminated each little colony of prey it found. During the 70 weeks before the predators died out and the experiment stopped, three complete predator–prey population cycles occurred (Figure 11.10).

Nearly 40 years after Gause's original work, new experiments using *Paramecium aurelia* as prey and *Didinium* as predator managed to stabilize

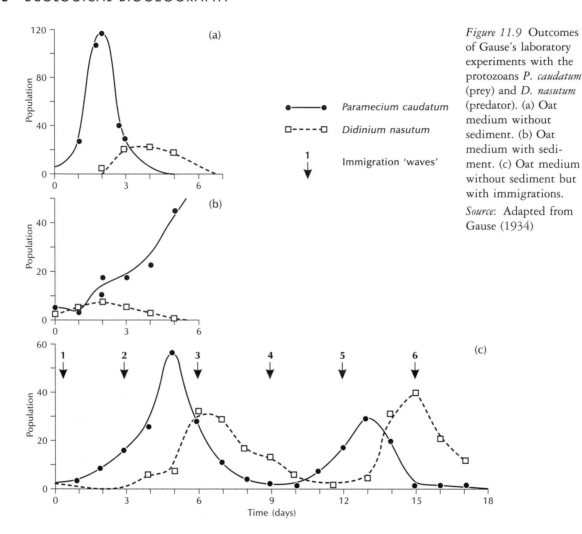

Figure 11.9 Outcomes of Gause's laboratory experiments with the protozoans *P. caudatum* (prey) and *D. nasutum* (predator). (a) Oat medium without sediment. (b) Oat medium with sediment. (c) Oat medium without sediment but with immigrations.

Source: Adapted from Gause (1934)

the system (Luckinbill 1973, 1974). When *Paramecium* was grown with *Didinium* in a 6 ml of standard cerophyl medium, *Didinium* ate all the prey in a few hours. When the medium was thickened with methylcellulose to slow down the movements of predator and prey alike, the populations went through two or three diverging oscillations lasting several days before becoming extinct. When a half-strength cerophyl medium was thickened with methylcellulose, the populations maintained sustained oscillations for 33 days before the experiment was concluded. A mathe-matical model of the system suggests that the stability arises because the increase in the cost–benefit ratio of energy spent searching to energy gained capturing prey apparently inhibits the predator searching at low prey densities (G. W. Harrison 1995).

An ambitious set of laboratory experiments studied competition and predation at the same time (Utida 1957). The system contained a bee-tle, the azuki bean weevil (*Callosobruchus chinensis*), as prey that was provided with an unlimited food supply. But this coleopteran paradise had a 'sting

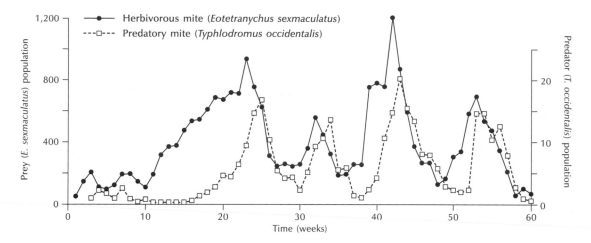

Figure 11.10 Population cycles in a laboratory experiment where a predatory mite, *Typhlodromus occidentalis*, feeds upon another mite, *Eotetranychus sexmaculatus*. The environment is heterogeneous, consisting of 252 oranges with one-twentieth of each orange available to the prey for feeding.

Source: Adapted from Huffaker *et al.* (1963)

in the tail' in the form of two competing species of predatory wasp – *Neocatolaccus mamezophagus* and *Heterospilus prosopidis*. The wasp species had similar life histories and depended upon the beetle as a food source. During the four years of the experiment, which represented 70 generations, all three populations fluctuated wildly but managed to coexist (Figure 11.11). The wasp population fluctuations were out of phase with one another. This was because *Heterospilus* was more efficient at finding and exploiting the beetle when it was at low densities, while *Neocatolaccus* was more efficient at higher prey densities. The competitive edge thus shifted between the two wasp species as the beetle density changed with time (owing to density-dependent changes in reproduction rate and the effects of the two wasp populations). The stability of the system was thus purely the result of predatory and competitive biotic interactions.

Mathematical 'experiments'

In theory, geography is a crucial factor in the coexistence of predator and prey populations. A simple mathematical model showed that, when predators and prey interact within a landscape, coexistence is rather easily attained (J. M. Smith 1974, 72–83). It is favoured by prey with a high migration capacity, by cover or refuge for the prey, by predator migration during a restricted period, and by a large number of landscape 'cells'. Furthermore, if the predator's migration ability is too low, it will become extinct. If the predator's migration ability is high, coexistence is possible if the prey is equally mobile.

Simple mathematical models of predator–prey interactions in a landscape have evolved into sophisticated metapopulation models. These, too, stress the vital role of refuges for prey species and migration rates between landscape patches. They also reveal the curious fact that some metapopulations may persist with only 'sink' populations, in which population growth is negative in the absence of migration. However, long-term persistence requires some local populations becoming large occasionally (Hanski *et al.* 1996a).

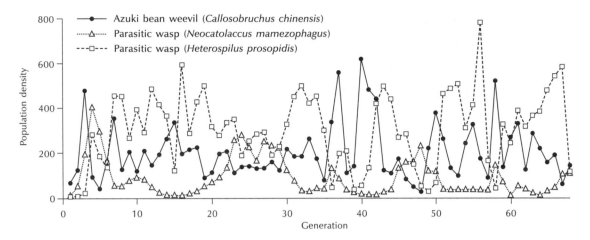

Figure 11.11 Population changes in a laboratory experiment with a beetle host, the azuki bean weevil (*Callosobruchus chinensis*), and two parasitic wasps, *Neocatolaccus mamezophagus* and *Heterospilus prosopidis*. The experiment ran for four years. All three populations fluctuated considerably, but they did survive.

Source: Adapted from Utida (1957)

LIFE AGAINST LIFE: ALTERNATIVES TO CHEMICAL CONTROL

Pests are organisms that interfere with human activities, and especially with agriculture. They are unwelcome competitors, parasites, or predators. The chief agricultural pests are insects (that feed mainly on the leaves and stems of plants), nematodes (small worms that live mainly in the soil, feeding on roots and other plant tissues), bacterial and viral diseases, and vertebrates (mainly rodents and birds feeding on grain and fruit). Weeds are a major problem for potential crop loss. A typical field is infested by 10–50 weed species that compete with the crop for light, water, and nutrients.

Pest control is used to reduce pest damage, but, even with the weight of modern technology behind it, pest control is not enormously successful. In the USA, one-third of the potential harvest and one-tenth of the harvested crop is lost to pests. A control operation is successful if the pest does not cause excessive damage. It is a failure if excessive damage is caused. Just how much damage is tolerable depends on the enterprise and the pest. An insect that destroys 5 per cent of a pear crop may be insignificant in ecological terms, but it may be disastrous for a farmer's margin of profit. On the other hand, a forest insect may strip vast areas of trees of their leaves, but the lumber industry will not go bankrupt.

Pests are controlled in several different ways. The blanket application of toxic chemicals called pesticides has inimical side-effects on the environment. Several other options for pest control are available (Table 11.3).

Biological control

Biological pest control pits predator species against prey species – parasites, predators, and pathogens are used to regulate pest populations. Two approaches exist – inundative biocontrol and classical biocontrol (Harris 1993). In *inundative control*, an organism is applied in the manner of a herbicide. Like a herbicide, the control agent is usually marketed by industry. In *classical biocontrol*, an organism (or possibly a virus) is established from another region. The pest is kept in check indefinitely, usually by government agencies acting in the public interest. Both biocontrol approaches, in the right circumstances, are effective means of pest control and bring few harmful environmental impacts, as the following examples will show.

Prickly-pear cactus in Australia

The prickly pear (*Opuntia stricta*), a cactus, is native to North and South America. It was brought to eastern Australia in 1839 as a hedge plant. It spread fast, forming dense stands, some 1–2 m high and too thick for anybody to walk through. By 1900, it infested 4,000,000 ha. Control by poisoning the cactus was not economically feasible – clearing infested land with poison was far more costly than the worth of the land. Searches in native habitats of the prickly pear were begun in 1912 to find a possible biological control agent. Eventually one was found – the cactus moth (*Cactoblastis cactorum*) native to northern Argentina, the larvae of which

burrow into the pads of the cactus, causing physical damage and introducing bacterial and fungal infections. Between 1930 and 1931, when the moth population had become enormous, the prickly pear stands were ravaged. By 1940, the prickly pear was still found here and there, but in very few places was it a pest.

A different control agent, the prickly pear cochineal (*Dactylopius opuntiae*), was used in an area of New South Wales, Australia, where *Cactoblastis cactorum* had not been a successful biological control agent (Hosking *et al.* 1994). *Dactylopius opuntiae* reduced a prickly pear (*Opuntia stricta* var. *stricta*) population in the central tablelands of New South Wales.

False ragweed in Australia and India

The false ragweed (*Parthenium hysterophorus*), a native of central America, is a problem weed of Australian rangeland, particularly in Queensland. Following field surveys in Mexico, the rust fungus, *Puccinia abrupta* var. *partheniicola*, was selected as a potential biological control agent

Table 11.3 Main pest control techniques

Control method	r-strategist pests	Intermediate pests	K-strategist pests
Pesticides (use chemical compounds to kill pests directly)	Early widespread application based on forecasting	Selective pesticides	Precisely targeted applications based on monitoring
Biological (use natural enemies, viruses, bacteria, or fungi)		Introduction or enhancement of natural enemies	
Cultural (change agricultural or other practices to alter the pest's habitat)	Timing, cultivation, and rotation	$\rightarrow$	$\leftarrow$ Changes in agronomic practice, destruction of alternative hosts
Resistance (breed animal and crop-plant varieties resistant to pests)	General, polygenic resistance	$\rightarrow$	$\leftarrow$ Specific monogenic resistance
Genetic (sterilize pest population to reduce its growth rate)			Sterile mating techniques

Source: Adapted from Conway (1981)

(Parker *et al.* 1994). One isolate was chosen for further investigation. Infection with the rust hastened leaf senescence, significantly decreased the life span and dry weight of false ragweed plants, and led to a tenfold reduction in flower production. Subsequent studies showed that the rust was sufficiently host-specific to be considered for introduction. In Bangalore, India, oneleaf senna (*Cassia uniflora*), a leguminous undershrub of some economic value, has replaced, over a five-year period, more than 90 per cent of the false ragweed on a 4,800-m² site (Joshi 1991). Leachates from oneleaf senna are allelopathic, inhibiting false ragweed seed germination and hampering the establishment of a summer false ragweed generation. The colonies of oneleaf senna are robust enough to prevent a false ragweed generation from forming below them in winter.

Pests in the Mediterranean

The whitefly, *Parabemisia myricae*, is one of the most serious citrus pests in the eastern Mediterranean region of Turkey. In 1986, a host-specific parasitoid of *P. myricae*, the aphelinid *Eretmocerus debachi*, was imported from California (Sengonca *et al.* 1993). In the following years *P. myricae* populations were rapidly reduced in all citrus orchards where the parasitoids were released. *E. debachi* was a good disperser, well adapted to the climatic conditions. Since its successful colonization, the whitefly is no longer a serious pest.

In Cyprus, the black scale (*Saissetia oleae*) is a pest primarily of the olive tree (*Olea europaea*). It attacks several other plants, including citrus trees and oleander (*Nerium oleander*). Two parasitoids (*Metaphycus bartletti* and *M. helvolus*) were imported, mass-reared, and permanently established in the island (Orphanides 1993). Following limited releases of these parasitoids, black scale populations fell from outbreak levels to almost non-existence. Black scale populations

have stayed low since parasitoid releases were discontinued.

Genetic control

Another alternative to chemical control of pests is *genetic control*. One method of genetic control is to breed sterile organisms. For example, gypsy moth (*Lymantria dispar*) pupae irradiated with a sterilizing dose of gamma radiation and mated with normal females produce a first generation of sterile male moths (Schwalbe *et al.* 1991). However, difficulties associated with large-scale deployment of partially sterile males render genetic control impractical on a large scale.

Genetic control is also achieved by breeding crop plants that are more resistant to pests. This control technique came to the rescue of the French wine industry. *Phylloxera* is an aphid native to America. It lives in galls on leaves and roots of vines, out of the reach of sprays. It multiplies prodigiously. *Phylloxera*-infected vines become stunted and die. In 1861, it was accidentally introduced into Languedoc, France. Two decades later, four-fifths of the Languedoc vineyards had been devastated and every wine-growing area of France was infected; no remedy had been found; and the outlook was bleak for wine drinkers. In 1891, it was discovered that the American vine (*Vitis labrusca*) was almost immune to *Phylloxera*. Scions of the European vine (*V. vinifera*) were grafted onto American rootstocks to produce a hybrid vine that, if not entirely immune, was affected far less seriously.

Integrated pest management

Modern pest control involves *integrated pest management*. This ecological approach to pest management brings together at least four techniques. First, it uses *natural pest enemies*, including parasites, diseases, competitors, and predators (biological control). Second, it advocates the planting of a greater *diversity of crops* to lessen the possibility that a pest will find a host. Third, it advocates *no or little ploughing* so that natural enemies of some pests have a chance to build up in the soil. Fourth, it allows the application of a set of *highly specific chemicals*, used sparingly and judiciously (unlike the old application method that tended to be profligate). Integrated pest management involves the use of chemicals, the development of genetically resistant stock, biological control, and land culture. Land culture is the physical management of the land – whether and how it is ploughed, what kind of crop rotation is used, the dates of planting, and basic means of handling crop harvests to reduce presence of pest in residues and products sold.

Integrated pest management has been used to tackle the oriental fruit moth (*Grapholitha molesta*), which attacks several fruit crops (Barfield and Stimac 1980). The moth is prey to a species of braconid wasp, *Macrocentrus ancylivorus*. The introduction of the wasp into fields and orchards helped to control the moth population. But an interesting discovery was made: the efficacy of the wasp in peach orchards was increased when strawberry fields lay nearby. The strawberry fields are an alternative habitat for the wasp and help it to overwinter.

SUMMARY

No population can exist in isolation – it needs to interact with others. Population interactions take several forms but fall into two categories – cooperation and competition. Cooperation occurs to varying degrees. Weak cooperation (protocooperation) is beneficial to both populations, but it is not obligatory – the interacting species would survive without one another. It includes mimicry. Mutualism is an extreme form of protocooperation in which the interaction is obligatory. Commensalism occurs when one of the cooperating species benefits and the other is unharmed by the association. Competition occurs where species interactions are detrimental to at least one of the interactants. The competitive exclusion principle states that no two species may occupy exactly the same niche. If they should occupy very similar niches, competition will occur. Competition takes the form of scramble competition (for resources) or contest competition (sometimes involving aggressive action between individuals). Resource partitioning, character displacement, and spatial heterogeneity allow the coexistence of competing species. Herbivory is the predation of plants by plant-eaters. The plant kingdom has evolved defences to foil their herbivorous adversaries. The defences include structural deterrents and chemical arsenals. Herbivores interact with plants in several characteristic ways. This is seen in patterns of seed predation and grazing. Carnivory is the predation of one animal by another. There are several carnivorous specializations, some rather bizarre. Evolution has finely tuned carnivore communities to promote coexistence through prey switching and prey selection. Predators and their prey sometimes display cycles in population numbers. Such cycles are sometimes the result of food availability, though chaotic behaviour in some predator–prey systems does appear to rest within the interacting populations themselves. Stability in predator–prey systems is more readily attained in heterogeneous environments. Life may be pitted against life in an attempt to control agricultural pests. Biological control is a useful alternative to pesticide application. Genetic control is another option. Biological control is usually applied within a broader system of integrated pest management.

ESSAY QUESTIONS

1 **What are the results of interspecific competition?**

2 **What are the advantages and disadvantages of species cooperation?**

3 **What are the pros and cons of biological control?**

FURTHER READING

Crawley, M. J. (1983) *Herbivory: The Dynamics of Animal–Plant Interactions* (Studies in Ecology 10). Oxford: Blackwell Scientific Publications.
A superb book, despite the mathematics.

Grover, J. P. (1997) *Resource Competition*. New York: Chapman & Hall.
Not easy but worth a look.

Kingsland, S. E. (1985) *Modeling Nature: Episodes in the History of Population Biology*. Chicago and London: The University of Chicago Press.
A first-rate account of the history of population biology. More exciting than it might sound.

Kormondy, E. J. (1996) *Concepts of Ecology*, 4th edn. Upper Saddle River, NJ: Prentice Hall.
Relevant sections well worth reading.

MacDonald, D. (1992) *The Velvet Claw: A Natural History of the Carnivores*. London: BBC Books.
All about carnivores with beautiful photographs.

12

COMMUNITIES

Communities consist of several interacting populations. Ecosystems are communities together with the physical environment that sustains them. This chapter covers:

- the nature of communities and ecosystems
- ecological roles in communities
- feeding relationships
- biological diversity

SOCIAL AND PHYSICAL CONNECTIONS: COMMUNITIES AND ECOSYSTEMS

An **ecosystem** is a space in which organisms interact with one another and with the physical environment. A *community* is the assemblage of interacting organisms within an ecosystem. Communities are sometimes called *biocoenoses*, with each part given a separate name – phytocoenose (plants), zoocoenose (animals), microbiocoenose (microorganisms) (Sukachev and Dylis 1964, 27). Where the physical environment supporting the community is included, the term biogeocoenose is used (and is a vowel-rich and somewhat awkward alternative to the term

ecosystem). Ecosystems and communities range in size from a cubic centimetre to the entire world.

A local ecosystem

The Northaw Great Wood, Hertfordshire, England, is a deciduous wood with a small plantation of conifers (Plate 12.1). It occupies 217 ha in the headwaters of the Cuffley Brook drainage basin. The eastwards-draining streams have cut into a Pebble Gravel plateau that sits at around 400 m. They have eroded valleys in London Clay, Reading Beds (sands), and, at the lowest points, chalk (Figure 12.1). A variety of soils have formed upon these rocks. The soils support many trees

Plate 12.1 Hornbeams (*C. betulus*) in the Northaw Great Wood, Hertfordshire, England. Photograph by Richard John Huggett.

(Figure 12.2). The main ones are oak (chiefly *Quercus robur*), hornbeam (*Carpinus betulus*), and silver birch (*Betula pendula*). The herb layer is varied, the main communities corresponding to the chief habitats – the forest floor, streams and other damp places, paths, clearings, the woodland edge, a marshy area, and a chalky area. It consists of flowering plants, ferns and horsetails, mosses and liverworts, and fungi. There are 283 plant species, at least 3 liverwort and 12 moss species, and 302 fungi species. A few lichens grow on tree trunks and buildings.

The plants and fungi support a great diversity of insects and woodlice. There are predators (ground beetles, burying beetles, ladybirds, and others), defoliators (leaf beetles and weevils), seed borers (weevils), bark borers (weevils and bark beetles), wood borers ('longhorn' beetles,

'ambrosia' beetles, and click beetles), root feeders (weevils and click beetle larvae – wireworms), and woodlice. There are butterflies and moths, snails and slugs. There are two amphibians – the common toad (*Bufo bufo*) and the common frog (*Rana temporaria*); and three reptiles – the warty newt (*Triturus cristatus*), the slow-worm (*Anguis fragilis*), and the ringed or grass snake (*Natrix natrix*). There are 50 species of breeding birds and some 22 visitors. In summer, the dominant species in the breeding community are chaffinch (*Fringilla coelebs*), willow warbler (*Phylloscopus trochilus*), blue tit (*Parus caerulus*), great tit (*P. major*), robin (*Erithacus rubecula*), blackbird (*Turdus merula*), wren (*Troglodytes troglodytes*), dunnock or hedge sparrow (*Prunella modularis*), blackcap (*Sylvia atricapilla*), garden warbler (*S. bocin*), chiffchaff (*Phylloscopus collybita*), greater

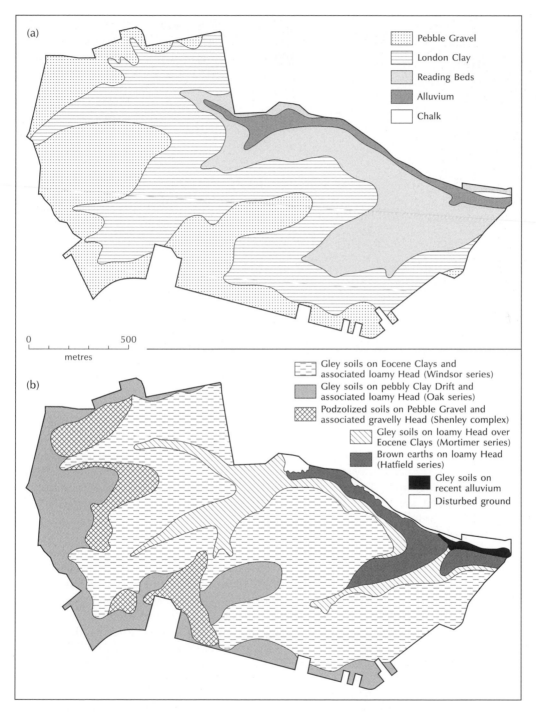

(a)

- ▦ Pebble Gravel
- ▤ London Clay
- ▨ Reading Beds
- ▨ Alluvium
- ▢ Chalk

0   500
metres

(b)

- ▤ Gley soils on Eocene Clays and associated loamy Head (Windsor series)
- ▨ Gley soils on pebbly Clay Drift and associated loamy Head (Oak series)
- ▩ Podzolized soils on Pebble Gravel and associated gravelly Head (Shenley complex)
- ▨ Gley soils on loamy Head over Eocene Clays (Mortimer series)
- ▨ Brown earths on loamy Head (Hatfield series)
- ■ Gley soils on recent alluvium
- ▢ Disturbed ground

Figure 12.1 Northaw Great Wood, Hertfordshire, England. (a) Geology. (b) Soils.
Sources: (a) Adapted from Sage (1966b). (b) Adapted from D. W. King (1966)

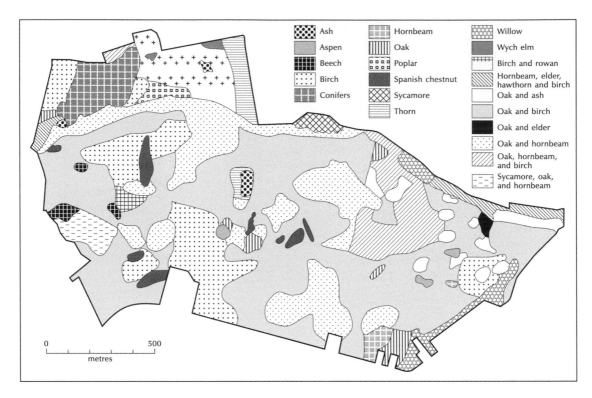

Figure 12.2 Tree distribution in the Northaw Great Wood, Hertfordshire, England.
Source: Adapted from Horsley (1966)

spotted woodpecker (*Dendrocopos major*), nuthatch (*Sitta europaea*), redstart (*Phoenicurus phoenicurus*), and tree pipit (*Anthus trivialis*). Twenty-three wild species of mammal are recorded (Table 12.1). Six mammalian orders are represented – Insectivora (shrews, hedgehogs, and voles), Chiroptera (pipistrelle and noctule bats), Lagomorpha (hares and rabbits), Rodentia (squirrels, rats, mice, and voles), Carnivora (foxes, badgers, stoats, and weasels), and Artiodactyla (fallow deer and muntjac deer).

Most ecosystems are heterogeneous and involve a mosaic of individuals, populations, and habitats. The Northaw Great Wood is mainly an oak–hornbeam woodland, with stands of silver birch on more acidic and better-drained soils. However, it contains patches and corridors. Paths and the streams form the corridors. The patches include cleared areas and areas where other tree species dominate. There are, for example, three stands of ash (*Fraxinus excelsior*), three stands of beech (*Fagus sylvatica*), three stands of aspen (*Populus tremula*), and seven stands of sweet chestnut (*Castanea sativa*).

Within the wood, several communities exist. A good example is the bracken (*Pteridium aquilinium*), wood anemone (*Anemone nemorosa*), and bluebell (*Hyacinthoides non-scripta*) community (Sage 1966a, 14). These species manage to live together as a plant association by dividing the habitat to avoid undue competition for resources. Bracken rhizomes may penetrate about 60 cm. The bluebell bulbs are not nearly so deep. The wood anemone taps the top 6 cm of soil. The wood anemone is a pre-vernal flowerer (March to May), whereas the bluebell is a vernal flowerer (April to

Table 12.1 Mammals in the Northaw Great Wood, Hertfordshire, England (to 1966)

Order	Family	Species
Insectivora	Erinaceidae	Hedgehog (*Erinaceus europaeus*)
	Talpidae	Mole (*Talpa europaea*)
	Soricidae	Common shrew (*Sorex araneus*), pygmy shrew (*Sorex minutus*), water shrew (*Neomys fodiens*)
Chiroptera	Vespertilionidae	Common pipistrelle (*Pipistrellus pipistrellus*), noctule (*Nyctalus noctula*)
Lagomorpha	Leporidae	Rabbit (*Oryctolagus cuniculus*), brown hare (*Lepus europaeus*)
Rodentia	Sciuridae	Grey squirrel (*Sciurus carolinensis*)
		Bank vole (*Clethrionomys glareolus*), field vole (*Microtus agrestis*), water vole (*Arvicola terrestris*)
		Wood mouse (*Apodemus agrestis*), yellow-necked mouse (*Apodemus flavicollis*), brown rat (*Rattus norvegicus*)
		Dormouse (*Muscardinus avellanarius*)
Carnivora	Canidae	Red fox (*Vulpes vulpes*)
		Stoat (*Mustela erminea*), weasel (*Mustela nivalis*), badger (*Meles meles*)
Artiodactyla	Cervidae	Fallow deer (*Dama dama*), Reeve's muntjac (*Muntiacus reevesi*)

May). Both complete assimilation before the bracken canopy develops in June, so avoiding competition for light. The bluebell is the only species that can withstand bracken, but if bracken growth is especially vigorous, the accumulations of fallen fronds exclude even the bluebell.

Communities possess vertical layers. This *stratification* results primarily from differences in light intensity. In some communities, distinct layers of leaves are evident – canopy, understorey, shrub layer, ground layer. Animal species are sorted among the layers according to their food and cover requirements. In the Northaw Great Wood, tree, shrub, field, and ground layers are present. Each layer acts as a kind of semitransparent blanket and modifies the conditions below. Light levels are reduced, as is precipitation intensity, from the canopy to the ground.

The horizontal distribution of species within communities is more complicated. Two extreme situations exist. First, a *gradient* is a gradual change in a species assemblage across an area. In the Northaw Great Wood, the dominant tree species change along a toposequence. Silver birch on summits gives way to oak and hornbeam in midslope positions and along the valley floor. Second, a *patch* is a clustering of species into somewhat distinctive groups. Although tree communities do gradually alter along environmental gradients in the Northaw Great Wood, there are also stands of trees that form distinct patches (Figure 12.2). Gradients and patches are two ends of a continuum and the subject of a lively debate in vegetation science.

The global ecosystem

All living things form the **biosphere**. The biosphere interacts with non-living things in its surroundings (air, water, soils, and sediments) to win materials and energy. The interaction creates the **ecosphere**, which is defined as life plus life-support systems. It consists of ecosystems – individuals, populations, or communities interacting with their physical environment. Indeed, the ecosphere is the global ecosystem (Huggett 1995, 8–11; 1997, 1999).

Communities and ecosystems possess properties that emerge from the individual organisms. These *emergent properties* cannot be measured in individuals – they are community properties. Four important community properties are biodiversity, production and consumption, nutrient cycles, and food webs. These community properties enable the biosphere to perform three important tasks (Stoltz *et al*. 1989). First, the biosphere harnesses energy to power itself and build up reserves of organic material. Second, it garners elements essential to life from the atmosphere, **hydrosphere**, and **pedosphere**. And third, it is able to respond to cosmic, geological, and biological perturbations by adjusting or reconstructing food webs.

PRODUCING AND CONSUMING: COMMUNITY ROLES

All organisms have a community and an ecological role. Some organisms produce organic compounds, and some break them down. The chief roles are:

1 **Producers (autotrophs)**. These include all photoautotrophic organisms: green plants, eukaryotic algae, blue-green algae, and purple and green sulphur bacteria. In some ecosystems, chemoautotrophs produce organic materials by oxidation of inorganic compounds and do not require sunlight.
2 **Consumers (heterotrophs)**. These are organisms that obtain their food and thus energy from the tissues of other organisms, either plants or animals or both – herbivores, carnivores, and top carnivores. Consumers occupy several **trophic** levels. Consumers that eat living plants are primary consumers or herbivores. Consumers of herbivores are the flesh-eating secondary consumers or carnivores. In some ecosystems, there are carnivore-eating carnivores; these are tertiary consumers or top carnivores.
3 **Decomposers (saprophytes) and detritivores (saprovores)**. Decomposers dissolve organic matter, while detritivores break it into smaller pieces and partly digest it.

Autotrophs produce organic material, the consumers eat it, and the decomposers and detritivores clean up the mess (excrement and organic remains).

Producer society: community production

Primary producers

Green plants use solar energy, in conjunction with carbon dioxide, minerals, and water, to build organic matter. The organic matter so manufactured contains chemical energy. **Photosynthesis** is the process by which radiant energy is converted into chemical energy. *Production* is the total biomass produced by photosynthesis within a community. Part of the photosynthetic process requires light, so it occurs only during daylight hours. At night, the stored energy is consumed by slow oxidation, a process called **respiration** in individuals and **consumption** in a community.

Sunlight comes from above, so ecosystems tend to have a *vertical structure* (Figure 12.3). The upper *production zone* is rich in oxygen. The lower *consumption zone*, especially that part in the soil, is rich in carbon dioxide. Oxygen is deficient in the consumption zone, and may be absent. Such gases as hydrogen sulphide, ammonia, methane, and hydrogen are liberated where reduced chemical states prevail. The boundary between the production zone and the consumption zone, which is known as the *compensation level*, lies at the point where there is just enough light for plants to balance organic matter production against organic matter utilization.

Phytomass is the living material in producers. It normally excludes dead plant material, such as tree bark, dead supporting tissue, and dead branches and roots. Where dead bits of plants are included, the term *standing crop* is applied. Evergreen trees in tropical forests are largely made of dead supporting structures (branches, trunks, roots); about 2 per cent is green plant biomass (leaves) (Figure 12.4). The relative proportion of

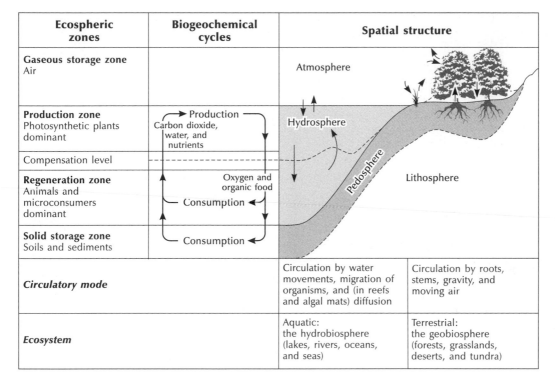

Ecospheric zones	Biogeochemical cycles	Spatial structure	
Gaseous storage zone Air		Atmosphere	
Production zone Photosynthetic plants dominant	Production Carbon dioxide, water, and nutrients	Hydrosphere	
Compensation level			
Regeneration zone Animals and microconsumers dominant	Oxygen and organic food Consumption	Pedosphere Lithosphere	
Solid storage zone Soils and sediments	Consumption		
Circulatory mode		Circulation by water movements, migration of organisms, and (in reefs and algal mats) diffusion	Circulation by roots, stems, gravity, and moving air
Ecosystem		Aquatic: the hydrobiosphere (lakes, rivers, oceans, and seas)	Terrestrial: the geobiosphere (forests, grasslands, deserts, and tundra)

Figure 12.3 Production zones, consumption zones, and biogeochemical cycles in ecosystems.
Sources: After Huggett (1997), partly adapted from Odum (1971)

leaves, branches, trunks, and roots is similar in temperate forests, though the absolute values are lower. Grassland plants are all leaf and root – there is negligible branch and trunk. Tundra and semi-desert plants are largely made of green plant biomass and contain little supporting tissue.

Primary production

The green plants that form the production zone, because they produce their own food from solar energy and raw materials, are called *photoautotrophs*. The amount of organic matter that they synthesize per unit time is **gross primary productivity**. Most of this matter is created in the plant leaves. Some of it is transported through the phloem to other parts of plants, and especially to the roots, to drive metabolic and growth processes.

Net primary productivity is the gross primary productivity less the chemical energy burnt in all the activities that constitute plant respiration. Net primary productivity is usually about 80–90 per cent of gross primary productivity. The mean net global primary productivity is 440 g/m²/yr (Vitousek *et al.* 1986), or about 224 petagrammes (1 Pg = 224 × 10¹⁵ grammes = 224 billion tonnes) (Figure 12.5). It is difficult to comprehend such vast figures. A cubic kilometre of ice (assuming the density is 1 g/cm³) would weigh 1 billion tonnes (1 petagramme). A block of ice 1 km high and resting on an area of 15 km × 15 km would have about the same mass as the global net primary production.

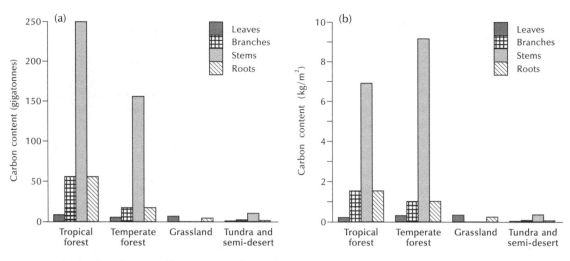

Figure 12.4 The distribution of biomass in different biomes. (a) Total carbon (gigatonnes, Gg) stored in leaves, branches, stems, and roots. (b) Carbon stored per unit area (kg/m²) in leaves, branches, stems, and roots. 'Tropical forest' includes tropical forest, forest plantations, shrub-dominated savannah, and chaparral. 'Temperate forest' includes temperate forest, boreal forest, and woodland.

Source: Compiled from data in Goudriaan and Ketner (1984)

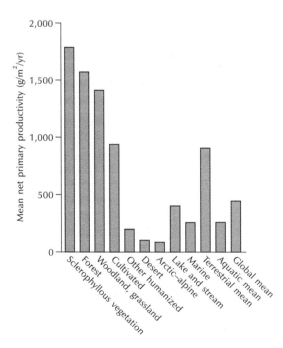

Figure 12.5 Mean net primary production for the biosphere and its component biomes.

Source: Data from Vitousek *et al.* (1986)

In terrestrial ecosystems, the main producers are green plants (megaphytes), with the lower plants playing a minor role. Land plants produce on average 899 g/m²/yr (Vitousek *et al.* 1986). That is a total of about 132 billion tonnes for the entire land area. In aquatic ecosystems, the main producers are autotrophic algae. These unicellular and colonial organisms are suspended in the water and are part of the plankton. Aquatic plants produce on average 225 g/m²/yr (Vitousek *et al.* 1986). That is a total of 92.4 billion tonnes for the entire water area. Of this total, marine plants produce 91.6 billion tonnes, and freshwater plants a mere 0.8 billion tonnes. Chemolithotrophic bacteria using chemical reactions around vents in the sea floor or in soils contribute a very tiny part global net primary production.

The world pattern of terrestrial net primary productivity is shown in Figure 12.6. It mirrors the simultaneous availability of heat and moisture. Production is high in the tropics, warm temperate zone, and typical temperate zone, and low in the arid subtropics, continental temperate regions, and polar zones.

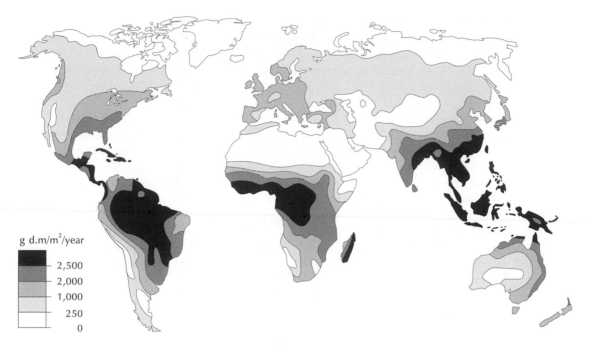

g d.m/m²/year

| 2,500 |
| 2,000 |
| 1,000 |
| 250 |
| 0 |

Figure 12.6 The world pattern of terrestrial net primary productivity. Units are grammes of dry matter per square metre per year (g d.m/m²/yr).

Sources: After Huggett (1997), adapted from Box and Meentemeyer (1991)

Consumer society: community consumption

Consumers

The material produced by photosynthesis serves as a basic larder for an entire ecosystem – it is used as an energy-rich reserve of organic substances and nutrients that is transferred through the rest of the system. The potential chemical energy of net primary production is available to organisms that eat plants, both living plant tissue and dead plant tissue, and indirectly therefore to animals (and the few plants) that eat other animals. All these organisms depending upon other organisms for their food are *heterotrophs* or *consumers*. They are browsers, grazers, predators, or scavengers. They include microscopic organisms, such as protozoans, and large forms, such as vertebrates. The majority are chemoheterotrophs, but a few specialized photo-

synthetic bacteria are photoheterotrophs. The stored chemical energy in consumers is called secondary production. There are two broad groups of consumers: *macroconsumers* or *biophages* eat living plant tissues; *microconsumers* or *saprophages* slowly decompose and disintegrate the waste products and dead organic matter of the biosphere.

The available phytomass in water is small, but the primary production (the amount of dry organic matter produced in a year) is relatively large because aquatic plants multiply fast. Animals eat the plants and incorporate much of the primary production in their bodies. For this reason, the *secondary production* or **zoomass** (animal biomass) in aquatic ecosystems is commonly more than 15 times larger than the phytomass. The situation is different in terrestrial ecosystems. Much of the phytomass consists of non-photosynthetic tissue, such as buds, roots, and trunks. For this reason, the standing phytomass is always very large, especially in woodlands. Primary production, on the other hand, is relatively small – a modest amount of dry organic matter is created each year. Animals living above the ground eat a mere 1 per cent, or thereabouts, of the phytomass. The zoomass is therefore small, being about 1 to 0.1 per cent of the phytomass. Biomass is the weight or mass of living tissue in an ecosystem – phytomass plus zoomass. It has an energy content, which may be thought of as bioenergy.

Human use plants for food, fuel, and shelter. This human harvest accounts for about 4.5 per cent of global terrestrial net production (Table 12.2). Land used in agriculture or converted to other land uses accounts for about 32 per cent of terrestrial net primary production. All human activities have reduced terrestrial net primary production by around 45 per cent. This probably amounts to the largest ever diversion of primary production to support a single species (Vitousek *et al.* 1986).

Ecosystems also contain *geomass* – soils (including litter and dead organic matter), sediments, air, and water that harbour a supply of water and nutrients (macronutrients, micronutrients, and other **trace elements**).

Decomposers and detritivores

Saprophages include decomposers (or saprophytes) and detritivores (or saprovores). *Decomposers* are

Table 12.2 Human appropriation of net primary production in the 1980s

Manner of consumption	Net primary production	
	Total mass (billion tonnes)	Percentage of terrestrial net primary production
Used by humans		
Plants eaten	0.8	0.48
Plants fed to domestic animals	2.2	1.67
Fish eaten by humans and domestic animals	1.2	–
Wood for paper and timber	1.2	0.91
Fuel wood	1.0	0.76
Total	7.2	4.54[a]
Used or diverted		
Cropland	15.0	11.35
Converted pastures	9.8	7.42
Other (cities, deforested)	17.8	13.48
Total	42.6	32.25
Used, diverted, or reduced		
Used or diverted	42.6	32.25
Reduced by conversion	17.5	13.25
Total	60.1	45.50

Source: Adapted from Diamond (1987)
Note: a Excluding fish

organisms that feed on dead organic matter and waste products of an ecosystem. They do so by secreting enzymes to digest organic matter in their surroundings, and soaking up the dissolved products. They are mainly **aerobic** and **anaerobic** bacteria, protozoa, and fungi. An example is the shelf fungus (*Trametes versicolor*). This grows on rotting trees and is important in decomposing wood. Some green plants lacking chlorophyll, such as Indian pipe (*Monotropa uniflora*) and pinesap (*M. hypopithys*), also obtain their food by diffusion from the outside.

Detritivores are organisms that feed upon detritus. They include beetles, centipedes, earthworms, nematodes, and woodlice. They are all microconsumers. Detritivores assist the breakdown of organic matter. By chewing and grinding dead organic matter before ingestion, they comminute it and render it more digestible. When egested, the carbon/nitrogen (C/N) ratio is a little lower, and the acidity (pH) a little higher,

than in the ingested food. These changes mean that the faeces provide a better substrate for renewed decomposer (and notably bacterial) growth. Successively smaller fragments of dead organic matter are passed through successively smaller detritivores, after having been subject to decomposer attack at each stage. The result is a *comminution spiral* (Figure 12.7).

Humification accompanies comminution and decomposition to produce a group of organic compounds called **humus**. Humus is in a sense the final stage of organic decomposition, but it is also the product of microbial synthesis and is always subject to slow microbial decomposition, the end product of which is stable humus charcoal. Humus decomposition in the temperate zones is incomplete and humus is brown or black. Decomposition is usually more advanced in the tropics and humus may be colourless. Mineral nutrients are released into the soil solution during decomposition, a process styled **mineralization**.

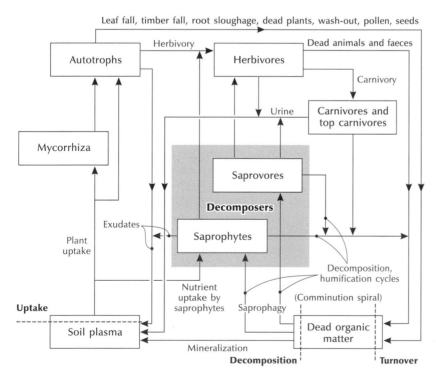

Figure 12.7 Grazing and detritus feeding relations in an ecosystem, showing comminution spirals.
Source: Adapted from Huggett (1980)

Recycling machines: ecosystem turnover

Biogeochemicals

The biosphere is made of three main elements – hydrogen (49.8 per cent by weight), oxygen (24.9 per cent), and carbon (24.9 per cent). Several other elements are found in the biosphere, and some of them are essential to biological processes – nitrogen (0.27 per cent), calcium (0.073 per cent), potassium (0.046 per cent), silicon (0.033 per cent), magnesium (0.031 per cent), phosphorus (0.03 per cent), sulphur (0.017 per cent), and aluminium (0.016 per cent). These elements, except aluminium, are the basic ingredients for organic compounds, around which biochemistry revolves. Carbon, hydrogen, nitrogen, oxygen, sulphur, and phosphorus are needed to build nucleic acids (RNA and DNA), amino acids (proteins), carbohydrates (sugars, starches, and cellulose), and lipids (fats and fat-like materials). Calcium, magnesium, and potassium are required in moderate amounts. Chemical elements required in moderate and large quantities are **macronutrients**. More than a dozen elements are required in trace amounts, including chlorine, chromium, copper, cobalt, iodine, iron, manganese, molybdenum, nickel, selenium, sodium, vanadium, and zinc. These are **micronutrients**. Functional nutrients play some role in the metabolism of plants but seem not to be indispensable. Other mineral elements cycle through living systems but have no known metabolic role. The biosphere has to obtain all macronutrients and micronutrients from its surroundings.

Biogeochemical cycles

There is in the ecosphere a constant turnover of chemicals. The motive force behind these chemical cycles is life. In addition, on geological timescales, forces in the **geosphere** producing and consuming rocks influence the cycles. **Biogeochemical cycles**, as they are called, involve the storage and flux of all terrestrial elements and compounds except the inert ones. Material exchanges between life and life-support systems are a part of biogeochemical cycles.

At their grandest scale, biogeochemical cycles involve the entire Earth. An exogenic cycle, involving the transport and transformation of materials near the Earth's surface, is normally distinguished from a slower and less well understood endogenic cycle involving the lower crust and mantle. Cycles of carbon, hydrogen, oxygen, and nitrogen are *gaseous cycles* – their component chemical species are gaseous for a leg of the cycle. Other chemical species follow *sedimentary cycles* because they do not readily volatilize and are exchanged between the biosphere and its environment in solution.

Minerals cycle through ecosystems, the driving force being the flow of energy. The circulation of mineral elements through ecosystems involves three stages – uptake, turnover, and decomposition. Green plants take up solutes and gases, the rate of uptake broadly matching biomass production rate, and incorporated into phytomass. Oxygen is released in photosynthesis. The remaining minerals either pass on to consumers or else return to soil and water bodies when plants die or bits fall off. The minerals in the consumers eventually return to the soil, sea, or atmosphere. Figure 12.8 summarizes some major *mineral cycles*.

The distribution of biogeochemicals within ecosystems varies among biomes. Figure 12.9 shows the amount of carbon stored in biomass, litter, humus, and stable humus charcoal in the major ecozones. This gives a good indication of how organic matter is apportioned in different ecosystems. Biomass carbon is 20–40 times greater in forests than in other ecosystems. Humus carbon is highest in temperate forests and grasslands. It is lower in tropical forests because it is rapidly decomposed in the year-round, hot and humid conditions. Stable-humus-charcoal carbon, which degrades exceedingly slowly, is highest in tropical forests, and is still significant in temperate forests and grasslands.

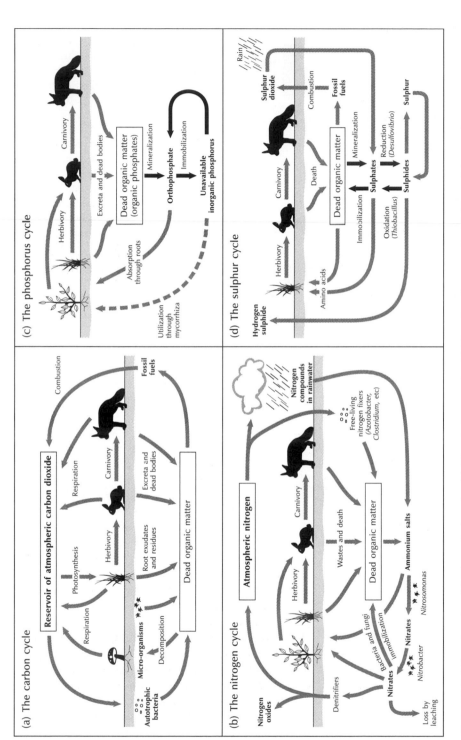

Figure 12.8 Mineral cycles. (a) The carbon cycle. (b) The nitrogen cycle. (c) The phosphorus cycle. (d) The sulphur cycle.

Source: Partly adapted from Jackson and Raw (1966)

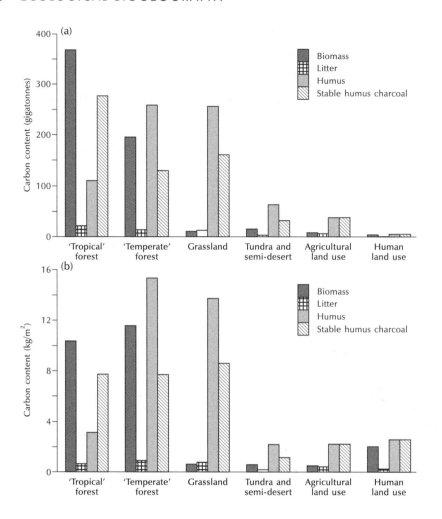

Figure 12.9 Carbon stored in biomass, litter, humus, and charcoal in the major ecozones. (a) Total carbon (gigatonnes, Gg). (b) Carbon per unit area (kg/m²). 'Tropical forest' includes tropical forest, forest plantations, shrub-dominated savannah, and chaparral. 'Temperate forest' includes temperate forest, boreal forest, and woodland.

Source: After data in Goudriaan and Ketner (1984)

EATERS AND THE EATEN: FOOD CHAINS AND FOOD WEBS

Food webs

An ecosystem contains two chief types of food web: a grazing food web (plants, herbivores, carnivores, top carnivores) and a decomposer or detritus food web (Figure 12.7).

Grazing food chains and webs

This simple feeding sequence – plant → herbivore → carnivore → top carnivore – is a **grazing** food chain. An example is leaf → caterpillar → bird → weasel. Usually though, food and feeding relations in an ecosystem are more complex because there is commonly a wide variety of plants available, different herbivores prefer different plant species, and carnivores are likewise selective about which herbivore they will consume. Complications also arise because some animals, the omnivores, eat both plant and animal tissues. For all these reasons, the energy flow through an ecosystem in most cases is better described as a **food web**.

Figure 12.10 shows the food web for Wytham Wood, Oxfordshire, England. The incoming solar energy supports a variety of trees, shrubs, and herbs. Oak (*Quercus* spp.) is the dominant plant. The plants are eaten by a variety of herbivores. Several invertebrates feed on plant material, and especially leaves. The winter moth (*Operophtera brumata*) and the pea-green oak twist (*Tortrix viridana*) are examples. The herbivores are prey to carnivores, including spiders, parasites, beetles, birds, and small mammals. *Cyzenis albicans* is a tachinid fly and a specific parasite of the winter moth. There are several predatory beetles in the litter layer, the commonest large ones being *Philonthus decorus*, *Feronia madida*, *Felonia melanaria*, and *Abax parallelopipedus*. Titmice feed partly on plants (e.g. beech mast), partly on insects, and partly on spiders. The main species are the great tit (*Parus major*) and the blue tit (*P. caeruleus*). Small mammals include the bank vole (*Clethrionomys glareolus*) and the wood mouse (*Apodemus sylvaticus*). Top carnivores include parasites and **hyperparasites**, shrews, moles, weasels, and owls. The common shrew (*Sorex araneus*), pygmy shrew (*S. minutus*), and mole (*Talpa europaea*) dominate the ranks of top carnivores. They are all eaten by the tawny owl (*Strix aluco*).

Decomposer (detritus) food chains and webs

Dead organic matter and other waste products generally lie upon and within the soil. As they are decomposed, minerals are slowly released that are reused by plants. There are complex food and feeding relations among the decomposers and a decomposer or detritus food chain is recognized, the organisms in which are all microconsumers (decomposers and detritivores).

On land, litter and soil organic matter support a *detritus food web*. In Wytham Wood, decomposers (mainly bacteria and fungi) slowly digest the litter. Detritivores – earthworms, soil insects, and mites – eat litter fragments. Predator beetles eat the detritivores, thus linking the grazing and detritus food webs. In aquatic ecosystems, the producers (mainly **phytoplankton**) live in the upper illuminated areas of water bodies. Animal microplankton and macroplankton eat the producers. In turn, fishes, aquatic mammals, and birds that take their prey out of water eat the animal plankton. All the organisms in the detritus aquatic food chain are eventually decomposed in the water and in sediments on lake, river, or sea bottoms.

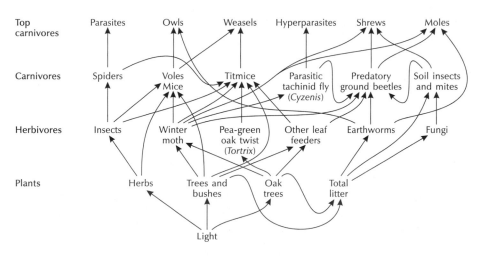

Figure 12.10 A food web, Wytham Wood, Oxfordshire.
Source: Adapted from Varley (1970)

Ecological pyramids

Energy and biomass are distributed among producers, herbivores, predators, top predators, and decomposers. Their distributions both resemble a pyramid – energy and biomass become less in moving from producer, to herbivore, to carnivore, to top carnivore. This is because, at each higher trophic level, there is successively less energy and biomass available to eat. In Cedar Bog Lake, Minnesota, the net production figures (in kcal/m²/yr) are: producers 879, herbivores 104, and carnivores 13 (Lindemann 1942). The biomass of each trophic level could be measured in the field, or could be derived by converting the production figures to biomass. Each gram of dry organic matter is equivalent to about 4.5 kcal, so each kilocalorie unit is equivalent to about 0.222 kg. The biomasses for Cedar Lake Bog (in kg/m²) are therefore approximately: producers 195, herbivores 23, and carnivores 3. This conversion to biomass does not change the shape of the pyramid.

Some ecosystems have an inverted pyramidal distribution of biomass – narrow at the bottom and wide at the top. In the English Channel, there are 4 g of phytoplankton per m² and 21 g of **zooplankton** per m². Clear-water aquatic ecosystems have a lozenge-shaped pyramid – narrow at the top and bottom and wide in the middle.

A *pyramid of numbers* represents the number of organisms at each trophic level – the number of plants, the number of herbivores, and the number of carnivores (Figure 12.11). This inform-ation can be troublesome when comparing two different ecosystems – it is not very informative to equate 'a diatom with a tree, or an elephant with a vole' (Phillipson 1966, 13). The typical pyramid of numbers applies when producers are small, as they are in aquatic ecosystems. In forests, the producers – mainly trees – are large, and a pyramid of the various consumer levels perches on a thin base. In plant–parasite–hyperparasite food chains, the pyramid of numbers is inverted.

Keystone species

Keystone species are species central to an ecosystem – species upon which nearly all other species depend. Several keystone species have been identified in the wild, but it is not easy to predict which species will be keystone because the connections between species in food webs are often complex and obscure. For instance, large cats act as keystone predators in neotropical forests. They limit the number of medium-sized terrestrial mammals, which in turn control forest regeneration. On Barro Colorado Island, Panama, jaguars (*Felis onca*), pumas (*F. concolor*), and ocelots (*F. pardalis*) have been removed. The populations of big-cat prey – including the red coati (*Nasua nasua*), the agouti (*Dasyprocta variegata*), and the paca (*Agouti paca*) – are about 10 times higher than on Cocha Cashu, Peru, where big cats still live (Terbough 1988). However, this increase may result from natural population variability rather than the lack of jaguars and pumas (Wright *et al*. 1994). The extreme removal of herbivores and

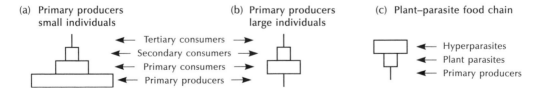

(a) Primary producers small individuals (b) Primary producers large individuals (c) Plant–parasite food chain

← Tertiary consumers →
← Secondary consumers →
← Primary consumers →
← Primary producers →

← Hyperparasites
← Plant parasites
← Primary producers

Figure 12.11 Ecological pyramids. (a) Primary producers are small organisms. (b) Primary producers are large organisms. (c) A plant–parasite–hyperparasite food chain.
Source: Adapted from Phillipson (1966)

frugivorous mammals would drastically affect forest regeneration, altering tree species composition, but the effects of modest changes in densities are less clear.

Several examples of keystone species will be considered, and then some general ideas about keystone-species removal will be examined.

Keystone predators

The sea otter (*Enhydra lutris*) is a keystone predator par excellence (Duggins 1980). A population of around 200,000 once thrived on the kelp beds lying close to shore from northern Japan, through Alaska, to southern California and Mexico (Figure 12.12). In 1741, Vitus Bering, the Danish explorer, reported seeing great numbers of sea otters on his voyage among the islands of the Bering Sea and the north Pacific Ocean. Some furs were taken back to Russia and soon this new commodity was highly prized for coats. Hunting began. In 1857, Russia sold Alaska to the USA for $7,200,000. This cost was recouped in 40 years by selling sea otter pelts. In 1885 alone, 118,000 sea otter pelts were sold. By 1911, when commercial hunting ceased, the sea otter was close to extinction, with a worldwide population of fewer than 2,000. It was hardly ever seen along the Californian coast from 1911 until 1938.

The inshore marine ecosystem changed where the sea otter disappeared. Sea urchins, which were eaten by the otters, underwent a population explosion. They consumed large portions of the kelp and other seaweeds. While the otters were present, the kelp formed a luxuriant underwater forest, reaching from the sea bed, where it was anchored, to the sea surface. With no otters to keep sea urchins in check, the kelp vanished. Stretches of the shallow ocean floor were turned into sea-urchin barrens, which were a sort of submarine desert.

Happily, a few pairs of sea otters had managed to survive in the outer Aleutian Islands and at a few localities along the southern Californian coast (Figure 12.12). Some of these were taken to intermediate sites in the USA and Canada where they were protected by strict measures. With a little help, the sea otters staged a comeback and the sea urchins declined. The lush kelp forest grew back and many lesser algae moved in, along with crustaceans, squids, fishes, and other organisms. Grey

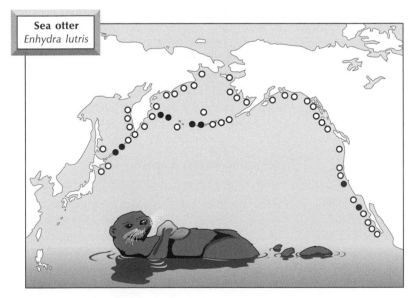

Sea otter
Enhydra lutris

Figure 12.12 Sea otter (*E. lutris*) distribution. Circles are its mid-eighteenth-century distribution. Black dots are its distribution around 1910.
Source: Partly adapted from Ziswiler (1967)

whales migrated closer to shore to park their young in breaks along the kelp edge while feeding on the dense concentrations of animal plankton. By the 1980s, some 55,000–74,000 sea otters lived in the Aleutian archipelago. However, a comparison between aerial surveys carried out in 1962, 1992, and 2000 revealed a sharp decline in the Aleutian archipelago sea otter population (Doroff *et al.* 2003). The decline started in the mid-1980s and reduced the population to an estimated 8,742 individuals by 2000. Predation of adults by killer whales (*Orcinus orca*), which probably shifted their diet to include sea otters after their preferred prey of harbour seals (*Phoca vitulina*) and Steller sea lions (*Eumetopias jubatus*) diminished, is implicated as the chief cause of the sea otters' decline. Sea otter and pinniped (sea lion, walrus, and seal) populations in the Commander Islands of Russia have not declined, which finding supports the predation hypothesis.

Keystone predators sometimes are more effective within certain parts of their range. The sea star (*Pisaster ochraceus*) is a keystone predator of rocky intertidal communities in western North America (Paine 1974). This starfish preys primarily on two mussels – the California mussel (*Mytilus californianus*) and the Pacific mussel (*M. trossulus*). A study along the central Oregon coast showed that three distinct predation regimes exist (Menge *et al.* 1994). Strong keystone predation occurs along wave-exposed headlands. Less strong predation by sea stars, whelks, and possibly other predators occurs in a wave-protected cove. Weak predation occurs at a wave-protected site regularly buried by sand.

Keystone herbivores and omnivores

Some keystone herbivores and omnivores are so dominant that they help to structure the ecosystems in which they live. Beavers, elephants, and humans are cases in point. The beaver (*Castor canadensis*) builds dams, so creating ponds and raising water tables (Naiman *et al.* 1988). The wetter conditions produce wet meadows, convert streamside forest into shrubby coppice areas, and open gaps in wood some distance away by toppling trees. The elongated area fashioned by a beaver family may exceed 1 km in length. Beavers also cause changes to the biogeochemical characteristics of boreal forest drainage networks (Naiman *et al.* 1994).

Elephants, rhinoceroses, and other big herbivores play a keystone role in the savannahs and dry woodlands of Africa (Laws 1970; Owen-Smith 1989). African elephants (*Loxodonta africana*) are relatively unspecialized herbivores. They have a diet of browse with a grass supplement. In feeding, they push over shrubs and small trees, thus helping to convert woodland habitats into grassland. They sometimes destroy large mature trees by eating their bark. As more grasses invade the woodland, so the frequency of fires increases, pushing the conversion to grassland even further. Grazing pressure from white rhinoceroses (*Ceratotherium simum*), hippopotamuses (*Hippopotamus amphibius*), and eland (*Taurotragus oryx*) then transforms medium-tall grassland into a mosaic of short- and tall-grass patches. The change from woodland to grassland is detrimental to the elephants, which begin to starve as woody species disappear, but beneficial to the many ungulates that are grassland grazers. Over millions of years, browsing and grazing by the **megaherbivores** of sub-Saharan Africa has created a mosaic of habitats and maintained a rich diversity of wildlife.

Some early human groups were keystone omnivores. Their invasion of North America may explain the mass extinction of megaherbivores (Owen-Smith 1987, 1988, 1989) (see p. 324).

Removing keystone species

What happens if a trophic level should be removed from an ecosystem? This is an enormously difficult question, to which there are at least four contradictory answers (Pimm 1991) (Box 12.1).

The effect of removing species depends upon the complexity of the food web (Figure 12.14).

Box 12.1

REMOVING TROPHIC LEVELS: THREE IDEAS

What happens if a trophic level is removed from a community? To answer this question it is helpful to ask what holds communities together. There are three competing ideas (Pimm 1991) (Figure 12.13).

The 'world is green' hypothesis

Carnivores, which have no predators above them to keep them in check, should keep herbivore populations low. In turn, the herbivores should have little impact on plants, which thus thrive. Consequently, removing herbivores should have a minor effect upon an ecosystem, whereas removing carnivores should have a major effect because it would allow herbivores to boom. In this model, competition is intense between plants and between predators, but not between herbivores. Removal of a predator will therefore have major consequences for other predators.

The 'world is prickly and tastes bad' hypothesis

Most plants possess defence systems, such as toxins and sharp spines, that limit herbivore numbers. The scarcity of herbivores in turn limits the number of carnivores. If either carnivores or herbivores should be removed, the effect upon the community as a whole would be limited because they are already kept in check by the plants at the base of the food chain. In this model, competition is between herbivores and between predators, and, if a species of either is removed, it should have consequences for the other species at the same level of the food chain.

The 'world is white, yellow, and green' hypothesis

This considers the likely effects of ecosystem productivity. 'White' ecosystems with low productivity, such as tundra, contain scant

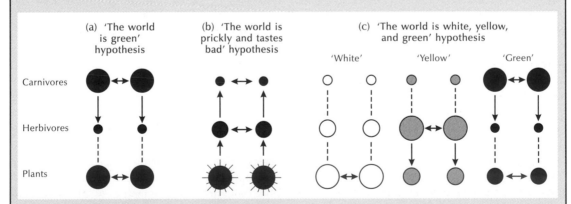

Figure 12.13 What holds communities together? Three possible answers. (a) 'The world is green' hypothesis. (b) 'The world is prickly and tastes bad' hypothesis. (c) 'The world is white, yellow, and green' hypothesis.
Source: Adapted from Pimm (1991)

plants that compete among themselves for limited resources. They do not produce enough food to support more than a few herbivores and even fewer carnivores. Removing herbivores and carnivores, which compete only weakly with their peers, from 'white' ecosystems will have little impact. 'Yellow' ecosystems, such as temperate forests and grasslands, have medium productivity. Enough plant material is produced to support a modest number of herbivores, certainly enough to keep plant numbers in check, but not to support a large number of carnivores. In 'yellow' ecosystems, carnivore removals will have little impact, but herbivore removals will have a major impact on plants. 'Green' ecosystems, such as tropical forests, are highly productive. There are enough carnivores to keep herbivores in check. In 'green' eco-systems, removing carnivores will affect herbivores numbers, which will in turn affect plants.

These hypotheses are simplistic, but the field evidence tends to favour the 'world is white, yellow, and green' hypothesis, which offers plausible explanations for certain features of ecosystems. It explains why carnivores are sometimes important in ecosystems, and sometimes they are not important. It explains why controlling carnivore numbers sometimes increases the population of a prey species, and sometimes it does not do so. 'White' ecosystems, such as the Arctic tundra, do seem resilient to major perturbations. 'Green' ecosystems, such as Yellowstone National Park, do suffer from the removal of carnivores − herbivore numbers increase greatly with an accompanying impact on vegetation.

In a complex food web, removal of a plant at the bottom has little effect through the rest of the ecosystem (Figure 12.14a). This is borne out by the limited impact of American chestnut (*Castanea dentata*) removal from the eastern forests owing to chestnut blight (p. 151). Seven species of butterfly that feed exclusively on the chestnut are probably extinct but 49 other species of butterfly that also fed on the chestnuts found alternative food sources, as did the insect predators that fed on all 56 butterflies. On the other hand, removal of keystone predators or herbivores is not so innocuous an event − a major shock cascades all the way down the food web and shakes the lowermost level (Figure 12.14b). Kangaroo rats (*Dipodomys* spp.) are keystone herbivores in the desert–grassland ecotone in North America. They have a major impact on seed predation and soil disturbance. Twelve years after their removal from plots of Chihuahuan Desert scrub, tall perennial and annual grasses increased threefold and rodent species typical of arid grassland had colonized (Brown and Heske 1990). Similarly, the cassowary is thought to be the sole disperser for over 100 species of woody tropical rain-forest plants in Queensland, Australia (Crome and Moore 1990). It usually inhabits large forests. Logging and habitat fragmentation have removed the bird from several areas, in which only small remnants of forest remain. A progressive and massive loss of trees is likely to follow, unless the cassowary adapts or adjusts its behaviour.

For simple food chains, the situation is reversed (Figure 12.14c). Highly specialized herbivores or carnivores are extremely vulnerable to the loss of their sole food source. The koala (*Phascolarctos cinereus*) is an arboreal folivore. It feeds almost exclusively on the foliage of gum trees (*Eucalyptus*). Although still widespread, the koala population is controlled in some areas where overpopulation would otherwise lead to defoliation, the death of scarce food trees, and an endangerment of the koala population (Strahan 1995, 198). The sabre-toothed cats were one of the main branches of the cat family (Felidae) throughout much of the Tertiary. They had enormous upper canine teeth and probably specialized in preying on large, slow-moving, thick-hided herbivores,

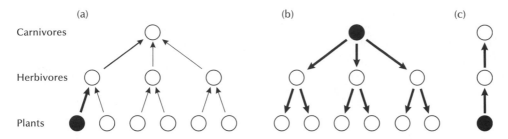

Figure 12.14 Simple and complex food webs: repercussions of removing trophic levels. (a) Removing a single plant from a food web has little impact on a predator, which draws upon a range of food sources. (b) Removing a predator from a food web creates a 'shock wave' that cascades down the trophic levels. (c) In a simple food chain, removing a single plant species may materially affect the predator.

Source: Adapted from Budiansky (1995)

such as mastodons and giant sloths. They became extinct during the Pleistocene, most likely because their prey was at first thinned by extinctions and finally vanished.

The examples suggest that introducing generalists should have a greater impact on an ecosystem than introducing specialists. Correspondingly, introductions into ecosystems containing generalist predators (which can limit the number of intruders) should have less of an impact than introductions of specialist predators.

Contaminating food webs

Humans are part of food webs. Their skills at hunting, and later farming, enabled them to exploit plants and animals in a manner quite unlike any other organism. They are super-omnivores. Hunting skills, 10,000 years ago, were equal to the task of driving large herbivores to extinction. This dubious ability has lasted and evolved into a new, and perhaps more subtle, form – farming has transformed the global land cover. In large tracts of continents, vast expanses of monoculture crops – wheat, maize, and rice – have replaced naturally diverse plant communities. Ecological efficiency runs at about 10 per cent, so it makes more sense for humans to eat plants than to let herbivores eat plants and then eat the herbivores. It makes little sense to let

carnivores eat the herbivores and then eat the carnivores. Most cultures do eat some meat and herbivores are the main source of animal protein. Nonetheless, carnivores do appear in some diets.

The human exploitation of food chains is an enormous topic. Three issues will be considered here because they impinge on biogeography – biological magnification, the long-range transport of radioactive isotopes, and the long-range transport of pesticides.

Biological magnification

Some substances, including pesticides, may be applied in concentrations that are harmless to all but the pests they are designed to eradicate. However, concentrations build up (or are magnified) as one organism eats another and the pesticide moves along a food chain. This process is called **biomagnification** or *food-chain concentration*. Rachel Carson brought its inimical effects to public notice in her book *Silent Spring* (1962). This book drew attention to the alarming build-up of long-lasting pesticides, mainly *DDT*, in the environment and the damage that they were causing to wildlife and humans near the top of food chains.

DDT is a chlorinated hydrocarbon. Othman Zeidler first prepared it in 1874. Paul Müller discovered its insecticidal properties in 1939 and, for doing so, received the Nobel Prize for Physiology. It was thought to be a panacea – a complete solution to pest control. By the early 1960s, its persistence in the environment and accumulation in the food chain were becoming apparent. In 1972, after years of forceful lobbying and petitioning in the USA, DDT was banned for all but emergency use by the Environmental Protection Agency. The biomagnification of DDT in the Long Island estuary, USA, clearly shows the increasing concentrations in moving up the trophic levels (Figure 12.15).

One reason that DDT accumulates along food chains is that it is firmly held in fatty animal tis-

sues. Some heavy metals are also stored in body tissues and are subject to biomagnification. In the Alto Paraguay River Basin, Brazil, large quantities of mercury, used in gold mining, are dispersed directly into the air and the rivers running into Pantanal, a wildlife reserve (Hylander *et al*. 1994). Local, commercially important catfish (*Pseudoplatystoma coruscans*) had a mercury content above the limit for human consumption, and significantly above the natural background level. Mercury content in bird feathers also indicated biomagnification. No statistically significant accumulation of mercury was found in soil and sediment samples. Evidently, organisms more

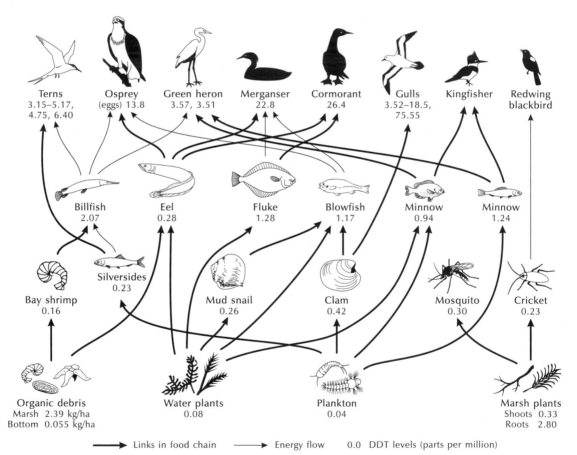

Figure 12.15 DDT biomagnification in the Long Island estuary food web, New York.
Source: Adapted from Woodwell (1967)

readily absorb mercury originating from the gold-mining process than they do mercury naturally present in soil minerals.

Ditches along busy roads are liable to pollution by heavy metals. In Louisiana, USA, animals and plants living in ditches have accumulated cadmium and lead (Naqvi *et al*. 1993). In the red swamp crayfish (*Procambarus clarkii*), the cadmium level was 32 times that in the water, and the lead level 12 times that in the water, giving bioaccumulation factors of 5.1 and 1.7, respectively.

Long-range transport of radioactive isotopes

Although southern and temperate biological systems have largely cleansed themselves of radioactive fallout deposited during the 1950s and 1960s, Arctic environments have not (D. J. Thomas *et al*. 1992). Lichens accumulate radioactivity more than many other plants because of their large surface area and long life-span; the presence and persistence of **radioisotopes** in the Arctic are of concern because of the lichen–reindeer (*Rangifer tarandus*)–human ecosystem.

Long-range transport of pesticides

Organochlorines (chlorinated hydrocarbons) include the best known of all the synthetic poisons – endrin, dieldrin, lindane, DDT, and others. They are widely used as **biocides**. *Polychlorinated biphenyls* (PCBs) are used in plastics manufacture and as flame retardants and insulating materials. A very worrying development is that high levels of organochlorines are recorded in Arctic ecosystems, where they are biomagnified and have inimical effects on consumers. The organochlorines are carried northwards from agricultural and industrial regions in the air. They then enter Arctic ecosystems through plants and start an upward journey through the trophic levels.

PCBs and DDT are the most abundant residues in peregrine falcons (*Falco peregrinus*) (D. J. Thomas *et al*. 1992). They reach average levels of 9.2 and 10.4 mg/g, respectively. These concentrations are more than 10 times higher than other organochlorines. They are also found in polar bears (*Ursus maritimus*) (Polischuk *et al*. 1995). A wide range of organochlorine pesticides and PCBs were measured in muscle tissue and livers of lake trout (*Salvelinus namaycush*) and Arctic grayling (*Thymallus arcticus*) from Schrader Lake, Alaska (R. Wilson *et al*. 1995). PCBs are recorded in marine Arctic ecosystems, too. Samples were collected near Cambridge Bay, at Hall Beach, and at Wellington Bay, all in Northwest Territories, Canada (D. A. Bright *et al*. 1995). Organisms studied included clams (*Mya truncata*), mussels (*Mytilus edulis*), sea urchins (*Strongylocentrotus droebachiensis*), and four-horned sculpins (*Myoxocephalus quadricornis*), which are fish. Some of the high concentrations in these samples were attributable to local organochlorine sources, but long-distance sources were also implicated.

FROM BACTERIA TO BLUE WHALES: BIODIVERSITY

Biological diversity, or **biodiversity** for short, incorporates three types of diversity – genetic, habitat, and species. *Genetic diversity* is the variety of information (genetic characteristics) stored in a gene pool. *Habitat diversity* is the number of different habitats in a given area. *Species diversity* is the number of species in a given area. It is also called species richness and species number.

Guestimates of the total number of species living on the Earth today vary enormously. They range from 4.4 million to 80 million. The number of known species is about 1,413,000. Therefore, the low estimate of 4.4 million would mean that 32 per cent of all species are known to date; the figure drops to a mere 2 per cent for the high estimate of 80 million.

The total species diversity disguises enormous differences between groups of organisms (Table 12.3). Insects are by far the most numerous group on the planet. Total diversity also disguises three important geographical diversity patterns — species–area relationships, altitudinal and latitudinal diversity gradients, and diversity hot-spots.

Species–area relationships

Count the species in increasingly large areas, and the species richness will increase. This, the **species–area effect**, is a fundamental biogeographical pattern. It applies to mainland species and to island species.

Species–area curves

A **species–area curve** describes an increasing number of species with increasing area. Figure 12.16 is a species–area curve for plants in Hertfordshire, England. To construct the curve, plant species records from 10-km grid squares were grouped into successively larger contiguous areas, until the entire county was covered. Figure 12.16a shows the data plotted on arithmetic coordinates. The result is a curvilinear relationship. The same data plotted on logarithmic coordinates (Figure 12.16b) produces a linear relationship between species recorded and area. The line is described by the equation

$$\log S = \log c + z \log A$$

where S is the number of species, A is area, c is the intercept value (the number of species recorded when the area is zero), and z is the slope of the line. This equation may also be written as

$$S = cAz$$

For the Hertfordshire plants, the equation is

$$S = 2.21A^{0.27}$$

The z-value (0.27) indicates that, for every 1 km² increase in area, an extra 0.27 plant species will be found.

Species also increase with area within island

Table 12.3 The diversity of selected living things

Group	Number of species
Insects	950,000
Flowering plants	258,650
Viruses	100,000
Fungi	100,000
Protozoa	60,000
Vertebrates	44,000
Fish	28,100
Amphibians	5,578
Reptiles	8,134
Birds	9,932
Mammals	4,842
Bacteria and cyanobacteria	8,500

Sources: Various

groups — large islands house more species than small islands. For amphibians and reptiles (the herpetofauna) living on the West Indian islands, the relationship, as depicted on Figure 12.17a, is

$$S = 2.16A^{0.37}$$

The line described by this equation fits the data for Sombrero, Redonda, Saba, Montserrat, Puerto Rico, Jamaica, Hispaniola, and Cuba very well. Trinidad lies well above the line. It probably carries more species than would be expected because it was joined to South America 10,000 years ago. For plant species on a selection of Scottish islands, the relationship (Figure 12.17b) is

$$S = 151.51A^{0.18}$$

A large number of studies have established the validity of this kind of relationship.

As a rule, the number of species living on islands doubles when habitat area increases by a factor of 10. For islands, values of the exponent z normally range from 0.24 to 0.34. In mainland areas, z-values normally fall within the range 0.12

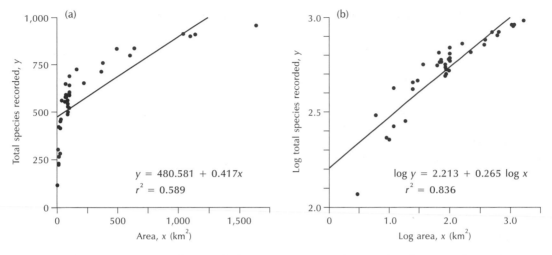

Figure 12.16 Species–area curve for Hertfordshire plants. (a) Arithmetic plot. (b) Logarithmic plot.
Source: Data from Dony (1963)

to 0.17 (the value of 0.27 for Hertfordshire plants may result from the high **geodiversity** of that county). This means that small areas contain almost as many species as large areas, but small islands contain fewer species than large islands. The difference may be partly attributable to relative isolation of islands, which makes colonization more difficult than on mainlands.

The habitat diversity and area-alone hypotheses

A crucial question is why there is such a good relationship between area and species richness. The answer is not simple. Two rival hypotheses have emerged (Gorman 1979, 24). The *habitat diversity hypothesis* suggests that larger areas have more habitats and therefore more species. The *area-alone hypothesis* proposes that larger areas should carry more species, regardless of habitat diversity. A study of **dicotyledonous plant species** on 45 uninhabited, unimproved, small islands off Shetland Mainland, Scotland, plus two similar headlands treated as islands, sought

to test the two hypotheses (Kohn and Walsh 1994). Species lists were complied during a systematic transect search covering each island. A second species list was derived by placing 50 cm × 50 cm-square quadrats randomly on 22 islands sufficiently vegetated for a reasonable number of samples. Habitats were classified according to physical characteristics assumed to be important for plants. Fourteen habitat types were identified (Table 12.4). The results show strong positive correlations between species per island, island area, and habitat diversity (Figure 12.18). Habitat diversity itself correlated with island size. A technique called path analysis was used to distinguish the effect of island size on species diversity from the direct effect of habitat diversity. The sum of the direct effects (area acting through area alone) and indirect effects (area acting through habitat diversity) of area were almost twice the overall effect of habitat diversity on species diversity. This finding strongly suggests a direct relationship between island area and species diversity, independent of habitat diversity.

The theory of island biogeography

The original MacArthur–Wilson model

When islands are colonized, the colonists seem to replace species that become extinct; in other words, there is a *turnover* of species. The **theory of island biogeography** combines geographical influences on species diversity with species change, and stresses the dynamism of insular communities. Frank W. Preston (1962), and Robert H. MacArthur and Edward O. Wilson (1963, 1967) proposed it independently. Preston stressed the idea that island species exist in some kind of equilibrium. MacArthur and Wilson explicitly set down an equilibrium model. Their central idea was that an equilibrium number of species (animals or plants) on an island is the outcome of a balance between *immigration* of new species not already on the island (from the nearest area of mainland) and *extinction* of species on the island. In other words, it reflects the interplay of species inputs (from colonization) and species outputs (from extinctions). The equilibrium is dynamic, the consequence of a constant turnover of species. In its simplest form, the MacArthur–Wilson hypothesis makes two key assumptions about what happens to immigration and extinction rates as the number of species living on the island mounts:

1 The rate of species immigration drops (Figure 12.19a). This happens because, on average, the more rapidly dispersing species would become established first, causing an initial rapid drop in the overall immigration rate, while the later arrival of slow colonizers would drop the overall rate to an ever-diminishing degree.
2 The rate of extinction of species rises (Figure 12.19a). It does so because the more species that are present, the more they are likely to go extinct in a unit time.

The point at which the lines for immigration rate and extinction rate cross defines the equilibrium

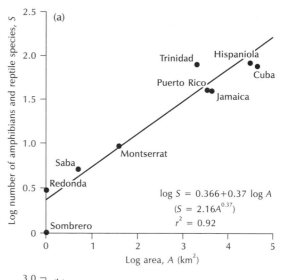

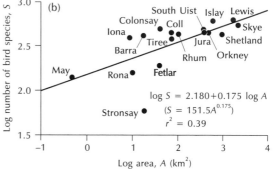

Figure 12.17 Species–area curve for island species. (a) Herpetofauna (amphibians plus reptiles) diversity on some West Indian islands. The equation for this classic best-fit line differs from the version presented in many books and papers because Trinidad and the tiny island of Sombrero were included in the computations. (b) Bird species on Scottish islands.

Sources: (a) From data in Darlington (1957, 483). (b) From data in M. P. Johnson and Simberloff (1974)

number of species for a given island (Figure 12.19a).

To refine their model a little, MacArthur and Wilson assumed that immigration rates decrease with increasing distance from source areas: immigration occurs at a higher rate on near islands than

Table 12.4 Habitat classification used in Shetland study

Class	Type		Number
Rocks	Sea cliff		1
	Scree		2
	Boulder field		3
Pasture	General, dry		4
	Rocky pasture		5
	Wet pasture:	Damp depressions, runnels	6
		Moss-dominated wet ground	7
		Mud	8
		Bog	9
		Open standing water	10
		Stream	11
	Grassy cliff		12
Shingle	Shingle		13
Sand	Sand		14
Total number of habitats			14

Source: After Kohn and Walsh (1994)

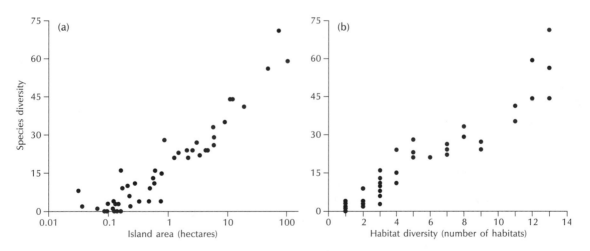

Figure 12.18 Results of the Shetland study. (a) Species diversity versus island area. (b) Species diversity versus habitat diversity.

Source: Adapted from Kohn and Walsh (1994)

on far islands (Figure 12.19b). For this reason, and all other factors being constant, the equilibrium number of species on a near island will be higher than that on a far island. In addition, MacArthur and Wilson assumed that extinction rates vary inversely with island size: extinction rates on small islands will be greater than extinction rates on large islands (Figure 12.19b). For this reason, and all other factors being constant, the equilibrium number of species on a small island will be lower than that on a large island. Several other refinements to the model were discussed in the monograph, *The Theory of Island Biogeography* (MacArthur and Wilson 1967).

Criticisms of the model

Several objections were raised to the original theory of island biogeography (for reviews see Williamson 1981, 82–91; Brown 1986; Shafer 1990, 15–18). Some claimed that it is so over-simplified as to be useless (e.g. Sauer 1969; Lack 1970; Gilbert 1980). This is the normal and, to some degree, understandable response of many field-based researchers to mathematical models that reduce the complex patterns and processes encountered in the field to a few 'simple' formulae. Admittedly, a fuller description of change in island-species richness, *S*, as a dynamic system could be written:

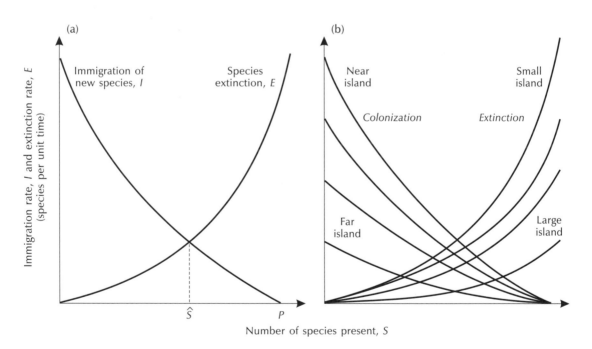

Figure 12.19 Basic relationships in the MacArthur–Wilson theory of island biogeography. (a) Equilibrium model of a biota on a single island. The equilibrial number of species is defined by the intersection of the curves for the rate of immigration of species not already resident on the island, *I*, and the rate of extinction of species from the island, *E*. (b) Equilibrium models of biotas of several islands lying at various distances from the principal source area of species and of varying size. An increase in distance (near to far) is assumed to lower the immigration rate, while an increase in area (small to large) is assumed to lower the extinction rate.

Source: Adapted from MacArthur and Wilson (1963)

$$\frac{dS}{dt} = \begin{matrix}(\text{Speciation} + \text{Immigration}) - \\ (\text{Extinction} + \text{Emigration})\end{matrix}$$

This equation is the equivalent to a population model in which population size changes because of births and deaths, immigration and emigration. Here are seven criticisms of the theory:

1   Autochthonous speciation (the evolution of species on the island itself) may be a significant factor. Within the Philippines, speciation on islands has contributed substantially to species richness, exceeding colonization by a factor of two or more as a supplier of new species (Heaney 1986). So, autochthonous speciation is important on some islands, and especially larger ones (Losos and Schluter 2000).

2   Immigration rate depends on island area as well as distance from mainland, since migrating animals are more likely to encounter a large island than a small one. Similarly, extinction rate may depend on distance from mainland as well as island area, since a dwindling species population may be saved from extinction by the arrival of immigrants. This 'rescue effect' is likely to be more efficacious on near islands than far islands, and, where dominant, can lead to a higher turnover rate of species on far islands than on near islands, a phenomenon often observed in nature (Brown and Kodric-Brown 1977).

3   The immigration curves for plants are unlikely to fall smoothly as species richness increases. In reality, the rate of immigration is likely to be low for pioneer species and only rise to high levels once pioneers have established themselves.

4   Species richness will be affected by habitat diversity, a factor that is not included in the theory. It does seem fair to suggest that the resources of an island will support a limited number of species; in other words, the species-carrying capacity will vary according to environmental factors and the heterogeneity of an island landscape.

5   Islands are characteristically impoverished in species, but the theory suggests that immigration becomes zero when the island houses the same number of species as in the mainland pool. In fact, the model explains the impoverishment of distant islands (compare near and far curves for a constant extinction rate in Figure 12.19b); but it is also possible that distant islands are impoverished in species because they are colonized more slowly and they have not yet been filled to a steady-state level.

6   The theory is uninformative about the mix of species on an island, which normally differs from the mix of species on neighbouring mainland owing to impoverishment. The poverty of island species becomes more pronounced towards the top of food chains. Differential impoverishment, combined with the fact that species have different dispersal abilities, leads to island biotas being disharmonious. Thus, oceanic islands tend to be populated by waif biotas and are characterized by **disharmonious communities**. The Pacific island faunas contain hardly any mammals (except bats and rats). Land birds are common, but most of them have a very restricted range, and only a few are migratory. Geckos and skinks that probably arrived on flotsam represent most of the reptiles, while snakes are found only on islands near continents. Amphibians and freshwater fishes are conspicuous by their absence. Land molluscs are well represented, and the dominant animals are insects.

7   The equilibrium number of species will vary with time in response to environmental changes. Indeed, insular faunas may not normally be in equilibrium because geological and climatic changes can occur as fast as colonization and speciation (Heaney 1986).

Despite these criticisms, many of which can be overcome by minor modifications of the basic model, the theory of island biogeography has engendered much valuable debate and a flood of field investigations into the effects of insularity. Indeed, many studies tend to vindicate the basic thesis that steady-state species numbers are a function of area and distance. For instance, mammalian faunas on 24 land-bridge islands in the Gulf of Maine, USA, could be explained by recurrent colonizations, mainly via ice bridges (Crowell 1986). Species richness varied directly with area and inversely with distance, though the richness was modulated by climatic change, range expansions, and human disturbance. Other studies of mammalian faunas on islands in Finnish lakes (Hanski 1986) and in the St Lawrence River, North America (Lomolino 1986) show that they are roughly in equilibrium with colonization rates balancing extinction rates.

The relaxation hypothesis

The theory of island biogeography has indisputably stimulated new ideas about species in true islands and habitat islands. An interesting idea is that the structure of insular mammalian faunas in a state of 'relaxation' may be described and explained by the *nested subset hypothesis*. First proposed by Bruce D. Patterson (1984), this hypothesis holds that the species in a depauperate insular fauna should consist of a proper subset of those in richer faunas, and that an archipelago of such faunas arranged by species richness should present a nested series (Patterson 1984, 1987, 1991; Patterson and Atmar 1986; Patterson and Brown 1991). A case in point is the debate over small mammal populations in montane islands in the American Southwest. It was originally thought that these montane islands contain 'relaxation' faunas, in which species richness decreased ('relaxed') through time as species demanding large resources (such as big specialist carnivores) became extinct and, owing to isolation, no immigrant species made good the losses (Brown 1971). The argument ran that the montane islands of forests and woodlands are fragments of a once continuous forest–woodland habitat that connected boreal habitats of the Sierra Nevada, Great Basin, and Rocky Mountains. After the Pleistocene epoch ended, climatic warming and drying broke up the forest–woodland that eventually contracted to higher elevations where relatively cool and moist local climates could support forest and woodland. Small, non-volant mammals living in these montane forest islands should thus be influenced by selective extinctions during relaxation, but not by Holocene immigrations of new species. The fact that mammalian species richness of these habitat islands in the Great Basin was correlated with island area (a measure of extinction probability), but not isolation (a measure of immigration potential) lent support to this hypothesis. A new hypothesis for the structure of mammalian communities on montane islands, which includes the possibility of Holocene immigration, was proposed when it was recognized that the great majority of montane islands in the American Southwest are now isolated by woodland and not grassland, chaparral, or desert scrub (Lomolino *et al.* 1989). Strong support for the hypothesis came from highly significant correlations between mammalian species richness and current isolation. Furthermore, when isolation was partitioned between distance to be travelled across grassland and chaparral habitats versus distance to be travelled across woodland, species richness was significantly correlated with the former but not with the latter, showing that woodlands do not present a major obstacle to immigration. Undoubtedly, historical processes of vicariance and subsequent selective extinction have helped to structure these communities, but the combined importance of immigration and extinction argues for a dynamically structured system on the lines set out by MacArthur and Wilson (1967).

New models of species–area relationships and island biogeography

A general model of species–area relationships takes into account the change in factors determined by species richness as geographical scale increases (Figure 12.20a) (Lomolino 2000a; Lomolino and Weiser 2001). The idea is that, on small islands, the dominant factors shaping species richness are such chance events as hurricanes. Beyond an 'ecological threshold', on larger islands, such deterministic factors as habitat diversity, carrying capacity, and extinction and immigration dynamics as envisioned by MacArthur and Wilson predominate. On large islands lying beyond an 'evolutionary threshold', *in situ* or autochthonous evolution of island endemics is a powerful force bolstering species numbers.

A species-based theory of island biogeography has also been presented (Lomolino 2000b). This theory is an alternative to the MacArthur–Wilson theory, taking as its premise the idea that general patterns of species assemblages on islands result from, not despite, non-random variation among

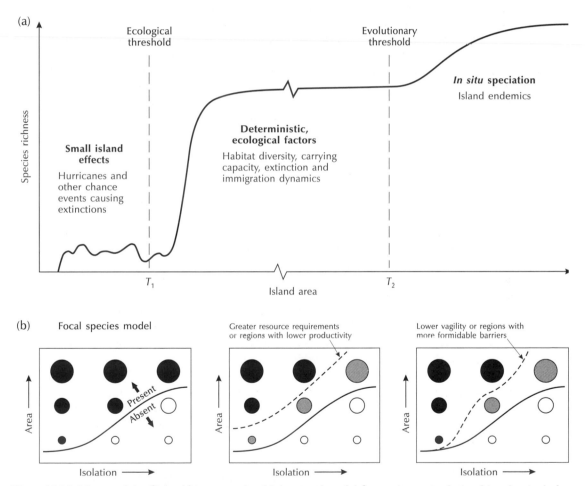

Figure 12.20 New models of island biogeography. (a) A general model for species–area relationships that includes scale-dependent changes in the chief factors shaping island communities. (b) Insular distribution functions delineating combinations of island area and island isolation where persistence time equals the time between immigration.

Sources: (a) Adapted from Lomolino (2000a). (b) After Lomolino (2000b)

species. It identifies 'insular distribution functions' that separate islands where a species should live from islands where it should not according to island size and island isolation (Figure 12.20b).

Diversity gradients and hot-spots

The latitudinal species diversity gradient

Many more species live in the tropics than live in the temperate regions, and more species live in temperate regions than live in polar regions. In consequence, a *latitudinal species diversity gradient* slopes steeply away from a tropical 'high diversity plateau'. Virtually all groups of organism exhibit the diversity gradient. An example is the species richness gradient of American mammals (Figure 12.21). The diversity falls from a tropical high of about 450 mammal species to a polar low of about 50 mammal species.

Why are there so many species (and genera, and families, and orders) in the tropics? Or, conversely, why are there so few species in temperate and polar regions? These are fundamental questions in biogeography and ecology and have exercised the minds of researchers for over a century. There is no shortage of suggested answers (Table 12.5). A recent survey of 20 hypotheses suggests that each one contains an element of circularity or else is unsupported by sufficient evidence (Rohde 1992). Hypotheses flawed by circular reasoning include biotic habitat diversity, competition, niche width, and predation. Hypotheses that lack an empirical base include climatic stability, climatic variability, latitudinal ranges, area, geodiversity, and primary production. It is possible that none of the proposed causes can alone account for the latitudinal species gradient. A universal explanation may lie in the changing impact of abiotic and

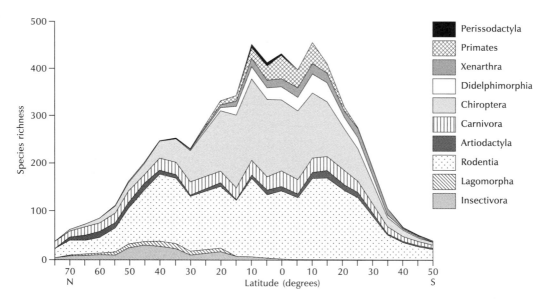

Figure 12.21 Latitudinal diversity gradient mammal species richness in the Americas. The species are grouped into Orders. Two Orders are omitted because they contain too few species. The Microbiotheria contains one species that ranges from 35°S to 45°S. The Paucituberculata contain three species found in bands from 15°N to 20°S and from 30°S to 45°S. Notice that species richness is fairly constant on the 'tropical species richness plateau' and declines steeply with increasing latitude outside the tropics.

Source: Adapted from Kaufman (1995)

Table 12.5 Some factors thought to influence species diversity gradients

Factor	Rationale
History	More time allows more colonization and the evolution of new species. In polar and temperate regions, diversity was greatly reduced during the ice ages and is now building up again
Climate	The warm, wet, and equable tropical conditions encourage a smaller niche breath, and therefore more species, than the colder and highly seasonal conditions elsewhere
Climatic stability	Tropical climatic stability is conducive to species specialization (smaller niche width) and therefore more species
Habitat heterogeneity	The greater the habitat diversity, the greater the species diversity. Forests contain more niches than grasslands. Tropical forests contain more niches than any other biome
Primary production	In food-deficient habitats, animals cannot be too choosy about their prey; in food-rich habitats, they can be selective. So food-rich environments allow greater dietary specialization and smaller niche width
Primary production stability	Habitats with stable and predictable primary production should allow greater dietary specialization (smaller niche width) than habitats with more variable and erratic primary production
Competition	Competition, which is most intense in the tropics, favours reduced niche width
Predation	Predation reduces competitive exclusion. Predators are therefore 'rarefying agents', reducing the level of competition between their prey species
Disturbance	Moderate disturbance mitigates against competitive exclusion
Energy	Species richness is limited by the partitioning of energy among species. In the energy-rich tropics, there is more energy to 'dish out' and it can be spread around a larger number of species than in temperate and polar regions
Latitudinal range size	Range size tends to increase towards the poles (Rapoport's rule), so fewer species can be packed into a given area
Area	The tropics cover more area than any other zone, which stimulates speciation and inhibits extinction (Rosenzweig 1992, 1995)

biotic factors in moving from equator to poles (Kaufman 1995).

The relative importance of abiotic and biotic factors in determining species distributions is shown in Figure 12.22. In polar regions, productivity is low and extreme fluctuations in abiotic conditions occur. These extreme environmental conditions result in physiological stress in organisms. Such adverse conditions must be countered by adaptations in structure, physiology, and behaviour, which all require energy expenditure. Abiotic factors are most likely to be severely limiting in polar regions and relax towards the equator. Contrariwise, biotic interactions are less limiting in polar regions because species diversity is low. They become severer towards the equator. As it becomes easier to cope with abiotic conditions, species can devote more resources to

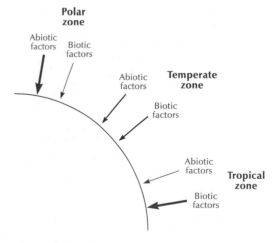

Figure 12.22 Controls on biotic diversity in polar, temperate, and tropical zones.

Source: Adapted from Kaufman (1995)

interacting with other species, that is, to competing. By these arguments, in passing along the gradient from the abiotically stressful poles to the more abiotically congenial tropics, biotic interactions should become increasingly important in limiting species distributions and influencing species diversity.

A general explanation for latitudinal gradients emerges from these ideas (Kaufman 1995). Abiotic conditions limit diversity by setting the higher latitude boundaries of species distributions, and permit only a few species to live near the poles. Biotic interactions become limiting where abiotic conditions are more favourable, set the lower latitude boundaries of species distributions, and allow many species to live in the tropics. Ultimately, the interplay of abiotic and biotic factors generates the latitudinal gradient in species diversity. Latitudinal diversity gradients occur in genera, families, and orders, as well as in species (Box 12.2).

Diversity hot-spots

Superimposed on latitudinal and altitudinal trends are *diversity hot-spots*. These are areas where large numbers of endemic species occur. Eighteen hot-spots have been identified globally (Figure 12.24). Fourteen occur in tropical forests and four in Mediterranean biomes. These hot-spots contain 20 per cent of the world's plant species on 0.5 per cent of the land area. All are under intense development pressure.

Box 12.2

LATITUDINAL DIVERSITY GRADIENTS, IN GENERA, FAMILIES, AND ORDERS

Animals within genera, families, and orders share a basic body plan (bauplan) and many other characteristics that constrain their distribution and diversity (Kaufman 1995). Members of the same genus are alike in many particulars of their morphology, physiology, behaviour, and ecology. The grass voles, genus *Microtus*, are the dominant voles in European grasslands. They feed on grasses and sedges, and have ever-growing cheek teeth with a sharply angular triangular pattern on the grinding surfaces. Members of the same Order share a basic suite of characteristics. All bats (Chiroptera), for instance, share traits associated with flight. Body plans are not limited by biotic interactions. Rather, they are constrained by abiotic factors – a body plan suitable for an aquatic existence is hardly suited to life in deserts (there are no desert-dwelling seals!). Abiotic factors have a top-down influence on the taxonomic hierarchy.

They influence geographical distributions most strongly at the level of orders, followed in decreasing strength by families, genera, and species. Just a few body plans can survive under the severe abiotic stresses of high latitudes, and those body plans tend to be generalized. So, abiotic factors act absolutely on diversity, limiting the number of body plans, and, because a species cannot exist where its body plan cannot exist, the number of species. The 17 bat species found north of 45°N belong to one family body plan – the Vespertilionidae. Likewise, the 8 xenarthran species found south of 35°S all belong to the family Dasypodidae. Conversely, biotic factors have a bottom-up influence on the taxonomic hierarchy, and their primary effects are felt at the species level. They do not limit the number of species per se, but they do limit each species individually by influencing population dynamics and niche width. Biotic factors

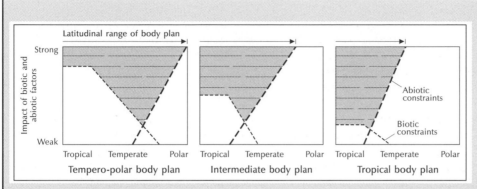

Figure 12.23 Latitudinal variations in abiotic pressure and biotic pressure acting on species of three hypothetical body plans: a tempero-polar body plan, an intermediate body plan, and a tropical body plan. The horizontal lines in the shaded areas represent latitudinal species ranges within a body plan.

Source: Adapted from Kaufman (1995)

lead to specialization and species packing. Biotic factors influence macrotaxonomic distribution only if the body has a feature that affects the interactions among species.

Each species and body plan trades off a resistance to abiotic pressures at its polar-most edge and to biotic pressures at its equatorial-most edge. The trade-off strategies of species and body plans produce different latitudinal distributions (Figure 12.23). Abiotic limitations operate to a significant effect only outside the tropics. Biotic limitations have a nearly constant effect in the tropics and then tail off towards either pole. Combined, abiotic and biotic factors limit species distributions within a body plan, and influence the body plan itself. Body plans occupy larger latitudinal ranges than species, because most body plans consists of many species. Body plans tend to be limited only by abiotic factors. Normally, they straddle the tropics and are limited by abiotic factors at each polar-most edge of their distribution. On the other hand, abiotic and biotic pressures potentially limit species. Therefore, they have reduced distributions, compared with their body plans, owing to the trade-offs in adapting to both these pressures.

SUMMARY

Communities are collections of interacting populations living in a particular place. Ecosystems are communities and their physical environment. Communities and ecosystems range in size from a few cubic centimetres to the entire biosphere and ecosphere. Organisms have roles within a community. Some are producers, making biomass from energy and mineral resources. Gross primary productivity is the amount of biomass produced in a unit time. Net primary productivity is net primary productivity less the energy burned by primary producers. Consumers eat the primary production. Macroconsumers are herbivores, carnivores, and top carnivores. Microconsumers are decomposers and detritivores. Energy flows through ecosystems. Biogeochemicals cycle around ecosystems, mineralization of dead organic remains providing food for renewed plant growth. Feeding relationships within communities produce food chains or food webs. Grazing food

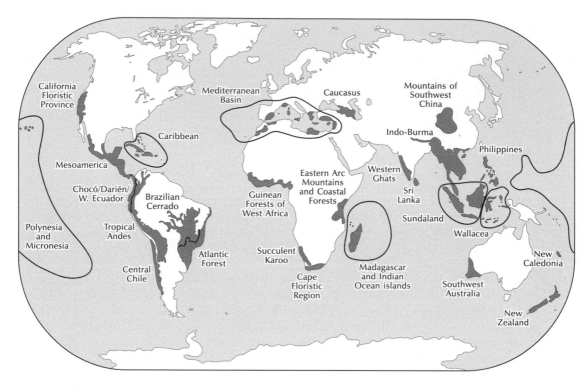

Figure 12.24 Diversity hot-spots. The circled areas circumscribe island hot-spots – Caribbean, Madagascar and the Indian Ocean islands, Sundaland, and Wallacea. The circle enclosing the Mediterranean Basin defines the Mediterranean hot-spot.

Source: Adapted from Mittemeier *et al.* (1999)

webs involve plant → herbivore → carnivore → top carnivore sequences. Decomposer food webs involve comminution spirals. Food chains have characteristic 'pyramids' of biomass, energy, and numbers. Keystone species are crucial to certain communities. Take them away and the community collapses or changes drastically. Humans have contaminated many food webs, largely because toxic substances become increasingly more concentrated as they move up the trophic levels. Biodiversity is a pressing concern. It has a consistent relationship with increasing area, as seen in species–area curves. There is a latitudinal diversity gradient – diversity decreases towards the poles from a tropical high-diversity plateau. There are also several diversity hot-spots.

ESSAY QUESTIONS

1 **What are the similarities and differences between grazing food chains and detritus food chains?**

2 **How global are global biogeochemical cycles?**

3 **How useful is the theory of island biogeography?**

FURTHER READING

Chameides, W. L. and Perdue, E. M. (1997) *Biogeochemical Cycles: A Computer-interactive Study of Earth System Science and Global Change*. New York: Oxford University Press.

If you demand much, this is the one for you.

Jeffries, M. L. (1997) *Biodiversity and Conservation*. London and New York: Routledge.

An excellent basic text.

Morgan, S. (1995) *Ecology and Environment: The Cycles of Life*. Oxford: Oxford University Press.

A good starting text.

Polis, G. A. and Winemiller, K. O. (1995) *Food Webs: Integration of Patterns and Dynamics*. New York: Chapman & Hall.

Have a look.

Quammen, D. (1996) *The Song of the Dodo: Island Biogeography in an Age of Extinctions*. London: Hutchinson.

Highly readable.

Reaka-Kudlka, M. L., Wilson, D. E., and Wilson, E. O. (1997) *Biodiversity II: Understanding and Protecting Our Biological Resources*. Washington, DC: National Academy Press.

Sure to become a classic.

Rosenzweig, M. L. (1995) *Species Diversity in Space and Time*. Cambridge: Cambridge University Press.

Fascinating, but not easy.

Schultz, J. (1995) *The Ecozones of the World: The Ecological Divisions of the Geosphere*. Hamburg: Springer.

A very good summary of the world's main ecosystems.

Whittaker, R. J. (1998) *Island Biogeography: Ecology, Evolution, and Conservation*. Oxford: Oxford University Press.

An excellent text.

Wilson, E. O. (1992) *The Diversity of Life*. Cambridge, MA: The Belknap Press of Harvard University Press.

A highly readable book.

COMMUNITY CHANGE

Communities seldom stay the same for long – community change is the rule. This chapter covers:

- communities and ecosystems in balance
- communities and ecosystems out of balance
- human-induced community change
- communities and global warming

As nature abhors a vacuum, so the biosphere abhors bare ground. Life soon colonizes land stripped clean by fire, flood, plough, or any other agency. What happens next is the subject of considerable debate. Proponents of two main schools of thought have their own views on the matter. These schools may be dubbed the 'climatic climax and balanced ecosystem' school, and the 'disequilibrium communities' school.

THE BALANCE OF NATURE: EQUILIBRIUM COMMUNITIES

Climatic climax

This influential school of thought had its heyday in the 1920s and 1930s. It stressed harmony, balance, order, stability, steady-state, and predictability within plant and animal communities. Its chief exponent was Frederic E. Clements, an American botanist. Clements (1916) reasoned that the first plant colonists are good at eking out a living under difficult conditions. By altering the local environment, they pave the way for further waves of colonization by species with a less pioneering spirit. And the process continues, with each new group of species changing the environment in such a way as to entice further colonists, until a stable community evolves whose equilibrium endures. The full sequence of change is called *vegetation succession*.

According to Clements, vegetation succession involves a predetermined sequence of developmental stages, or *sere*, that ultimately leads to a self-perpetuating, stable community called *climatic climax vegetation*. He recognized six stages in any successional sequence:

1 *Nudation* – an area is left bare after a major disturbance.
2 *Migration* – species arrive as seeds, spores, and so on.
3 *Ecesis* – the plant seeds establish themselves.
4 *Competition* – the established plants complete with one another for resources.
5 *Reaction* – the established plants alter their environment and so enable other new species to arrive and establish themselves.
6 *Stabilization* – after several waves of colonization, an enduring equilibrium is achieved.

Clements believed in the idea of *monoclimax*. He argued that a *prisere* (primary sere) was a developmental sequence that, for given climatic conditions, would always end in a reasonably permanent stage of succession – climatic climax vegetation. However, he realized that climatic climax formations contain patent exceptions to this rule. He termed these aberrant communities proclimax states. There are four forms of proclimax state (Figure 13.1): subclimax, disclimax, preclimax, and postclimax (anticlimax is something very different):

1 *Subclimax* is the penultimate stage of succession in all full primary and secondary seres. It persists for a long time but is eventually replaced by the climax community. In eastern North America, an early Holocene coniferous forest eventually gave way to a deciduous forest. Clementsians would therefore conclude that the coniferous forest was a subclimax community.
2 *Disclimax* (*plagioclimax*) partially or wholly replaces or modifies the true climax after an environmental disturbance. Cattle and sheep grazing maintain most of the so-called natural grasslands in the British Isles lying below 500 m. If grazing were to cease, the grasslands would revert to a deciduous forest. In the arid and semi-arid USA, an annual grass – cheatgrass (*Bromus tectorum*) – introduced from the Mediterranean permanently alters vegetation succession around some ghost towns (Knapp 1992).
3 *Preclimax* and *postclimax* are caused by local conditions, often topography, producing climatic deviations from the regional norm. Cooler or wetter conditions promote postclimax vegetation; drier or hotter conditions promote preclimax vegetation. In the Glacier Bay area, southeast Alaska, the climax vegetation is spruce (*Sitka sitchensis*)–hemlock (*Tsuga* spp.) forest, but at wetter sites bog mosses (*Sphagnum* spp.) dominate with occasional shore pine (*Pinus contorta*) trees forming a postclimax vegetation (see p. 258).

A *polyclimax hypothesis* supplanted Clements' monoclimax hypothesis (e.g. Tansley 1939). According to the polyclimax hypothesis, several different climax communities could exist in an area with the same regional climate owing to differences in soil moisture, nutrient levels, fire frequency, and so on. The *climax-pattern hypothesis* (R. H. Whittaker 1953) was a variation on the polyclimax theme. As in the polyclimax hypothesis, natural communities are adapted to all environmental factors, but, according to the climax-pattern hypothesis, there is a continuity of climax types that grade into one another along environmental gradients, rather than forming discrete communities that change through very sharp ecotones.

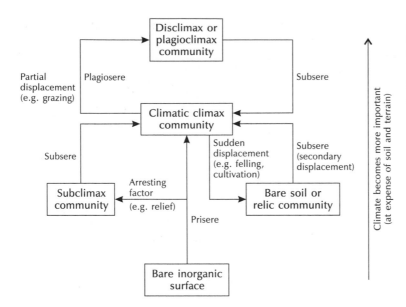

Figure 13.1 Types of climax community. The subclimax community is also called 'deflected succession'.
Source: Adapted from Eyre (1963)

Balanced ecosystems

The idea of *'balanced ecosystems'* emerged in the 1940s and persisted until the late 1960s. It did not differ radically from the idea of Clements' enduring climaxes. But it did switch the emphasis from plant formations to ecosystems and related ideas of energy flow, trophic levels, homeostasis, and *r*- and *K*-selection. An ecosystem was defined as a biotic community together with its immediate life-support system (soil, water, and air) (Tansley 1935). During the next four decades, the ecosystem idea was promoted and refined by Raymond L. Lindeman, G. Evelyn Hutchinson, and, most persuasively, Eugene P. Odum. To Odum, each ecosystem has a strategy of development that leads to mutualism and cooperation between individual species; in other words, that leads to a balanced ecosystem. The **Gaia hypothesis** is, in part, an extension of this idea. It sees an overall **homeostasis** in the ecosphere, suggesting that life and its supporting environment are a globally balanced ecosystem (e.g. Lovelock 1979).

From bare soil to climax

Facilitation, tolerance, and inhibition are the three mechanisms thought to drive succession (Horn 1981). Researchers disagree about the relative importance of these mechanisms, so there are three main models of succession.

Facilitation model

The pioneer species make the habitat less suitable for themselves and more suitable for a new round of colonists. The process of reaction (p. 255) continues, each group of species facilitating the colonization of the next group. This is the classical model of succession, as expounded by Henry C. Cowles and Clements. Reaction helps to drive successional changes. On a sandy British beach, the first plant to colonize is marram-grass (*Ammophila arenaria*). After a rhizome fragment takes hold, it produces aerial shoots. The shoots impede wind flow and sand tends to accumulate around them. The plant is gradually buried by the sand and, to avoid 'suffocation', grows longer shoots. The shoots keep growing and the mound of sand keeps growing. Eventually, a sand dune is produced. This is then colonized by other species, including sand fescue (*Festuca rubra* var. *arenaria*), sand sedge (*Carex arenaria*), sea convolvulus (*Calystegia soldanella*), and two sea spurges (*Euphorbia paralias* and *E. portlandica*), which help to stabilize the sand surface.

Tolerance model

According to this model, late successional plants, as well as early successional plants, may invade in the initial stages of colonization. In northern temperate forests, for example, late successional species appear almost as soon as early successional species in vacant fields. The early successional plants grow faster and soon become dominant. But late successional species maintain a foothold and come to dominate later, crowding out the early successional species. In this model, succession is a thinning out of species originally present, rather than an invasion by later species on ground prepared by specific pioneers. Any species may colonize at the outset, but some species are able to outcompete others and come to dominate the mature community.

Inhibition model

This model takes account of chronic and patchy disturbance, a process that occurs when, for example, strong winds topple trees and create forest gaps (p. 146). Any species may invade the gap opened up by the toppling of any other species. Succession in this case is a race for uncontested dominance in recent gaps, rather than direct competitive interference. No species is competitively superior to any other. Succession works on a 'first come, first served' basis – the species that happen to arrive first become established. It is a disorderly process, in which any directional changes are due to short-lived species replacing long-lived species.

Allogenic and autogenic succession

The members of a community propel **autogenic** succession. In the facilitation model, autogenic succession is a unidirectional sequence of community, and related ecosystem, changes that follow the opening up of a new habitat. The sequence of events takes place even where the physical environment is unchanging. An example is the heath cycle in Scotland (Watt 1947). Heather (*Calluna vulgaris*) is the dominant heathland plant. As a heather plant ages, it loses its vigour and is invaded by lichens (*Cladonia* spp.). In time, the lichen mat dies, leaving bare ground. Bearberry (*Arctostaphylos uva-ursi*) invades the bare ground, and in turn heather invades the bearberry. The cycle takes about 20–30 years.

Fluctuations and directional changes in the physical environment steer **allogenic** succession. A host of environmental factors may disturb communities and ecosystems by disrupting the interactions between individuals and species. For example, deposition occurs when a stream carries silt into lake,. Slowly, the lake may change into a marsh or bog, and eventually the marsh may become dry land.

Clements distinguished between primary succession and secondary succession.

Primary succession

Primary succession occurs on newly uncovered bare ground that has not supported vegetation before. New oceanic islands, ablation zones in front of glaciers, developing sand dunes, fresh river alluvium, newly exposed rock produced by faulting or volcanic activity, and such human-made features as spoil heaps are all open to first-time colonization. The full sequence of communities forms a primary sere or prisere. Different priseres occur on different substrates: a *hydrosere* is the colonization of open water; a *halosere* is the colonization of salt marshes; a *psammosere* is the colonization on sand dunes; and a *lithosere* is the colonization of bare rock.

A hydrosere is recognized in the fenlands of England (e.g. Tansley 1939). Aquatic macrophytes, and then reeds and bulrushes, colonize open water. Decaying organic matter from these plants accumulates to create a reed swamp in which water level is shallower. Marsh and fen plants establish themselves in the shallower water. Further accumulation of soil leads to even shallower water. Such shrubs as alder then invade to produce 'carr' (a scrub or woodland vegetation) and, ultimately, so the classic interpretation claimed, carr would change into mesic oak woodland.

Glacier Bay, Alaska

A classic study of primary succession was made in Glacier Bay National Park, southeast Alaska (Cooper 1923, 1931, 1939; Crocker and Major 1955). The glacier has retreated considerably since about 1750 (Figure 13.2; Plate 13.1). Roadside rock moss (*Racomitrium canescens*) and hoary rock moss (*R. lanuginosum*), broad-leaved willow herb or river beauty (*Epilobium latifolium*), northern scouring rush (*Equisetum variegatum*), yellow mountain avens (*Dryas drummondii*), and the Arctic willow (*Salix arctica*) characterize the pioneer stage. In the next stage, the Barclay willow (*S. barclayi*), Sitka willow (*S. sitchensis*), feltleaf willow (*S. alaxensis*), undergreen willow (*S. commutata*), and other willows appear. They start as prostrate forms but eventually develop an erect habit, forming dense scrub. Later stages of succession vary from place to place. They involve three main changes in vegetation. First, green or mountain alder (*Alnus crispa*) establishes itself in some areas. On the east side of Glacier Bay (Muir Inlet) and within 50 years, it forms almost pure thickets, some 10 m tall, with scattered individuals of black cottonwood (*Populus trichocarpa*). Second, Sitka spruce (*Picea sitchensis*) invades and after some 120 years forms dense, pure stands. The spruce finds it difficult to establish itself in those areas dominated by alder thickets. Third, western hemlock (*Tsuga heterophylla*) enters the

community, arriving soon after the spruce. Eventually, after a further 80 years or so, spruce–hemlock forest is the climax vegetation, at least on well-drained slopes. In wetter sites, where the ground is gently sloping or flat, *Sphagnum* species invade the forest floor. The more *Sphagnum* there is, the wetter the conditions become. Trees start to die and wetland forms. *Sphagnum* and occasional shore pine individuals (*Pinus contorta*), which are tolerant of wet habitats, dominate the wetland.

Krakatau Islands, Indonesia

The eruptions of 20 May to 27 August 1883 largely or completely sterilized and greatly reshaped the islands in the Krakatau group. There are four islands in the group – Rakata, Sertung, Panjang, and Anak Krakatau. They lie in the Sunda Strait roughly equidistant 44 km from 'mainland' Java and Sumatra (Figure 13.3). Two large 'stepping stone' islands – Sebesi and Sebuku – connect them with Sumatra. Rakata, Sertung, and Panjang were sterilized in 1883. Anak Krakatau emerged in 1930 and has a disturbed history of colonization.

Community succession on each of the three main islands follows a similar pattern for the first 50 years. The coastal communities were established rapidly. Typical dominant strandline species were among the first to colonize, producing a one- or two-phase succession. In the interior lowlands and upland Rakata, an early phase of highly dispersive ferns, grasses (carried by winds), and a few Compositae slowly diversified. By 1897 they had gained such species as the heliophilous (sun-loving) terrestrial orchids – bamboo orchid (*Arundina graminifolia*), nun orchid (*Phaius tankervilliae*), and ground orchid (*Spathoglottis plicata*). A scattering of shrubs and trees, most of which were animal-dispersed colonists that were interspersed among the savannah vegetation, soon spread. Forest covered Rakata by the end of the 1920s. The forest closure led to the loss of a number of early colonists that

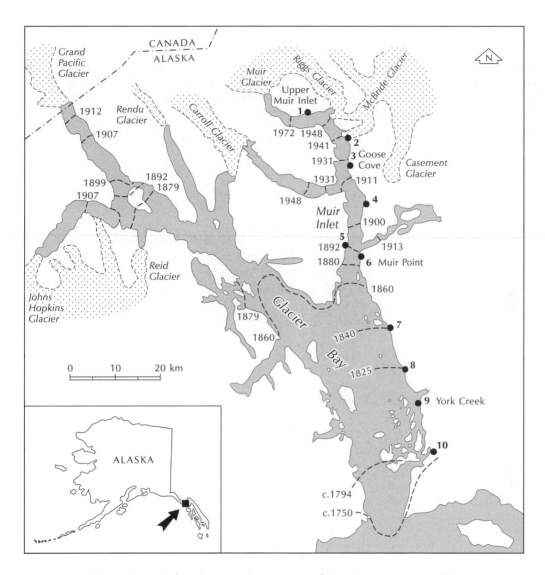

Figure 13.2 Glacier Bay, Alaska, showing the positions of the glacier termini and Fastie's (1995) study sites (which are referred to on p. 263).

Sources: Adapted from Crocker and Major (1955) and Fastie (1995)

Plate 13.1 Plant succession at Glacier Bay, Alaska. (a) Upper Muir inlet (Site 1 in Fig. 13.2), 20 years old: glacial till surface with young plants of yellow mountain avens (*D. drummondii*), black cottonwood (*P. trichocarpa*), and Sitka spruce (*S. sitchensis*). (b) Goose Cove (Site 3 in Fig. 13.2), 60 years old: alder thicket overtopped by black cottonwood. (c) Muir Point (Site 6 in Figure 13.2), 105–110 years old: alder and willow thicket over-topped by scattered Sitka spruce. (d) York Creek (Site 9 in Fig. 13.2), 165 years old: Sitka spruce forest (with western hemlock) that was not preceded by a dense alder thicket or cottonwood forest.

Photographs by Christopher L. Fastie.

were open-habitat species. Additionally, open areas were reduced to relatively small gaps and fauna dependent on open areas suffered. Losses included birds such as the red-vented bulbul (*Pycnonotus cafer*) and the long-tailed shrike (*Lanius schach bentet*).

Secondary succession

Secondary succession occurs on severely disturbed ground that previously supported vegetation. Fire, flood, forest clearance, the removal of grazing animals, hurricanes, and many other factors may inaugurate secondary succession.

Abandoned fields in Minnesota

Before the 1880s, upland habitats in the Cedar Creek Natural History Area, Minnesota, USA, were a mosaic of oak savannah (open oak wood-land), prairie openings, and scattered stands of oak forest, pine forest, and maple forest (Tilman 1988). Farming converted some of the land to fields. Some fields were abandoned at different

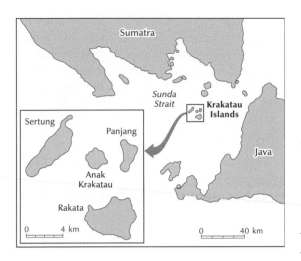

Figure 13.3 The location of the Krakatau island group. Rakata, Panjang, and Sertung are remnants of the pre-1883 volcano; Anak Krakatau, which appeared in 1930, is the only volcanically active island.

Source: Adapted from Whittaker and Bush (1993)

times and secondary succession started. A set of 22 abandoned fields, placed in chronological order, provides a picture of the successional process (Figure 13.4). The initial dominants are annuals and short-lived perennials, many of which, including ragweed (*Ambrosia artemisiifolia*), are agricultural weeds. These are replaced by a sequence of perennial grasses. After 50 years, the dominant grasses are the little bluestem (*Schizachyrium scoparium*) and big bluestem (*Andropogon gerardi*), both native prairie species. Woody plants – mainly shrubs, vines, and seedlings or saplings of white oak (*Quercus alba*), red oak (*Q. rubra*), and white pine (*Pinus strobus*) – slowly increase in abundance. After 60 years, they account for about 12 per cent of the cover. Many of the successional changes are explained by an increase of soil nitrogen over the 60-year period. Much of the pattern reflects the differing life histories and time to maturity of the component species.

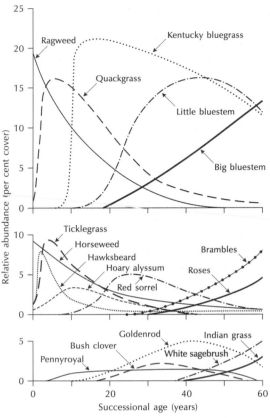

Figure 13.4 Secondary succession in Cedar Creek Natural History Area, Minnesota, USA. The diagram is based on an old field chronosequence and observations within some permanent plots. The species are quackgrass (*Agropyron repens*), ragweed (*Ambrosia artemisiifolia*), big bluestem (*Andropogon gerardi*), Kentucky bluegrass (*Poa pratensis*), little bluestem (*S. scoparium*), ticklegrass (*Agrostis scabra*), hoary alyssum (*Berteroa incana*), hawksbeard (*Crepis tectorum*), horseweed (*Erigeron canadensis*), red sorrel (*Rumex acetosella*), brambles (*Rubus* spp.), roses (*Rosa* spp.), white sagebrush (*Artemisia ludoviciana*), pennyroyal (*Hedeoma hispida*), bush clover (*Lespedeza capitata*), goldenrod (*Solidago nemoralis*), and Indian grass (*Sorghastrum nutans*).

Source: Adapted from Tilman (1988)

Ghost towns in the western Great Basin

Terrill and Wonder, which are in central-western Nevada, USA, are abandoned gold- and silver-mining camps. Terrill was abandoned in about 1915 and Wonder in about 1925. Secondary succession following abandonment was studied for two levels of disturbance – greatly disturbed (abandoned roads) and moderately disturbed (within 5 m of building foundation edges) – and at 'undisturbed' control plots (Knapp 1992). Twenty-seven species were found at Terrill. Cheatgrass (called drooping brome-grass in the UK) (*Bromus tectorum*), Shockley's desert thorn (*Lycium shockleyi*), indian ricegrass (*Oryzopsis hymenoides*), and Bailey's greasewood (*Sarcobatus baileyi*) dominated the abandoned road. Shadscale (*Atriplex confertifolia*), cheatgrass, and *Halogeton glomeratus* (an introduced Asiatic annual) dominated the foundation edges. The control plot was dominated by greasewood, shadscale, and cheatgrass. At Wonder, just seven species were found. Big sagebrush (*Artemisia tridentata*) was the dominant species in the abandoned road and control site. Big sagebrush and cheatgrass comprised 90 per cent of the cover around foundation edges.

In both towns, a striking feature of succession is the higher percentages of therophytes around foundation peripheries. This is seen in the high percentage of cheatgrass at Terrill and Wonder, and in the high percentages of *Halogeton* and tumbleweed (*Salsola kali*) at Terrill. They are not so common on highly disturbed abandoned roads because soil bulk density is higher there, which hinders the growth of tumbleweed's and cheatgrass's extensive root systems, and because phosphorus and nitrogen levels are low, which hampers the establishment of cheatgrass. Vegetation has reverted to something approaching its original state (salt desert shrub in the case of Terrill and sagebrush–grass in the case of Wonder), but complete recovery to a 'climax' state is unlikely. Therophytes, especially cheatgrass, are likely to persist. Cheatgrass excels in disturbed areas. It establishes itself easily on disturbed sites and creates a self-enhancing cycle by promoting the occurrence of additional disturbances – fire intensity and frequency and grazing by small mammals. So, cheatgrass, which is a Mediterranean introduction (p. 255), will probably stay dominant at the expense of other, native, species. The secondary succession has produced new 'climax' vegetation, a sort of permanent disclimax.

EGOCENTRIC NATURE: DISEQUILIBRIUM COMMUNITIES

Henry Allan Gleason and Alexander Stuart Watt first raised the idea that communities and ecosystems might not be in equilibrium, at least not in Clements' sense of stable climax communities or Odum's sense of balanced ecosystems, in the 1920s. The *disequilibrium view* rose to stardom in the early 1970s, when some ecologists dared to suggest that succession leads nowhere in particular – that there are no long-lasting climatic climaxes (e.g. Drury and Nisbet 1973). Instead, each species 'does its own thing', communities are ever-changing and temporary alliances of individuals, and succession runs in several directions. This disequilibrium view emphasizes the *individualistic behaviour* of species and the evolutionary nature of communities. It stresses imbalance, disharmony, disturbance, and unpredictability in nature. And it focuses on the geography of ecosystems – landscape patches, corridors, and matrixes replace the climax formation and ecosystem, and landscape mosaics replace the assumed homogeneous climaxes and ecosystems.

The individualistic comings and goings of species influence community change in a profound way. Communities change because new species arrive and old species are lost. New species appear in speciation events and through immigration. Old species vanish through local extinction (**extirpation**) and through emigration. Some species increase in abundance and others decrease in abundance, thus tipping the competitive balance within a community. Each species has its own propensity for dispersal, invasion, and population expansion. **Community assembly** is an unceasing process of species arrivals, persistence, increase, decrease, and extinctions played out in an individualistic way. Evidence that communities assemble (and disassemble) in this manner is seen in multidirectional succession, as revealed in recent vegetation **chronosequences** and discussed below, and in community impermanence, as revealed in palaeobotanical and palaeozoological studies, which are discussed in Chapter 16.

Multidirectional succession

A result of individualistic community assembly is that succession may continue along many pathways, and is not necessarily fenced into a single predetermined path. Some field studies support this idea.

Hawaiian montane rain forest

Montane rain forest on a windward slope of Mauna Loa, Hawaii, displays a primary successional sequence on lava flows with ages ranging from 8 years to 9,000 years (Kitayama *et al*. 1995). Both downy (pubescent) and smooth (glabrous) varieties of the lehua (*Metrosideros polymorpha*) dominated the upper canopy layers on all lava flows in the age range 50 to 1,400 years. The downy variety was replaced by the smooth variety on the flows more than 3,000 years old. Lower forest layers are dominated by a matted fern, uluhe (*Dicranopteris linearis*), for the first 300 years, and then by tree ferns (*Cibotium* spp.). The *Cibotium* cover declined slightly after 3,000 years, while other native herb and shrub species increased. A 'climax' vegetation state was not reached – biomass and species composition changed continuously during succession. Such divergent succession may be unique to Hawaii, where the flora is naturally impoverished and disharmonic due to its geographical isolation (Kitayama *et al*. 1995). Against this view, Hawaii

is the sort of place where succession should have the best chance of following the classical model (R. J. Whittaker personal communication). It is very interesting that montane forest on Hawaii does not follow the classical model. Rather, it may be an example of a general pattern of non-equilibrium dynamics of the kind revealed by new work at Glacier Bay and the Krakatau Islands.

Glacier Bay revisited

A re-evaluation of the succession in Glacier Bay, Alaska, used reconstructions of stand development based on tree-ring records from 850 trees at 10 sites of different age (Fastie 1995). The findings suggested that succession there is multidirectional. The three oldest sites were deglaciated before 1840. They differ from all younger sites in three ways. First, they were all invaded early by Sitka spruce (*Picea sitchensis*). Second, they all support western hemlock (*Tsuga heterophylla*). Third, they all appear to have had early shrub thickets. Sitka alder (*Alnus sinuata*) is a nitrogen-fixing shrub. Only at sites deglaciated since 1840 has it been an important and long-lived species. Black cottonwood (*Populus trichocarpa*) has dominated the overstorey only at sites deglaciated since 1900. The new reconstruction of vegetation succession in the Glacier Bay area suggests additions or replacements of single species. These single-species changes distinguish three successional pathways that occur in different places and at different times. This re-evaluation of the evidence dispels the idea that communities of different age at Glacier Bay form a single chronosequence describing unidirectional succession.

The multiple successional pathways seem to result from the number and timing of woody species arriving on the deglaciated surfaces. They do not appear to stem from spatial differences in substrate texture and lithology. For example, site proximity to a seed source at the time of deglaciation accounts for up to 58 per cent of the variance in early Sitka spruce recruitment. Nitrogen-fixing is another factor influencing successional pathways. Long-lived alder thickets produce soil nitrogen pools that tend to reduce conifer recruitment and prevent the rapid development of spruce–hemlock forest.

Krakatau Islands revisited

The latest studies of primary succession on Krakatau favour a disequilibrium interpretation (e.g. R. J. Whittaker *et al*. 1989, 1992; Bush *et al*. 1992; R. J. Whittaker and Jones 1994). With the exception of the strandline species, all components of the Krakatau fauna and flora are still changing. A lasting equilibrium does not prevail because dispersal opportunities and disturbance events ranging from continuing volcanism to the fall of individual trees continually alter the faunal and floral diversity. The patchy nature of these processes helps to explain why the forest-canopy architecture on Panjang and Sertung changed materially from 1983 to 1989, while the pace of change on Rakata over the same period was comparatively slow. It also accounts for significant differences within and between forests on the four islands. Overall, the forest dynamics of Krakatau have been highly episodic, with stands experiencing times of relatively minor change punctuated by pulses of rapid turnover and change; and all the while, new colonists species have arrived and have spread (R. J. Whittaker personal communication).

On Rakata, for example, four common successional pathways have produced distinct forest communities (R. J. Whittaker and Bush 1993; Plate 13.2). The first occurs on coastal communities, where either tropical almond (*Terminalia catappa*)–sea poison tree (*Barringtonia asiatica*) woodland is rapidly established, or else the same woodland is established after a stage of Australian pine (*Casuarina equisetifolia*) woodland. Path two is followed inland (but not in uplands) and runs from ferns, to grass savannah, to parasol leaf tree (*Macaranga tanarius*)–fig (*Ficus* spp.) forest, to hooded bur-flower tree (*Neonauclea calycina*) forest. Path three is the same as path two, but the current

stage is kajeng sampeyan (*Ficus pubinervis*)– hooded bur-flower tree forest. Path four occurs in the upland. It runs from ferns, to grass savannah, to *Cyrtandra sulcata* scrub, to *Schefflera polybotrya–Saurauia nudiflora–Ficus ribes* submontane forest.

The vegetation of Rakata is quite different from the vegetation on Panjang and Sergung (Tagawa 1992). Hooded bur-flower tree seeds successfully formed forest on Rakata, but failed to do so on Panjang and Sertung. Factors in seed dispersal, soil conditions, and disturbance by volcanic activity on Anak Krakatau all help to explain these differences. The *Timonius compressicaulis* and red dysox (*Dysoxylum gaudichaudianum*) forests found on Panjang and Sertung developed on both islands after the appearance of Anak Krakatau. Both *T. compressicaulis* and particularly red dysox have larger seeds than hooded bur-flower tree and they were able to germinate and grow under the regenerated mixed forest canopy, but it was not so easy for them to colonize the hooded bur-flower tree forest. Red dysox invasion progresses only gradually on Rakata.

Chaotic communities

In the 1990s, the notion of non-equilibrium was formalized in the *theory of chaotic dynamics*. The main idea is that all nature, including the communities and ecosystems, is fundamentally erratic, discontinuous, and inherently unpredictable. This view of communities and ecosystems arose from mathematical models. The models confirmed what ecologists had felt for a long time but had been unable to prove – emergent community properties, such as food webs and a resistance to invasion by alien species, arise from the host of individual interactions in an assembling community. In turn, the emergent community properties influence the local interactions.

All evolving ecosystems, from the smallest pond to the entire ecosphere, possess emergent properties and appear to behave like super-

Plate 13.2 Successional pathways on Rakata. (a) Path one: *Casuarina equisetifolia* woodland (at north end of Anak Krakatau). (b) Path two: lowland *Neonauclea calycina* forest. (c) Path three: lowland *Ficus pubinervis–Neonauclea calycina* forest. (d) Path four: upland forest. Photographs by Robert J. Whittaker.

oganisms. But this superorganic behaviour is the result of a continuing two-way feedback between local interactions and global properties. It is not the outcome of some mystical global property determining the local interactions of system components, as **vitalists** would contend. Nor is it the cumulative result of local interactions in the system, as **mechanists** would hold. No, the whole system is an integrated, dynamic structure that is powered by energy and involving two-way interaction between all levels.

Experiments with 'computer communities' showed that species-poor communities were easy to invade (Pimm 1991). Communities of up to about 12 species offered essentially open access to intruding species. Beyond that number, in species-rich communities, there were two results. First, newly established species-rich communities were more difficult to invade than species-poor communities. Second, long-established communities were even harder to invade than newly established species-rich communities. Other mathematical experiments started with a 125-species pool of plants, herbivores, carnivores, and omnivores (Drake 1990). Species were selected one at a time to join an assembling community.

Second chances were allowed for first-time failed entrants. An extremely persistent community emerged comprising about 15 species. When the model was rerun with the same species pool, an extremely persistent community again emerged, but this time with different component species than in the first community. There was nothing special about the species in the communities: most species could become a member of a persistent community under the right circumstances; the actual species present depended on happenstance. The dynamics of the persistent communities was very special: a persistent community of 15 species could not be reassembled from scratch using only those 15 species. This finding suggests that communities cannot be artificially manufactured from a particular set of species – they have to evolve and to create themselves out of a large number of possible species interactions.

UNDER THE AXE AND PLOUGH: LAND COVER TRANSFORMATION

In almost all parts of the world, humans are changing biomes on a massive scale. Much of this *land-cover transformation* has taken place in the last two centuries (Buringh and Dudal 1987) (Figure 13.5). It resulted from the western European core region pushing out successive waves of exploitation into peripheral regions of the globe. As the frontiers expanded, so isolated subsistence economies fell irresistibly into a single world market. Basic commodities – crops, minerals, wood, water, and wild animals – were traded globally. Combined with steam power, medicine, and advances in agricultural technology, these economic changes set in motion the Great Transformation – a worldwide alteration of land cover. Figure 13.6 shows the extent to which humans now dominate much of the land surface.

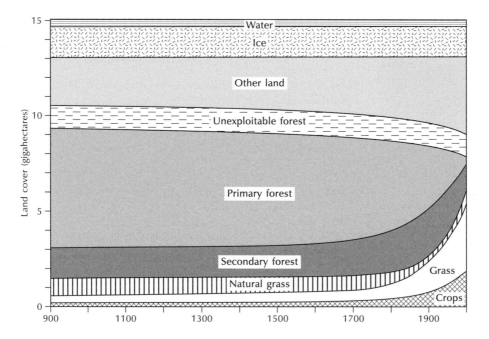

Figure 13.5 Land-cover transformation, AD 900–1977. The greatest transformation took place in the last two centuries.

Source: Adapted from M. Williams (1996)

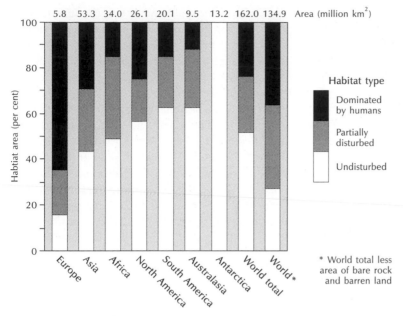

Figure 13.6 Human disturbance of habitats by continent.
Source: From data in Hannah *et al.* (1994)

The land-cover transformation has changed biomes. From 1860 to 1978, 8,517,000 km² of land were converted to cropping (Revelle 1984). The converted land was originally forest (28.5 per cent), woodland (18.3 per cent), savannah (13.8 per cent), other grassland (34.9 per cent), wetland (2.5 per cent), and desert (2.0 per cent). A major problem with this ecosystem conversion is its patchy nature – a field here, a clearing there – that has broken the original habitat into increasingly isolated habitat fragments. Five processes cause land-cover change – perforation, dissection, fragmentation, shrinkage, and attrition (Forman 1995, 406–15). Forests, for example, are being perforated by clearings, dissected by roads, broken into discrete patches by felling, and the newly created patches are shrinking and some of them disappearing through attrition. The overall effect of land-cover change is habitat fragmentation.

Habitat fragmentation

This is the breaking up of large habitats or areas into smaller parcels. It is enormously significant for wildlife and poses a global environmental problem. Several community changes occur as habitat patches become smaller and more isolated. The numbers of generalist species, species that can live in more than one habitat, edge species, and exotic species all rise. The nest predation rate increases, populations fall, and extinctions become more common. The numbers of species specialized for life within a habitat interior, and species with a large home range, both decrease, as does the richness of habitat interior species. Habitat fragmentation also changes ecosystem processes. It disturbs the integrity of drainage networks, affects water quality in **aquifers**, alters the disturbance regime to which species have adapted, and influences other ecosystem processes.

The effects of fragmenting habitats on wildlife are evident in the examples of the silver-spotted skipper butterfly in Britain, the malleefowl in Australia, and the reticulated velvet gecko in Australia.

The silver-spotted skipper butterfly in Britain

In Britain, the silver-spotted skipper butterfly (*Hesperia comma*) lives in heavily grazed calcareous grasslands (Plate 13.3) (C. D. Thomas and Jones 1993). Rabbits (*Oryctolagus cuniculus*) do most of the grazing. During the mid-1950s, myxomatosis devastated the rabbits and the habitat became overgrown. The silver-spotted skipper population shrank and became a metapopulation, occupying 46 or fewer localities in 10 refuges (Figure 13.7). The rabbits had recovered by 1982, and many former grassland sites, now neatly cropped, appeared suitable for recolonization by the silver-spotted skipper. Indeed, from 1982 to 1991, the number of populated habitat patches increased by 30 per cent in the South and North Downs, with most of the increase taking place in East Sussex. Most of the recolonized habitat patches were fairly large and close to other populated patches. But, even by 1991, many suitable habitat patches had not been reoccupied and it was thought that, in the fragmented landscape, further spread during the twenty-first century was unlikely, except in East Sussex, because bands of unsuitable habitat, which are more than 10 km wide, would halt the spread. Were the habitat not fragmented, the silver-spotted skipper might be expected to fill its previous range in southeast England within 50 to 75 years. However, since the early 1990s the range of the silver-spotted skipper has further expanded, thanks to the restoration of suitable habitat, and climate warming making more of the landscape suitable (e.g. less restricted to south-facing slopes) (C. D. Thomas *et al.* 2001). Nonetheless, there is still evidence that 'gaps' in the landscape are preventing full recolonization of its original range.

Plate 13.3 A silver-spotted skipper (*H. comma*), on the North Downs, Surrey, England. Photograph by Owen Lewis.

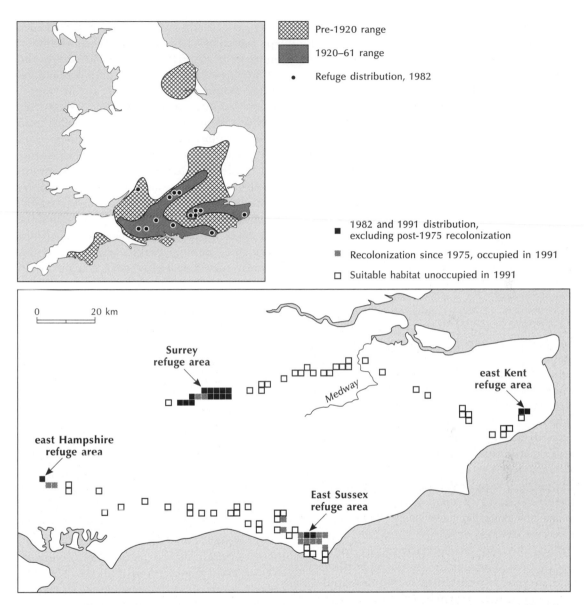

Legend:

- Pre-1920 range
- 1920–61 range
- • Refuge distribution, 1982

- ■ 1982 and 1991 distribution, excluding post-1975 recolonization
- ▨ Recolonization since 1975, occupied in 1991
- □ Suitable habitat unoccupied in 1991

0   20 km

Surrey refuge area

Medway

east Kent refuge area

east Hampshire refuge area

East Sussex refuge area

Figure 13.7 Silver-spotted skipper (*H. comma*) decline in England during the twentieth century. The three published maps show the skipper butterfly beyond the range shown, but each of the five records occurs on just one of the maps and they are excluded. The bottom map shows the results of a 1991 survey.

Source: Adapted from C. D. Thomas and Jones (1993)

The malleefowl in Australia

Mallee is an Australian sclerophyllous shrub formation (equivalent to the maquis in the Mediterranean region) characterized by high bushes of shrubs and small trees. Loss, fragmentation, and degradation of mallee habitat within the New South Wales wheatbelt have caused a marked decline in the range and local abundance of malleefowl (*Leipoa ocellata*) (Priddel and Wheeler 1994). The malleefowl is a large, ground-dwelling bird that makes a mound for its nest and keeps the temperature tightly controlled. Small, disjunct populations of malleefowl now occupy small and isolated remnants of mallee habitat. Several of these populations have recently become locally extinct. As an experiment, young malleefowl (8–184 days old) reared in captivity were released in March and June 1988 into a 55-ha remnant of mallee vegetation that contained a small but declining population of malleefowl. From the first day after release, malleefowl were found dead. Deaths continued until, within a relatively short time, no malleefowl remained alive. The main cause of their demise was predation, which accounted for 94 per cent of the deaths. Raptors accounted for 26–39 per cent of the loss, and introduced predators, chiefly the red fox (*Vulpes vulpes*), accounted for 55–58 per cent. Helpings of supplementary food made no difference to survival. Young malleefowl rely principally on camouflage for safety. They have no effective defence or escape behaviour to evade ground-dwelling predators. Foxes are imposing severe predation pressure on young malleefowl, and are probably curtailing recruitment into the breeding population. Foxes are thus a major threat to the continuance of malleefowl remnant populations in the New South Wales wheatbelt.

An Australian gecko

The reticulated velvet gecko (*Oedura reticulata*) is restricted in distribution to southwest Western Australia (Sarre 1995; Plate 13.4a). Within this region, it occurs mainly in smooth-barked eucalypt woodland habitat, mainly consisting of gimlet gum (*Eucalyptus salubris*) and salmon gum (*E. salmonophloia*). Much of the woodland has become fragmented by clearing and remains as islands within a sea of wheat (Plate 13.4b). For example, about 40 per cent of the Kellerberrin region contained smooth-barked eucalypt woodland suitable for reticulated velvet gecko populations before land clearance. The same habitat now occupies just 2 per cent of the region. Only four remnants, including three nature reserves, contain stands of more than 570 gimlet gum and salmon gum trees. Demographic characteristics of nine reticulated velvet gecko populations that currently survive in Kellerberrin eucalypt woodland remnants were assessed and compared with reticulated velvet gecko populations in three nature reserves. The woodland remnants vary in area from 0.37 to 5.40 ha, but are similar in age, tree species composition, and vegetation structure. Reticulated velvet gecko population sizes vary considerably among remnants. They are poorly correlated with remnant area and the number of smooth-barked eucalypts. Population size does not appear to be limited by habitat availability in most remnants. The number of adults of breeding age is small in most populations suggesting that they may be susceptible of stochastic extinction pressures. The species is a poor disperser – interaction between local populations is small or zero, even over distances of 700 m or less. This poor dispersal ability means that the possibility of remnant recolonization after an extinction event is unlikely. The occupancy rate of reticulated velvet gecko in remnant woodland is thus likely to decline, and the species is likely to become restricted to a few large

Plate 13.4 (a) Reticulated velvet gecko (*O. reticulata*). (b) A remnant patch of gimlet gum (*E. salubris*) in a wheat crop, near Kellerberrin, Western Australia. Photographs by Stephen Sarre.

(a)

(b)

remnants. It may therefore be futile to direct conservation effort into protecting small woodland remnants of a few hundred hectares.

Loss of wetlands

The world's *wetlands* are the swamps, marshes, bogs, fens, estuaries, salt marshes, and tidal flats. They cover about 6 per cent of the land surface and include some very productive ecosystems. But this figure is falling as wetlands are reclaimed for agricultural or residential land, drowned by dam and barrage schemes, or used as rubbish dumps. Sea-level rise during this century also poses a major threat to coastal wetlands. The following two cases illustrate the complexity of changes in wetland habitats.

Bottomland forests in the Santee River floodplain

The coastal plain drainage of the Santee River, South Carolina, USA, was drastically modified in 1941. Almost 90 per cent of the Santee River discharge was diverted into the Cooper River as part of a hydroelectric power project (Figure 13.8). However, this diversion of water has led to silting in Charleston Harbor, the main navigation channels which require year-round dredging. To alleviate this costly problem, authorization was granted in 1968 to redivert most of the water back into the Santee River through an 18.5 km-long canal feeding off Lake Moultrie (Figure 13.8). An additional hydroelectric power plant would be constructed on this canal. Fears have been voiced that this rediversion would inundate

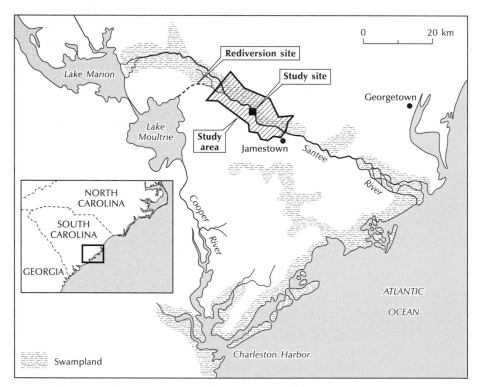

Figure 13.8 Location of the Santee River and proposed river diversion.
Source: Adapted from Pearlstine *et al.* (1985)

much of the Santee floodplain and cause a substantial decline in the bottomland forest. After the diversion, the Santee River would return to 80 per cent of its pre-diversion flow rate with seasonal spates of floods and stages of low flows. A crucial change would be a reduction of flow early in the growing season.

A forest growth model, called FORFLO, was developed to quantify the effects of the changed hydrological regime on the bottomland forests (Pearlstine *et al*. 1985) (Box 13.1). It was used to simulate the effect of the proposed river rediversion, and a modified version of it, along a 25 km reach of the Santee forested floodplain from the rediversion site downstream to Jamestown (Figure 13.8). In the proposed rediversion, flow from Lake Marion to the Santee River stays the same, while flow to the Cooper River is reduced to 85 m³/s (the level which prevents silting in

Charleston Harbor), increasing the annual flow to the Santee River via the rediversion canal to 413 m³/s. The modified rediversion was the same as the proposed rediversion, except that, during the early growing season (April to July), flow through the rediversion canal would be kept to a level that could be handled by just one of the three turbines at the power station. Although this would mean that the Cooper River would exceed the critical flow of 85 m³/s for the four months of the growing season, and thus cause some silting at the coast, it might promote the preservation of the bottomland forest.

The simulated responses of the forests to the rediversion are shown in Figure 13.9a and b. The annual duration of flooding is a crucial factor in determining the course of vegetational change. With an annual flood duration of more than 30–35 per cent (Figure 13.9a), bottomland

Box 13.1

KEY FEATURES OF FORFLO

The FORFLO model allows hydrological variables to influence tree species composition through seed germination, tree growth, and tree mortality. Key features of the model are the assumed relationships between flooding and various aspects of forest growth succession. First, for all tree species, save black willow (*Salix nigra*) and eastern cottonwood (*Populus deltoides*), seeds will not germinate when the ground is flooded. If the plot should be continuously flooded during that period of the year when a species would germinate, then the germination of that species fails. Black willow and eastern cottonwood can germinate whether the land is flooded or not. Second, after having germinated, the survival of seedlings depends on environmental conditions. A notable determinant of the seedling survival rate is the duration of the annual flood. Each

species has a tolerance to flooding, and will survive if its range of tolerance should fall within the flood duration for the plot. Third, the optimum growth of trees was reduced by, among other things, a water-table function that models floodplain conditions. The water-table function modifies the tree-growth equation to account for the tolerance of species to the level of water on the plot during the growing season. The model computes the height of water for each half month during the growing season. It is assumed that all trees will fail to grow during the half months when they are more than three-quarters submerged by floodwater. At lower levels of submergence, tree growth was related to water level by a curvilinear function in which the optimum water-table depth for each species is taken into account.

hardwood forest is replaced by cypress–tupelo forest, bald cypress (*Taxodium distichum*) and water tupelo (*Nyssa aquatica*) being the only species that would manage to regenerate. When annual flood duration was above 65–70 per cent, no species was able to survive and the forest was replaced by a non-forest habitat; this happened more rapidly in the subcanopy.

The effects of the proposed rediversion and the modified rediversion version on the frequency of habitat types are indicated in Figure 13.10. In both cases there is a large loss of bottomland forest: a 97 per cent loss in the case of the proposed rediversion and a 94 per cent loss in the modified rediversion. The saving grace of the

modified rediversion plan is that a forest cover is maintained: bottomland forest changes to cypress–tupelo forest, rather than to open water. If these predictions of sweeping changes in the bottomland forest along the Santee River should be trustworthy, then plainly it would be advisable to rethink the plans for rediverting the flow from the Cooper River.

A coastal ecosystem in southern Louisiana

Ecosystems occupying coastal locations are under threat from a variety of human activities: gas and oil exploration, urban growth, sediment diversion, and greenhouse-induced sea-level rise, to

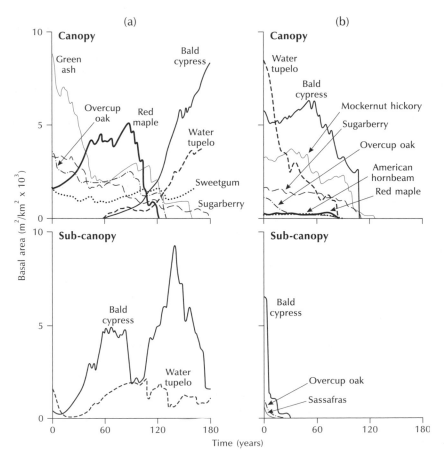

Figure 13.9 Results of simulations. (a) Bottomland forest community subjected to an annual flood duration of 45 ± 4 per cent. (b) Bottomland forest community subjected to an annual flood duration of 72 ± 5 per cent. The flood duration is the percentage of a year during which a plot is flooded. The plant species are bald cypress (*T. distichium*), water tupelo (*N. aquatica*), and sassafras (*Sassafras albidum*).

Source: Adapted from Pearlstine *et al.* (1985)

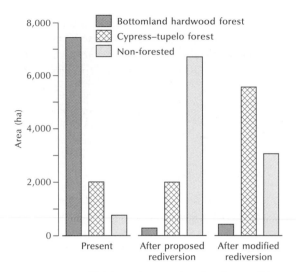

Figure 13.10 Habitat changes in the Santee River study area.

Source: Adapted from Pearlstine *et al.* (1985)

name but a few. To protect and preserve these ecosystems, it is valuable to know what the effects of proposed human activities are likely to be, and how these effects differ from natural changes. These questions were addressed in the Atchafalaya Delta and adjacent Terrebonne Parish marshes in southern Louisiana, USA, using a mathematical model to simulate likely changes (Figure 13.11) (Costanza *et al.* 1990). This landscape, part of the Mississippi River distributary system, is one of the most rapidly changing landscapes in the world. The Atchafalaya River is one of the two principal distributaries of the Mississippi River. It carries about 30 per cent of the Mississippi discharge to the Gulf of Mexico. Since the mid-1940s, sediments transported by the Atchafalaya River have been laid down in the bay area. In consequence, the bay area has gradually become filled in, and in 1973 a new subaerial delta appeared that has since grown to about 50 km². Other changes are taking place in the area. The western Terrebonne marshes are becoming less salty, while the eastern part of the area is becoming more salty: the boundary between fresh and brackish marshes has shifted closer to the Gulf in the western marshes, and farther inland in the east (Figure 13.12). As a whole, the study region is losing wetland, but rate of loss in the Terrebonne marshes has slowed and reversed since the mid-1970s owing to river deposition. The hydrology of the area has been greatly altered by dredging of waterways and the digging of access canals for petroleum exploration.

A dynamic spatial model was developed and christened the Coastal Ecological Landscape Spatial Simulation model (CELSS model for short). It divided the marsh–estuary complex into 2,479 square grid-cells, each with an area of 1 km². The model was used to predict changes of habitats under a range of climatic, management, historical, and boundary scenarios. Table 13.1 summarizes the results of several scenario analyses, which predicted changes to the year 2033. Climate scenarios take 'climate' to mean all the driving variables including rainfall, Atchafalaya River flow, wind, and sea-level. They address the impact of climatic changes on the study area. Management scenarios look at the effect of specific human manipulations of the system. Historical scenarios consider what the system would have been like had not the environment been altered by human action, and if climatic conditions had been different. Boundary scenarios delve into the potential impacts of natural and human-induced variations in the boundary conditions of the system, such as sea-level rise. Running the management and boundary scenarios by restarting the model with the actual habitat map for 1983, rather than the predicted habitat map for that year, added more realism. Climate and historical scenarios were run starting in 1956, so that the full impacts of climate and historical variations could be assessed. After 1978, the rate of canal and levee construction for oil and gas exploration slowed appreciably compared with the 1956–78 period, so all scenarios assumed that no canals or levees were built after 1978, except those specifically mentioned in the scenarios.

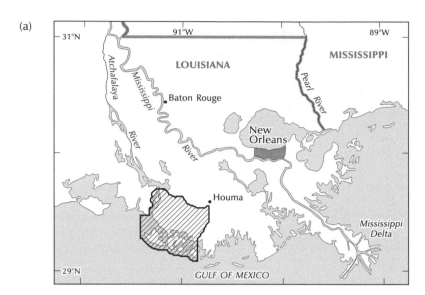

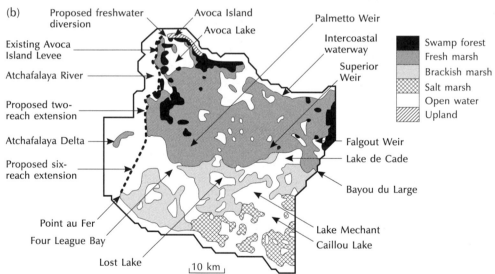

Figure 13.11 The Atchafalaya Delta and Terrebonne Parish marshes study area in southern Louisiana, USA.
(a) General location map.
(b) Major geographical features, types of habitat in 1983, and management options considered in the simulations.

Source: Adapted from Costanza *et al.* (1990)

It is clear from Table 13.1 that the assumptions made about climatic change have a big influence on habitat distribution by the year 2033. The mean climate scenario, wherein the long-term average for each variable was used for all weeks, produced a modest loss in land area. On the other hand, the weekly average climate scenario, wherein each weekly value of each climatic variable for the entire run from 1956 to 2033 was set to the average value for that week in the 1956–83 data, produced a drastic loss of land area. This finding indicated that the annual flood cycle and other annual cycles in climatic variables are important to the land-building process, but that chance events, such as major storms and floods, tend to have a net erosional

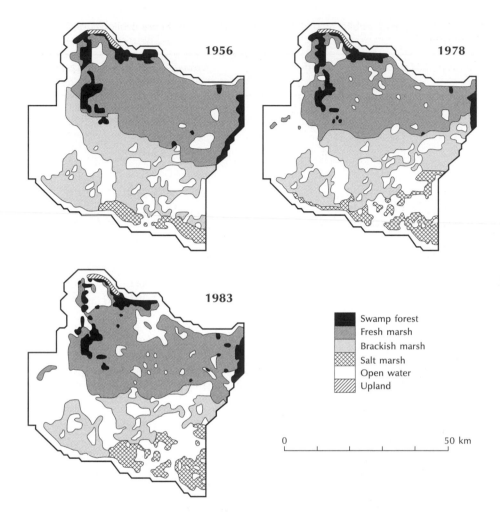

Figure 13.12 Observed distribution of habitats in the Atchafalaya–Terrebonne study area.
Source: Adapted from Costanza *et al.* (1990)

effect on marshland. If the global climate should become less predictable in the future as a result of global warming, then the stability of coastal marshes may be in jeopardy.

Several management scenarios were run. As the data in Table 13.1 indicate, the largest loss of land would arise from the full six-reach levee extension scheme that had been considered at one time. With this scheme, 48 km^2 of brackish marsh and 6 km^2 of fresh marsh would be lost by 2033, largely because the extended levees prevent sediment-laden water reaching the brackish marsh bordering Four League Bay, where most of the loss occurs. Boundary scenarios considered the effects of projected rates of sea-level rise, both high and low projections, on the area. The results

Table 13.1 Area (km²) occupied by each habitat type for three years for which data are available, and for 2033, under various scenarios

	Swamp	Fresh marsh	Brackish marsh	Saline marsh	Upland	Total land	Open water
Survey data							
1956	130	864	632	98	13	1,737	742
1978	113	766	554	150	18	1,601	878
1983	116	845	347	155	18	1,481	998
Climate scenarios							
Base case	84	871	338	120	10	1,423	1,056
Mean climate	94 (+10)[a]	974 (+103)	402 (+64)	136 (+16)	11 (+1)	1,617 (+194)	862 (−194)
Weekly average climate	128 (+44)	961 (+90)	813 (+475)	300 (+180)	11 (+1)	2,213 (+790)	266 (−790)
Management scenarios							
No levee extension[b]	100	796	410	123	15	1,444	1,035
Full six-reach levee extension	103 (+3)	790 (−6)	362 (−48)	122 (−1)	15 (0)	1,393 (−52)	1,087 (+52)
Freshwater diversion	103 (+3)	803 (+7)	404 (−6)	123 (0)	15 (0)	1,448 (+4)	1,031 (−4)
Boundary scenarios							
Low sea-level rise[c]	104 (+4)	800 (+4)	411 (+1)	124 (+1)	15 (0)	1,454 (+10)	1,025 (−10)
High sea-level rise[d]	89 (−11)	794 (−2)	396 (−14)	131 (+8)	15 (0)	1,425 (−19)	1,054 (+19)
Historical scenarios[e]							
No original Avoca levee	84	951	350	126	13	1,524	955
No effects	130	863	401	144	12	1,550	929

Source: Adapted from Costanza *et al.* (1990)

Notes: a Brackets indicate changes from the base case. b This is the base case for the management scenarios. c 50-cm rise by the year 2100. d 200-cm rise by 2100. e No comparisons with a base case are given for the historical scenarios because these runs started in 1956 rather than 1983

were unexpected. Surprisingly, doubling the rate of eustatic sea-level rise from 0.23 to 0.46 cm/yr caused a net gain in land area of 10 km² relative to the base case. This was probably because, so long as sediment loads are high, healthy marshes can keep pace with moderate rates of sea-level rise.

Two historical scenarios were tested. The first probed the changes in the system that might have taken place had not the original Avoca Island levee been built. The second considered the changes that might have ensued had not the Avoca levee nor any of the post-1956 canals been constructed. The results shown in Table 13.1 suggest that the original levee and the canals had a major influence on the development of the system, causing a far greater loss of land than would have occurred in their absence.

GETTING WARMER: COMMUNITIES IN THE TWENTY-FIRST CENTURY

Individuals and communities, by processes of natural selection, become adapted to the environment in which they live. If that environment should change, life systems must adapt, move elsewhere, or perish.

Species under pressure

The fate of many species in the twenty-first century, as the world warms and habitats shrink or vanish, is worrying (Peters 1992a, b). It seems that extinction will continue apace and global biodiversity will drop. But which species are the most vulnerable to global warming? The safest species are mobile birds, insects, and mammals who can track their preferred climatic zone. In the British Isles, the white admiral butterfly (*Ladoga camilla*) and the comma butterfly (*Polygonia c-album*) expanded their ranges over the last century as temperatures rose by about 0.5°C (Ford 1982). However, even mobile species would need a food source, and their favourite menu item might be extinct or have wandered elsewhere. Experimental work suggests that insects do not respond uniformly to global warming: the population density of phloem feeders increases with elevated carbon dioxide levels, but the population densities of free-living and mining and chewing insects show no change or even a reduction, probably because compensatory feeding (finding alternative food plants) is common in these groups (J. B. Whittaker 2001).

Five kinds of species appear to be most at risk:

1 *Peripheral species*. Populations of animals or plants that are at the contracting edge of a species range – peripheral populations – are vulnerable to extinction. The surviving California condor (*Gymnogyps californianus*) population is an example.
2 *Geographically localized species*. Many currently endangered species live in alarmingly limited habitats. The golden lion tamarin (*Leontopithecus rosalia*) is a squirrel-sized primate that lives in lowland Atlantic forest, Brazil. Lowland Atlantic forest is one of the most endangered rain forests in the world. It once covered 1,000,000 km² along the Brazilian coastline. Today, a mere 7 per cent of the original forest remains, putting the golden lion tamarin under enormous threat (Figure 13.13). An estimated 1,000 golden lion tamarins live in the wild, some 220 in the Poço das Antas Biological Reserve, 130 in the Fazenda União Biological Reserve, 400 in several reintroduced forests, and about 250 scattered outside protected areas. Additionally, some 480 golden lion tamarins live in 150 zoos around the world. During the 1990s, a combination of relocating tamarin families from vulnerable areas into the protection of the Poço das Antas Biological Reserve, and reintroducing captive-bred tamarins into the wild, has put the little primate on the road to recovery. Predators, probably 3–6-kg weasel-like tayra (*Eira barbara*), extirpated all but one of 15 tamarin study groups between 1996 and 2000. Fortunately, golden lion tamarins soon recolonized some the territories occupied by the extirpated groups. Even so, for the first time in 17 years, some areas with suitable habitat in the reserve no longer contain golden lion tamarins and the population size in the reserve has dropped by 36 per cent.
3 *Highly specialized species*. Many species have a close association with only one other species. An example is the Everglade (or snail) kite (*Rostrhamus sociabilis*) that feeds exclusively on the large and colourful Florida apple snail (*Pomacea paludosa*) in Florida wetland. Swamp drainage has already robbed the kite of its prey in some areas.
4 *Poor dispersers*. Many trees have heavy seeds that are not broadcast far. Plants with limited dispersal abilities have knock-on effects along the food chain. Some birds, mammals, and insects are closely associated with specific forest trees, which cannot migrate rapidly. These species would be hard pressed to survive. The endangered Kirtland's warbler (*Dendroica*

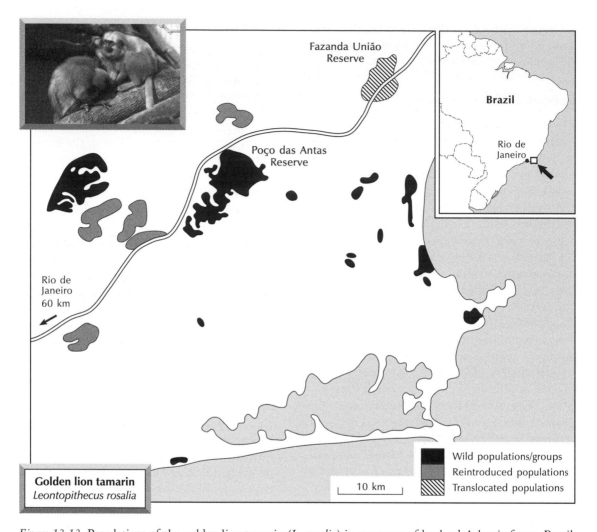

Figure 13.13 Populations of the golden lion tamarin (*L. rosalia*) in remnants of lowland Atlantic forest, Brazil. The inset shows a golden lion tamarin.

Sources: Map adapted from Frankham *et al*. (2002); photograph by Pat Morris

kirtlandii), for example, only breeds in and nests upon the ground under jack-pine (*Pinus banksiana*) forest on well-drained sandy soil in north-central Michigan, USA. But not any jack-pine forest will do – it has to be young, secondary growth forest that emerges in the aftermath of a forest fire. As the global thermometer rises, the jack pines may move north-wards onto less well-drained soils, and the Kirtland's warbler may find itself without any suitable nesting sites (Botkin *et al*. 1991). This species could be an early casualty of global warming.

5 *Climatically sensitive species.* Species in climatically sensitive communities are vulnerable to global warming. Communities in this class

include wetlands, montane and alpine biomes, Arctic biomes, and coastal biomes. Wetlands will dry out, with grave consequences for amphibians (Box 13.2), mountains will become warmer towards their tops, tundra regions will warm up, and coastal biomes will be flooded. Climatic warming is predicted to be greatest at high latitudes. Animals and plants in these regions, including polar bears (*Thalarctos maritimus*), will have to cope with the most rapid changes. Tundra vegetation may be pushed northwards as much as 4 degrees of latitude. This could mean that, for a climatic warming of 3°C, 37 per cent of present tundra will become forest. The same degree of climatic warming would fuel local temperature increases on mountains. Animals and plants would need to shift their ranges upwards by about 500 m. Animal and plant populations on mountains may retreat upwards as climate warms, but eventually some of them will have nowhere to go. This would probably happen to the pikas (*Ochotona* spp.) living on alpine meadows on mountains of the western USA. Pikas feed on grasses and do not range into boreal forests. As the boreal forest habitat advances up the mountainsides, so the pikas will retreat into an ever-decreasing area that may one day vanish. This process is already happening to Edith's checkerspot butterfly (*Euphydryas editha*), which lives in western North America (Parmesan 1996). Records of this species suggest that its range is shifting upwards and northwards, and extinctions have occurred at some sites. Similarly, the lower altitudinal limit of the gelada baboon (*Theropithecus gelada*), a medium-sized African primate living in high montane grassland of the Ethiopian plateau, is predicted to rise about 500 m for every 2°C increase in global mean temperature (Dunbar 1998). A 7°C temperature rise would confine the species to a small number of isolated mountain peaks.

Communities under pressure

Community composition and structure change in the face of environmental perturbations. During the current century, *global warming* will perturb many communities. This will affect the geographical location of biomes and the species composition of communities.

Prairie wetlands in North America

Prairie wetland habitat occupies a broad swathe of land across the American Midwest and southern Canadian prairie provinces (Figure 13.14) (Poiani and Johnson 1991). It is the main breeding area for waterfowl on the North American continent. The habitat consists of relatively shallow-water temporary ponds (holding water for a few weeks in spring), seasonal ponds (holding water from spring until early summer), semi-permanent ponds (holding water throughout much of the growing season in most years), and large permanent lakes (Plate 13.5). For breeding purposes, the waterfowl require a mix of open water and emergent vegetation. The temporary and seasonal ponds provide a rich food source in the spring and are used by dabbling ducks. The semi-permanent ponds provide food and nesting areas for birds as seasonal wetlands dry. As temperatures rise over the next decades, water depth and the numbers of seasonal ponds in these habitats should decrease. The lower water levels would favour the growth of emergent vegetation and reduce the amount of open water. Waterfowl may respond by migrating to different geographical locations, relying more upon semi-permanent wetlands but not breeding, or failing to re-nest as they do at present during droughts.

Mires in the Prince Edward Islands, Indian Ocean

Temperatures in the Prince Edward Islands have increased by approximately 1°C since the early 1950s, while precipitation has decreased (Chown

Box 13.2

AMPHIBIANS AND GLOBAL WARMING

Amphibians need water. They must live in it, near it, or else in very humid conditions. Warmer and drier conditions could be disastrous for them. In places where global warming is associated with increasing aridity, they are likely to suffer. Evidence of this comes from endemic golden toad (*Bufo periglenes*) and harlequin frog (*Atelopus varius*) populations in the Monteverde Cloud Forest Preserve, Costa Rica (Pounds 1990; Pounds and Crump 1994). In May and April 1987, there were thousands of golden toads thriving along streams in the reserve. The males live in underground burrows near the water table; they emerge only to mate. There was also a considerable number of harlequin frogs living in wet stream-bank patches. These amphibians retreat into crevices or gather in damp pockets during the dry season. Four years later, the golden toad and the harlequin frog had vanished. Rainfall during May 1987 was 64 per cent less than average, the lowest ever recorded for that month. Normally, early rains in May and June help the amphibians to recover from the November and December dry season. The

1987 dry season was extra dry. During the drought, water tables fell, springs all but dried up, and streams fed by aquifers, which would usually give enough water to keep the mossy riverbanks wet, dwindled. As the amphibians disappeared suddenly, it may be that the extreme drought killed the adults and their tadpoles or eggs. This catastrophe augurs ill for amphibians during the twenty-first century.

The Costa Rican toads are part of a global amphibian decline that has caused much public concern and a frenzy of scientific research (e.g. Carey and Alexander 2003; Carrier and Beebee 2003; Green 2003; Kats and Ferrer 2003; Storfer 2003). The conclusion is that there is no plain and simple answer as to declining amphibian numbers. Complex interactions of several human factors seem the most likely explanation. The six prime suspects are: alien species, overexploitation, land-use change, global change (including global warming), increased use of pesticides and other toxic chemicals, and emerging infectious diseases (Collins and Storfer 2003).

and Smith 1993). The changing water balance has led to a reduction in the peat-moisture content of mires and higher growing season 'warmth'. In consequence, the temperature-sensitive and moisture-sensitive compact hook-sedge (*Uncinia compacta*) has increased its aerial cover on Prince Edward Island. However, harvesting of seeds by feral mice (*Mus musculus*), which can strip areas bare, has prevented an increase in sedge cover on Marion Island. Such extensive use of resources suggests that prey switching may be taking place at Marion Island. Mice are not only eating ectemnorhinine weevils to a greater extent than found in previous studies of populations at Marion

Island, but they also prefer larger weevils. A decrease in body size of preferred weevil prey species (*Bothrometopus randi* and *Ectemnorhinus similis*) has taken place on Marion Island (1986–92), but not on Prince Edward Island. This appears to be a result of increased predation on weevils. Adults of the prey species, *E. similis*, are relatively more abundant on Prince Edward Island than adults of the smaller congener, *E. marioni*, and could not be found on Marion Island in the late southern summer of 1991. Results not only provide support for previous hypotheses of the effect of global warming on mouse–plant–invertebrate interactions on the Prince Edward

Figure 13.14
Location of North American prairie wetlands. The prairie wetlands are relatively shallow, water-holding depressions of glacial origin.

Source: Adapted from Poiani and Johnson (1991)

Plate 13.5 Semi-permanent prairie wetland, Stutsman County, North Dakota. This wetland is part of the Cottonwood Lake site where long-term hydrological and vegetation data have been collected by the US Geological Survey (Water Resources and Biological Resources Division). The plants in the foreground are cattails (*Typha* spp.).

Photograph by Karen A. Poiani.

Islands, but also provide limited evidence for the first recorded case of predator-mediated speciation.

Biome and ecotone shifts

Global warming fuelled by a doubling of carbon dioxide levels could produce large shifts in the distribution of biomes. One study, which explored five different scenarios for vegetation redistribution with a doubling of atmospheric carbon dioxide levels, predicted large spatial shifts, especially in extratropical regions (Neilson 1993a). Large spatial shifts in tropical and boreal vegetation were predicted (Figure 13.15). Boreal biomes (taiga and tundra) retreat northwards and decrease in size by 62 per cent (the range is 51–71

per cent, depending on the scenario used). Boreal and temperate grassland increase in area under two scenarios. The increase is 36 per cent in one case and 82 per cent in the other. Temperate forest area alters little. Tropical forests show a slight reduction in area in all but one scenario, which predicts a slight increase. Maps of predicted change in the **leaf–area index** (which reflects the maximum rate of transpiration) imply drought-related biomass losses in most forested regions, even in the tropics. The areas most sensitive to drought-induced vegetation decline are eastern North America and eastern Europe to western Russia.

In detail, spatial shifts of biomes should produce discrete 'change' zones and 'no-change' zones (Figure 13.16) (Neilson 1993b). Such

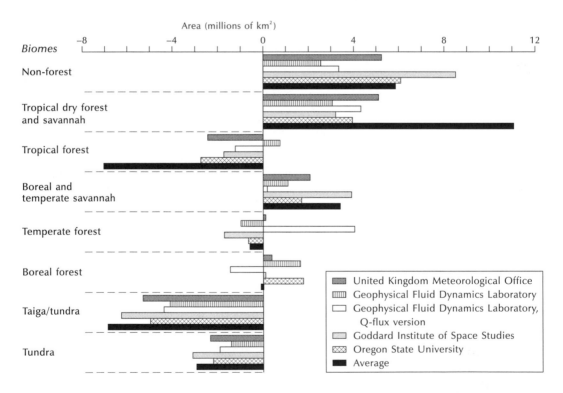

Figure 13.15 Changes in biome area predicted by the MAPPS (Mapped Atmosphere–Plant–Soil System) model under various scenarios.

Source: Adapted from Neilson (1993a)

changes alter the pattern of vegetation. Large, relatively uniform biomes are reduced in size, while new biomes, which combine the original biome and the encroaching biomes, emerge. Biomes and ecotones alike would probably be affected by drought followed by infestations and fire if climate were to change rapidly. Ecotones would be especially sensitive to climatic change, whether it were slow or fast. The pattern of ecotone change would be sensitive to water stress (Figure 13.17). With global warming unaccompanied by water stress, habitats should not fragment as under water stress, but should display a wave of high habitat variability as the ecotone gradually tracks the climatic shifts. Under extreme drought stress, the entire landscape should become fragmented as every variation in topography and soil become important to site water-balances and the survivorship of different organisms. The ecotone disappears for a while and reappears at a new location. It does not visibly shift geographically, as it would with no water stress, but disassembles and then re-establishes itself later.

Forest change in eastern North America

It is probably common for species to move in the same general direction. However, that does not mean that entire communities move together. Quite the contrary – species move at different rates in response to climatic change. The result is that communities disassemble, splitting into their component species and losing some species that fail to move fast enough or cannot adapt. Mathematical models are helpful in understanding the possible changes in communities as the world warms up. Clearly, profound changes in the composition and geography of communities would occur. This is evident in the following case studies.

An early model simulated forest growth at 21 locations in eastern North America, as far west as a line joining Arkansas in the south to Baker Lake, Northwest Territories, in the north (Solomon 1986). All forests started growing on a clear plot and grew undisturbed for 400 years under a modern climate. After the year 400, climate was changed to allow for a warmer

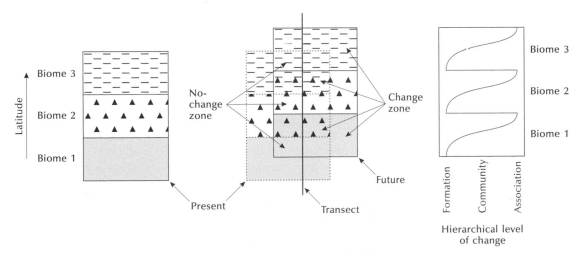

Figure 13.16 Geographical shift in biomes induced by global warming. Discrete 'change' and 'no-change' zones are produced. Large, uniform biomes shrink. New biomes, which combine characteristics of original biomes and encroaching biomes, emerge.

Source: Adapted from Neilson (1993b)

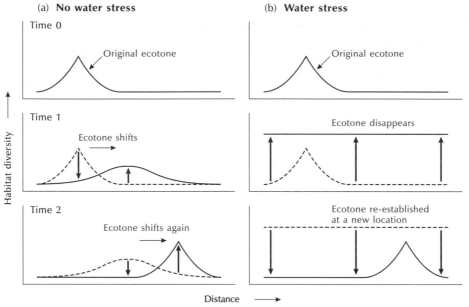

(a) **No water stress**   (b) **Water stress**

Time 0
Original ecotone

Time 1
Ecotone shifts
Ecotone disappears

Time 2
Ecotone shifts again
Ecotone re-established at a new location

Habitat diversity

Distance

Figure 13.17
Ecotone changes induced by global warming. (a) With no water stress. (b) With water stress.
Source: Adapted from Neilson (1993b)

atmosphere. A linear change of climate between the years 400 and 500 was assumed, the new climate at the year 500 corresponding to a doubling of atmospheric carbon dioxide levels. Climatic change continued to change linearly after the year 500, so that, by the year 700, the new climate corresponded to a quadrupling of atmospheric carbon dioxide levels. At the year 700, climate stabilized and the simulation went on for another 300 years. Simulation runs at each site were repeated 10 times and the results averaged. Validation of the results was achieved by testing with independent forest composition data, as provided by pollen deposited over the last 10,000 years and during the last glacial stage, 16,000 years ago.

Results for some of the sites are set out in Figure 13.18. At the tundra–forest border, the vegetation responds to climatic change in a relatively simple way (Figure 13.18a). With four times the carbon dioxide in the atmosphere, the climate supports a much-increased biomass and,

at least at Shefferville in Quebec, some birches, balsam poplar, and aspens. The response of northern boreal forest to warming is more complicated (Figure 13.18b). With a fourfold increase in atmospheric carbon dioxide levels, the climate promotes the expansion of ashes, birches, northern oaks, maples, and other deciduous trees at the expense of birches, spruces, firs, balsam poplar, and aspens. In all northern boreal forest sites, the change in species composition is similar, but takes place at different times. This underscores the time-transgressive nature of vegetational response to climatic change. The southern boreal forest and northern deciduous forest (Figure 13.18c and d) also have complicated responses to climatic change. In both communities, species die back twice, the first time around and just after 500 years, and the second time from about 600 to 650 years. In the southern boreal forest, the change is from a forest dominated by conifers (spruces, firs, and pines) to a forest dominated initially by maples and basswoods, and later by northern oaks and hickories. At some sites in the northern deciduous forest, species appear as climate starts to change then vanish once the stable climate associated with a quadrupled level of atmospheric carbon dioxide becomes established. Species that do this include the butternut (*Juglans cinerea*), black walnut (*J. nigra*), eastern hemlock (*Tsuga canadensis*), and several species of northern oak. In western and eastern deciduous forests, the response of trees to climatic warming is remarkably uniform between sites (Figure 13.18e and f). Biomass declines everywhere, generally as soon as warming starts in the year 400. The drier sites in the west suffer the greatest losses of biomass, as well as a loss of species; this is to be expected as prairie vegetation would take over. The decline of biomass in the eastern deciduous forest probably results chiefly from the increased moisture stress in soils associated with the warmer climate. The increased moisture stress also gives a competitive edge to smaller and slower-growing species, such as the southern chinkipin oak (*Quercus muehlenbergii*), post oak (*Q. stellata*), and live oak (*Q. virginiata*), the black gum (*Nyssa sylvatica*), sugarberry (*Celtis laevigata*), and the American holly (*Ilex opaca*), which come to dominate the rapid-growing species such as the American chestnut (*Castanea dentata*) and various northern oak species.

Soil water regimes and forest change

Forest growth is sensitive to changes in soil water regimes. Global warming is sure to alter water regimes and this will have an inevitable impact upon forests. A forest-growth model was used in conjunction with a soil model to assess the impact on forests of a doubling of atmospheric carbon dioxide levels (Pastor and Post 1988). The model was run for several sites in the northeastern USA, including a site in northeastern Minnesota. The climate of this region is predicted to become warmer and drier. The major changes will occur at the boundary between the boreal forest and cool temperate forest. Forest growth was modelled on two soil types: soils with a high water-holding capacity and soils with a low water-holding capacity. The simulations began in 1751. Bare plots were 'sown' with seeds of the tree species common in the area. The model forest was then allowed to grow for 200 years under present climatic conditions. Next, carbon dioxide concentrations were gradually increased until, after a century, their present value had been doubled. After having reached that value, carbon dioxide levels were held constant for the next 200 years.

The model predicted that, on soils with a high water-holding capacity, where there will be enough water available to promote tree growth, productivity and biomass will increase. On soils with a low water-holding capacity, productivity and biomass will decrease in response to drier conditions. In turn, raised productivity and biomass will up levels of soil nitrogen, whereas lowered productivity and biomass will down levels of soil nitrogen. The vegetation changes are, therefore, self-reinforcing. On water-retentive soils, global warming favours the expansion of northern hardwoods (maples, birches, basswoods) at the expense of conifers (spruces and firs); on well-drained, sandy soils, it favours the expansion of a stunted oak–pine forest, a relatively unproductive vegetation low in nitrogen, at the expense of spruce and fir.

Ecological lessons from simulation models

Species show a wide range of responses to climate. In consequence, the response of different animal and plant species to climatic change will be varied. This should mean that natural communities are likely to disassemble and that habitats restructure in a transient, non-equilibrium fashion as climatic change unfolds. Validated models that help forecast these events are needed to aid scientists in better understanding the

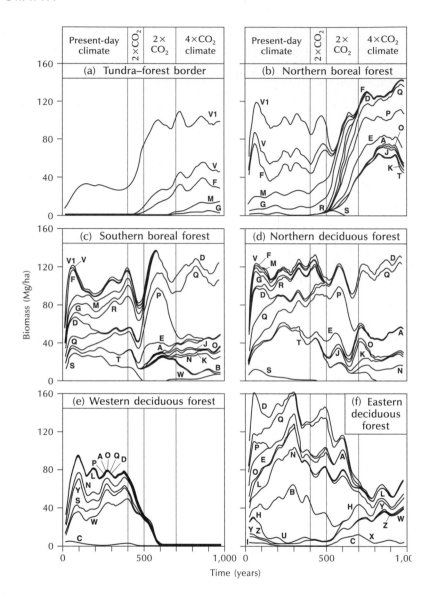

ecological ramifications of global climatic change. Also, and perhaps more important for conservation biology, such validated models can help provide probabilities for the occurrence of these events, which will allow policy makers to make better, informed decisions (T. L. Root and Schneider 1993).

SUMMARY

Communities change. The nature of this change is debatable. The classic view saw climatic climaxes and balanced ecosystems resulting from unidirectional succession. Succession is a complex process and is explained by at least

Figure 13.18 Simulations of forest biomass dynamics over one millennium in response to climatic change induced by increasing levels of carbon dioxide in the atmosphere at six sites in eastern North America. (a) Shefferville, Quebec (57°N, 67°W). (b) Kapuskasing, Ontario (49°N, 83°W). (c) West upper Michigan (47°N, 88°W). (d) North central Wisconsin (45°N, 90°W). (e) South central Arkansas (34°N, 93°W). (f) Central Tennessee (36°N, 85°W). The tree species are as follows: A. American beech (*Fagus grandifolia*); B. American chestnut (*Castanea dentata*); C. American holly (*Ilex opaca*); D. ashes: green ash (*Fraxinus pennsylvanica*), white ash (*F. americana*), black ash (*F. nigra*), blue ash (*F. quadrangulata*); E. basswoods: American basswood (*Tilia americana*) and white basswood (*T. heterophylla*); F. birches: sweet birch (*Betula lenta*), paper birch (*B. papyrifera*), yellow birch (*B. alleghaniensis*), and grey birch (*B. populifolia*); G. balsam poplar (*Populus balsamifera*), bigleaf aspen (*P. grandidentata*), trembling aspen (*P. tremuloides*); H. black cherry (*Prunus serotina*); I. black gum (*Nyssa sylvatica*); J. butternut (*Juglans cinerea*) and black walnut (*J. nigra*); K. eastern hemlock (*Tsuga canadensis*); L. elms: American elm (*Ulmus americana*) and winged elm (*U. alata*); M. firs: balsam fir (*Abies balsamea*) and Fraser fir (*A. fraseri*); N. hickories: bitternut hickory (*Carya cordiformis*), mockernut hickory (*C. tomentosa*), pignut hickory (*C. glabra*), shagbark hickory (*C. ovata*), shellbark hickory (*C. laciniosa*), and black hickory (*C. texana*); O. hornbeams: eastern hornbeam (*Ostrya virginiana*) and American hornbeam (*Carpinus caroliniana*); P. maples: sugar maple (*Acer saccharum*), red maple (*A. rubra*), and silver maple (*A. saccharinum*); Q. northern oaks: white oak (*Quercus alba*), scarlet oak (*Q. coccinea*), chestnut oak (*Q. prinus*), northern red oak (*Q. rubra*), black oak (*Q. velutina*), bur oak (*Q. macrocarpa*), grey oak (*Q. borealis*), and northern pin oak (*Q. ellipsoidalis*); R. northern white cedar (*Thuja occidentalis*), red cedar (*Juniperus virginiana*), and tamarack (*Larix laricina*); S. pines: jack pine (*Pinus banksiana*), red pine (*P. resinosa*), shortleaf pine (*P. echinata*), loblolly pine (*P. taeda*), Virginia pine (*P. virginiana*), and pitch pine (*P. rigida*), and, T. white pine (*P. strobus*); U. yellow buckeye (*Aesculus octandra*); V. spruces: black spruce (*Picea mariana*), red spruce (*P. rubens*); and v1 white spruce (*P. glauca*); W. southern oaks: southern red oak (*Quercus falcata*), overcup oak (*Q. lyrata*), blackjack oak (*Q. marilandica*), chinkipin oak (*Q. muehlenbergii*), Nuttall's oak (*Q. nuttallii*), pin oak (*Q. palustris*), Shumard's red oak (*Q. shumardii*), post oak (*Q. stellata*), and live oak (*Q. virginiana*); X. sugarberry (*Celtis laevigata*); Y. sweetgum (*Liquidambar styraciflua*); Z. yellow poplar (*Liriodendron tulipifera*).

Source: Adapted from Solomon (1986)

three models – the facilitation model, the tolerance model, and the inhibition model. It may also be driven by factors external to the community (allogenic factors). Primary succession is the colonization of land or submarine surfaces that have never existed before. Secondary succession is the invasion of newly created surfaces resulting from removal of pre-existing vegetation. A modern view of community change stresses the disequilibrium behaviour of communities. It sees succession going in many possible directions, and sees communities as temporary collections of species that assemble and disassemble as the environment changes. Much community change has been caused by land cover transformation over the last 200 years. Two important aspects of this transformation are habitat fragmentation, and its attendant effects on wildlife, and the loss of wetlands. Global warming during the present century and beyond is likely to put peripheral, geographically restricted, highly specialized, poorly dispersive, and climatically sensitive species under intense pressure. It will also cause significant changes in many communities. Wetlands, tundra, and alpine meadows are especially vulnerable.

ESSAY QUESTIONS

1 **Why has Clements' unidirectional view of succession been revised?**

2 **Why does habitat fragmentation pose a threat to many species?**

3 **What community changes are likely to occur because of global warming?**

FURTHER READING

Gates, D. M. (1993) *Climate Change and Its Biological Consequences*. Sunderland, MA: Sinauer Associates.

A clear account.

Huggett, R. J. (1993) *Modelling the Human Impact on Nature: Systems Analysis of Environmental Problems*. Oxford: Oxford University Press.

If you like modelling but are not too mathematically minded, this might be of interest.

Huggett, R. J. (1997) *Environmental Change: The Evolving Ecosphere*. London: Routledge.

Provides a broad perspective.

Jackson, A. R. W. and Jackson, J. M. (1996) *Environmental Science: The Natural Environment and Human Impact*. Harlow: Longman.

A basic textbook with several relevant sections.

Matthews, J. A. (1992) *The Ecology of Recently Deglaciated Terrain: A Geoecological Approach to Glacier Forelands and Primary Succession*. Cambridge: Cambridge University Press.

An excellent case study and lots more.

Peters, R. L. and Lovejoy, T. E. (eds) (1992) *Global Warming and Biological Diversity*. New Haven, CT, and London: Yale University Press.

A host of examples in this one.

PART III

HISTORICAL BIOGEOGRAPHY

DISPERSAL AND DIVERSIFICATION IN THE DISTANT PAST

Animal and plant species originate at a particular place and may then spread elsewhere, commonly branching out to fill new niches as they do so. This process of dispersal occurs today and occurred in the past. This chapter covers:

■ how to restock the world from a single point of origin
■ how to account for endemic regions in a world of dispersing organisms
■ evidence of dispersal in the biogeographical history of continents

A CLASSIC BIOGEOGRAPHICAL DEBATE: HOW DID THE ANIMALS ON NOAH'S ARK RESTOCK THE WORLD?

The roots of modern biogeography date back to at least the seventeenth century (Browne 1983). They are worth exploring a little as they illustrate how thinkers approach the seemingly simple biogeographical question – why do animals and plants live where they do? – from very different angles at different times.

During the Renaissance, the Scriptures underwent a reinterpretation: taken as allegories and metaphors in the Middle Ages, they became regarded as a literal tract, describing actual past events rather than being stories with a strong moral message. Scholars therefore took Noah's Flood as a real event and his ark as a real ark. This raised an interesting question to the inquiring mind – how could the ark accommodate all the animals then known? Athanasius Kircher (1602–80) took the size and shape of the ark described in Genesis and concluded that accommodating a pair of all species was feasible. But, at that time, all species meant 130 species of mammals, 30 pairs of snakes, and 150 different kinds of birds. Foreign travel started to reveal an increasing number of new species and the difficulty of squeezing them all in the ark started to strain the argument. But writers such as John Ray (1627–1705) and Tancred Robinson (d. 1748) remained undeterred, arguing that the Bible said a flood had occurred, and indeed tangible evidence for a universal flood was coming to light (e.g. fossil seashells found in mountains), so all species must have found a berth. Scholars also considered another logical outcome of the Flood: how did animals and plants restock the world from the disembarkation point on the flanks of Mount Ararat?

Several species were introduced into Europe in the seventeenth century and seemed to have little trouble establishing themselves. So, it seemed reasonable to opine that animals and plants have a marked facility for migration. The notion that animals and plants dispersed from a single centre – Mount Ararat – thus held no intellectual problems. Indeed, the tenth and eleventh chapters of the Book of Genesis tell how all men were descended from Noah, and how they made their way from Armenia to their present countries – the story was one of multiplication and diffusion under God's guiding hand. Animal dispersal was explained as synchronous with the dispersal of man. José de Acosta (1540–1600), noting that many species are found only in the Americas, suggested that man had taken them there by boat. A critic, Justus Lipsius (1547–1606), was quick to point out that no one would willingly ride with rattlesnakes and bears. Rather, he said, animals got to the Americas via Africa, which was at the time connected to the Americas by dry land (the notion of land bridges had arrived). Various other possiblilites for making the Atlantic crossing were proposed, such as crossing in the winter when the north seas were frozen. Dissenting views began to mount. Isaac de la Peyrère (1596–1676) doubted that organisms had a propensity to migrate and championed the view that the Flood was a European and Middle Eastern event, and elsewhere in the world man lived on with a full complement of birds, beasts, and plants; this view was condemned in Catholic France. Abraham van der Myl (1563–1637) seized on Peyrère's idea and suggested that there had been more than one Creation. This was rejected, along with Peyrère's

thesis, on the grounds that, as humans increase in number every generation, a family of two would grow to eight in 34 years. Looking back, every generation must be smaller than the one that followed. Logically, the process of diminution must end in a single couple – Adam and Eve – for whom there could be no natural origin, and a Creator must be assumed. The same argument must apply to animal species.

By the eighteenth century, two big changes in thought had occurred. First, a mechanistic worldview promulgated by Isaac Newton saw a divinely contrived balance in nature. Second, the new philosophers who, in the Age of Reason or Enlightenment, stressed the value of clear and rational thinking, played down the role of God. In biogeography, these changes helped to convert a literal ark into a metaphorical ark. By the time that Carl Linnaeus (1707–78) was writing, the number of known species had proliferated enormously. Linnaeus himself listed 14,000, including some 5,600 animals, during his lifetime. He believed that the ark described in Genesis was not an actual boat but a metaphor, suggesting that it was a mountain – Mount Ararat seemed a likely candidate – that stood above the highest floodwaters as a world in miniature. He argued that a mountain contains all the climatic zones and so species could be housed according to their climatic needs (he argued convincingly of the close link between organisms and their environment). The biogeographical problem did not change: how to explain worldwide dispersal from a small centre-of-origin. But now there was the added problem of trying to explain how a cold-loving species such as the reindeer could travel from the chilly upper slopes of Mount Ararat across inhospitably hot terrain to reach Lapland. A new line of thought came from George Leclerc, Comte de Buffon (1707–88), who observed that the Old and New World tropics have no mammal species in common. This observation was found to apply to plants, insects, and reptiles and was generalized as 'Buffon's law'. To Buffon, his law suggested that all North

American species originated in the Old World but underwent alterations (what would now be called evolution) in dispersing, which idea has become the mainstay of the dispersal biogeographers' argument. By way of explanation, Buffon contended that animals originated in the North Pole of the Old World in a warmer period and moved south as temperatures fell little by little.

Eberhardt Zimmerman in 1777 finally laid the literal ark to rest in a zoological tract that was the first book to describe in the detail the distribution of mammals. Zimmerman pointed out that the proliferation of a single pair of each species would be impossible – the first pair of lions would eat the first pair of sheep, followed by the first pairs of goats and all the other herbivore species in quick succession; the lions would then die from hunger having eaten everything in sight. (Zimmerman's own solution to the problem of animal distribution was the thesis that every animal was created in the area where it now lives, under the same climate that it now enjoys, with the same food rations already in abundant supply – in short, the entire global ecosystem was called into existence in one fell swoop.)

WANDERLUST: CENTRES-OF-ORIGIN AND DISPERSAL

The classic dispersal model

Advances in mapping the distribution of species as more and more of the world was explored were a key factor in shaping nineteenth-century biogeography. Indeed, the Darwin–Wallace notion of centre-of-origin and dispersal grew out of efforts to map the distribution of plants, mammals, insects, and reptiles. All the maps showed areas of endemism. Auguste-Pyramus de Candolle (1820) identified 20 botanical regions, which he later extended to 40. Buffon's law applied to these regions and not just to the Old and New Worlds – they each were home to particular species. To Charles Lyell, an implication of

endemism was that, as different regions with the same physical conditions house different species, new species must somehow have appeared, the process involved being a mystery to him. Wallace followed up Lyell's idea on the creation of species, and suggested that they must have come from pre-existing species. Darwin and Wallace discovered a mechanism by which species might change during dispersal – natural selection. Darwin's thesis was that barriers to dispersal should delay movement long enough for natural selection to modify dispersing species. Wallace (1876, 1880) laid the foundations of modern biogeography, many of which have endured into the twenty-first century (Box 14.1).

Modern dispersal biogeography

Refinements to the dispersal model came from Ernst Mayr, George Gaylord Simpson, Philip J. Darlington, and others. Taken in conjunction with evolutionary theory, two schools of thought emerged within the centre-of-origin–dispersal paradigm. Darlington (1957) made three assertions: (1) All groups tend to speciate most actively in their centre-of-origin, which is a limited geographical area. (2) When ancestral species produce two daughter species (a 'sister' pair), one is always advanced (apomorphous) and one primitive (plesiomorphous). (3) Once speciation creates the sister pair, the advanced form stays in the centre-of-origin and the primitive form moves to the periphery. Members of the other school, led by Willi Henning and Lars Brundin, agreed with Darlington's first two assertions, but not the third. Instead, they opined that the primitive form stays in the centre-of-origin and the advanced form moves away. This is Henning's progression rule.

Related to the process of dispersal is the phenomenon of taxon pulses. Identified by Darlington in the early 1940s, named by Edward O. Wilson in 1961, and championed by Terry L. Erwin in the 1970s, a *taxon pulse* is an adaptive shift in a group of organisms as they spread

Box 14.1

ALFRED RUSSEL WALLACE AND THE FOUNDATIONS OF BIOGEOGRAPHY

These are some of Wallace's conclusions from a long lifetime's study of biogeography (Brown and Lomolino 1998, 28):

Distance

The similarity of two biogeographical regions is not necessarily a function of the distance between them.

Climate

Strongly affects the similarity between two biogeographical regions.
Palaeoclimates, and especially those in the recent past, strongly influence the distribution of biotas.

Dispersal

Fossil record provides positive evidence for this process.
Affected by competition, predation, and other biotic factors.
Long-distance dispersal is the likely means of colonizing remote islands.

Reuniting of landmasses

When two landmasses are joined (by a land bridge), competition will lead to extinctions.

Speciation

Occurs through geographical isolation and subsequent modification to local climates and habitats.

Extinction

Affected by biotic factors.
Discontinuous ranges and disjunctions.
Discontinuous ranges result from extinctions in intervening areas or from patchy habitats.
Disjunctions of genera are older than disjunctions of species, and so forth for higher taxa.

Islands

Island biotas important because relationships between distribution, speciation, and adaptation are more readily observed in them.
May be classified as continental islands recently separated from mainlands, continental islands more distantly separated from mainlands, and remote oceanic islands. Each type fosters a different sort of biota.

from one habitat to another while dispersing (e.g. Erwin 1981). Erwin, studying carabid beetles, found that the first shift is from a generalist species living in tropical wetlands to specialist species occupying zones away from the waterside. This shift may produce forest-floor specialist, under-canopy specialists, climate specialists (those that move out of the tropical zone into higher latitudes), upland specialists, and so on. A subsequent adaptive shift creates superspecialists, such as canopy superspecialists, steppe and desert superspecialists, high altitude superspecialists, and tundra superspecialists. The original wetland generalists tend to persist and generate several pulses before being replaced. Each pulse may overtake and replace previous ones. Evolutionary forces, including competition and predation, keep the pulses flowing, which may help to account for the stunning diversity of the carabid beetles through time.

THE PAST AT PRESENT: DISPERSAL, DIVERSIFICATION, AND FAUNAL STRATA

The current floras and faunas of the continents, and their fossil faunas and floras, provide clues to biogeographical history. Different schools of biogeographers interpret these clues in unlike ways. The next chapter will focus on interpretations from a vicariance biogeogaher's perspective; this one will examine continental patterns from a dispersalist's angle. The mammal faunas of South America and Australia will serve to show how the present compliment of mammals on those continents appears to consist of several layers or *faunal strata* that have built up through

successive *waves of invasion* over the last 100 million years or so. Each wave of invasion has triggered adaptive radiations, producing shining examples of parallel and convergent evolution, with similar environments on opposite sides of the globe shaping remarkably similar forms (phenotypes) from animals of remotely related stock.

South American invasions

A classic example of a dispersal model is the *mammalian invasions* of South America (Simpson 1980). Although South America now joins North America, it was an island-continent for most of the last 65 million years or so. The classic interpretation is that, on two occasions during that time, a land connection with North America, probably through a chain of islands, developed for a few million years and was then lost. During these times, and in recent times, mammals invaded South America from the north. Four phases of invasion are recognized (L. G. Marshall 1981a) (Figure 14.1):

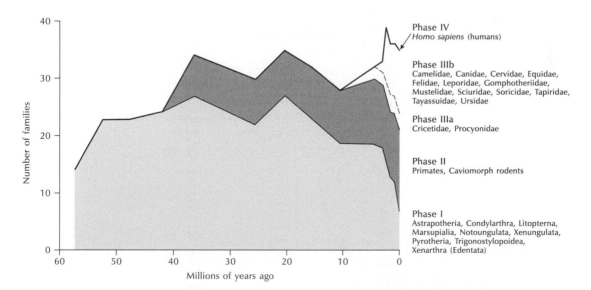

Figure 14.1 The history of South American mammals.
Source: Adapted from L. G. Marshall (1981a)

1 *Phase I*, which occurred in the Late Cretaceous period and earliest **Palaeocene epoch**, was an invasion by 'old timers' or 'ancient immigrants'. Three orders of mammal invaded – only two of which are represented in South America today – the marsupials, the xenarthrans (edentates in some classifications), and the condylarthrans. Once isolated again, these ancestral stocks evolved independently of mammals elsewhere and some unique forms arose. The first marsupial invaders were didelphoids and included ancestors of the modern opossums (Box 14.2). The borhyaenids were dog-like marsupial carnivores. They bore a strong resemblance to the Australian thylacine or Tasmanian wolf, the likeness being the result of convergent evolution. *Thylacosmilus*, a marsupial equivalent of the sabre-toothed tigers, was also a splendid example of convergent evolution (Figure 14.3). The xenarthrans ('strange-jointed mammals') were ancestors of living armadillos, sloths, and anteaters, and extinct glyptodonts (Figure 14.4). The condylarthrans produced a spectacular array of endemic orders of hoofed mammals, all of which are now extinct – the condylarthra, litopterns, notoungulates, astrapotheres, trigonstylopoidea, pyrotheres, and xenungulates (Figure 14.5).

2 *Phase II* involved a brief 'window of dispersal opportunity' arising in the Late **Eocene** and Early **Oligocene epochs**, from about 40 million to 36 million years ago. During this time, a series of islands linked the two American continents. According to Simpson, two groups of mammal invaded South America – primates and ancestors of the caviomorph rodents. These were the 'ancient island hoppers'. After having arrived in South America, both groups underwent an impressive adaptive radiation to produce the great variety of rodents found in South America today and the New World or platyrrhine monkeys. The rodents include some interesting extinct forms – the Miocene

Box 14.2

A CURIOUS SOUTH AMERICAN MARSUPIAL

The yapok or water opossum (*Chironectes minimus*) is a curious living didelphid marsupial. It is an otter-like aquatic carnivore with webbed hind feet, living along riverbanks and eating mostly crayfish, fish, and frogs (Figure 14.2). It is the only aquatic marsupial in the world. A sphincter muscle closes the female's pouch when diving to prevent her babies from drowning.

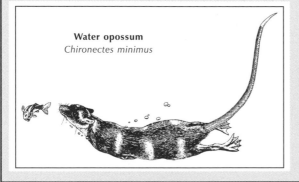

Water opossum
Chironectes minimus

Figure 14.2 The yapok or water opossum (*C. minimus*).

Source: Adapted from Rodríguez de la Fuente (1975)

Phoberomys pattersoni was a 608-kg, buffalo-sized relative of the guinea pig and the late Pleistocene *Telicomys gigantissimus* was nearly as large as a rhinoceros. However, the origin of monkeys (and perhaps even the caviomorph rodents) in South America is unsettled. They may have come from North America, from Africa, or even from Antarctica. Current opinion seems to favour an African homeland, despite the fact that the two continents separated during the Mesozoic era. The oldest fossil primates in South America come from late Oligocene deposits, so a middle Oligocene arrival time seems possible. During the late Oligocene, South America was about the same distance from North America. The late Oligocene ocean currents would have assisted a crossing from Africa to South America but hindered a crossing from North America; and a middle Oligocene fall of sea-level may have permitted rafting (Fleagle 1988). In addition, the platyrrhines and the African parapithecids share a morphological feature – postorbital closure – that is missing in North American primates (Fleagle and Kay 1997). Interestingly, the South American caviomorph rodents did not appear until the Oligocene, and their closest relatives are

African porcupines, which fact shows that other animals may have rafted across the Atlantic (Hoffstetter and Lavocat 1970).

3 *Phase III* began in the Late Miocene epoch, some 6 million years ago. Then, the Bolivar Trough connected the Caribbean Sea with the Pacific Ocean and deterred the passage of animals (Figure 14.6). Phase IIIa saw the first mammals rafting across the seaway on clumps of soil and vegetation. These 'ancient mariners' were members of two families – the 'field mouse' family (Cricetidae) and the racoon and its allied family (Procyonidae). By 3 million years ago, a complete land connection – the Panamanian land bridge – furnished a gateway for faunal interchange between North and South America. Phase IIIb began as a flood of mammals simply walked into South America. Members of many families were involved – deer (Cervidae), camels (Camelidae), peccaries (Tayassuidae), tapirs (Tapiridae), horses (Equidae), mastodons (Gomphotheriidae), rabbits (Leporidae), squirrels (Sciuridae), shrews (Soricidae), mice (Muridae), dogs (Canidae), bears (Ursidae), weasels (Mustelidae), and cats (Felidae). The traffic was two-way – many South American species travelled northwards and entered North America.

Figure 14.3 Convergent evolution of the placental sabre-tooth *Smilodon* and the marsupial sabre-tooth *Thylacosmilus*.

Source: After L. G. Marshall (1981a)

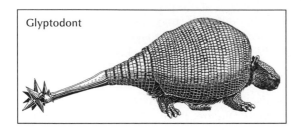

Figure 14.4 A glyptodont. This tanklike Pleistocene herbivore was descended from old xenarthran invaders.
Source: After L. G. Marshall (1981a)

4 *Phase IV* started about 20,000 years ago (the exact date is debatable) as humans spread into South America.

The outcome of these waves of invasion is distinct *faunal strata* in the present mammals of South America (Figure 14.1). Anteaters, sloths, armadillos, and opossums are survivors of the first invasion. The caviomorph rodents and New World monkeys are survivors of the second invasion. All other South American mammals (save recent introductions) are survivors of the Great American Interchange.

Australian invasions

Australia was isolated from other continents for much of the **Cenozoic era**. This isolation led to its having an impoverished fauna, a high degree of endemism, many examples of convergent evolution, and a magnificent adaptive radiation in the pouched mammals. As in South America, new groups of mammals arrived in Australia at different times and radiated adaptively. These invading groups produced faunal strata that can be seen in Australia today. George Gaylord Simpson (1953, 1961) identified four successive waves of invaders, which he styled the 'archaic immigrants', 'old island hoppers', 'middle island hoppers', and 'late island hoppers', to which should be added the 'recent island hoppers'.

Figure 14.5 Unique South American mammals that evolved from Late Cretaceous condylarthran invaders.
Source: After Simpson (1980)

Writing well before plate tectonics was thought of, Simpson argued that all the invading groups came from Asia. A subsequent change of ruling theory in geology and fieldwork on fossil mammals in Australia has meant this view is no longer accepted for two of the groups – the monotremes and the marsupials.

Here are the five phases in the colonization of Australia by mammals:

1 *Phase I*. The 'archaic immigrants' were early monotremes (a subclass of mammals found only in Australia). Fossil finds at Lightning Ridge in northern New South Wales suggest that 110 million years ago Australia supported a number of different monotremes, but no marsupials.
2 *Phase II*. The 'old island hoppers' were marsupials and bats. The first evidence of marsupials in Australia comes from an Eocene fossil site

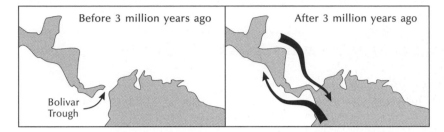

Figure 14.6 The Bolivar Trough and Panamanian Isthmus. Before 3 million years ago, a marine barrier separated central America and South America. When the barrier dried up, the Panamanian land bridge was formed. Animals ventured northwards and southwards in the Great American Interchange.

Source: Adapted from L. G. Marshall (1981b)

at Murgon, near Kingaroy in southeast Queensland. This 55 million-year-old Murgon site has yielded a range of marsupial fossils, many with strong affinities to South American marsupials. Marsupials appear to have evolved in the northern hemisphere about 95 million years ago, then spread to South America, Antarctica, and Australia (Box 14.3). The Murgon site includes the teeth of condylarthrans (see p. 298) that suggest a connection with the Americas. The presence of condylarths in Australia suggests that marsupials and placentals were in Australia at the same time and that the marsupials could have outcompeted the placentals. The remains of a platypus (a Monotreme) in 63 million-year-old rocks of Patagonia, South America, also suggests a connection between Australia and South America. The Australian marsupials underwent a splendid adaptive radiation (Figure 14.7). As well as the extant marsupials, the adaptive radiation produced a number of large extinct forms including a 130–260-kg marsupial 'lion' (*Thylacoleo carnifex*), a 1–2-tonne hippopotamus-like *Diprotodon optatum*, a 200-kg giant wombat (*Phascolonus gigas*), and giant kangaroos, with *Procoptodon goliah* the biggest at 3 m tall.

3 *Phase III*. The 'middle island hoppers' were murid rats that crossed from Asia in the Pliocene epoch (e.g. Godthelp 1997). Fossil monotremes from Pleistocene deposits in Mammoth Cave, Western Australia, include *Zaglossus hacketti*, a sheep-sized echidna and probably the largest-ever monotreme.

4 *Phase IV*. The 'late island hoppers' were the ancestors of aborigines and dingo (*Canis dingo*), which may have come with them some 40,000 to 60,000 years ago. The water gap at the time, when sea-levels were lower, would have been about 60 km (Fagan 1990).

5 *Phase V*. The 'recent island hoppers' were white settlers who from the end of the eighteenth century brought a long catalogue of exotic species with them.

The outcome of these waves of invasion, like their counterparts in South America, is distinct faunal strata in the current mammalian fauna of Australia. The echidnas (or spiny anteaters) and the duck-billed platypus are survivors of the first invasion, although they may well have evolved in Australia. The pouched mammals and bats are survivors of the second invasion. All other Australian mammals – save the recent humans and the animals they brought with them – are descendants of individuals that island-hopped from Polynesia and Asia.

Box 14.3

HISTORICAL BIOGEOGRAPHY OF MARSUPIALS

The pouched mammals or marsupials live today in Australia, South America, and North America. Eurasia, North Africa, and Antarctica yield fossil forms. Marsupials probably evolved in Asia, which houses a 125 million-year-old marsupial fossil (*Sinodelphys szalayi*), from an ancestor that also sired the placental mammals (Luo *et al*. 2003). The marsupial and placental mammals probably split about 130 million years ago, early in the Cretaceous period, the marsupials then moving into North America. Albian deposits in Utah, USA, which are about 101 million years old, have yielded a probable marsupial (Cifelli 1993). There are several rival explanations for the current distribution of marsupials (L. G. Marshall 1980; Springer *et al*. 1997).

The classic explanation of marsupial distribution, proposed before the acceptance of continental drift, assumed stationary continents. It argued that marsupials dispersed from a Cretaceous North American homeland to other continents. Some time in the Late Cretaceous period, marsupials hopped across islands linking North and South America. During the Eocene epoch, they moved into Asia and Europe across a land bridge spanning the Bering Sea between Alaska and Siberia. From there they spread into Europe and, using Indonesia as an embarkation point, into Australia. Several variations on this 'centre-of-origin followed by dispersal' hypothesis played out on a stationary land surface were forthcoming. The variations involved different centres-of-origin (South America or Antarctica) and different dispersal routes.

Revised explanations of the marsupial history followed the acceptance of continental drift. Some of these new hypotheses still invoked centre-of-origin and dispersal. They were similar to the hypothesis developed for stationary continents, but they did not need to invoke fanciful land bridges between widely separated continents. Other hypotheses lay emphasis on the fragmentation of Pangaea, the Triassic supercontinent, and stressed vicariance events rather than dispersal over pre-existing barriers.

It now seems likely that marsupials appeared and went through an initial phase of diversification in Mesozoic Laurasia, and they did not range into Gondwana (the combined landmasses of South America, Antarctica, Australia, Africa, and India) until after the Cretaceous period (Springer *et al*. 1997). The existing marsupial orders, which almost all occur on the remnants of Gondwana, do not stem from this initial diversification. Rather, they are the product of a second bout of diversification that occurred in the Late Cretaceous and Palaeocene. The ancestor of this radiation may have reached western Gondwana (South America) from North America in the late Cretaceous period, and then diversified entirely on Gondwana, perhaps filling the vacant adaptive zones left by the terminal Cretaceous mass extinctions, to produce the existing marsupial orders before South America, Antarctica, and Australia broke apart. To be sure, marsupials appear to have arrived on Australia about 55 million years ago. The secondary diversifications of marsupials in South America and Australia (and possibly Antarctica too) were thus roughly contemporaneous and involved closely related ancestors (Springer *et al*. 1997). The Asian marsupials spread to America, Europe, and North Africa during late Cretaceous, Palaeocene, and Eocene times, but by the end of the Miocene epoch, North American, Asian, European, and North African marsupials were extinct. South American marsupials invaded North America in the Pleistocene epoch. Current explanations of marsupial biogeography thus call on dispersal and vicariance events.

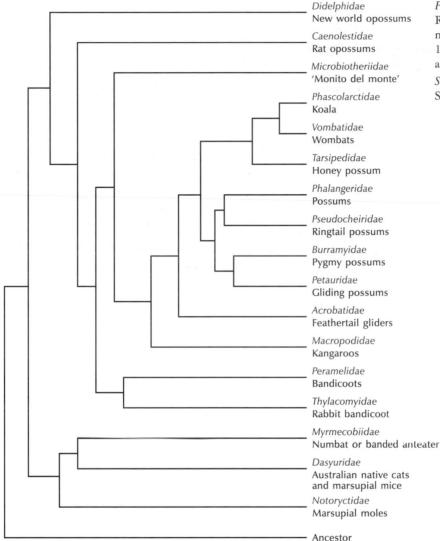

Didelphidae
New world opossums

Caenolestidae
Rat opossums

Microbiotheriidae
'Monito del monte'

Phascolarctidae
Koala

Vombatidae
Wombats

Tarsipedidae
Honey possum

Phalangeridae
Possums

Pseudocheiridae
Ringtail possums

Burramyidae
Pygmy possums

Petauridae
Gliding possums

Acrobatidae
Feathertail gliders

Macropodidae
Kangaroos

Peramelidae
Bandicoots

Thylacomyidae
Rabbit bandicoot

Myrmecobiidae
Numbat or banded anteater

Dasyuridae
Australian native cats
and marsupial mice

Notoryctidae
Marsupial moles

Ancestor

Figure 14.7
Relationships between marsupials based on 102 morphological and anatomical characters.

Source: Adapted from Springer *et al.* (1997)

SUMMARY

Historical biogeography began as a scholarly debate about the restocking of the Earth from Mount Ararat after Noah's Flood. The debate evolved into the classical theory of centres-of-origin and dispersal, first proposed by Charles Robert Darwin and Alfred Russel Wallace. This theory contends that all species have a centre-of-

origin from which they disperse, and that, because barriers impede their movement, species spread slowly enough for natural selection to cause them to change while dispersing. Modern biotas of continents and islands, such as South America and Australia, are the product of successive waves of invasion that have built up faunal strata over geological timescales. Remnants of the early invasion waves survive in the modern faunas.

ESSAY QUESTIONS

1 Is the history of biogeography relevant to modern debates in the subject?

2 What problems arise in reconstructing the biogeographical history of South America?

3 How unique is the fauna of Australia?

FURTHER READING

Briggs, J. C. (1995) *Global Biogeography* (Developments in Palaeontology and Stratigraphy 14). Amsterdam: Elsevier.

An advanced text but worth dipping into.

Darlington, P. J., Jr (1957) *Zoogeography: The Geographical Distribution of Animals*. New York: John Wiley & Sons.

A classic dispersalist interpretation of biogeographical history in its day, which was before the revival of continental drift.

Flannery, T. F. (1994) *The Future Eaters: An Ecological History of the Australasian Lands and People*. New York: George Brazillier.

A great read.

Simpson, G. G. (1980) *Splendid Isolation: The Curious History of South American Mammals*. New Haven, CT, and London: Yale University Press.

An exciting case study, but needs updates on the arrival of primates and caviomorph rodents.

Udvardy, M. D. F. (1969) *Dynamic Zoogeography, with Special Reference to Land Animals*. New York: Van Nostrand Reinhold.

Rather old, but still deserves to be read.

VICARIANCE IN THE DISTANT PAST

Processes in the physical environment, such as continental drift, may break animal and plant distributions into fragments, which then evolve independently of each other. Drifting fragments may collide and fuse and their biotas mingle. This chapter covers:

- vicariance biogeography
- biotic fragmentation
- biotic fusion

SPLITTING ASUNDER: WHAT IS VICARIANCE?

Vicariance biogeography arose as an antidote to the hegemony of dispersal biogeography. Anti-dispersalist grumblings were heard early in the twentieth century. Léon Croizat, a Franco-Italian scholar, directed the first major critical onslaught (Croizat 1958, 1964; see also Humphries 2000). Croizat tested the Darwinian centre-of-origin–dispersal model by mapping the distributions of hundreds of plant and animal species. He found that species with quite different dispersal propensities and colonizing abilities had the same pattern of geographical distribution. He termed these shared geographical distributions *generalized* or *standard tracks*. Croizat argued that standard tracks do not represent lines of migration. Rather, they are the present distributions of a set of ancestral distributions, or a biota of which individual components are relict fragments. His reasoning was that widespread ancestral taxa had been fragmented by tectonic, sea-level, and climatic changes. He termed the fragmentation process 'vicariism' or 'vicariant form-making in immobilism'. Fortunately, it has become known as *vicariance*. Of course, to become widespread in the first place, a species must disperse. But vicariance biogeographers claim that ancestral taxa achieve widespread distribution through a mobile phase *in the absence of barriers*. They allow that some dispersal across barriers does occur, but feel that it is a relatively insignificant biogeographical process.

BREAKING UP: CONTINENTAL FISSION

Continents break up, drift, and collide. In doing so, they make and break dispersal routes and alter species distributions. Overall, continental drift, and processes associated with it (such as mountain building), help to explain several features in animal and plant distributions for both living and fossil forms.

The tearing asunder of previously adjoining landmasses causes the separation of ancestral populations of animals and plants. Once parted, the populations evolve independently and diverge. Eventually, they may become quite distinct from the ancestral population that existed before the landmasses broke apart. There are several good examples of this process.

Reptiles and mammal-like reptiles on Pangaea

The breakup of Pangaea, and in particular the southern part of it known as **Gondwana**, split several animal and plant populations into separate groups. In consequence, members of these ancient populations, and their living descendants, have disjunct distributions.

One of the first pieces of fossil evidence used in support of the idea that Africa and South America were formerly joined was the distribution of a small reptile called *Mesosaurus* that lived about 270 million years ago, in the **Permian period** (Figure 15.1). *Mesosaurus* was about a metre long, slimly built with slender limbs and paddle-like feet. At the front end it had extended, slender jaws carrying very long and sharp teeth, probably used to catch fish or small crustaceans. At the rear end was a long and deep tail, admirably suited to swimming. *Mesosaurus* surely spent much time in the water. The sediments in

which it has been found suggest that it inhabited freshwater lakes and ponds. Its remains have been found only in southern Brazil and southern Africa. If, as the evidence suggests, it was a good swimmer, then it would probably have had a wider range than it apparently did have. The most parsimonious explanation is that Brazil and Africa abutted in the late **Palaeozoic era**. As Alfred Sherwood Romer (1966, 117) said, 'although *Mesosaurus* was obviously a competent swimmer in fresh waters, it is difficult to imagine it breasting the South Atlantic waves for 3,000 miles'.

Another disjunct distribution is displayed by *Lystrosaurus*, a squat, powerful, mammal-like reptile. *Lystrosaurus* lived in the Early Triassic period, about 245 million years ago. It was about a metre long, with a pair of downwards-pointing tusks. The position of its eyes, high on its head, suggests that it spent much time submerged with all but the top of its head below water. Specimens of *Lystrosaurus* have been found at localities in India, Antarctica, South Africa (all formerly part of Gondwana), and China (Figure 15.2). The distribution of *Lystrosaurus* on Gondwana accords with modern ranges of terrestrial reptiles, such as the snapping turtle (*Chelydra serpentina*) (Colbert 1971). Its presence in China, in a place normally assumed to have been part of eastern Laurasia in Triassic times, presents a problem. One explanation, though an unlikely one, is that some individuals migrated from Gondwana, all the way

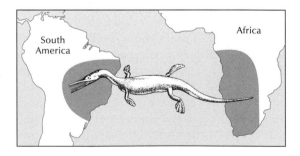

Figure 15.1 The distribution of *Mesosaurus* in the Permian period.

round the head of the Tethys embayment, to eastern Laurasia. Another explanation is that Gondwana was bigger than is normally supposed, and extended beyond the northern edge of the Indian plate to include a large segment of what is now China (Crawford 1974). If this were the case, all the *Lystrosaurus* faunas would have lived on Gondwana, and the Chinese members of the population would not be anomalous. Another possibility, suggested by recent work on plate tectonics in the southeast Asian region, is that, in Late Permian times, or even before, large fragments of Australia (terranes) appear to have broken off and drifted northwards, colliding with Asia. These terranes could have carried, in the manner of Noah's arks (p. 313), *Lystrosaurus* with them, or could have acted as a series of stepping-stones making dispersal possible (C. B. Cox 1990, 125).

Cenozoic land mammals

During Cenozoic times, continental drift greatly affected land-animal evolution (Kurtén 1969). Fossil Cenozoic faunas, and particularly the mammal faunas, are different on each continent. The most distinctive faunas are those of South

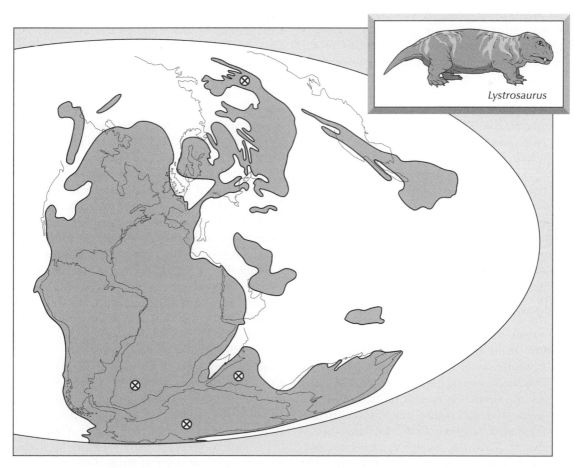

Figure 15.2 The distribution of *Lystrosaurus* in the Early Triassic period.
Source: Base map adapted from A. G. Smith *et al.* (1994)

America, Africa, and Australia – the southern continents. There are some 30 orders of mammal, almost two-thirds of which are alive today. All the mammalian orders probably had a common origin – an ancestor that lived in the Mesozoic era. The Cenozoic divergence of mammals may be due to the late Mesozoic fragmentation of Pangaea. Interestingly, the land-dwelling animals (mostly reptiles) that lived in the Cenozoic era displayed far less divergence than the mammals. By the end of the Cretaceous period, after about 75 million years of evolution, some 7–13 orders of reptiles had appeared, far fewer than the 30 or so produced by mammalian evolution over 65 million years. A possible explanation for this difference lies in the palaeogeography of the Mesozoic continents. For much of the Age of Reptiles, the continents were not greatly fragmented. There were two supercontinents: Laurasia lay to the north and Gondwana to the south. Rifts between the continents existed as early as the Triassic period, but they were not large enough to act as barriers to dispersal until well into the Cretaceous period. In Late Cretaceous and Early Tertiary times, when the mammals began to diversify, the rapid break-up of the former Pangaea, coupled with high sea-levels, led to the separation of several land-masses and genetic isolation of animal populations. The result was divergence of early mammalian populations.

Pangaea and plant distributions

The break-up of Pangaea accounts for many disjunct plant distributions. Some plants have seeds unsuitable for jump dispersal. An example is the genus *Nothofagus* (the southern beeches) that consists of about 60 species of evergreen and deciduous trees and shrubs. Its present distribution is disjunct, being found on remnants of Gondwana – South America, New Zealand, Australia, New Caledonia, and New Guinea, but not Africa. Fossil *Nothofagus* pollen of Oligocene age has been found on Antarctica. The conclusion

normally drawn is that the modern disjunct range of the genus has resulted from the break-up of Gondwana. The same explanation applies to plants of the protea, banksia, and grevillea family – the Proteaceae (p. 62).

Big, flightless birds

Continental break-up helps to explain the enigma of the large flightless birds. This group forms an avian superorder – the *ratites*. There are five families of living ratites and two extinct ones, all of which are thought to have arisen early in bird evolution from a common ancestor and thus form a **monophyletic group**. The moas (Dinornithidae) lived on New Zealand until a few hundred years ago; the kiwis (Apterigidae) still live on New Zealand (Plate 15.1). Emus (Dromaiidae) and cassowaries (Casuariidae) both live in Australia, and the cassowaries are also found on New Guinea (Plate 15.2). The extinct dromorthinids are also Australian ratites. Ostriches (Struthionidae) live in Africa and Europe. The elephant birds (Aepyornithidae) lived on Madagascar and possibly Africa (e.g. Senut *et al.* 1995), but went extinct around the mid-seventeenth century. Rheas (Rheidae) live in South America.

With the exception of the chicken-sized kiwis, most of the ratites are well-built cursorial birds with huge hind limbs for fast and sustained running. The ostrich is the world's largest living bird, standing about 2.5 m high and weighing in at about 140 kg. The largest of the moas, *Diornis maximus*, stood about 3.5 m high and weighed around 240 kg; it was the tallest bird known to have lived. The largest elephant-bird, *Aepyornis maximus*, was a ponderous giant, with elephantine-style legs; it stood 3 m tall and weighed about 450 kg. How could such large birds have spread through the southern continents if they could not fly, or even if they could? A plausible hypothesis involves the early evolution of flightlessness and continental drift (Cracraft 1973, 1974). A flying ancestor of all the ratites probably lived in west Gondwana, or what is

Plate 15.1 Brown or spotted kiwi (*Apterix australis*).
Photograph by Pat Morris.

now South America. The flightless tinamous, partridge-like birds that have their own super-order (the Tinamae), may be descendants of this extinct bird that stayed at home in South America. Flightlessness may have evolved during the Cretaceous period, before continental separation was far advanced. The flightless birds could then have dispersed through the southern continents by walking to other parts of Gondwana. The ancestors of the New Zealand moas and kiwis must have walked through west Antarctica before New Zealand drifted away from the main landmass. Interestingly, a fossil ratite has recently been discovered in the Palaeogene La Meseta Formation, Seymour Island, Antarctica (Tambussi *et al.* 1994). The recently extinct elephant birds of Madagascar could have walked directly from South America. The emus and cassowaries may have walked through east Antarctica. The ostriches of Africa and Europe may have walked directly from South America, and the South American rheas presumably evolved from less adventurous South American flightless relatives.

More recent work challenges the Gondwanan origin of the ratites (e.g. Houde 1988). Forms ancestral to modern ratites appear to have evolved in North America and Europe from the Late Palaeocene to the Middle Eocene epochs. This

Plate 15.2 Common cassowary (*Casuarius casuarius*).
Photograph by Pat Morris.

would mean that living ratites are southern hemisphere relics of a widespread group that probably originated in the early Tertiary period.

Greater Antillean insectivores

Vicariance events in the geological history of the Caribbean region may explain the distribution of *relict insectivores* in the Greater Antilles. There are two genera of such insectivores – *Nesophontes* and *Solenodon*. Several species of *Nesophontes* are known only from skull and skeletal remains in owl pellets – the St Michel nesophontes (*N. paramicrus*) and Haitian nesophontes (*N. zamicrus*) both lived on Haiti; the west Cuban nesophontes (*N. micrus*) lived on Cuba and Haiti; the atalaye nesophontes (*N. hypomicrus*) lived on the Dominican Republic and Haiti. There were also the east Cuban nesophontes (*N. longirostris*), the Puerto Rican nesophontes (*N. edithae*), and an undescribed species on the Cayman Islands. They all seem to have gone extinct by the sixteenth century at very latest, though a few researchers think a few may have survived into the early twentieth century. Two species of *Solenodon* – the Cuban solendon (*S. cubanus*) and Haitian solendon (*S. paradoxus*) – still live on Cuba and Hispaniola, respectively (Plate 15.3). These insectivores were, and in the solenodons' case are, shrew-like animals but larger than

a normal shrew. Solenodons are about 15–16 cm and weigh 40–46 g. They are nocturnal and live in caves, burrows, and rotten trees.

One explanation of the relict insectivore distribution runs as follows (MacFadden 1980). In the Late Cretaceous, North American and South American landmasses were joined by a landmass that was to become the Antilles (Figure 15.3a). At this time, ancestral insectivores lived on North America and the proto-Antillean block. Early in the Cenozoic era, the proto-Antilles moved eastwards, relative to the rest of the Americas, carrying a cargo of insectivores with it (Figure 15.3b). Later in the Cenozoic era, the proto-Antilles had reached their present position, and the gap between North and South America was filled by the lower central American landmass. However, the mainland relatives of *Nesophontes* and *Solenodon* on North America were by now extinct, leaving their Antillean relatives as relics of a once widespread distribution (Figure 15.3c).

The history of the Greater Antillean insectivores is interpretable in other ways. The traditional view is that they colonized the islands by over-water dispersal from the Americas – a sweepstakes route. Support for this interpretation came from a study of Caribbean tectonics, the composition of the fauna, and the fossil record in the Americas (Pregill 1981). The evidence suggested

Plate 15.3 Solenodon, a relict Antillean insectivore.
Photograph by Pat Morris.

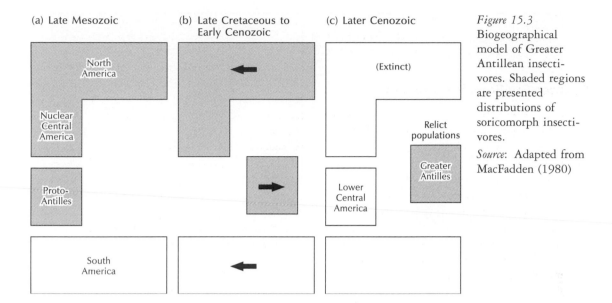

(a) Late Mesozoic

(b) Late Cretaceous to Early Cenozoic

(c) Later Cenozoic

North America

Nuclear Central America

Proto-Antilles

South America

(Extinct)

Relict populations

Greater Antilles

Lower Central America

Figure 15.3 Biogeographical model of Greater Antillean insectivores. Shaded regions are presented distributions of soricomorph insectivores.

Source: Adapted from MacFadden (1980)

that the Antilles started life as a volcanic archipelago in the Late Cretaceous epoch. Modern terrestrial vertebrates probably started arriving by over-water dispersal during the Oligocene epoch, when most of the living genera in the West Indies first appeared, and continued to do so through the Quaternary. The various living genera and species are unevenly distributed throughout the islands and remarkably few orders, families, and genera represent the vertebrates. This composition is consistent with an island biota built up though dispersal, rather than vicariance.

A recent reconstruction of geological history in the Caribbean region may offer a compromise between the dispersal and vicariance explanations of the Antillean insectivores and other animals and plants (Perfit and Williams 1989). About 130 million years ago, the proto-Antilles was a chain of volcanic islands lying along a subduction zone at the Pacific Ocean rim (Figure 15.4). They stretched 2,000 km between the west coast of Mexico and the coast of Ecuador. Some 100 million years ago, the North American and South American plates started to move westwards over the proto-Caribbean sea floor, and the islands drifted relatively eastwards, at the leading edge of the Caribbean plate. By 76 million years ago, Cuba struck the Yucatán region of Mexico. The remaining islands suffered uplift and deformation as they squeezed through the gap between the Yucatán and Colombia. A land bridge between the Americas (the proto-Costa Rica–Panama land bridge) had begun to grow along a new subduction line formed where the Pacific Ocean dived underneath the Caribbean plate. On occasions, the proto-Antillean islands may have formed a complete dry land connection between the Americas around this time. The proto-Antilles stayed close to the North American continent for the next 20 million years, and indeed was often connected to it. About 55 million years ago, Cuba hit the Bahamas bank (a large limestone platform) and became wedged there. The remaining islands kept moving eastwards, causing considerable shear against the beached Cuba. The shear stress caused the islands to break up and adopt their modern configuration. Puerto Rico broke away from Hispaniola about 35 million years ago; Hispaniola separated from Cuba about 23 million years ago.

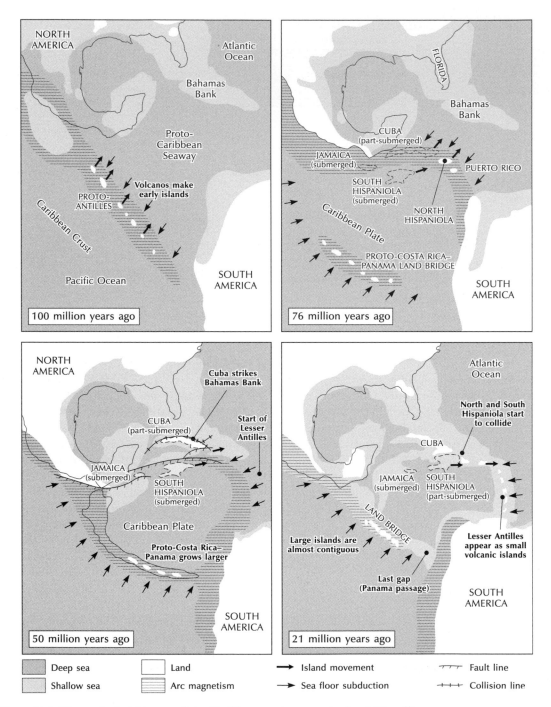

Figure 15.4 The geological history of the Caribbean region over the last 100 million years.

Sources: Adapted from Perfit and Williams (1989) and Reddish (1996)

These tectonic changes affected the biogeography of the area. Frogs of the genus *Eleutherodactylus*, which are all small (one from eastern Cuba is at 0.9 cm the smallest four-legged beast in the world), originally came from South America (Hedges 1989). They appear to have moved along the proto-Antillean island chain around 70–80 million years ago. They disembarked when the island reached the Yucatán, and established a foothold on the North American continent. Today, there are 68 species in Mexico. The frogs that chose not to disembark have produced the 139 species found on the Greater Antillean islands today. The traffic was two-way. While the *Eleutherodactylus* frogs went ashore onto North America, several North American species – including pines, butterflies, and todies (a kind of bird) – boarded the islands while they were docked. The presence of pines on the Caribbean islands today is difficult to explain by dispersal because their heavy seeds are ill designed for crossing hundreds of kilometres of salt water. Similarly, many of the butterflies found on the Greater Antilles, including the fragile glasswings (family Ithomiidae) are not good dispersers. Todies are a family of birds endemic to the Caribbean, and distantly related to the kingfishers. They are not powerful fliers. A 30 million-year-old tody from Wyoming at least allows the possibility that ancestors of the present birds could have boarded the islands early in their eastward travels, perhaps flying the short distance across to Cuba. At same time, ancestors of the primitive insectivores could have crossed over, by either walking or swimming the short distance. Most mammals had not evolved when the islands docked against North America, but ancestors of these insectivores are known from fossils in North America to be at least 55–60 million years old.

COMING TOGETHER: CONTINENTAL FUSION

Drifting continents eventually collide and fuse. When they do so, they unite to form a new single landmass. Uplift or a fall of sea-level may also forge a land bridge between two formerly separated landmasses. A moving plate may carry species (indeed, entire faunas and floras) over vast distances, acting like a gigantic raft or Noah's ark; it may also carry a cargo of fossil forms, and is like a Viking funeral ship (McKenna 1973). The **biota** on these more-or-less isolated landmasses should attain a steady state in which origination and extinction (turnover) will be roughly in balance. These biota should be saturated – all available niches should be filled.

Faunal mixing

When two continents collide, or a land bridge forms between them, the fauna and flora of the two continents are free to mix. The mixing process has four possible outcomes (L. G. Marshall 1981a):

1 *Active competition* occurs between species or genera occupy the same or very similar niches on the different landmasses. If the native form should be more efficient at exploiting resources, the invader will be expelled or kept out. Of course, such unsuccessful invasion attempts are seldom, if ever, recorded in the fossil record. The alternative outcome is that an invader is successful and ousts the native vicar.
2 *Passive replacement* may occur when, by chance, a native species sharing a niche with an invading species happens to go extinct. There is no competition involved – the surviving species was merely lucky to be in the right place at the right time.
3 Intruders may come across a biota with unexploited ecological niches and are able to insinuate themselves into the pre-existing community without having an overt affect on it. The *insinuators* are ecologically unique and have no discernible counterparts in the native biota. Insinuating species will boost the biotic diversity.

4   An invader may have a similar niche to a native species and character displacement (p. 198) takes place so that both species can live together. These invaders are *competitors-cum-insinuators*.

As the faunas and floras mix, so they come to resemble one another, though they may still retain distinctive features. The Great American Interchange is perhaps the most spectacular and best-documented faunal mixing event. Resulting from the fusion of the North and South American continents, it demonstrates many of the possible outcomes of faunal mixing.

The Great American Interchange

The faunas of North and South America began to mix at the close of the Miocene epoch, about 6 million years ago (p. 299). The *Great American Interchange* began as a trickle. It became a flood around 3 million years ago, when the Panamanian Isthmus formed, peaked around 2.5 million years ago, and is still going on today.

It was once widely believed that the invasion of South America by North American species caused the extinction of many native taxa. The placental carnivores appeared and outcompeted marsupial carnivores, which became extinct. Likewise, South American 'ungulates' suffered heavy losses in competition with the northern invaders. Recent interpretations of the evidence paint a more complicated picture of events. Two cases illustrate this point (L .G. Marshall 1981a; see also Lessa *et al*. 1997).

Before the invasions, South American 'ungulates' included litopterns (protutheres and macrauchenids) and notoungulates (toxodons, mesotheres, and hegetotheres). To these should be added five families of xenarthrans and even two of rodents that occupied the large herbivore niche. Once the Bolivar Trough was closed, many families of northern ungulates (mastodons, horses, tapirs, peccaries, camels, and deer) travelled south. There is some evidence that the disappearance of native 'ungulates', especially the notoungulates, began before the appearance of northern rivals. The factors causing this decline were thus not related to the interchange. This conclusion is supported by the fact that no invading northern species were vicars (ecological equivalents) of southern species – they occupied slightly different niches. The most likely vicars were invading 'camels' and the camel-like macrauchenids. However, these two groups co-existed for 2.5 million years.

It was originally claimed that the South American dog-like borhyaenids (marsupials) declined to the point of extinction when in competition with invading placental carnivores from the north. However, it now seems clear that the borhyaenids were extinct before dogs, cats, and mustelids moved south. Some large omnivorous borhyaenids declined and fell when the waif members of the Procyonidae arrived in Phase IIIa (p. 299). For instance, *Stylocynus*, a large, omnivorous, bear-like borhyaenid, had a vicar in *Chapalmalania*, a large, bear-like procyonid. In turn, *Chapalmalania* vanished when members of the bear family (Ursidae) arrived. It is also possible that the placental sabre-tooths (felids) replaced the native marsupial sabre-tooths (thylacosmilids).

When the Panamanian Isthmus triggered the Great American Interchange, a large majority of land-mammal families crossed reciprocally between North and South America around 2.5 million years ago, in Late Pliocene times. Initially, there was an approximate balance between northward traffic and southward traffic. During the Quaternary, the interchange became decidedly unbalanced (S. D. Webb 1991). Groups of North American origin continued to diversify at exponential rates. In North America, extinctions more severely affected South American immigrants – six South American families were lost in North America, while two North American families were lost in South America.

SUMMARY

Vicariance is a rival idea to dispersal. It emphasizes the splitting of ancestral and widespread species distributions by geological or ecological factors. A good example is the break-up of Pangaea, which appears to have dictated the current distributions of several animal and plant groups. Drifting continental blocks tend to collide and fuse. Continental fusion leads to faunal mixing, a splendid example of which is the Great American Interchange.

ESSAY QUESTIONS

1 **To what extent was the distribution of Mesozoic and Cenozoic animals shaped by vicariance events?**

2 **To what extent is the present distribution of plant groups fashioned by the break-up of Pangaea?**

3 **What impact did the Great American Interchange have on the faunas of South and North America?**

FURTHER READING

Nelson, G. and Rosen, D. E. (eds) (1979) *Vicariance Biogeography: A Critique* (Symposium of the Systematics Discussion Group of the American Museum of Natural History, May 2–4). New York: Columbia University Press.

Only for the brave.

Simpson, G. G. (1980) *Splendid Isolation: The Curious History of South American Mammals*. New Haven, CT, and London: Yale University Press.

Contains a discussion of the Great American Interchange.

Woods, C. A. (ed.) (1989) *Biogeography of the West Indies: Past, Present, and Future*. Gainesville, FL: Sandhill Crane Press.

Dip into this for a case study.

PAST COMMUNITY CHANGE

Communities change of their own volition or through changes in climate and other environmental factors. This chapter covers:

- historical changes
- Pleistocene changes
- pre-Quaternary changes

Community change is the rule. Seldom do communities remain stable for long periods. Environmental changes and processes within the communities themselves drive them to new states. It is expedient to consider community changes in three separate blocks of time – the last 10,000 years or so (the Holocene epoch), during the Pleistocene ice ages, and before the Quaternary.

HISTORICAL TIMES: COMMUNITY CHANGE IN THE HOLOCENE

Climatic and other environmental changes during the Holocene epoch in northwest Europe are traditionally divided into a series of time zones (Table 16.1). Broadly speaking, rapid climatic warming occurred once the ice sheets withdrew, leading to an early to mid-Holocene 'Climatic Optimum' (also called Hyspithermal, Altithermal, and other names) with temperate latitude temperatures some 2–3°C warmer than now. After that episode, temperatures fell and precipitation increased in temperature latitudes. These climatic changes, in conjunction with cultural and land-use events in some parts of the world (Table 16.1), led to readjustments and adaptations of animal and plant populations and communities. Three examples serve to illustrate these faunal and flora repercussions: the changing size of some animal species; the disassembling and reassembling of Holocene communities; and the migration of plants in a changed environment.

Table 16.1 Climatic and environmental changes over the last 14,000 years in the Peak District, England

Zone (I to VIII)	Date (years BC)	Climate	Vegetation	Culture and land-use events
Glacial	Pre-18,000	Arctic	Dwarf willow and arctic herbs	
Late glacial I Older Dryas	12,380–10,000	Tundra	Arctic–alpine herbs, including mountain avens (*Dryas octopetala*) with arctic willows (*Salix* spp.) and dwarf birch (*Betula nana*)	
II Allerød	10,000–8,800	Warmer	Birches – hairy birch (*Betula pubescens*) and silver birch (*B. pendula*), aspen (*Populus tremula*), and pine open woodland with herbs	
III Younger Dryas	8,800–8,300	Arctic	Arctic–alpine herbs, including mountain avens (*D. octopetala*)	
Postglacial IV Pre-Boreal	8,300–7,600	Warmer (sub-Arctic)	Open birch woodland	
V Boreal	7,600–5,500	Warmer and drier	Pine and hazel replace birch; elm and oak towards end with small amounts of alder	Mesolithic: burning for improved grazing
VI				
VII(a) Atlantic	5,500–3,000	Warmer and wetter (oceanic)	[English Channel forms] Alder increases. Deciduous woodland with oak, elm, hazel, lime	Neolithic
VII(b) Sub-Boreal	3,000–600	Drier (more continental)	Elm declines, ash expands. First weeds appear	Bronze Age: deforestation, pastoralism, cultivation
VIII Sub-Atlantic	600 to present	Colder and wetter	Heather spreads. Birch increases, lime declines	Iron Age, Romans, Saxons, Normans: continual disturbance of vegetation

Source: Adapted from Anderson and Shimwell (1981)

Bergmann's rule and Holocene faunas

Fossil assemblages sometimes reveal geographical variations in morphological features, including the size, of mammals. Nine species from the New Paris No. 4 fauna in Pennsylvania, for instance, tend to increase in size towards the north in agreement with Bergmann's rule, although four species tend to get larger in the opposite direction (Lundelius *et al.* 1983).

Bergmann's rule also seems to operate in the changing size of members of a population during time – the size adjustments may track climatic change. In Missouri, archaeological specimens of the grey squirrel (*Sciurus carolinensis*) increase in size from the early to middle Holocene epoch; specimens of the eastern cottontail (*Sylvilagus floridans*) decrease in size during the same time interval and then increase in size to modern proportions during the late Holocene epoch (Purdue 1980). James R. Purdue (1989) believes that changes in the size of the white-tailed deer (*Odocoileus virginianus*) in central Illinois during Holocene times were strongly influenced by insolation-driven summer climate acting through food resources, in particular summer forage. Similar size changes occurred in other animals during the Quaternary (e.g. Klein 1986).

Disassembly and reassembly of Holocene communities

Communities, like individuals, are impermanent. Species abundances and distributions constantly change, each according to its own life-history characteristics, largely in response to an ever-changing environment.

Orbital changes appear to have harmonized the broad shifts of climate during the Holocene epoch. These shifts varied from region to region. Climatic change in western Europe, for instance, was (and still is) sensitive to changes in the north Atlantic, and especially to changes that influenced winter conditions on the west of the continent. Holocene climatic changes forced animal and plant communities to disassemble and reassemble. In Europe, pollen data for the last 13,000 years indicate individualistic changes in species distributions and evanescent community composition (Huntley 1990; Huntley and Prentice 1993). Figure 16.1 illustrates these patterns.

Singular climatic events produced by volcanoes interrupt the broad sweep of Holocene climatic swings. These caused temporary disruption of the climate system and left their mark in the growth rings of trees. Tree rings in European oaks (*Quercus*), for example, responded to a volcanic eruption (or possibly two) in AD 536 (Baillie 1994).

Migrating species

Following the retreat of the Pleistocene ice sheets, new land was available for plants and animals to colonize and new habits became available as the environment altered. Migration rates vary considerably between species. Three basic migration strategies seem to exist and are analogous to opportunist, equilibrium, and fugitive population strategies (Delcourt and Delcourt 1991, 57–60) (see p. 176):

1 *Opportunist species*. Fast migrators typically push forwards along a steep migration front, but fail to maintain large populations in their wake. Spruce (*Picea*) is an *r*-migration strategist and displays these characteristics (Delcourt and Delcourt 1987, 306–12). With the retreat of the Laurentide Ice Sheet, spruce spread rapidly onto the newly exposed ground. The rate of migration reached 165 m/yr 12,000–10,000 years ago. Only the ice halted the spruce's northwards advance. Behind its migration front, populations diminished as climates warmed. Black spruce (*Picea mariana*) was the first conifer species to invade northern Quebec immediately after the ice had melted 6,000 years ago (Desponts and Payette 1993).

2 *Equilibrium species*. Slow migrators rise more slowly to dominance after an initial invasion, the highest values of dominance lying well behind the 'front lines' and reflecting an unhurried build-up of populations. Oak (*Quercus*) is a *K*-migration strategist. It was a minor constituent of late-glacial forests in eastern North America. It did reach the Great Lakes in late-glacial times, but its rise to dominance was a slow process that reached a ceiling in the Holocene epoch in the region now occupied by the eastern deciduous forest (Delcourt and Delcourt 1987, 313–16).

3 *Fugitive species*. These are the 'weeds' of the animal and plant kingdoms. They colonize temporary, disturbed habitats, reproduce rapidly, and soon depart before the habitat

disappears or before competition with other organisms overwhelms them. A fugitive-migration strategist occupies a patchwork of habitat islands. The tamarack (*Larix laricina*) is a fugitive-strategist (Delcourt and Delcourt 1987, 319–22). Its changing distribution mirrors the availability of wetland habitats, and any snapshot of its distribution shows low dominance in most areas, with dominance hot-spots in local wetland sites (e.g. Payette 1993).

A differential migration rate of species helps to disassemble communities and to make the species composition and abundance in reassembled communities virtually impossible to predict. However, in some cases, communities retain their essential character and shift bodily in the wake of shifting climatic zones. In Africa, for instance, vegetation belts swung swiftly northwards 9,000 years ago, after an increase in the Atlantic monsoon led to greater rainfall.

ICES AGES: COMMUNITY CHANGE IN THE PLEISTOCENE

Glacial–interglacial cycles and Pleistocene communities

During the Pleistocene epoch, *orbital variations* generated a sequence of glacial–interglacial switches that marine and terrestrial communities were obliged to track. Within an interglacial episode, the Croll–Milankovitch astronomical parameters (changes in the tilt of the Earth's axis, the eccentricity of its orbit, and the precession of the equinoxes that alter the seasonal and latitudinal pattern of solar radiation receipt) gradually change. The result is a cycle of climatic change involving an increase in temperature and possibly humidity followed by decreasing temperature. Plants and animals responded to this climatic change in a four-stage (later revised to a five-stage) interglacial cycle (Iversen 1958; Andersen

1966; H. J. B. Birks 1986). The five stages are as follows:

1 *Cryocratic stage*. Skeletal mineral soils support open herb communities.
2 *Protocratic stage*. Increasing temperature and possibly humidity favours the formation of unleached calcareous soils with basic grassland and woodland.
3 *Mesocratic stage*. Temperatures peak. Fertile brown soils form and support mixed deciduous forest.
4 *Oligocratic stage*. During the first part of the oligocratic stage, leaching of brown soils produces acid podzols that favour coniferous woodlands and heaths.
5 *Telocratic stage*. This sets in as temperatures begin to fall.

Orbital signals are present in records of Quaternary vegetation change. Pollen in three cores from the Grand Pile bog, Vosges Mountains, northeastern France, registers the 23,000-year precessional cycle and its harmonics (Molfino *et al.* 1984). The abundances of oak (*Quercus*), grass (*Poaceae*), wormwood (*Artemisia*), and fir (*Abies*) pollen in a core from the Tyrrhenian Sea spanning 55,000–9,000 years ago are locked in phase with global ice volume, as measured in oxygen isotope ratios in marine cores (Rossignol-Strick and Planchais 1989).

Orbital variations did not drive all climatic changes during the Pleistocene. They did not force *Heinrich events*, which were massive periodic advances of ice-streams from the eastern margin of the Laurentide ice sheet. Heinrich events were significant enough climatic fluctuations to leave traces in the pollen record. In Lake Tulane, south-central Florida, sediment cores show alternating peaks of vegetation dominated by pine (*Pinus*) and vegetation dominated by oak (*Quercus*) and ragweed (*Ambrosia*) (Grimm *et al.* 1993; Watts and Hansen 1994). The pine peaks correlate neatly with the first five Heinrich events.

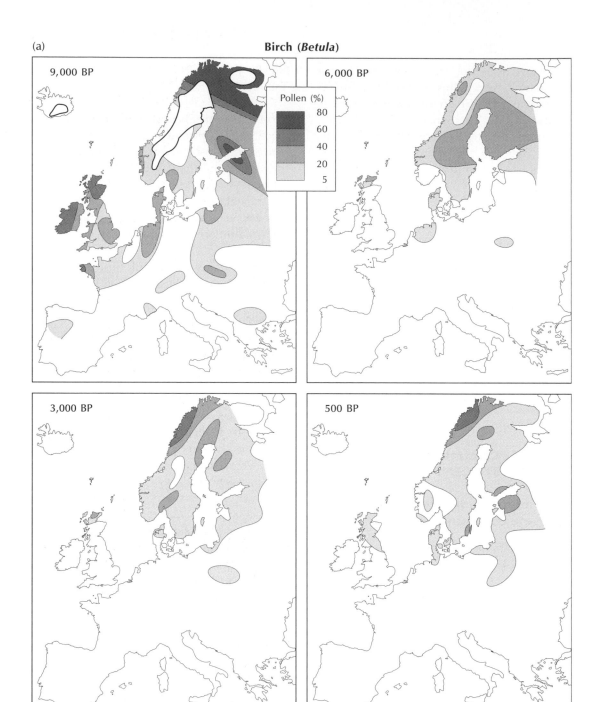

Figure 16.1 Isopoll maps for two tree species in Europe, 9,000 years ago to the present. (a) Birch (*Betula*) was abundant over much of unglaciated northern Europe 9,000 years ago. Its southern limit shifted northwards between 9,000 and 6,000 years ago. It readvanced slightly towards the present. This change is consistent with a general increase in growing-season warmth over central and northern Europe from 9,000–6,000 years ago,

Hazel (*Corylus*)

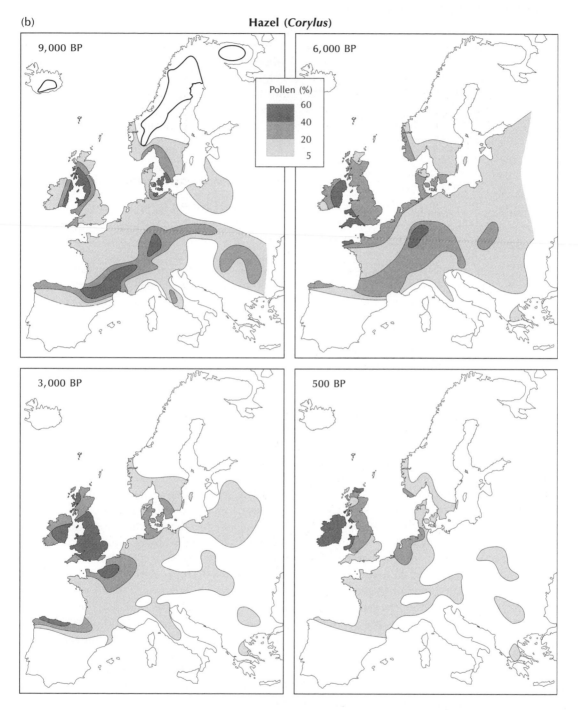

followed by a decline. (b) Hazel (*Corylus*) was abundant in a broad band across central Europe 9,000–6,000 years ago, but then, as steppe vegetation expanded into eastern Europe, it retreated westwards.

Source: Adapted from Huntley and Prentice (1993)

Glacial refugia

What happened to the temperate fauna and flora under the glacial climates that prevailed for some four-fifths of the last 2 million years? The assumption is that they survived in *refugia*, spots where they could endure the inclement conditions. The exact location of these more favourable pockets is still under investigation (e.g. Willis and Whittaker 2000). One study using macroscopic charcoal, in conjunction with analyses of pollen and molluscs, showed that at least seven tree species grew in Hungary between about 32,500 and 16,500 radiocarbon years BP (Willis *et al.* 2000). They probably eked out a living in favorable microenvironments that provided an important cold-stage refugium for the European fauna and flora. Indeed, such refugia for temperate species appear to have exerted a strong influence over current patterns of biodiversity in that continent. For instance, the distinctive patterns of genetic variation in present-day populations of grasshoppers, hedgehogs, the brown bear, house mice, voles, oaks, common beech, black alder, and silver fir seemed to be linked to isolation in cold-stage refugia in Spain, southern Italy, and the Balkans (Willis and Whittaker 2000). In short, the cold-stage isolation of these species has left a mark on present-day biodiversity.

Interestingly, work on the tropics saw refugia in the Amazon Basin – isolated pockets of tropical forest surrounded by lowland savannah – acting as 'species pumps' that generated new species, a model that seemed to account for the high levels of endemicity in many tropical regions (Haffer 1969). However, while the refugia hypothesis is gaining acceptance as an explanation of temperate latitude diversity patterns, it is going out of favour as an explanation of tropical diversity. Pollen evidence suggests that lowland tropical forests were not widely replaced by savannah during glacial stages, and that forests persisted over much of the lowland Amazon Basin (Colinvaux *et al.* 1996). Sea-level rises of around 100 m during interglacial stages are a possible cause of the patterns of endemic species in the Amazon (Nores 1999). Sea-level rises of that magnitude would have fragmented the region into two large islands and several smaller archipelagoes. The insular populations would have been isolated so promoting allopatric speciation during warm stages.

Pleistocene megafaunal extinctions

During the Pleistocene epoch, from about 1.6 million to 10,000 years ago, the fauna of North America contained species of large size (over about 1 tonne) and huge populations – a megafauna. Herbivorous members of the **megafauna** included mammoths, mastodons, camels, horses, stag-mooses (a member of the deer family), giant ground sloths, and beavers the size of modern black bears. The species were prey to a range of carnivores – sabre-toothed cats, savage short-faced bears, cheetahs, maned lions, and dire wolves. Around 13,000 years ago they all vanished within about four centuries. Similar mass extinctions of megafauna occurred on other continents, but not all at the same time. The extinctions at the end of the Pleistocene epoch affected much of the world's terrestrial fauna, and particularly its megafauna. In North America, 43 genera died, including 73 per cent of the megafauna; in South America, at least 46 genera became extinct, including 80 per cent of the megafauna; and in Australia, 26 genera disappeared, including 86 per cent of the megafauna. There were few true extinctions in Europe, Africa, and Asia, though there were many extirpations. Many taxa survived the extinction event but did not come through unscathed: surviving animals and plants changed their geographical ranges and abundances and reassembled into new communities.

The cause of these Pleistocene megafaunal mass extinctions is the subject of a red-hot debate that was begun some 150 years ago. Three rival hypotheses have emerged: the climatic change hypothesis, the overkill hypothesis, and the disease hypothesis. Some authorities favour a mixture of climate change and overkill. Until recently, the argument was always divided between climatic change and human impact as the causative agent (Owen 1846; Lyell 1863), though middle-of-the-road positions have also been taken wherein environmental change weakens populations rendering them vulnerable to human hunters who delivered the final blow (e.g. Kurtén and Anderson 1980, 363).

Climatic change hypothesis

Donald K. Grayson (1987, 1991) and Ernest L. Lundelius Jr (1987) are the chief champions of the climatic hypothesis. Climatic change associated with *deglaciation* would have altered the megafaunas' habitats. In North America, some mammals adapted to the changing conditions, those with small ranges being particularly successful at doing so by adjusting to the new conditions where they lived or seeking suitable conditions elsewhere. Others, notably those with large ranges, failed to adapt and became extinct. To be sure, late Cenozoic extinction episodes do correlate with bouts of rapid climatic change, particularly those associated with rapid glacial terminations (S. D. Webb 1984). Also, climatic change on a regional scale is implicated in late Pleistocene extinctions where human activity is unlikely to have been involved.

Perhaps the best-documented regional relationship between climatic change and mass extinctions comes from the mid-Appalachian Mountain area (Guilday 1984) and the north-central Great Plains (Wendland *et al.* 1987) of the USA. The vegetation of the mid-Appalachian region, comprising much of Pennsylvania, West Virginia, western Virginia, western Maryland, and parts of Ohio, Kentucky, and Tennessee, changed through the late Pleistocene epoch into the Holocene epoch in response to the amelioration of climate. From 18,000 to 10,500 years ago, the northernmost parts of the region resembled *periglacial tundra*, while the rest was a parkland with spruce (*Picea*), jack pine (*Pinus banksiana*),

fir (*Abies*), birch (*Betula*), and an understorey of woody shrubs, grasses, sedges, and herbs. The composition of the vegetation has no exact modem counterpart, but is indicative of boreal conditions. By 10,000 years ago, the vegetation began to shift from open coniferous forest to present-day closed-canopy deciduous forest in response to *postglacial warming*. This climatically driven change in the vegetation precipitated a radical change in the fauna. Eighteen large and one small species became extinct in the region, while three large and ten small species were extirpated. Some mammals survived the change: four species are now rare and local boreal relicts; nine species have become less common or have undergone sweeping ecological readjustment as reductions of ranges (Guilday 1984, 254). The same pattern of faunal change occurred in the north-central Great Plains region (Graham and Mead 1987). From about 18,000 to 10,500 years ago, the fauna of Iowa consisted of 70 per cent boreal species and 20 per cent steppe or deciduous species. Around 10,500 years ago, the figures were 30 per cent boreal species and 50 per cent steppe or deciduous species, and have stayed more or less at those values ever since (Wendland *et al.* 1987). In Illinois and Missouri, the change in faunal composition was less dramatic but still evident.

Climatic changes 'postdicted' by computer simulation models may account for the North American extinctions and range disruptions. From about 18,000 to 15,000 years ago, adiabatic warming of air flowing off the Laurentide ice sheet, coupled with the particular combination of the Croll–Milankovitch variables, gave rise to a unique climate. It was cooler than at present on average, but with less extreme differences between summer and winter: July temperatures were 7–10°C cooler than today, and January temperatures 5–10°C cooler. By 9,000 years ago, the ice had retreated and, owing to changes in the Croll–Milankovitch variables, the climate had become more seasonal, with warmer summers and cooler winters than those experienced today.

These changes would have deeply affected the faunas (Barnosky 1989). As climate became more seasonal, communities should have disassembled and then reassembled as animals not adapted to cold winters moved south and those not adapted to hot summers moved north. This is what did happen as the late Pleistocene disharmonious floras and faunas (p. 327) broke up. In general, the large herbivores that survived the extinctions were ruminants such as bison, deer, moose, and sheep, which could live on vegetation of low diversity. They could follow their preferred food plants despite the restructuring of plant communities. On the other hand, herbivores requiring a broad range of food plants within their normal grazing range – mammoths, mastodons, horses, camels, sloths, and peccaries – became extinct or were extirpated over wide areas. Herbivores, such as cervids with enormous antlers, whose physiology was locked into the 'old' pattern of **seasonality**, also became extinct. In turn, the downfall of the large herbivores led to the extinction of the large carnivores and scavengers that preyed on them. To be sure, the disappearance of the proboscideans is likely to have disrupted intricate grazing food webs, thus adding to the change of community composition and then extirpation and extinction.

An interesting twist to the climatic change hypothesis is the suggestion that species evolve to withstand environmental perturbations caused by frequently occurring combinations of the Croll–Milankovitch pulses, such as those that forced glacial–interglacial cycles during the late Pleistocene, but not infrequent combinations of them (Bartlein and Prentice 1989). The rare combinations of forcings, perhaps coupled with other causes of climatic change such as epeirogeny, may therefore lead to widespread extinction. This might have happened at the close of the Pleistocene, when an infrequent combination of Croll–Milankovitch forcings induced a rather extreme climatic change, involving the maximum change that would be expected in insolation and ice volume, and far greater than typical transitions from glacial to interglacial regimes.

As plausible as the climatic change hypothesis may seem, it is difficult to test – in some cases, climatic change coincides with the arrival of humans, so making it difficult to point the finger at the guilty party. Only where climatic change did not coincide with the appearance of humans, or vice versa, can the rival hypotheses be tested. In Australia, major extinctions of the megafauna occurred around 20,000 years ago when climates became less equable, some 20,000 years after the first human occupancy of the area. Critics of the climatic change hypothesis also argue that climatic change is not sufficient to explain the different intensities of the megafaunal extinctions on different continents – they were heavy in America and Australia but light in Asia and Africa.

Overkill hypothesis

Paul S. Martin (1984) promulgated the overkill hypothesis. He argued that the extinctions followed in the wake of the megafauna's first contact with migrating modern humans, and believed that humans hunted the large animals to extinction in a veritable blitzkrieg. The hunting was made easy by naive beasts with no fear of humans armed with spears and cunning hunting strategies. In North America, the Clovis hunters, who migrated through an ice-free corridor east of the Canadian Rockies some 11,500 years ago, would have precipitated the extinctions (Grayson 1991). They would have slain the animals for food. If the death rate were too high, then the species would fail to reproduce and replenish their numbers fast enough. The same pattern of overhunting would apply to the megafaunas on other continents. Tim F. Flannery (1994, 2001) believes it explains the megafaunal extinction in Australia and North America (see also Haynes 2002).

Against the overkill hypothesis is a paucity of kill or butcher sites – stone tools seldom accompany megafaunal remains. Admittedly, mammoth

PAST COMMUNITY CHANGE   325

kills are quite common, but mastodon kills are rare and horse and camel kills non-existent, even though these animals were an abundant component of the megafauna (Grayson 1991). Some critics simply fail to see how primitive hunters armed with spears could annihilate a continent bursting with over 130 species of mammals (MacPhee and Marx 1997). In addition, humans and megafaunas coexisted in Africa for millions of years without overhunting taking place.

Another possibility, still involving hunters, is that big carnivores shifted their attention to other kinds of animals as humans killed their preferred prey, thereby helping to press the substitutes to extinction. Once their prey became extinct for whatever reason, many of the predator species vanished as well.

Disease hypothesis

Ross D. E. MacPhee and Preston A. Marx (1997) favour a radically different explanation of late Pleistocene extinctions – the *disease hypothesis*. They contend that the immune system of the megaherbivores and megacarnivores were vulnerable to the lethal pathogens carried by the dogs, rats, and birds that came with the migrating humans. Disease would have moved across the continent in the wake of humans, either reducing animal populations to levels from which they could not recover or springing up again and again to infect new generations. To achieve this, the lethal pathogen would need several credentials. First, it would have to kill rapidly and affect all age groups in a given animal population (something hunters do not do). Second, it would have to have an independent host, either people or creatures that arrived with people. Third, it would require a reservoir of immune carriers from which the disease could spread. Fourth, it would have to affect a broad array of species without causing epidemics in humans. The only modern diseases that meet all these criteria are leptospirosis, a bacterium spread in rat urine, and the rabies virus. It is doubtful if either of these diseases actu-

ally caused the megafaunal extinctions, but they do attest to the existence of pathogens that could have done the job.

Many researchers are sceptical of the disease hypothesis, although accept that it is a possibility. To be sure it is testable – genetic traces of the disease organism should be left in the megafaunal remains.

Multicausal hypothesis

Anthony J. Stuart (1991, 1993) prefers a mixed explanation of late Pleistocene extinctions, arguing that *multiple causes* seem the likely answer. He maintains that climatic change reduced some populations of the Eurasian megafauna to vulnerable levels, and that hunting pressure then supplied the final blow. He also points out that a set of extinctions coincided with the arrival of modern humans in Europe about 40,000 years ago, but others occurred later and corresponded with the climatic disruptions that occurred at the end of the last ice age.

Irish elk extinction

The overkill and disease hypotheses may work for some terminal Pleistocene extinctions, but in many parts of the world individual extinctions can be shown to relate to climatic change, both on a local scale and on a regional scale. The correlation between local changes of climate and the extinction of individual species has been demonstrated for the American mastodon (*Mammut americanum*) (J. E. King and Saunders 1984) and the Irish elk (*Megaloceras giganteus*) (Barnosky 1986). The Irish elk makes an interesting case study.

In Ireland, that magnificent beast and largest of the European deer, the Irish elk, became extinct when the Younger Dryas cold spell took hold some 10,600 years ago, but 1,000 years before the first humans arrived in Ireland (Barnosky 1985, 1986). Careful survey disclosed no signs of butchering and hunting on hundreds of known

Irish specimens of the giant elk (Barnosky 1985). Investigation of pollen data support climatic change as the root cause of the Irish elk's extinction. It suggests that the quantity and quality of forage, as well as a shortened feeding season, decreased during the Nahanagan Stadial (roughly equivalent to the Younger Dryas). The teeth of the giant elk suggest that it was an opportunistic browser, supplementing its diet with large amounts of grass. Browse plants, such as juniper (*Juniperus*), crowberry (*Empetrum*), and birch (*Betula*), were common before the Nahanagan Stadial when they all but disappeared. Grasses declined, too. Compared with the environment immediately before the Nahanagan cold spell, fewer nutrient-rich plants were available for fewer weeks in spring and summer, and fewer browse plants were available in winter. Studies of living artiodactyls indicate that these changes would have put the elk populations under stress because the energy intake required to sustain the large bodies and build up fat reserves for the next winter would have been increasingly difficult to maintain. Eventually, deaths, caused chiefly by winterkill, would have outnumbered births and extinction would have ensued. Taphonomic data support this conclusion: attritional age-frequency distributions, the presence of antlers, and bone weathering in most of the Irish specimens, and the disproportionately high number of young adults found in a site at Balybetagh, attest to death during the winter, probably owing to malnutrition. Thus climatic deterioration, acting through changes in vegetation, seems to have led to the expulsion of the Irish elk from Ireland.

Stephen Jay Gould's (1974, 216) suggestion that the growth of woodland impaired the free travel of the large-antlered animals seems less believable, especially as the pollen record indicates a decrease, rather than an increase, of tree cover at the time that the Irish elk waned (Barnosky 1986, 133). If a shorter feeding season did cause the demise of the Irish elk, then it may also explain why the elk never returned to Ireland once food plants were again plentiful. The argu-ment that the elk could not have returned because the Irish Sea stood in its path is not entirely satis-factory because *Megaloceras giganteus* was unable to survive anywhere by early Holocene times. It seems more plausible that, by 10,000 years ago, summer insolation at latitude 50° N was begin-ning to decrease towards its present values, and the length of the spring 'green-up', the time when plants contained maximum levels of nutrients so vital to the fitness of large artiodactyls with large antlers, became shorter than it had been in the late Pleistocene epoch. It is probably no coinci-dence that the Irish elk's presence in Ireland, and its maximum abundance in Britain, was associ-ated with a time of maximum summer and minimum winter insolation from about 12,000 to 10,000 years ago (Barnosky 1986, 133).

Community comings and goings in the Pleistocene

Community impermanence

If each species 'does its own thing', it follows that communities, both local ones and biomes, will come and go in answer to environmental changes (e.g. Graham and Grimm 1990). This argument leads to a momentous conclusion: there is nothing special about present-day communities and bio-mes. However, this bold assertion should be tem-pered with a cautionary note – insect species and communities have shown remarkable constancy in the face of Quaternary climatic fluctuations (Coope 1994).

Community impermanence is evidenced in two aspects of communities – in communities with no modern analogues and in communities that are disharmonious.

No-modern-analogue communities

Some modern communities and biomes are simi-lar to past ones, but most have no exact fossil coun-terparts. Contrariwise, many fossil communities and biomes have no precise modern analogues. This finding supports the disequilibrium view of communities. The list of past communities that lack modern analogues is growing fast. In the Missouri–Arkansas border region, USA, from 13,000 to 8,000 years ago, the eastern hornbeam (*Ostyra virginiana*) and the American hornbeam (*Carpinus caroliniana*) were significant plant community components (Delcourt and Delcourt 1994). These communities, which were found between the Appalachian Mountains and the Ozark Highlands, bore little resemblance to any modern communities in eastern North America. They appear to have evolved in a climate characterized by heightened seasonality and springtime peaks in solar radiation. Farther north, in the north-central USA, a community rich in spruce and sedges existed from about 18,000 to 12,000 years ago. This community was a boreal grassland biome (Rhodes 1984). It occupied a broad swathe of land south of the ice sheet and has no modern counterpart, though it bore some resemblance to the vegetation found in the southern part of the Ungava Peninsula, in northern Quebec, Canada, today.

Disharmonious communities

The fauna and flora of communities with no modern analogues are commonly described as disharmonious communities. This inapt name inadvertently conjures an image of animal and plants struggling for survival in an alien environment. It is meant to convey the idea that these communities had evolved in, and flourished under, climatic types that no longer exist anywhere in the world (Graham and Mead 1987). In the southern Great Plains and Texas, USA, present-day grassland or deciduous forest species – including the least shrew (*Cryptotis parva*), the bog lemming (*Synaptomys cooperi*), the prairie vole (*Microtus ochrogaster*), and short-tailed shrews (*Blarina* spp.) – lived cheek-by-jowl with present-day boreal species – including the long-tailed shrew (*Sorex cinereus*), white-tailed jack rabbit

(*Lepus townsendii*), ermine or stoat (*Mustela erminea*), and meadow vole (*Microtus pennsylvanicus*) (Lundelius *et al.* 1983). During the late Pleistocene epoch, disharmonious animal communities were found over all the USA, except for the far west where vertebrate faunas bore a strong resemblance to their modern day equivalents, and date from at least 400,000 years ago to the Holocene epoch. These disharmonious communities evolved from species responding individually to changing environmental conditions during late Pleistocene times (R. W. Graham 1979).

At the end of the Pleistocene, new environmental changes led to the disassembly of the communities. The climate became more seasonal and individual species had to readjust their distributions. Communities of a distinctly modern mark emerged during the Holocene epoch. This is seen in the changing distributions of three small North American mammals since the late Quaternary (Figure 16.2). The northern plains pocket gopher (*Thomomys talpoides*) lived in southwestern Wisconsin around 17,000 years ago and persisted in western Iowa until at least 14,800 years ago. Climatic change associated with deglaciation then caused it to move west. The same climatic change prompted the least shrew (*Cryptotis parva*) to shift eastwards. The collared lemming (*Dicrostonyx* spp.), which lived in a broad band south of the Laurentide ice sheet, went 1,600 km to the north.

North America does not have a monopoly in disharmonious communities. In Australia, an Early Pliocene fauna from Victoria – the Hamilton local fauna – contains several extant genera whose living species live almost exclusively in rain forest or rain-forest fringes (Flannery *et al.* 1992). The indication is, therefore, that the Pliocene fauna lived in a rain-forest environment, but a more complex rain forest than exists today. Modern representatives of four genera (*Hypsiprymnodon, Dorcopsis, Dendrolagus,* and *Strigocuscus*) are almost entirely rain-forest dwellers, but they live in different kinds of rain forest. Living species of the New Guinean forest wallabies (*Dorcopsis*) live in high mountain forests, lowland rain forests, mossy

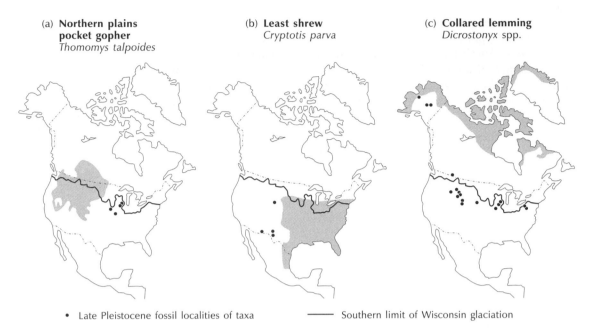

(a) **Northern plains pocket gopher** *Thomomys talpoides*

(b) **Least shrew** *Cryptotis parva*

(c) **Collared lemming** *Dicrostonyx* spp.

• Late Pleistocene fossil localities of taxa   —— Southern limit of Wisconsin glaciation

Figure 16.2 Individualistic response of some small North American mammals since the late Quaternary. (a) Northern plains pocket gopher (*T. talpoides*). (b) Least shrew (*C. parva*). (c) Collared lemming (*Dicrostonyx* spp.).
Source: Adapted from Graham (1992)

montane forests, and mid-montane forests. Living species of tree kangaroos (*Dendrolagus*) live mainly in montane rain forests. The musky rat-kangaroo (*Hypsiprymnodon moschatus*) is restricted to rain forest where it prefers wetter areas. The modern New Guinea species, the ground cuscus (*Strigocuscus gymnotis*), is chiefly a rain-forest dweller, though it also lives in areas of regrowth, mangrove swamps, and woodland savannah. Living species of pademelons (*Thylogale*) live in an array of environments, including rain forests, wet sclerophyllous forests, and high montane forests. Other modern relatives of the fossils in the Hamilton fauna, which includes pseudocheirids, petaurids, and kangaroos and wallabies, live in a wide range of habitats. It contains two species, brushtail possums (*Trichosurus*) and cuscuses (*Strigocuscus*), whose ranges do not overlap at present. It is thus a disharmonious assemblage. Taken as a whole, the Hamilton mammalian assemblage suggests a

diversity of habitats in the Early Pliocene. The environmental mosaic consisted of patches of rain forest, patches of other wet forests, and open area patches. Nothing like this environment is known today.

THE DISTANT PAST: PRE-QUATERNARY CHANGES

The fossil and sedimentary record reveals several key features in the history of life: it has become increasingly complex and more diverse; in doing so, climatic cycles have affected it; and it has gone through times of crisis.

Long-term trends in biodiversity and complexity

Life on Earth first appeared at least 3.8 billion years ago. Since then, it has followed secular trajectories of increasing complexity and diversity. The first living things were prokaryotes; they were anaerobic, fermenting heterotrophic bacteria. The first autotrophic bacteria evolved by about 3.5 billion years ago, and the first aerobic photoautotrophic bacteria and the nitrate-reducing and sulphate-reducing bacteria by 2.5 billion years ago. Eukaryotes had evolved by around 1.5 billion years ago, and metazoans by 600 million years ago. During the Phanerozoic aeon, key events in the biosphere were the appearances of the following:

- calcareous and siliceous skeletons (570 million years ago)
- the origin of vertebrates (510 million years ago)
- land plants (440 million years ago)
- wingless insects (420 million years ago)
- winged insects (310 million years ago)
- mammals (225 million years ago)
- birds (150 million years ago)
- flowering plants (140 million years ago)
- humans (3 million years ago).

Communities and ecosystems, as well as individuals, have become more complex, in so far as they contain more, and a greater variety of, species. And they have become more diverse, mainly because the abiotic and biotic environments have become increasingly patchy – geodiversity and biodiversity have both increased. The increase in geodiversity is partly caused by the action of organisms, especially by those that, after their death, form the material of such sedimentary rocks as limestone.

Life has always interacted with its environment, but occasionally unusual events have caused spurts of evolution and biodiversity. An example of such an event is the late Palaeozoic oxygen pulse (mid-Devonian, Carboniferous, and Permian). This involved a marked rise (possibly to a hyperoxic 35 per cent) and then fall (possibly to 15 per cent) in atmospheric oxygen and associated changes in atmospheric carbon dioxide. It

was probably caused by bottlenecks in lignin cycling, and in the cycling of other refractory compounds synthesized by the newly evolved land plants (J. M. Robinson 1990, 1991). Its effect was to quicken the terrigenous organic-carbon cycle and to enable terrestrial production to increase with concomitant rise in atmospheric oxygen levels. The oxygen pulse influenced diffusion-dependent features of organisms (including respiration and lignin biosynthesis), and may have fuelled diversification and ecological radiation, permitting greater exploitation of aquatic habitats and the newly evolving terrestrial biosphere (J. B. Graham et al. 1995).

An increase in biodiversity has paralleled the evolutionary burgeoning of life-forms. This is seen in the rising number of fossil taxa recorded in successive stages during the Phanerozoic aeon (Figure 16.3). An initial upsurge marks the Cambro-Ordovician explosion. A decline sets in towards the close of the Palaeozoic era and ends in the rapid drop in the late Permian period. From the Triassic period the trend is consistently upward. This curve is thought to record real changes in the diversity of life, and is not an artefact of sampling (Valentine et al. 1978).

The causes of these long-term diversity changes are not known for sure. One suggestion is that, after the Cambro-Ordovician explosion, which might have filled the available niches, diversity tracked changes in the disposition of the continents. Moderately high diversity was associated with moderately separated continents during the Palaeozoic era; diversity dropped as the continents came together to form Pangaea; and diversity rose after the Permian period as the continents broke up. A more recent study of Phanerozoic diversity used five major and essentially independent estimates of lower taxa (trace fossil diversity, species per million years, species richness, generic diversity, and familial diversity) in the marine fossil record (Sepkoski et al. 1981). Strong correlations between the independent data sets indicated that there is a single underlying pattern of taxonomic diversity during the

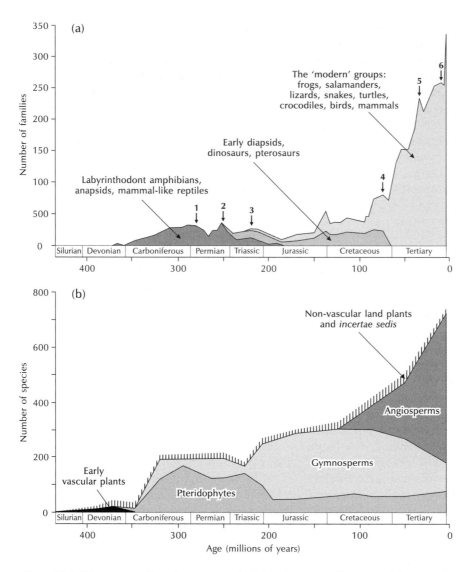

Figure 16.3 Phanerozoic diversity changes. (a) Diversity curves for terrestrial tetrapod families. The upper curve shows the total diversity with time. Six apparent mass extinctions are indicated by points immediately preceding drops in diversity and numbered 1–6. The mass extinctions were produced by a slightly elevated extinction rate and a reduced origination rate. (b) Diversity curves for vascular plant species. Each group comprises plants sharing a common structural grade, a common reproductive grade, or both.

Sources: (a) Adapted from Benton (1985). (b) Adapted from Niklas *et al.* (1983)

Phanerozoic. The pattern is low diversity during the Cambrian period; a higher but not steadily increasing diversity through the Ordovician, Silurian, Devonian, Carboniferous, and Permian periods; low diversity during the early Mesozoic era, notably in the Triassic period; and increasing diversity through the Mesozoic era culminating in a maximum diversity during the Cenozoic era (see Signor 1994). Congruent patterns have been discerned in the record of marine vertebrates (Raup and Sepkoski 1982), non-marine tetrapods (Figure 16.3a), vascular land plants (Figure 16.3b), and insects (Labandeira and Sepkoski 1993).

It is tempting to suppose that there is an upper limit to biodiversity, or a global carrying capacity. However, this diversity ceiling is cranked up by evolutionary innovations and by climatic and geological changes. Evolutionary innovations occasionally have led to the raising of this global carrying capacity and biodiversity can be expected to have risen. Climatic and geological processes incessantly increase the complexity of the physical environment – they drive geodiversity to ever-greater levels. The biosphere has always striven to reach this ever-rising biodiversity ceiling, but it has been hindered by the major disturbances that have led to mass extinctions. Thus, the biodiversity ceiling is seldom reached by the biota: it is a theoretical maximum and the biosphere strives towards it between perturbations (see Kitchell and Carr 1985). For these reasons, the biodiversity is unlikely ever to attain a true steady state; rather, it will increase through time, tracking the increasing diversity of the physical environment, with occasional setbacks caused by mass extinctions (Cracraft 1985). The physical complexity that is possible on the Earth may have its limits (Valentine 1989) – geodiversity will then limit biodiversity.

Prolonged phases of steady-state biodiversity are found within the overall biodiversity increase through geological time (Rosenzweig 1995, 52). Over the last 5 million years of the Ordovician period, the species diversity of muddy benthos organisms from Nicolet River Valley, Quebec, was stable – it fluctuated around a mean of about 32 species. This was a steady state, and represented an approximate balance between speciation events and extinctions. Cenozoic mammal faunas display similar steady-state diversities for long periods. The diversity of large herbivores and carnivores in North America fluctuated around a steady value for 45 million years (Van Valkenburgh and Janis 1993).

Providing the environment changes slowly enough, populations can simply ride with the tide of change. That does not mean that communities will stay intact, but there is evidence to suggest that communities stay stable for lengthy periods, say a few million years, when environmental changes are small and slow. During the Cenozoic era, for example, land mammals in North America consisted of stable sets of taxa (chronofaunas) that evolved slowly and persisted for some 10 million years (S. D. Webb and Opdyke 1995).

Hothouse–icehouse cycles

The long-term **hothouse–icehouse** and **warm-mode–cool-mode** climatic shifts drove past communities and ecosystems through protracted successional sequences (see Huggett 1997, 133–5). Particularly well documented is the Cenozoic succession from the Cretaceous hothouse to the Quaternary icehouse. Floral and faunal changes in many parts of the world point to a cooling and, in some regions, drying of climate during the Miocene epoch. In the broad expanses of the North American mid-continent, the shift to cooler and drier climatic conditions in Miocene times directly drove a change from forest to savannah and from savannah to steppe, and indirectly drove parallel changes in the diversity of browsing and grazing ungulate taxa (S. D. Webb 1983).

An excellent record of the long descent into the current icehouse Earth comes from the Siwalik Group sediments, in the Potwar Plateau region, northern Pakistan (Quade and Cerling 1995).

These floodplain sediments span the last 18 million years. Carbon isotope ratios in palaeosols suggest the following changes in the floodplain ecosystem:

1 *From 17–7.3 million years ago.* A pure, or almost pure, C_3 biomass, probably a mosaic of closed canopy forest and grassland, dominated the floodplains.

2 *From 7.3–6 million years ago* (Late Miocene). A gradual expansion of C_4 plants (essentially grassland) on the floodplain occurred, probably in response to monsoon intensification. Two lines of evidence support this interpretation. First, concurrent changes in soils – depth of leaching decreased, the colour of leached zones changed from dominantly strong reds and oranges to yellower hues, and humic soil horizons, possibly owing to the appearance of grassland, became commoner. Second, a major faunal turnover occurred at about the same time. This turnover saw grazing animals, such as a new tragulid with high-crowned dentition, replace larger browsing animals (tragulids, suids, okapi-like giraffes, low-crowned bovids, and others); and it saw rodents undergoing a rapid change (Flynn and Jacobs 1982; Barry *et al.* 1985; Barry and Flynn 1990).

3 *From 6–0.4 million years ago.* C_4 grasslands remained dominant throughout that time.

In eastern Africa, as in Asia, aridification led to grassland expanding at the expense of forests. The marked floral changes at around 2.5 million years ago had a drastic effect upon mammals (e.g. Vrba *et al.* 1989). Many forest antelope species became extinct and new savannah-dwelling species evolved, most of which survive as elements of the modern African fauna. The climatic and floral changes might also have nudged gracile australopithecines (*Australopithecus afarensis* and *A. africanus*) in an evolutionary direction that produced the first members of the human genus – *Homo* (Stanley 1992, 1995).

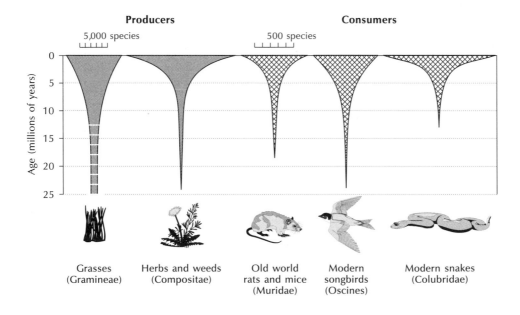

Figure 16.4 A relay of adaptive radiations in terrestrial ecosystems that were favoured by global aridification during the last 25 million years.

Source: Adapted from Stanley (1990)

The trend towards drier climates (*aridification*) over the last 25 million years led to several adaptive radiations (Figure 16.4). The radiations displayed a relay effect. The relay started at the food-web base with the diversification of grasses. Herbs and weeds diversified during the spread of grasses. Many songbirds and Old World rats and mice, which fed on the seeds of the expanding plant groups, then underwent a radiation. Lastly, modern snakes (many of which belong to the family Colubridae), expanded as the rats and mice, and eggs and chicks of songbirds, appeared on the menu.

A switch from an icehouse to a hothouse climate leads to many environmental changes. In the seas, warmer waters develop in polar regions. Deep ocean waters become warm, sluggish, and contain little or no oxygen (dysaerobic or anoxic). Such a change in the hydrosphere would cause widespread extinctions in the deep sea. During the mid-Cretaceous high-stand of sea-level, when seas stood about 300 m above present levels, anoxic conditions in the ocean deeps touched the deeper portions of epicontinental seas. At this time, environmental perturbations created pulses of biotic destruction that comprise the Cenomanian–Turonian mass extinction (Kauffman 1995). Marine diversity was at a peak for the Cretaceous period immediately before the extinction events. Many lineages had evolved narrow tolerance limits in the hothouse conditions. This meant that marine life would be prone to extinction if environmental conditions should change. Tropical reef communities suffered heavy losses near the beginning of the early and middle Cenomanian boundary. Within 520,000 years of the Cenomanian–Turonian boundary, non-tropical biotas suffered a loss of species in the range 45–75 per cent, depending on the group. These losses were stepped, most of the species disappearing in short-term events. They started in the subtropical and warm temperate biotas and progressed to the cool temperate biotas. The extinction events were closely connected with abrupt, large-scale changes in the ocean and atmosphere that are recorded in wild fluctuations of trace element, stable isotope, and organic carbon levels. The expanded anoxic zone of the deep ocean floor spread to deeper continental shelf and epicontinental sea habitats, initiating trace element advection and chemical stirring of the ocean. These changes might have been the result of ocean impacts of asteroids and comets during a Cenomanian impact storm.

Biotic crises

A biotic crisis, or *mass extinction*, occurs when species biodiversity falls to low levels. It may arise from a higher than normal extinction rate, from a lower than usual speciation rate, from species loss through net outward migration, or from a combination of all these.

There are varying levels of crisis. *Mild crises* involve an elevated turnover of species. *Severe crises* involve a loss of 20 per cent or more of all species. When such severe crises act globally, they are called mass extinctions. But crises can occur at all geographical scales. Landscape patches commonly suffer species loss after a disturbance, and explosions of volcanic islands cause local mass extinctions. The biospheric stresses that cause extinction events and the post-crisis recovery processes have parallels at local, regional, and global scales.

Extinction is the shared fate of all species. Like the death of an individual, it is an inevitability. The fossil record points to a continuum of extinction events that range from everyday background levels to mass extinctions. However, it also suggests that mass extinctions are not the chance coincidence of independent extinction events, but regular episodes of mass killings. Additionally, extinction is to some extent selective, but in many cases it appears to act randomly in that the survivors are apparently not better adapted to life in the post-crisis environment than the victims – they are simply luckier.

A few key questions need addressing here. First, why should background extinction levels suddenly 'go critical'? In other words, what causes mass extinctions? Second, are mass extinction events grand global dyings occurring within days, weeks, months, or years? Or are they clusters of independent extinction episodes occurring over hundreds of millennia? And third, what happens to the biosphere in the aftermath of a mass extinction? The first two questions are too closely allied to separate and will be considered together.

Times of crisis

Data on marine genera covering the last 270 million years show peaks where the extinction rate has risen above background levels 12 times (Figure 16.5). Eight of these peaks match well-known extinction events, and nine are roughly periodic, occurring once every 26 million years (Sepkoski 1989). But are these 'mass extinctions'? When does a surge in extinction rate (or sharp drop in speciation rate) become a mass extinction? Three mass extinctions seem indisputable, for they involved an estimated loss of more than 63 per cent of all species. They are the Late Permian extinction event, the upper Norian (Triassic–Jurassic) extinction event, and the Maastrichtian (Cretaceous–Tertiary) extinction event. The Late Permian event is the mother of all mass extinction events: it entailed an estimated loss of at least 80 per cent (Stanley and Yang 1994), and perhaps 93 per cent (Sepkoski 1989) of all species (see also Benton 2003). The other six events run at extinction rates about twice or thrice the background level.

It is difficult to read rates of extinction directly from the fossil record (Benton 1994). The existing evidence gives a mixed message about the rate at which mass extinctions occurred: some evidence points to protracted extinction episodes; other evidence suggests sudden extinction. That statement begs an obvious but tricky question: how sudden is sudden? Some researchers think that sudden means sudden – a year or so (McLaren 1988). This view of suddenness was given strong support by the discovery of a marker horizon at the Cretaceous–Tertiary boundary (Alvarez *et al.*

1980). Some geologists took the marker horizon as concrete evidence that the terminal Cretaceous extinction event was indeed geologically instantaneous, and was caused by an asteroid colliding with the Earth. The association of impact-event signatures within boundary layer sediments lends much weight to the view that impacts did occur contemporaneously with boundary-layer clay formation. Some boundary clays contain organic chemicals with a composition highly suggestive of a cosmic origin, and some contain glass spherules of probable impact origin. To be sure, some geochemical signatures do change suddenly at boundary events. An example is the carbon-isotope ratio at the Permo-Triassic boundary in British Columbia, Canada (Wang *et al.* 1994). That does not necessarily mean that the impacts were the primary cause of the extinctions; they might simply have been the knockout blows.

The fossil record almost invariably does record long-lasting extinction episodes. Investigations of many boundary sites show that mass extinctions occurred in a series of discrete steps spread over a few million years (stepwise extinction), and not in an instant. Mass extinctions during Late Ordovician, Late Devonian, and Late Permian times were long affairs in which tropical marine biotas, including stenothermal calcareous algae, declined greatly, and reef communities were decimated (Stanley 1988a, 1988b). The detailed pattern of late Cretaceous extinctions suggests a relatively gradual extinction-rate increase for many groups of organisms, followed by a catastrophe lasting a few tens of thousands of years. In the marine realm, the extinction of planktonic foraminiferal species spanned 300,000 years below, and some 200,000 to 300,000 years above, the Cretaceous–Tertiary boundary (Keller 1989). The dinosaurs might have suffered a gradual extinction (M. E. Williams 1994), but the Hell Creek Formation in eastern Montana and western North Dakota suggests a sudden extinction (Sheenan *et al.* 1991; but see Hurlbert and Archibald 1995).

(a)

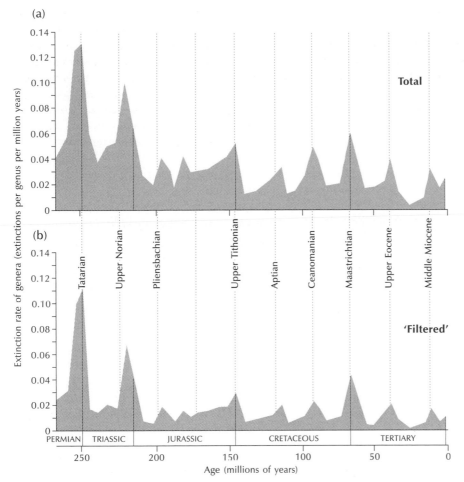

Figure 16.5 Extinction rate per genus for 49 sampling intervals from the mid-Permian (Leonardian) to Recent. A periodicity of 26 million years is indicated by the vertical lines. (a) Time series for the entire data set of 17,500 genera. (b) 'Filtered' time series for a subset of 11,000 genera, from which genera confined to single stratigraphical intervals are excluded.

Source: Adapted from Sepkoski (1989)

In the aftermath of crises

Even the severest of mass extinctions were no match for the biosphere. If they had been, you would not be reading this book. The biosphere has managed to recover from all biotic crises in its long history. Recovery after the heavy crises is commonly delayed, but it does take place. Mass extinctions have regulated reef diversity during the Phanerozoic aeon: reef development after extinction events is negligible and diversity stays low, sometimes almost zero, for long periods (Kauffman and Fagerstrom 1993). Recovery may be delayed by the persistence of post-crisis inimical conditions, or else the slow restoration of vulnerable taxa (Stanley 1990). Adverse conditions in the post-crisis environment may reduce ecosystem primary productivity. In British

Columbia, Canada, a marine sequence through the Permo-Triassic mass-extinction event shows an abrupt drop in the carbon-isotope ratio after the event indicating a reduction in surface-water productivity (Wang *et al.* 1994). A similar drop occurred at Botomian time (Lower Cambrian) in Siberia, where it coincided with the mass extinction of archaeocyathan reef biota (Brasier *et al.* 1994).

The species surviving a biotic crisis have one thing in common – they are lucky. If they can grasp the opportunity for expansion and diversification, they may be spectacularly successful. The Norian (Upper Triassic) mass extinction saw the demise of the mammal-like reptiles and rhynchosaurs. This event was quickly followed, still within the Norian age, by the adaptive radiation of the dinosaurs. It is unlikely, therefore, that the dinosaurs outcompeted the mammal-like reptiles, as was traditionally claimed (Tucker and Benton 1982). No, the original success of the dinosaurs was simply due to opportunism – they happened to be in the right place at the right time. The same argument applies to the rise of the mammals after the dinosaur extinctions (e.g. Sheenan and Russell 1994).

Extinction cycles?

One of the more contentious issues about mass extinctions is the assertion that they recur at regular intervals through geological time. This claim can at least be tested against the fossil record. Most of the evidence for periodicity in the fossil record has come from biostratigraphic data on marine organisms. Marine genera and families display a very dominant period of 26 million years (Sepkoski 1989; Rampino and Caldeira 1993). Extinctions in the Upper Permian, Upper Norian, and Maastrichtian were truly massive and bouts of widespread volcanism or episodes of heavy bombardment seem likely causes. However, many of the periodic extinction events occurred over several stratigraphical stages or substages. Generic extinctions follow two patterns, both of which

suggest a gradualistic mechanism behind the periodicity (Sepkoski 1989). First, most of the extinction peaks, especially for filtered data, have almost identical amplitudes, well below the amplitudes of the three major events. Second, the widths of most of the peaks span several stages. Taken with the three massive extinction events, two distinct causes of mass extinctions are indicated. The first cause is a 26 million-year oscillation of the Earth's oceans or climates, or both, that leads to higher extinction rates over long periods, either continuously or in high-frequency, stepwise episodes. The second cause involves independent agents or constraints upon extinction, including bombardment episodes, volcanic episodes, and sea-level changes, that boost the periodic oscillation of extinction rates when they happen to occur at times of increasing extinction. In other words, large impacts and massive outpourings of lava may trigger mass extinctions under some circumstances. And it would be wrong to surmise that all extinctions run to a rigid 26 million-year timetable. Cycles of extinction in mid-Palaeozoic reef communities failed to 'run on time', and appear to relate to climatic change (and possibly ocean anoxic events) associated with plate tectonic processes (Copper 1994).

Long-term life cycles

Communities and ecosystems may display directional changes over millions of years. Such *megatrends* are sometimes a response to secular changes in environmental factors, and sometimes a response to the crossing of internal community and ecosystem thresholds. For example, the pulse of Cenozoic mammal communities displays a 'syncopated equilibrium', with long-lasting and stable chronofaunas separated by rapid turnover episodes involving radical reorganizations of terrestrial ecosystems (S. D. Webb and Opdyke 1995). The result is a large-scale and long-term succession of terrestrial ecosystems. The same process has operated throughout the Phanerozoic (Sheenan and Russell 1994). Biotic crises abruptly ended long periods of stability in major varieties of dominant organisms. After each crisis, rapid diversification and ecological reorganization ushered in a new period of stability.

Protracted cycles of 150 million and 300 million years appear to be locked into climatic warm-mode–cool-mode and hothouse–icehouse cycles. For instance, the sites of greatest carbon seques-tering switch from low latitudes during icehouse times to higher latitudes (poleward of 40°) during hothouse times (Spicer 1993). Marine palaeonto-logical data hint at a 30 million-year cycle (Fischer 1984). This cycle is expressed in the global diversity of planktonic and nektonic taxa, including globigerinacean foraminifers and ammonites, and in the episodic development of superpredators with body lengths of 10–18 m (Fischer 1981). Long-lasting phases of increasing diversity ('oligotaxic' phases), when oceans were cool, were punctuated roughly every 30 million years by high-diversity crises ('polytaxic' phases), when the oceans were warm. Each polytaxic pulse brought a new group of superpredators – ichthyosaurs, pliosaurian plesiosaurs, mosasaurs, whales, and sharks successively filled the super-predator niche opened up after each biotic crisis.

SUMMARY

Species and communities changed in the past. During the last 10,000 years, they have displayed at least three interesting patterns of change. The size of some species has tracked climatic changes, becoming larger under cooler climates and smaller under warmer climates. Holocene communities have disassembled and reassembled as climates have changed. Species have migrated at different rates, with opportunist, equilibrium, and fugi-tive migrators being recognized. During the Pleistocene epoch, species and communities have tracked glacial–interglacial climatic swings, tem-perate species survived cold stages in refugia, and the continental megafaunas suffered mass extinc-tions. In addition, species have 'done their own thing', causing often disharmonious communities with no modern counterparts to disassemble and reassemble with a new composition. Geological changes of species and communities include long-term trends in diversity and complexity as life burgeoned, cyclical swings associated with switches between hothouse and icehouse climates, and biotic crises.

ESSAY QUESTIONS

1 How sensitive are animals and plants to Croll–Milankovitch cycles?

2 What is the significance of communities with no modern analogues?

3 How did species and communities respond to late Cenozoic climatic cooling?

FURTHER READING

Benton, M. J. (2003) *When Life Nearly Died: The Greatest Mass Extinction of All Time*. London: Thames and Hudson.

A fascinating read about the Permian extinction.

Huggett, R. J. (1997) *Environmental Change: The Evolving Ecosphere*. London: Routledge.

An overview.

Ingrouille, M. (1995) *Historical Ecology of the British Flora*. London: Chapman & Hall.

Interesting coverage of all timescales.

Martin, P. S. and Klein, R. G. (eds) (1984) *Quaternary Extinctions: A Prehistoric Revolution*. Tucson, AZ: The University of Arizona Press.

Still a mass of information and ideas in the Pleistocene extinctions debate.

Roberts, N. (1998) *The Holocene: An Environmental History*, 2nd edn. Oxford: Blackwell.

Excellent introduction to the Holocene in Britain.

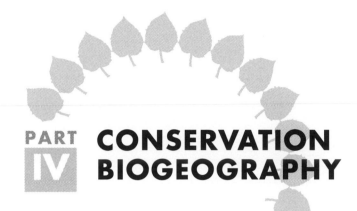

PART IV CONSERVATION BIOGEOGRAPHY

17

CONSERVING SPECIES AND POPULATIONS

Biological diversity is plummeting, mainly due to habitat degradation and loss, pollution, overexploitation, competition from alien species, disease, and changing climates. There is an urgent need to preserve species and populations. This chapter covers:

■ justifying species conservation
■ conservation bodies and categories
■ conservation strategies

Biodiversity responds to changes in the physical and biological environment in a maze of complex ways. Many scientists think the recent steep fall in global biodiversity levels alarming and fear that the human species is single-handedly manufacturing a mass extinction far speedier than any past mass extinctions. Two big worries about current biodiversity decline are that it is an irrevocable process set to undermine the basis of human existence, and that it will deny hosts of species their right to exist.

WHY CONSERVE SPECIES?

Underpinning *conservation biogeography* is the assumption that humans should protect species, communities, and ecosystems wherever necessary. The justification for this assumption is that the environment has value and deserves protecting. Conservationists present at least five types of justification: economic, ecological, aesthetic, moral, and cultural (Botkin and Keller 2003, 263–4):

1 *Economic* or *utilitarian justification* for conservation stems from a need of individuals or societies for an environmental resource to gain an economic benefit or even to survive. For instance, farmers make their living from the land and need a supply of crops or livestock to do so. A common argument upheld under the economic justification banner is that it is

unwise to destroy species that may prove useful to humans, say as anticancer agents (Table 17.1). Another powerful economic force for conservation is ecotourism, which breeds on exotic communities and particularly **flagship species**.

2 *Ecological justification* for conservation is rooted in the necessity of sustaining life-support functions of the ecosphere that, although they have no direct benefit to individuals, are essential to the healthy functioning of regional and global ecosystems. A malfunction or failure or these life-support processes would imperil human survival. For example, the profligate production of greenhouse gases by humans may change the global climate and affect the entire ecosphere. Similarly, soil bacteria are of no apparent immediate benefit to humans but some of them fix nitrogen, converting to a reactive form that is then available to other living things. Ecological and utilitarian justification for conservation may overlap. Coral reefs provide a habitat for marine fish, and although humans do not eat coral, they may eat the reef-dwelling fish.

3 *Aesthetic justification* for conservation comes from the human perception of the natural world. The argument is that biological and landscape diversity add to the quality of human existence. Many people, for instance, regard wilderness as beautiful and would prefer to preserve wilderness areas than to lose them. Similarly, people deem many organisms – birds, large land mammals, flowering plants, many insects and ocean animals – as beautiful and celebrate them in books, poems, paintings, sculptures, and plays. The human appreciation of beauty in the natural world appears to be a compelling motive for conserving endangered species, communities, and landscapes.

4 *Moral justification* for conservation rests in the belief that elements of the environment – including species, communities, and landscapes – have a right to exist independently of humans, and that humans have a moral responsibility to preserve them. The United Nations General Assembly World Charter for Nature, signed in 1982, declared that species have a moral right to exist. Likewise, the US

Table 17.1 Species useful to humans

Species used	Examples
Wild strains of grains and other crops	Fresh genetic material from wild strains of wheat and corn is needed to aid disease resistance
New crops, especially from the tropics	Twenty new tree species identified in Peru as worth exploiting for national and international markets, including andiroba (*Carapa guianensis*) and yacushapana (*Terminalia amazonica*)
Chemical compounds from wild organisms	Colchiline (anti-inflammatory) from the glory lily (*Gloriosa superba*); digitalis from foxglove; aspirin from willow bark; vincristine and vinblastine (anti-cancer agents) and raubasine (anti-hypertension drug) from rosy periwinkle (*Catharanthus roseus*)
	Non-addictive painkillers from snake venom; arthritis treated with bee venom; squalene from shark-liver oil is a bactericide
Species used in medical research	Armadillo can contract leprosy and helps in finding cure for the disease
Species used in pollution control	Plants, fungi, and bacteria remove toxins from the environment: carbon dioxide and sulphur dioxide are removed by vegetation; carbon monoxide is reduced and oxidized by soil fungi and bacteria

Endangered Species Act includes statements that express the rights of organisms to exist. Moral questions about the natural world are the subject of **environmental ethics** – a new and fast-growing discipline. Deep ecology, begun by the Norwegian Arne Naess, has contributed greatly to the debate over environmental ethics.

5 *Cultural justification* rests on the need of many indigenous peoples for specific species, which may include those in endangered and threatened categories. For many indigenous peoples, their local ecosystems provide food, shelter, tools, fuel, materials for clothes, and medicines. A reduction in the biodiversity of such ecosystems may render the indigenous peoples much poorer and their survival may require continual assistance from outside agencies.

CONSERVATION BODIES AND CATEGORIES

Conservation organizations

A large number of national and international organizations devote themselves to conservation. Many of these have highly informative websites and resources (Table 17.2). The big international players are the International Union for the Conservation of Nature and Natural Resources (IUCN), the World Wildlife Fund (WWF), and the World Conservation Monitoring Centre (WCMC).

Conservation categories of species

Not all species are equally vulnerable to extinction. The IUCN (1994) devised 10 conservation categories (Figure 17.1; Table 17.3). These

Table 17.2 Selected international and national conservation organizations

Organization	Role
International	
International Union for the Conservation of Nature and Natural Resources (IUCN) [www.iucn.org]	The foremost coordinating body for international conservation efforts. More than 10,000 internationally recognized scientists and experts from more than 180 countries volunteer their services to its six global commissions. Its 1,000 staff members in offices around the world are working on some 500 projects
World Wildlife Fund (WWF) [www.wwf.org]	A leading conservation organization with branches worldwide. Active in research and in the management of national parks. Has 52 offices working in more than 90 countries
United Nations Environment Programme-World Conservation Monitoring Centre (UNEP-WCMC) [www.unep-wcmc.org.uk]	Provides information for policy and action to conserve the living world. Activities include assessment and early warning studies in forest, dryland, freshwater, and marine ecosystems. Research on endangered species and biodiversity indicators provide policy-makers with vital knowledge on global trends in conservation and sustainable use of wildlife and their habitats
National	
Zoological Society of London [www.zsl.org]	Worldwide conservation of animals and their habitats
Center for Plant Conservation, Missouri Botanical Garden [www.centreforplant conservation.org] [www.mobot.org]	To conserve the rare native plants of the USA
Royal Botanic Gardens, Kew [www.rbgkew.org.uk]	To enable better management of the Earth's environment by increasing knowledge and understanding of the plant and fungal kingdoms

Table 17.3   IUCN conservation categories of taxa

Category name	Category code	Definition
Extinct	EX	A taxon for which there is no reasonable doubt that the last individual has died. Exhaustive surveys in known or expected habitat at appropriate times (diurnal, seasonal, annual) throughout its past range fail to record an individual
Extinct in the wild	EW	A taxon that survives only in cultivation, in captivity, or as a naturalized population (or populations) well outside its past range
Critically endangered	CR	A taxon facing an extremely high risk of extinction in the wild. Specifically, a taxon with an observed, estimated, inferred, or suspected population-size (number of individuals in the taxon) reduction of 90 per cent or more over the last 10 years or three generations, whichever is the longer, where the causes of the reduction are plainly reversible, understood, and no longer operative. Where the causes of the reduction may not have ceased, may not be understood, and may be irreversible, the reduction is 80 per cent over 10 years or three generations (in some cases to a maximum of 100 years) Or a taxon with a restricted geographical range (extent of occurrence less than 100 km^2; area of occupancy less than 10 km^2) Or a taxon with a population of fewer than 250 mature individuals
Endangered	EN	A taxon facing a very high risk of extinction in the wild. Specifically, a taxon with an observed, estimated, inferred, or suspected population-size (number of individuals in the taxon) reduction of 70 per cent or more over the last 10 years or three generations, whichever is the longer. Where the causes of the reduction may not have ceased, may not be understood, and may be irreversible, the reduction is 50 per cent over 10 years or three generations (in some cases to a maximum of 100 years) Or a taxon with a restricted geographical range (extent of occurrence less than 5,000 km^2; area of occupancy less than 500 km^2) Or a taxon with a population of fewer than 2,500 mature individuals
Vulnerable	VU	A taxon considered to be facing a high risk of extinction in the wild. Specifically, a taxon with an observed, estimated, inferred, or suspected population-size (number of individuals in the taxon) reduction of 50 per cent or more over the last 10 years or three generations, whichever is the longer. Where the causes of the reduction may not have ceased, may not be understood, and may be irreversible, the reduction is 30 per cent over 10 years or three generations (in some cases to a maximum of 100 years) Or a taxon with a restricted geographical range (extent of occurrence less than 20,000 km^2; area of occupancy less than 2,000 km^2) Or a taxon with a population of less than 10,000 mature individuals
Near threatened	NT	A taxon that, after evaluation, does not qualify for Critically endangered, Endangered, or Vulnerable categories, but is close to qualifying for or is likely to qualify for a threatened category in the near future
Least concern	LC	A taxon that, after evaluation, does not qualify for Critically endangered, Endangered, or Vulnerable categories
Data deficient	DD	A taxon for which information data is inadequate to make a direct or indirect assessment of its risk of extinction based on its distribution or population status or both. A taxon in this category may be well studied, and its biology well known, but biologists lack appropriate data on its abundance or distribution or both. Data deficient is therefore not a category of threat: taxa in this category indicate that more information is required and acknowledges the possibility that future research will show that threatened classification is appropriate
Not evaluated	NE	A taxon not yet evaluated against the criteria

Source: Adapted from *The IUCN Red List of Threatened Species*™, 2001 Categories and Criteria (v. 3.1), www.redlist.org/info/categories_criteria2001.html

categories are useful nationally and internationally in: (1) focusing attention around species of concern; and (2) identifying species threatened with extinction for protection through international agreements, such as the Convention on International Trade in Endangered Species (CITES).

CONSERVATION STRATEGIES

The best strategy for safeguarding the long-term survival of individual species is to preserve wild populations and the communities in which they live. The main reasons for this are twofold. First, only wild species can maintain the diversity of the species gene pool. Second, only wild species are open to the evolutionary interactions within a natural community that let them adapt to changing environmental conditions, such as changes in pest populations or climate. However, with mounting human disturbance, *in situ* or *on-site preservation* is often not a workable option. In remnant populations too small to persist, or in which remaining individuals live outside protected areas, on-site preservation may be useless. The only alternative is then *ex situ* or *off-site preservation*, in which human supervisors maintain individuals under artificial conditions. This section will explore these strategies of species conservation.

On-site conservation

On-site conservation, which usually takes place in protected areas, aids the maintenance of wild animal and plant diversity, as well as the diversity of traditional agricultural species and varieties.

Wild populations

In an ideal world, an endangered species would be conserved by protecting as many individuals as possible in the largest possible reserves of suitable habitat. In practice, managers must usually make do with a restricted number of individuals on reserves of limited size. The big question is how many individuals should be preserved and how they can be preserved given conflicting demands on the resources within a reserve. *Population viability analysis* (PVA) is a process used by conservation biologists to help answer these questions. PVA is a process for identifying the threats faced by a species, and for evaluating the chances of its surviving for a specified time into the future (Akçakaya *et al.* 1999). It extends the demographic models mentioned in Chapter 10 to assess a species' capacity to survive in an area (Soulé 1990). The Chicago Zoological Society distributes VORTEX, software for performing PVA, free of charge (www.brookfieldzoo.org).

Population viability analysis mainly focuses on the conservation and management of rare and

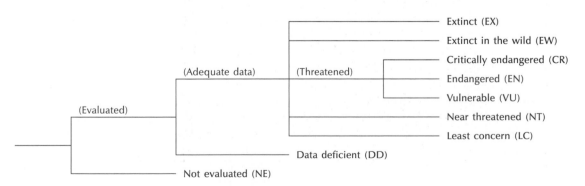

Figure 17.1 Conservation categories established by the IUCN.

threatened species, in the hope that their chances can be improved. Threatened species' management has short-term and longer-term objectives. The short-term objective is to lessen the risk of extinction. The longer-term objective is to encourage conditions in which species retain their potential for evolutionary change without intensive management. Against this background, PVA may be used to address three aspects of threatened species management (Akçakaya *et al.* 1999):

1 *Planning research and data collection.* PVA may reveal that population viability is insensitive to particular parameters, and research can then target factors that have a significant impact on extinction probabilities or on the rank order of management options.
2 *Assessing vulnerability.* In conjunction with cultural priorities, economic imperatives, and taxonomic uniqueness, PVA may be used to set policy and priorities for allocating scarce conservation resources.
3 *Ranking management options.* PVA may be used to predict the likely response of species to reintroduction, captive breeding, approved burning, weed control, habitat rehabilitation, or different designs for nature reserves or corridor networks.

There is no single recipe that applies to all applications of PVA, as each case tends to differ in some respects. Two concepts are important within PVA: minimum viable population and minimum area of suitable habitat (minimum dynamic area). The *minimum viable population* (*MVP*) defines the number of individuals required to guarantee the survival of a species. It is formally defined as the smallest isolated population having a 99 per cent chance of remaining extant for 1,000 years in the face of foreseeable effects of demographic, environmental, and genetic fluctuations, and natural catastrophes (Shaffer 1981). The *minimum area of suitable habitat* (*MASH*) or *minimum dynamic area* (*MDA*) of a species is the area required to sustain the MVP.

It is estimated from the home-range sizes of individuals and groups. As would be anticipated, MASH increases with the size of individuals. Small mammal populations need 10,000–100,000-ha reserves; grizzly bears in Canada require 49,000 km^2 for 50 individuals and 2,420,000 km^2 for 1,000 individuals (Noss and Cooperrider 1994).

A prime example of PVA is a study of an endangered forest primate, the Tana River crested mangabey (*Cercocebus galeritus galeritus*) from eastern Kenya (Kinnaird and O'Brien 1991). The Tana River crested mangabey lives in fragmented floodplain forest patches within the Tana River National Primate Reserve. Between about 1970 and 1989, agriculture reduced the area occupied by the primate and its population size fell by a half to 700 (although, owing mainly to a large portion of non-reproductive individuals, the effective population was nearer 100); the number of Tana River crested mangabey groups also declined. PVA predicted a 40 per cent chance of extinction within a century. The danger with such a small effective population size is a loss of genetic variability. To preserve an effective population of 500 Tana River crested mangabeys, which should be enough to maintain genetic variability, would require a population of 5,000 individuals. However, to raise the survival probability over a century to 95 per cent, a population of 8,000 individuals would be needed. Unfortunately, the genetic and demographic analyses paint a bleak future for the Tana River crested mangabeys. The subspecies has a restricted range and a restricted habitat that is under increasing pressure from the growing human population in the area. Maintaining 5,000–8,000 individuals is probably unrealistic. The best management plan may be to increase the protected area, plant more mangabey food plants within existing forests, and build corridors to help movement between forest fragments.

Another example of PVA is a study of the Atlantic Forest spiny rat (*Trinomys eliasi*) in Brazil (Brito and de Souza Lima Figueiredo 2003).

Habitat loss and a restricted geographical distribution threaten this species of spiny rat. VORTEX software predicted that 200 animals could form a demographically stable population, but that 2,000 animals would be needed to buffer against declines in genetic variability. Minimum areas of suitable habitat were 250 ha for 200 animals, and 2,500 ha for 2,000 animals. The sensitivities of the parameters in the model were tested for five scenarios, ranging from very optimistic, through baseline, to very pessimistic (Figure 17.2).

Agricultural populations

On-site conservation is a vital method of maintaining diversity in agricultural systems where varieties have evolved alongside particular farming systems. On-site conservation is very important in the conservation of landraces, of germplasm, and of forest trees.

A *landrace* is a crop cultivar or an animal breed that evolved in association with traditional farming practices and remains unadulterated by modern breeding techniques. The traditional practices genetically improved the landraces, which carry genes resistant to particular diseases or pests. Such genetic variability is often crucial to the agricultural industry as it strives to maintain and boost the high productivity of modern crops in the face of an environment changing through drought, soil salinization, and so on. Landraces commonly occur in small areas: for instance, potatoes in parts of the Andes, rubber and chocolate in central Brazil, corn in central America, citrus fruits in India, and rice in India and China. The genetic variability provided by landraces is dwindling as traditional farmers around the world swap their local crop and animal breed varieties for standard, high-yielding ones. In Sri Lanka, for example, farmers grew 2,000 different varieties of rice until the late 1950s. Since then, they have grown just five high-yielding varieties. Researchers are combating this genetic erosion of traditional varieties. One solution is to designate village-level 'landrace

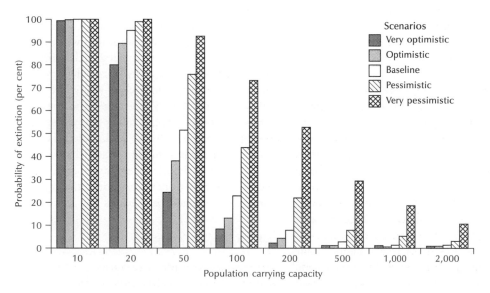

Figure 17.2 Sensitivity analysis of mean litter size in Atlantic Forest spiny rat (*T. eliasi*) populations predicted by a PVA. Mortality rate and sex ratio gave similar results.
Source: After Brito and de Souza Lima Figueiredo (2003)

custodians'. The idea is to pay farmers to grow a sample of the endangered native landraces at perhaps a few hundred key sites in areas where diseases, pests, and pathogens strongly influence the evolution of local crop varieties. These farms would bolster off-site conservation efforts, maintain the potential for further evolution in important landraces, preserve knowledge of traditional farming systems, and provide regional education on the importance of biodiversity conservation. The economic costs of subsidizing such on-site maintenance of traditional landraces are modest compared with the huge cost of subsidizing, for example, modern fertilizer and pesticide application, which speeds up the loss of traditional landraces.

A number of plant species possess *germplasm* (seeds) that are unsuited to long-term storage in seed banks (p. 352). These *recalcitrant seeds* belong to about 15 per cent of the world's flora. They either lack dormancy or are intolerant of low-temperature storage conditions. Better storage techniques for such recalcitrant seeds may eventually emerge, but, in the short-term, on-site conservation offers an appealing and practical approach. Similarly, the off-site conservation of wild crop relatives is not always practicable, but on-site conservation provides an attractive alternative. An advantage of on-site conservation of wild crop relatives is that it preserves a host of other species that share the same habitat with the target species. Many species have wild relatives that are key candidates for on-site conservation, including relatives of groundnut, oil palm, banana, rubber, coffee, cocoa, members of the onion family, citrus fruits, mango, cherries, applies, pears, and many forage species. In short, protected areas serve to safeguard the germplasm of recalcitrant seeds and of wild crop relatives. Such on-site conservation is in its infancy, but pilot schemes are up and running. In India, a gene sanctuary protects *Citrus* species. In Zambia, two reserves protect Zambesi teak (*Baikiaea plurijuga*). In Canada, a protected area maintains jack pine (*Pinus banksiana*). The WCMC listed 12,754 protected areas larger than 1,000 ha in the late 1990s, many of which will house many wild crop relatives, though the actual numbers are unknown.

The conservation of the genetic diversity of *forest tree species* suffers from sad neglect. Populations of the couple of dozen intensively bred tree species survive in various protected areas and their genetic base is unlikely to contract. On the other hand, populations of the several hundred wild or semi-domesticated tree species of economic value are vulnerable to extinction in all or part of their ranges. In 1986, the FAO (Food and Agriculture Organization of the United Nations) identified 86 such threatened species, and the number is growing. About 50 per cent of the populations threatened in 1989 grew in protected areas. The practice of *high grading*, which removes the best trees during each harvest cycle so that the less useful trees become the breeding stock, renders wild forest species conspicuously susceptible to genetic erosion. High grading has converted several formerly vigorous timber-producing forests in Japan, Korea, Turkey, and the Himalayas to stands of stunted and malformed trees. A combination of on-site conservation, off-site storage, and the use of living collections of plants (known as *off-site field gene banks*) is the best safeguard against the erosion of forest genetic diversity. However, it is uncertain how long the seeds of many tree species will last in storage. To be sure, the long-term storage of tropical species may prove difficult and field gene banks become indispensable. Even species with seeds suitable for seed bank storage will need occasional regeneration – as forest species have long generation times, the continual maintenance of field gene banks will be required.

Off-site conservation

Off-site preservation is expensive. The costs of maintaining African elephants and black rhinos in zoos is 50 times greater than protecting the same number of individuals in east African national parks. It costs about a billion dollars a year to maintain US zoos. The costs of off-site conservation reduce with the size of animals, so it is easier to justify off-site conservation on economic grounds for small-bodied species (Balmford *et al*. 1995). In terms of population size, the cost of off-site programmes rises markedly as numbers increase, whereas on-site conservation is relatively insensitive to the size of the population that it supports. Thus, for large mammals, the most effective use of available funds appears to be in protected areas, rather than in captive breeding programmes. It can be argued that zoos should be more selective in choosing threatened species for captive breeding programmes, and purely on financial grounds should concentrate on smaller-bodied species (Balmford *et al*. 1995). However, much of a zoo's revenue comes from visitors, and the principal attractions are the larger mammals. Thus zoos are going to be reluctant to concentrate solely on the smaller candidates for conservation. With zoos tending to favour displaying flagship species (charismatic or keystone megafauna) – pandas, giraffes, and elephants – huge numbers of insects and other invertebrates are ignored. Encouragingly, recent off-site conservation efforts include invertebrates as well – butterflies, beetles, dragonflies, spiders, and molluscs. These efforts are imperative, since most invertebrates have very restricted distributions.

On-site and off-site preservation are not necessarily mutually exclusive and an intermediate strategy exists that involves elements of both. It involves intensive monitoring and management of rare and endangered species in small, protected areas. Such populations are still to some extent wild, but human intervention on occasions may prevent population decline. Furthermore, on-site and off-site preservation practices often go hand-in-hand. Individuals for off-site populations may be released into the wild to help on-site programmes. Research conducted on captive programmes may furnish insights into the basic biology of a species and suggest better conservation measures in an on-site population. Self-maintaining off-site populations reduce the need

to collect individuals from the wild for display or research purposes. Displaying individuals from endangered species may help to educate the public about the necessity of preserving species, and so save other members of the species in the wild from harm. In turn, the on-site preservation of species is vital to the survival of species that are difficult to sustain in captivity (e.g. rhinoceros), as well as to the continued ability of zoos, aquaria, and botanical gardens to display new species.

Off-site facilities for animal preservation include zoos, game farms, aquaria, and captive breeding programmes. Off-site facilities for plant preservation include botanical gardens, arboreta, and seed banks. These are discussed below.

Zoos and aquaria

Zoological gardens

Zoos help with the conservation of biodiversity in many ways (e.g. Rabb 1994). Along with affiliated universities, government wildlife departments, and conservation organizations, zoos maintain over 700,000 individuals from some 3,000 species of mammals, birds, reptiles, and amphibians.

Most zoos now operate *captive breeding programmes*, which is perhaps their major contribution to conservation. Nonetheless, a tiny proportion of mammal species kept by zoos have self-sustaining populations big enough to preserve their genetic variation. In the USA, for example, zoos contain self-sustaining populations of just 96 species. Worldwide, just 26 of 274 species of rare mammals in captivity maintain self-sustaining populations. To overcome this problem, zoos and affiliated organizations are making strenuous efforts to build even more facilities to establish breeding colonies of rare and endangered species (e.g. the snow leopard and orang-utan) and to develop methods and programmes for re-establishing species in the wild.

Techniques to increase the reproductive rates of captive species include *cross-fostering* (using mothers of common species for young of rare species – mainly for the incubation of eggs by domestic birds), *artificial incubation*, *artificial insemination*, and *embryo transfer* (using surrogate mothers from different species). *Head-starting* is the protection of wild eggs and aiding hatchlings of, for instance, turtle and crocodiles, in an effort to avoid heavy mortality rates just after hatching.

Captive breeding is designed to produce enough individuals to consider release back into the wild. Such releases are *translocations* and take several forms. *Augmentation* or *restocking* is the bolstering of a wild population with animals bred in captivity or caught elsewhere in the wild. *Reintroduction* (also called re-establishment, restoration, and repatriation) establishes a new population within the historical range of a species, as described for the European beaver (p. 183). *Introductions* are releases outside the original range of a species in new territory.

Zoos are home to the sole remaining representatives of some species, including the California condor (*Gymnogyps californianus*) and possibly the black-footed ferret (*Mustela nigripes*). The California condor once ranged in coastal regions from British Columbia to Baja California, east to Florida, and north to New York. By 1987, the California Condor Recovery Team had controversially caught all surviving wild California condors – just 27 individuals – in the wild and placed them in Los Angeles Zoo and San Diego Wild Animal Park. (The very last captured condor had a brief sojourn at San Diego Zoo, owing to protestors at the gates of Los Angeles Zoo.) By 2000, there were 155 condors, 56 of which were back in the wild. The black-footed ferret, which lived in the prairies of the central USA, is the most endangered mammal in North America. Thought to be extinct until 1981, a small population was found living in Meeteetse, Wyoming. Starting in 1985, biologists collected wild black-footed ferrets and by 1987 had a surviving captive population of 18. The success of the captive breeding

programme – some 2,600 ferrets were born between 1987 and 1998 – enabled reintroduction to the wild. The wild population numbers about 80, with releases in the Shirley Basin in Wyoming, UL Bend National Wildlife Refuge in Montana, the Fort Belknap Reservation in Montana, the Badlands National Park in South Dakota, Buffalo Gap National Grasslands in South Dakota, and Aubrey Valley in Arizona.

Captive breeding programmes have led to the *reintroduction* of at least 18 species into the wild. In the cases of Père David's deer (*Elaphurus davidianus*), Przewalski's horse (*Equus przewalski*), the red wolf (*Canis rufus*), the Arabian oryx (*Oryx leucorx*), the American bison (*Bison bison*), the Guam kingfisher (*Halcyon cinnamomina cinnamomina*), and the Guam rail (*Rallus owstoni*), the species were extinct in the wild at the time of reintroduction (Box 17.1). Many other species, although not extinct in the wild, have benefited from reintroduction to portions of their range from which they had vanished. These include the peregrine falcon (*Falco peregrinus*), the alpine ibex (*Capra ibex ibex*), the black-footed ferret (*M. nigripes*), the red wolf (*C. rufus*), the Guam rail (*R. owstoni*), the Puerto Rican parrot (*Amazona vittata*), the Mauritius pink pigeon (*Nesoenas mayeri*), and the whooping crane (*Grus americana*).

Aquaria

Globally, aquaria sustain some 600,000 individual fish. Aquaria have lagged behind zoos in instituting captive breeding programmes for threatened species. With an increasing number of freshwater species joining the threatened list, aquaria have come to assume a vital role in off-site fish conservation – witness the work of the Captive Breeding Conservation Group of the IUCN. This group develops captive breeding programmes for endangered fish species. Species benefiting from these endeavours include the fishes of Lake Victoria in Africa, and the desert fishes and the Appalachian stream fishes of North America. The programmes restore natural habitat

Box 17.1

PÈRE DAVID'S DEER

Père David's deer (*E. davidianus*) disappeared from the wild about 1200 BC. It lived on in hunting reserves kept by Chinese royalty, first coming to the attention of westerners in 1865, when it was observed by the missionary Père Armand David in the gardens of the Chinese emperor, near Beijing. Some specimens were sent to Europe, where they prospered in captive herds. The Chinese populations all perished during the Boxer Uprising of 1898–1900. After the Second World War, many of the world's zoos received breeding stock from England. In 1960, the species was re-established in China. Père David's deer now lives in captive herds. Its natural habitat is unknown, but it may have lived in swampy plains of China until agriculture changed the landscape.

and breed species in aquaria, with the aim of safeguarding against the loss of wild species and educating the public on threats posed to fishes.

Botanical gardens and arboreta

Worldwide, some 1,600 botanical gardens grow about 4 million plants from at least 35,000, and maybe as many as 80,000, species. This represents 14 per cent or 32 per cent of the world flora. Additional species grow in greenhouses, subsistence gardens, and hobby gardens, usually with very few individuals per species. The largest botanical gardens are the Royal Botanic Gardens at Kew, England. These gardens grow 27,000 plant species under cultivation (11 per cent of the world's flora), of which 2,700 are rare, endangered, or threatened.

Botanic gardens are essential tools for maintaining the diversity of plant species and their gene pools. Some 800 of them claim to be active in conservation. Captive plant populations are easier to manage than captive animal populations. Plants do not need cages, they demand less tending than animals, and their natural habitats are generally easy to recreate. Moreover, individual plants can be crossed more readily than individual animals can; many plants can be self-pollinated or vegetatively propagated; and most plants are bisexual, which means that only half the number of individuals is needed to retain their genetic diversity. In addition, seed banks offer an efficient way of storing many plants in their dormant stage (see p. 352). Even if only 300 to 400 of the world's botanic gardens were to house major conservation collections and only 250 of these were to maintain seed banks, by one estimate it would still be possible to save viable populations of up to 20,000 plant species from extinction in them.

Many botanical gardens specialize in certain groups of plants. The Arnold Arboretum of Harvard University has hundreds of temperate tree species. In California, a specialized pine arboretum grows 72 of the world's 110 pine species, and one garden boasts one-third of the state's native species. In South Africa, a botanic garden contains roughly a quarter of the country's flora. Botanical gardens are the last hope for some plants – just one known original individual of the Murray birch (*Betula murrayana*) survives in the University of Michigan's demonstration forest. Other plants survive only in botanical gardens and arboreta (Box 17.2).

All botanical gardens need to increase the number of individuals per species to protect the range of genetic variability.

Box 17.2

THE FRANKLIN TREE AND PRESIDIO CLARKIA

The Franklin tree (*Franklinia altamaha*) is a plant that is probably extinct in the wild but survives in arboreta and other cultivated gardens. Botanists John and William Bartram first discovered it growing along the banks of the Altamaha River, near Fort Barrington, McIntosh County in Georgia, USA, in 1765. It was named after Benjamin Franklin. No one has seen the Franklin tree growing in its natural habitat since 1790; indeed, many people think that it is extinct in the wild. What caused its extinction is a mystery. Bartram's original collection appears to be the source of all specimens grown and sold today.

Similarly, in California, Presidio clarkia (*Clarkia franciscana*), a slender annual herb with lavender-pink flowers, once extinct in the wild, survived in a botanic garden. Later it was reintroduced to its native habitat – areas with serpentine substrate at the Presidio of San Francisco and in the Oakland Hills of Alameda County. One population at the Presidio is increasing and another decreasing, probably owing to competition from exotic species and to human disturbance (trampling, bicycling, and grass mowing), but more Presidio habitat is being created by removing exotic shrubs and over 100 species of non-native trees. The populations in Alameda County are under pressure from such exotic species as pampas grass (*Cortaderia selloana*) and French broom (*Genista monspessulanus*).

Germplasm and seed banks

Various methods allow the off-site preservation of genetic diversity in hi-tech storage facilites. In animals, cryo-preservation of gametes and embryos, or *Genome Resource Banks* (*GRBs*), can serve as genetic reservoirs. Such 'frozen zoos' are possibly helpful to species conservation, but they must take second place to techniques that are more conventional. Living populations pass on non-genetic, learned behaviour patterns – deep-frozen genetic material may be largely worthless in the absence of live animals.

Worldwide, more than 50 major *seed banks*, many in developing countries, store the seeds collected from a range of plant species. The Consultative Group on International Agricultural Research (CGIAR) coordinates seed-bank activities (Fuccilo *et al.* 1998). Seed banks store seeds in cold, dry conditions or are frozen. Cryogenic storage in dry, low-temperature, vacuum containers is possible in plants bearing so-called 'orthodox seeds'. A storage life of over a century is attainable for some of these species using extremely low (below −130°C) temperatures. The preservation of species with recalcitrant seeds, which must germinate immediately or die, can only occur on-site or in field collections, botanic gardens, and arboreta. Species with recalcitrant seeds are very common in the tropics and many economically important tropical fruit trees, timber trees, and plantation crops (e.g. cocoa and rubber) cannot be stored. Some recalcitrant seed species, such as banana or taro, will grow from cuttings. *In vitro* culture, which is the growing of plant tissue or plantlets under specific conditions in glass or plastic vessels, helps to 'store' these species.

Although widely employed, problems beset some off-site technologies such as seed storage. Even in the best-tended seed banks, the long-term integrity of the germplasm is difficult to guarantee. On a mechanical level, equipment or power failure may damage or ruin a frozen collection. Even without mishaps, seeds stored in cold conditions eventually lose their ability to germinate because their energy reserves diminish and harmful mutations accumulate. The occasional germination of seed samples, maturation of adult plants, and collection of fresh seeds is necessary to avoid this gradual erosion of seed quality. Such revitalization of seed samples is a formidable task in large collections. Another problem with seed banks is that botanists may unintentionally select individuals or accidentally cross them with other varieties. Plants stored *in vitro* display relatively high mutation rates. Furthermore, under any off-site storage conditions, species evolution is impossible and no adaptation to pests or environmental change takes place. For this reason, off-site storage preserves the germplasm but fails to conserve the species. Another difficulty is identifying the source of the germplasm. Some 50 per cent of the 2 million 'accessions' (collections of seed from a specific locality) to gene banks worldwide lack a record of the plant's characteristics and the place of its collection. In addition, off-site collections, and especially very-low-temperature seed banks, are expensive to run. If the economic climate should become adverse, some seed banks will suffer cutbacks or will even close. And the money pumped into storage costs often leaves scant funds for describing the germplasm stored in the seed banks, which diminishes the bank's value to plant breeders.

Surmounting all the above-named difficulties is the absence of some important species from seed banks. In particular, seed banks often lack species of regional importance (especially species of significant value in tropical countries), species with recalcitrant seeds, wild species, and livestock (Box 17.3).

Box 17.3

SPECIES DEFICIENT IN SEED BANKS

Crops of regional importance

Germplasm collections initially focused on food crops of greatest value in world commerce. Because many of the most important subsistence crops in developing countries are not widely traded in world markets and because many tropical species possess recalcitrant seeds, many regionally important crops are poorly represented in germplasm banks.

Species with recalcitrant seeds

Many important crops are poorly represented in off-site collections because their seeds are hard to store or because the species normally propagates vegetatively. Crops such as rubber, cacao, palms, many tuber crops, and many tropical fruits and other tree species can be conserved only on-site or in off-site field gene banks.

Wild species

Wild crop relatives are a useful source of genes conferring resistance to parasites and pests. With only two exceptions – wheat and tomatoes – the wild relatives of crops are poorly represented in off-site collections and in very few instances has their on-site conservation been attempted. Thus, many wild relatives of crops of economic importance face the same threat of extinction as other wild species do.

Livestock

Controlled breeding and the development of livestock varieties suitable for modern commercial production have eroded the genetic diversity in livestock. There is as yet no coordinated international effort for conserving the genetic resources of livestock. Because far fewer species and varieties are involved, less effort than that needed to conserve crop genetic resources is demanded, and, though the cost per species will be higher than in plants, the long-term benefits that these genetic resources could provide will be substantial.

The chief benefit of off-site preservation is in providing breeders with ready access to a wide range of genetic materials already screened for useful traits. Off-site preservation may also represent a last resort for many species and varieties that would otherwise die out as their habitat is destroyed or modern varieties of plants or animals take their place. On-site conservation is often less expensive than off-site techniques, it insures against loss of off-site collections, and it allows the continuing evolution of the crop varieties. On-site conservation also preserves knowledge of the farming systems with which local varieties evolved. Thus, the off-site and on-site techniques complement each other and are better used mutually.

SUMMARY

Scientists and conservationists advance arguments on several grounds to justify conserving species – economic, ecological, aesthetic, moral, and cultural. Conservation is 'managed' by national and international organizations, the biggest of which is probably the IUCN. The IUCN recognizes 10 conservation categories, ranging from 'extinct' and 'extinct in the wild' to 'least concern'. Two main conservation strategies are on-site conservation and off-site conservation. On-site conservation involves assessing possible management plans for species using, for example, population viability analysis. On-site conservation of landraces, germplasm (in seed banks), and forest trees tries to stem the loss of animals and plants used in agriculture. Off-site conservation has saved several species that are extinct in the wild. However, it is an expensive enterprise. Zoos and aquaria house captive animals, while botanical guards, arboreta, and seed banks house plants.

ESSAY QUESTIONS

1 **Is the cost of conserving threatened species justified?**

2 **How useful is population viability analysis?**

3 **Is off-site conservation a realistic alternative to on-site conservation?**

FURTHER READING

Akçakaya, H. R., Burgman, M. A., and Ginzburg, L. R. (1999) *Applied Population Ecology: Principles and Computer Exercises Using RAMAS®EcoLab*, 2nd edn. Sunderland, MA: Sinauer Associates.

If you are interested in hands-on modelling, try this.

Frankham, R., Ballou, J. D., Briscoe, A. D., and McInnes, K. H. (illustrator) (2002) *Introduction to Conservation Genetics*. Cambridge: Cambridge University Press.

An excellent introductory-level text.

Jeffries, M. L. (1997) *Biodiversity and Conservation*. London and New York: Routledge.

An excellent basic text.

Primack, R. B. (2000) *A Primer of Conservation Biology*, 2nd edn. Sunderland, MA: Sinauer Associates.

Highly readable and informative.

CONSERVING COMMUNITIES AND ECOSYSTEMS

Not just species, but the communities and ecosystems in which they live are altering in the face of current environmental change. There is a pressing need to preserve biodiversity and restore damaged communities and ecosystems. This chapter covers:

■ The nature, design, and management of protected areas
■ The value of conservation in unprotected areas
■ Restoring ecosystems

The conservation of species discussed in the previous chapter must go hand-in-hand with the conservation of communities and ecosystems – species are part of communities and ecosystems and they cannot be conserved in isolation. Much effort goes into conserving communities and ecosystems within protected areas. Since the 1980s, conservationists have acknowledged the vital role of protecting communities outside protected areas. In addition, the theory and practice of restoring ecosystems has emerged as a thriving discipline. This final chapter will explore the efforts of conservation biogeographers in conserving biological diversity within protected areas, in conserving it outside protected areas, and in ecosystem restoration.

PROTECTED AREAS

Categories of protected areas

In 1998, 6.4 per cent of the land area was occupied by nature reserves, national parks, wildlife sanctuaries, and protected landscapes. This figure is unlikely ever to exceed 7–10 per cent, owing to human demand for resources. The IUCN (1994) designated six categories of protected areas (Table 18.1). These are:

I   Strict nature reserves and wilderness areas
II   National parks
III National monuments and landmarks
IV   Managed wildlife sanctuaries and natures reserves

Table 18.1 IUCN system of protected areas

Category and description	Management objectives
I Strict Nature Reserve: protected area managed mainly for science	
Area of land and/or sea possessing some outstanding or representative ecosystems, geological or physiological features and/or species, available primarily for scientific research and/or environmental monitoring	Preserve habitats, ecosystems and species in as undisturbed a state as possible Maintain genetic resources in a dynamic and evolutionary state Maintain established ecological processes Safeguard structural landscape features or rock exposures Secure examples of the natural environment for scientific studies, environmental monitoring and education, including baseline areas from which all avoidable access is excluded Minimize disturbance by careful planning and execution of research and other approved activities Limit public access
Ia Wilderness Area: protected area managed mainly for wilderness protection	
Large area of unmodified or slightly modified land, and/or sea, retaining its natural character and influence, without permanent or significant habitation, which is protected and managed so as to preserve its natural condition	(Same as for Category I)
II National Park: protected area managed mainly for ecosystem protection and recreation	
Natural area of land and/or sea, designated to (a) protect the ecological integrity of one or more ecosystems for present and future generations, (b) exclude exploitation or occupation inimical to the purposes of designation of the area, and (c) provide a foundation for spiritual, scientific, educational, recreational, and visitor opportunities, all of which must be environmentally and culturally compatible	Protect natural and scenic areas of national and international significance for spiritual, scientific, educational, recreational, or tourist purposes Perpetuate, in as natural a state as possible, representative examples of physiographic regions, biotic communities, genetic resources, and species, to provide ecological stability and diversity Manage visitor use for inspirational, educational, cultural, and recreational purposes at a level which will maintain the area in a natural or near natural state Eliminate and thereafter prevent exploitation or occupation inimical to the purposes of designation Maintain respect for the ecological, geomorphological, sacred, or aesthetic attributes which warranted designation Take into account the needs of indigenous people, including subsistence resource use, in so far as these will not adversely affect the other objectives of management
III Natural Monument: protected area managed mainly for conservation of specific natural features	
Area containing one (or more) specific natural or natural/cultural feature which is of outstanding or unique value because of its inherent rarity, representative or aesthetic qualities, or cultural significance	Protect or preserve in perpetuity specific outstanding natural features because of their natural significance, unique or representational quality, and/or spiritual connotations An extent consistent with the foregoing objective, to provide opportunities for research, education, interpretation, and public appreciation Eliminate and thereafter prevent exploitation or occupation inimical to the purpose of designation Deliver to any resident population such benefits as are consistent with the other objectives of management

IV Habitat/Species Management Area: protected area managed mainly for conservation through management intervention

Area of land and/or sea subject to active intervention for management purposes so as to ensure the maintenance of habitats and/or to meet the requirements of specific species

- Secure and maintain the habitat conditions necessary to protect significant species, groups of species, biotic communities, or physical features of the environment where these require specific human manipulation for optimum management
- Facilitate scientific research and environmental monitoring as primary activities associated with sustainable resource management
- Develop limited areas for public education and appreciation of the characteristics of the habitats concerned and of the work of wildlife management
- Eliminate and thereafter prevent exploitation or occupation inimical to the purposes of designation
- Deliver such benefits to people living within the designated area as are consistent with the other objectives of management

V Protected Landscape/Seascape: protected area managed mainly for landscape/seascape conservation and recreation

Area of land, with coast and sea as appropriate, where the interaction of people and nature over time has produced an area of distinct character with significant aesthetic, ecological, and/or cultural value, and often with high biological diversity. Safeguarding the integrity of this traditional interaction is vital to the protection, maintenance, and evolution of such an area

- Maintain the harmonious interaction of nature and culture through the protection of landscape and/or seascape and the continuation of traditional land uses, building practices, and social and cultural manifestations
- Support lifestyles and economic activities which are in harmony with nature and the preservation of the social and cultural fabric of the communities concerned
- Maintain the diversity of landscape and habitat, and of associated species and ecosystems
- Eliminate where necessary, and thereafter prevent, land uses and activities which are inappropriate in scale and/or character
- Provide opportunities for public enjoyment through recreation and tourism appropriate in type and scale to the essential qualities of the areas
- Encourage scientific and educational activities which will contribute to the long-term well-being of resident populations and to the development of public support for the environmental protection of such areas
- Bring benefits to, and to contribute to the welfare of, the local community through the provision of natural products (such as forest and fisheries products) and services (such as clean water or income derived from sustainable forms of tourism)

VI Managed Resource Protected Area: protected area managed mainly for the sustainable use of natural ecosystems

Area containing predominantly unmodified natural systems, managed to ensure long-term protection and maintenance of biological diversity, while providing at the same time a sustainable flow of natural products and services to meet community needs

- Protect and maintain the biological diversity and other natural values of the area in the long term
- Promote sound management practices for sustainable production purposes
- Protect the natural resource base from being alienated for other land-use purposes that would be detrimental to the area's biological diversity
- Contribute to regional and national development

Source: Adapted from IUCN (1994)

V   Protected landscapes and seascapes
VI   Managed resource protected areas.

Areas in Categories I to V are truly protected areas with their habitats managed first and foremost to preserve biological diversity, although it could be argued that only the first three categories are truly protected. The managed areas, though at the bottom of the list, are by no means insignificant. Quite the contrary, they are often much larger than protected areas and many contain their original species. Indeed, a matrix of managed areas often surrounds protected areas.

Examples of protected areas

Strict nature reserves and *wilderness areas* are designed to protect species and natural processes in as undisturbed a state as possible. An example is the *Bolshoi Arkticheskiy* (*Great Arctic*) *State Nature Reserve* in Russia, which is one of the biggest wildernesses on Earth. It covers 4.2 million hectares of tundra in the north Siberian Taymyr peninsula. Some 4,000 indigenous Dolgans and Nganasans inhabit it. They subsist from hunting, fishing, and herding reindeer. The Reserve is to be maintained in its natural condition, unchanged by humans. To this end, development, including vehicle access, is forbidden, although scientific research and low-density tourism are endorsed, as is the traditional practice of reindeer husbandry.

National parks are large areas of scenic and natural beauty designed to protect one or more ecosystems and to provide for scientific, educational, and education users. Commercial extraction is normally banned. The *Kruger National Park* in South Africa is 1,948,528 ha of arid and semi-arid habitat types. The park's natural environment is protected from disturbance and supports an active research programme with first-rate facilities. The park caters for large-scale recreation. Conservation, education, and recreation services occupy zoned areas. Fences around the park boundary guarantee minimal threats from

outside. The *Yellowstone National Park*, USA, covers 898,349 ha of mainly pristine wilderness. It was the world's first national park, established by an act of Congress in 1872. Its stupendous scenery and natural resources are conserved, while allowing access to the public, of which 3 million visit every year. The 300 people involved in scientific research and park management live within the park. A fire management plan lets natural fires burn through 70 per cent of the area.

National monuments are smaller reserves set aside to preserve unique biological, geological, and cultural features of special interest. A good example is *Metéora* in Greece. It has exceptional geological formations. Impressive sandstone rock pinnacles stand over 400 m above the Thessalian Plain. Atop the pinnacles are caves and medieval monasteries. The site holds 50 pairs of Egyptian vultures (*Neophron percnopterus*), the largest population in Greece. Local laws protect Metéora. Construction and rock extraction are prohibited or limited. The 375-ha site is managed primarily for conservation of the monasteries and the natural surroundings. Owing chiefly to the inaccessibility of the cliffs, the relict flora and fauna have remained comparatively undisturbed.

Managed wildlife sanctuaries and *nature reserves* are like strict nature reserves, but some human manipulation – setting managed fires, removing exotic species, and some controlled harvesting for example – is permitted. *Lüneburger Heide Nature Reserve* in Lower Saxony, Germany, is one of the best surviving examples of lowland heath in western Europe; it is also one of the most threatened by human impacts. The reserve comprises 19,720 ha of human-modified landscape, which is thought to have originated about 5,000 years ago. It is protected and activities that could alter or eliminate heathland are proscribed. *Verein Naturschutzpark*, a nature conservation society, has privately acquired much of the land and manages the site, involving local groups and the public in the process. A comprehensive system of controlled grazing, use of chemicals, and scything actively sustains the heathland. Fire is not used to manage the heathland as it would damage the juniper. Three million visitors a year are allowed access to 14 per cent of the site, which restriction minimizes ecological damage and maintains the seral vegetation communities.

Protected landscapes and *seascapes* allow non-destructive traditional uses of the environment by residents, as in fishing villages and orchards. These areas often have special opportunities for tourism and recreation. The 91,300-ha *Dartmoor National Park* in the UK is mostly owned privately, although the public has access rights to over half the area. Three national nature reserves, two forest nature reserves, and 25 sites of special scientific interest lie within the park and cover 29 per cent of its area.

Managed resource protected areas allow sustained production of natural resources (water wildlife, grazing for livestock, timber, tourism, fishing) in a way that safeguards biodiversity. The 25,000-ha *Kiunga Marine National Reserve*, Kenya, protects fairly pristine tropical coastal habitats, and is managed in a sustainable way for the benefit of the local community. It consists of a narrow mainland strip, some 50 small offshore islands that support seabird colonies, and the surrounding waters. Human interference in the area is very low, with just a limited exploitation of local wildlife. Tourism, non-destructive activities such as swimming and sailing, and fishing by specified traditional means are endorsed. However, fishing using poison, spear-guns, or dynamite is illegal, as is the collecting of shells and corals. Permits control the passage and anchorage of boats, diving, and access by non-locals.

Biosphere reserves

The United Nations Educational, Scientific, and Cultural Organization's (UNESCO's) Man and the Biosphere Programme (MAB), which began in 1971, created an international network of Biosphere Reserves. The number of reserves has proliferated, although some areas (e.g. New Guinea, much of India, and Amazonia) still worryingly lack reserves (Figure 18.1). The Biosphere Reserves are meant to be models demonstrating the compatibility of conservation efforts with sustainable development for the benefit of the local people by integrating human activities, research on the natural environment, and tourism. The general pattern of a reserve is a central protected area in which communities and ecosystems are strictly protected; a surrounding buffer zone in which traditional human activities (such as the collection of thatch, medicinal plants, and small fuelwood) are monitored and managed and in which non-destructive research is undertaken; and a transition zone in which sustainable development (e.g. small-scale farming), some extraction of natural resources (e.g. selective logging), and experimental research are carried out (Figure 18.2). This three-zone strategy has at least three benefits. First, it encourages local people to support the goals of the protected area. Second, it ensures the maintenance of desirable landscape features fashioned by humans. And third, the buffer zones may aid animal dispersal and gene flow between firmly protected conservation areas and human-dominated and unprotected areas.

DESIGNING AND MANAGING PROTECTED AREAS

The distribution of parks and conservation areas is haphazard, largely owing to the availability of land and money. The location and size of protected areas depends on several factors, including the distribution of people, land values, the political efforts of conservation bodies, and historical factors. Protected areas commonly occupy out-of-the-way, uninhabited, infertile, and resource-poor land unwanted for any commercial activity. In large metropolitan areas, small conservation areas are common on land bought by local governments or conservation organizations, or donated by rich citizens.

The rather random acquisition and creation of most parks and conservation areas has not

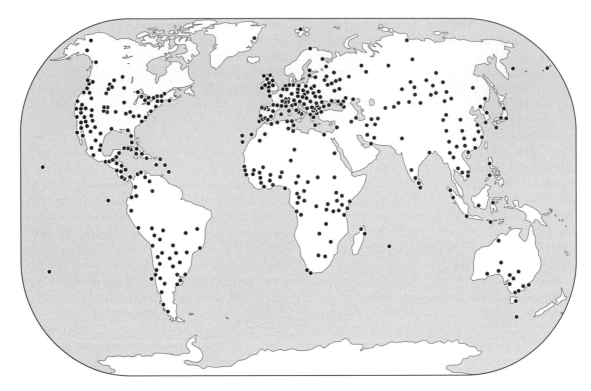

Figure 18.1 UNESCO's Biosphere Reserves in 2003.

deterred ecologists proposing several guidelines for designing reserves to capitalize on the preservation of biodiversity. Landscape ecology, and especially the theory of island biogeography (p. 242), informs conservation practice in parks and reserves (Figure 18.3). Six major themes in landscape ecology have relevance to conservation plans: reserve number and size, reserve shape, reserve edges, conservation corridors, reserve functioning, and regional settings.

Reserve number and size

A major conservation debate concerns two seemingly simple questions. First, is it better to have a *large reserve or a small reserve* (*LOS*)? Second, is it better to have a *single large or several small reserves* (*SLOSS*)? This long-running debate started in the mid-1970s (see Shafer 1990, 79–82). To an

extent, arguments over the connectivity of nature reserves, the worth of wildlife corridors, and the problems of scale have superseded the debate, but its arguments still have currency.

Large reserves have several qualities of ecological value (Forman 1995, 47):

- they safeguard water quality of aquifers and lakes;
- they aid fish and terrestrial mammal populations by the connectivity of a low-order stream network;
- they sustain patch-interior species;
- they support vertebrates with large home ranges by affording core habitat and ample escape cover;
- they furnish a source of species dispersing into the surrounding matrix;
- they maintain near-natural disturbance

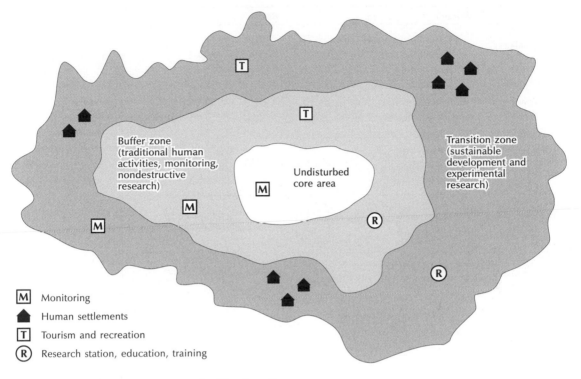

M Monitoring
▲ Human settlements
T Tourism and recreation
Ⓡ Research station, education, training

Figure 18.2 The three-zone pattern of a Biosphere Reserve.

regimes, which are advantageous to those species that have evolved with, and require, disturbance;

- they provide a buffer against extinction during environmental change;
- in suburban areas, they may house a limited number of interior species, but these patches can play a major role in ameliorating microclimates, making the downwind neighbourhood cooler and moister. They also soak up rainwater, so reducing floods.

Small reserves also have qualities of ecological value (Forman 1995, 47):

- they provide much-needed habitats and stepping-stones for species dispersal and for recolonization after local extinction of interior species;

- they support high densities of species and large populations of edge species;
- they augment matrix heterogeneity, so providing escape cover for prey species and decreasing wind fetch and erosion;
- they provide habitats for small-patch-restricted species (p. 136);
- they preserve scattered small habitats and rare species.

On balance, large benefits tend to accrue from large reserves, and small benefits from small reserves. Oddly, few benefits arise in middling size reserves. Disadvantages, such as an increase in the proportion of edge species, occur mainly in small reserves.

Accepting that large reserves are ordinarily preferable to small reserves, the question then becomes how many large reserves are needed to

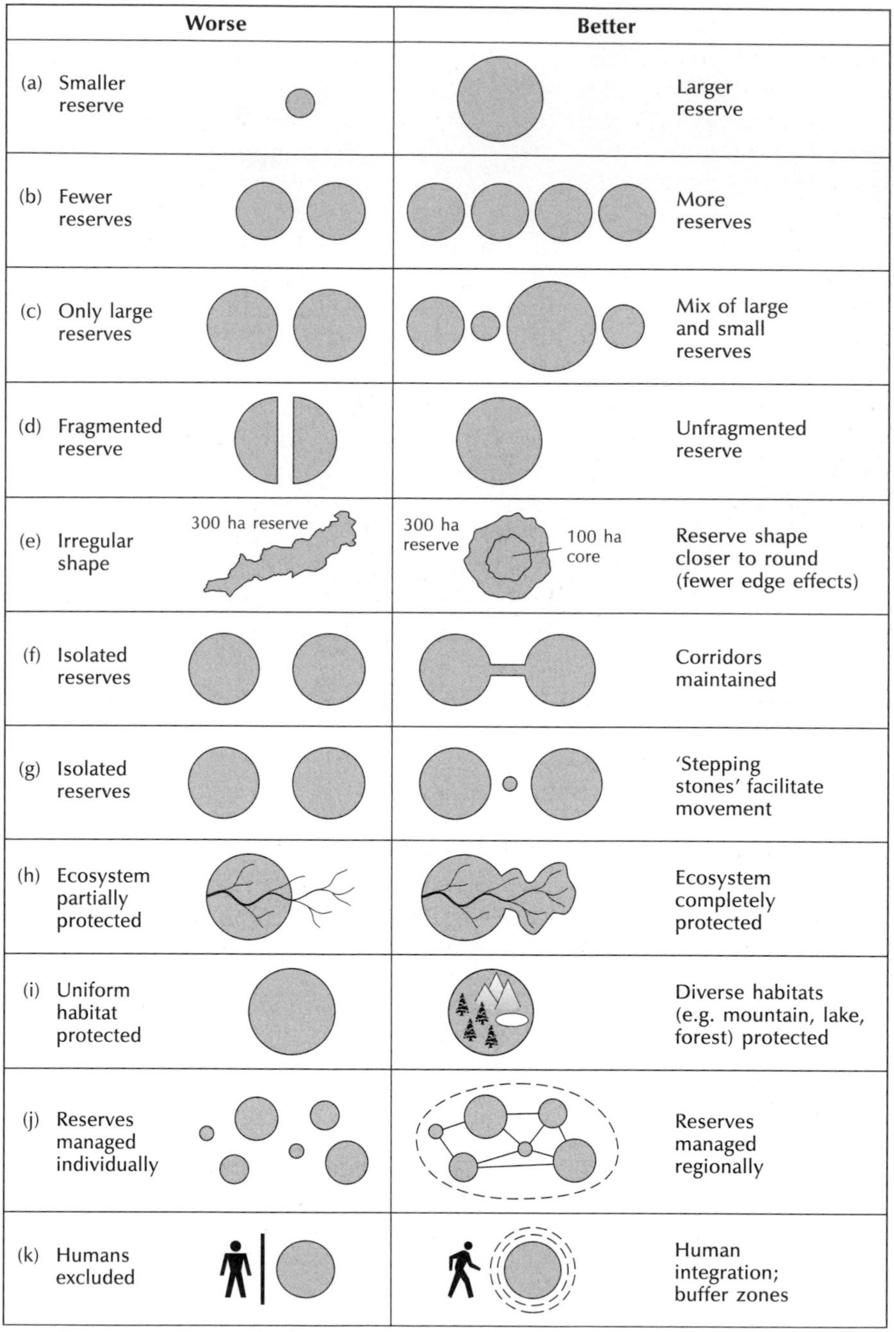

Worse		Better
(a) Smaller reserve		Larger reserve
(b) Fewer reserves		More reserves
(c) Only large reserves		Mix of large and small reserves
(d) Fragmented reserve		Unfragmented reserve
(e) Irregular shape	300 ha reserve / 300 ha reserve / 100 ha core	Reserve shape closer to round (fewer edge effects)
(f) Isolated reserves		Corridors maintained
(g) Isolated reserves		'Stepping stones' facilitate movement
(h) Ecosystem partially protected		Ecosystem completely protected
(i) Uniform habitat protected		Diverse habitats (e.g. mountain, lake, forest) protected
(j) Reserves managed individually		Reserves managed regionally
(k) Humans excluded		Human integration; buffer zones

Figure 18.3 Design principles for reserves as suggested by the theory of island biogeography.
Source: Adapted from Shafer (1997)

achieve the desired ecological outcomes? This question is not well researched, but there are snippets of studies that offer some clues. In New Jersey woodlots, USA, more than three 24-ha woods would be needed to maximize bird species diversity at the landscape scale (Forman *et al*. 1976). Another way of approaching the question is to look at the species pool (the total species living in a habitat type within a landscape). If a large patch should contain the entire pool, then no more patches would be required since they would add no extra species. If a large patch contains 90 per cent of the species pool, then a second patch may add the other 10 per cent, though it is more likely that three or more patches would be needed to include the 'tail-end', rare species. The *minimum number point* is useful as a conservation target. It is the number of patches beyond which the species diversity–patch number curve gains less than 5 per cent of the species pool with the addition of another patch. It thus includes nearly all the characteristic species of the patch-type in a landscape. Plainly, the lower the number of species present in the first patch, the more patches that are required. In northern Sweden since 1945, large-scale clear-cuttings have hit boreal forests. To boost biodiversity in this region, the preservation of small patches of old forest, less than 1 ha in size, has been advocated. However, a study of birds in Grandanlet, a 280-km^2 landscape mosaic of virgin boreal forest and mire, suggested that such small forest patches would provide some habitat for more generalist species, but to maintain a diverse bird fauna requires larger forest patches, preferably larger than 10 ha, to be set aside (Edenius and Sjöberg 1997).

Recent fieldwork on reserve size reveals the problem with generalizations. For instance, the finding that patch-size–species richness relationships vary with species even in the same area has management implications. A recent study tackled the SLOSS question head-on, and also considered the effect of reserve shape (Virolainen *et al*. 1998). The subject of the study was vascular plant species living on boreal spruce and pine mires in Finland. Species richness, species rarity, and taxonomic diversity in the reserves were all investigated. The results (Table 18.2) showed that species richness was not related to the spruce mire size, but it did increase in relation to the pine mire size. In contrast, the species-rarity score increased in relation to the area of spruce mires, but it was not related to the area of pine mires. Taxonomic diversity was unrelated to reserve size in the case of spruce-mire reserves, but it increased with the size of pine-mire reserves. By comparing large and small reserves, it was found that several small mire reserves contained more vascular plant species than a large mire reserve of equal area. Several small mire reserves also had higher rarity scores and higher taxonomic diversities than a single large mire reserve. Indeed, species richness, rarity score, and taxonomic diversity all increased in relation to the number of small mires in both spruce and pine groups. Mire shape had no influence upon species richness, rarity score, and taxonomic diversity, which brings us to the next theme.

Reserve shape

It is reasonable to think that reserve shape, particularly in smaller reserves, will affect population

Table 18.2 Summary of Finland results

Ecological measure	Spruce mire size	Pine mire size
Species richness	No relationship	Positive relationship
Rarity score	Positive relationship	No relationship
Taxonomic diversity	No relationship	Positive relationship

Source: Adapted from information in Virolainen *et al*. (1998)

dynamics on a reserve. Three possible effects come to mind. Convoluted shapes:

- have a greater length exposed to surrounding matrix, and this increased exposure may raise susceptibility to external stresses or disturbances arising from domesticated animals, poaching, and so on;
- have proportionately more edge than less-convoluted shapes; this difference has several attendant consequences, including the occurrence of more pronounced edge effects in convoluted patches compared with less-convoluted patches of the same size, which may explain some area-related extinctions;
- contain less interior habitat than less-convoluted reserves of the same size, and this may interfere with reserve dynamics.

Despite these reasons for supposing that reserve shape would play a role in reserve dynamics, research into the effects of reserve shape has produced a mixed message. A study of species richness in English farm woodlands showed, after controlling for the effects of patch area, patch shape explained no significant variation in species richness (Usher *et al.* 1992). Conversely, in urban woodlots in four adjacent cities in southern Ontario, Canada, edge length (which is related to shape) was a superior predictor of bird species richness than patch area (Gotfryd and Hansell 1986). The relationship between reserve shape and species richness may be inconsistent, but the relationship between reserve shape and the abundance of specific species adds another and little-researched dimension to the problem (Kupfer 1995). It is likely that small reserves with convoluted edges will house mainly edge species, whereas more circular reserves will house a larger number of rare, interior species. However, far more work needs doing on reserve shape and its effect upon species composition and dynamics before reaching secure conclusions.

Reserve edges

As well as reducing the size of patches, habitat fragmentation also increases the proportion of edge to interior. Although wildlife managers once considered forest edges a boon, mainly because they often are species-rich and attract many game birds, current thinking is less upbeat. Edge habitats are not necessarily desirable. For conservation purposes, for example, their propensity to attract weedy species, rather than species that need protecting, is hardly a blessing. With these drawbacks in mind, conservation schemes try to build in edge processes. In addition, edges tend now to be treated as dynamic, functioning habitats that interact with the matrix on either side. The pressing needs are to re-evaluate edges and achieve a fuller comprehension of their roles in landscape processes, and to examine the significance of reserve boundaries to reserve functioning and viability (Kupfer 1995).

Conservation corridors

Corridors have captured the public imagination and a range of 'greenways', 'greenbelts', 'linear reserves', 'lifelines', 'buffer strips', and 'wildlife corridors' are built into management plans and planning strategies (Bennett 1999). The idea that corridors may negate the isolation of populations resulting from habitat fragmentation carries an intuitive appeal. And the building of corridors to ease the movement of endangered animals is often within the scope and resources of conservation-minded groups.

Corridors possess several advantages in conservation, but there are potential disadvantages, too. On the positive side, corridors:

- are thought to boost dispersal between reserves, which compensates for the damaging effects of small populations and restricted gene-pools by expediting interbreeding between populations; inter-reserve dispersal also influences species demography positively by increasing population growth, facilitating recolonization of reserves following local extinctions, and promoting the survival of metapopulations by raising the chances of 'rescue';

- increase the foraging area of wide-ranging species;
- provide temporary refuges from predators for species moving between patches;
- offer a mix of accessible habitats and successional stages for species demanding a diversity of habitats for different activities or stages of their life cycles;
- afford alternative shelter from large distances, such as a refuge from fire;
- create 'greenbelts' to limit urban sprawl, abate pollution, furnish recreational opportunities, and enhance scenery and land values;
- may assist large-scale species migrations necessitated by climatic change.

Corridors may have the following negative effects:

- they may hasten the spread of diseases, exotic predators, and other disturbance while failing to enhance the survival or dispersal of the targeted species;
- they may increase species flow between reserves, so impoverishing genetic variability through the loss of local genotypes;
- they may permit the spread of fire and other 'contagious' disturbances;
- they may increase the exposure of wildlife to hunters, poachers, and other predators;
- in the case of riparian strips, used as conduits, they may block the dispersal or survival of upland species;
- they conflict with strategies to preserve the habitat of endangered species, when the inherent quality of a corridor habitat is low.

Despite the arguments against them, the consensus is that well-designed conservation corridors are efficacious in managing species within fragmented landscapes. Moreover, there is evidence that several reserves linked by corridors helps to maintain regional-scale landscapes and conserve many species. In 14 western North American parks, 13 had lost 43 per cent of their historical lagomorph (rabbits, hares, and pikas), carnivore, and ungulate species (Newmark 1987). The Kootenay–Banff–Jasper–Yoho park system, which has significant connections between reserves, maintained its entire original mammal fauna. Nonetheless, it is well to bear in mind the makeshift nature of conservation corridors:

> Corridors, regardless of how effective they may be, can never replace large reserves for the protection of ecosystems and species. This is because corridors, by definition, contain little habitat, and because they are intrinsically 'edgy'. This edginess of corridors means that these landscape links are hazardous for edge-sensitive and predation-sensitive species, but very suitable habitat indeed for many weedy species and pathogens. Corridors are bandages for a wounded natural landscape, and at best can only partly compensate for the denaturing activities of humans.
>
> (Soulé and Gilpin 1991, 8)

Road corridors often serve as conservation areas. The paradox of road systems is that:

- on the one hand, they split and isolate wildlife populations, cause disturbance and pollution in the adjacent environment, and lead to a heavy toll of dead animals;
- on the other hand, roadside habitats attract a wide range of animals and plants, may preserve the only remnants of natural habitat in a region, and may act as conduits for animal passage through forbidding terrain (Bennett 1991).

To resolve this paradox, road systems must be viewed in their regional setting. New road proposals should consider impacts on the regional landscape. For instance, where a road is to pass through forest, wildlife management priorities should be to minimize the efficiency of the road as a barrier, to limit road kills, and to keep noise and pollution to the lowest possible levels. And when a road is to pass through agricultural landscapes, efforts should be made to retain and expand wildlife habitats along the roadside, to enhance the continuity of roadside vegetation in an effort to build natural corridors across the landscape, and to monitor road kills (Bennett 1991). In some agricultural areas of the world, such as the central wheatbelt of Western Australia, native vegetation survives as small remnants, the use of which depends on connections through road verges (Fortin and Arnold 1997). This underscores the importance of roadside habitats in conservation.

Reserve functioning

The animal and plant species preserved within a reserve do not live in isolation. They are components of the reserve ecosystem and interact with the reserve's physical environment. Traditionally, reserves were managed to protect a particular species or to sustain species richness. Reserves for single species are inefficient and philosophically questionable. Piecemeal species management and piecemeal land management may pervert ecosystem integrity and precipitate severe ecosystem changes (see Merriam 1991). Management plans now recognize that some kind of ecological completeness is desirable and they commonly include the conservation of ecosystem processes within a reserve.

Upsets of reserve ecosystems arise from many processes, but habitat fragmentation comes high on the list, if not at the top. Effects of upsets are varied and fall into biotic and abiotic categories:

1 *Biotic effects* resulting from habitat fragmentation include 'extinction' cascades'. If a top carnivore goes extinct because there is not enough of its habitat left to support it, repercussions may cascade right down the food web. Fragmentation also leads to the incursion of edge habitats into interior habitats. Concomitant changes in ecosystem processes ensue, including higher nest predation rates. Human modification of habitats in reserves may alter the age-structure of habitat patches. Logging, for instance, tends to induce a faster turnover of forest patches and maintain a young forest. This change of forest age, with a drastic reduction in old trees, has injurious effects on such species as the northern spotted owl (*Strix occidentalis caurina*) and the red-cockaded woodpecker (*Picoides borealis*).

2 *Abiotic effects* resulting from disruption of a reserve ecosystem involve changes in sediment budgets and the land-surface water cycle. Both geomorphic and hydrological changes may have knock-on effects in animal and plant communities on a reserve. Microclimates are also altered by ecosystem modifications. Near-surface wind, temperature, and relative humidity all change following deforestation.

Disturbance is a major ecological factor that needs to be included in reserve design. Soon after disturbance became a popular topic of research in the 1970s, reserve functioning often focused on the connection between reserve size and the scale of the natural disturbance regime (Kupfer 1995). Particular attention was paid to minimum dynamic area (p. 346), defined as the smallest area under a natural disturbance regime that keeps sources for recolonization and so minimizes extinctions (e.g. Pickett and Thompson 1978). Subsequently, researchers have questioned the value of the minimum dynamic area concept (e.g. W. L. Baker 1989, 1992). Despite these problems, the notion of reserve functioning now recognizes the important relationships between changing landscape structures and altered landscape functioning (Kupfer 1995).

Regional reserve settings

The regional setting of a reserve affects its local dynamics. For this reason, ecologists, planners, and wildlife managers should heed large-scale ecological processes and look beyond the bounds of an individual reserve (e.g. Noss 1983). Of course, the process is two-way: local dynamics affect regional processes. A danger in metapopulations is that local extinctions in patchy landscapes will accumulate into landscape extinctions, and even regional and subcontinental extinctions (see Merriam 1991). The acknowledgement of regional processes and their interaction with local ecosystem dynamics has at least two implications for conservation (Kupfer 1995):

1 It encourages a switch toward landscape-level population management and strengthens the case for studying large-scale population dynamics within spatially heterogeneous landscapes. Reserves do not stand in isolation; they interact with surrounding ecosystems, the interaction affecting species within the reserve.
2 The recognition of regional processes suggests a need to include spatial factors and scale effects in conservation plans. A study of factors affecting the distribution of Leadbeater's possum (*Gymnobelideus leadbeateri*), an endangered arboreal marsupial, and their implications for animal conservation, is a superb example of this practice (Box 18.1).

Several other conservation biologists stress the value of regional perspectives (e.g. Sexton *et al.* 1999; Poiani *et al.* 2000). An example is the major transition zone between temperate and subtropical species in Florida, USA, which is threatened by climatic warming (Crumpacker *et al.* 2001a, b, c). Predictions suggest that, with temperature rises of 1–2°C, the zone will move northwards. A likely consequence is the loss of live oak (*Quercus viginiana*) and bald cypress (*Taxodium distichum*), the latter being a very important dominant species in south Florida swamps. Other woody species in Florida are at risk from global warming. Mitigation efforts include the building of an extensive system of connected conservation areas in the upper two-thirds of the Florida peninsula and keeping the disturbance of the southern cypress forests to a minimum.

MANAGING UNPROTECTED AREAS

Over 90 per cent of the world's land area will stay outside protected areas. In consequence, it is vital that conservation strategies preserve biodiversity outside protected areas, as well as inside them. Species and communities cannot be saved in parks and reserves alone. If areas around parks are degraded, the parks themselves will suffer. This fact was recognized for the large herbivore species in Kenyan and Tanzanian national parks (Western 1989). Surprisingly, about three-quarters of Kenya's 2 million large animals live outside the national parks. The figures for living outside are giraffes 89 per cent, impalas 72 per cent, Grevy's zebras 99 per cent, oryx 73 per cent, and ostriches 92 per cent. Only rhinoceroses, elephants, and wildebeest live largely inside the parks. If the parks were sealed off, then their large herbivore biodiversity would probably fall by between about 24 and 48 per cent.

Management strategies

The trick in managing species, communities, and ecosystems in unprotected areas is to develop strategies for reconciling human needs and conservation interests. Several examples of such efforts have yielded promising results. They come under the categories of privately owned land, agricultural practices, restricted access areas, sparsely populated areas, and government-owned land designated for multiple use:

• *Privately owned land.* Much unprotected land is privately owned. The key to preserving rare species on such land is to persuade private

Box 18.1

MANAGING LEADBEATER'S POSSUM

David B. Lindenmayer's research into Leadbeater's possum has been running for over a decade (e.g. Lindenmayer 1991, 1997, 2000; Lindenmayer *et al.* 1993). It has involved field and modelling studies at four spatial scales – tree level, stand level, landscape level, and regional-climate level. Leadbeater's possum is largely confined to ash-type eucalypt forests in the central highlands of Victoria, southeastern Australia. Most of these forests are designated for timber harvesting and pulpwood production. The possums live in colonies of up to 12 individuals that shelter and nest in hollows within large trees. They eat arthropods, sap from wattle (*Acacia* spp.) and eucalypt manna, and honeydew from phloem-feeding insects. The total population is predicted to fall by over 90 per cent by 2020 to 2030, owing to the natural collapse of existing large trees containing hollows, low recruitment of new hollow-bearing trees, and timber extraction. Studies on the possums at the four scales cover a variety of themes (see Lindenmayer 2000, Table 1):

1 *Tree-level studies* include models of den-tree attributes and tree occupancy, performance tests of nest tree models, den-tree use, long-term occupancy of nest sites, cavity development in trees, and falls rate of hollow trees.
2 *Stand-level studies* involve habitat requirement models, performance tests of habitat models, and applications of habitat models.
3 *Landscape-level studies* consider corridor use and model population viability.
4 *Regional-scale studies* focus on possum-based climate analysis and tree-distribution climate analysis.

The research findings strongly suggest that the conservation of Leadbeater's possum must embody the needs of the animals at different spatial scales. Plans must include large reserves, landscape-scale strategies such as wildlife corridors, and stand-level operations, such as maintaining structural attributes and vegetation complexity. By adopting strategies at several ecological and management scales, it should be possible to ensure that appropriate conditions for the possum will exist somewhere in part of its range, including areas that are used for wood production.

Although the possum work has focused on one species, the implications of its findings have universal pertinence. Lindenmayer identifies five generic implications: (1) Data gathered at each scale can inform processes and conservation strategies at other scales. (2) The conservation implications for different scales may be disparate. (3) Conservation strategies may be more readily acceptable if the scales required for their employment are congruent with the scales used in the management and planning of natural resources. (4) Even for a single species, there is a need for an array of conservation strategies that cannot stem from a single study at one spatial scale. (5) Implementing a range of conservation strategies across the scales spreads the risk involved and means that, if a strategy at one scale should prove to be faulty, other strategies are in place that should work and so not fatally jeopardize the overall success of the conservation exercise.

landowners to adopt apt conservation measures, and perhaps even compensate them for doing so. Another measure is for government agencies to ensure that builders and developers know the location of rare species or habitats so that they can take steps to avoid harming them.

• *Agricultural practices*. Agricultural practices may be manipulated to sustain biodiversity in unprotected areas. Forests in which selective logging over a long cutting cycle or in which traditional shifting cultivation is practised maintain a large portion of their original biota (Chapman *et al*. 2000). In western Panama, coffee is grown under shade trees and the bird and insect diversity of these plantations is in some cases comparable to that in nearby natural forest. Dina L. Roberts and her team (2000) studied mixed-species parties of birds that follow swarm raids of army ants that flush out more palatable insects from the leaf litter. They found that all shade-coffee and forested areas supported many bird species. Open-sun areas, where coffee grows without shade trees, had no army ant swarms and, consequently, no birds following them. However, many forest-dwelling birds were absent from shade-coffee areas located some distance from continuous forests, perhaps because they do not nest in disturbed habitats or because they are unwilling to fly across brightly lit areas.

• *Restricted access areas*. Areas cordoned off or managed for some purpose that does not harm the ecosystem may support native species. Security zones around government installations, military sites, and nuclear power stations are examples. In England, the Ministry of Defence acquired Salisbury Plain for military training early in the twentieth century and this large expanse of lowland grassland has consequently not been subject to intensive farming methods. Similarly, the Defence Science and Technology Laboratory range at Porton Down has remained unaffected by modern farming. Some 14,000 ha of chalk downland remains and supports 13 species of nationally rare and scarce plants, 67 species of rare and scarce invertebrates, and is a site of international importance for birds. A fifth of the rare stone curlew (*Burhinus oedicnemus*) population breeds at Salisbury Plain, and neighbouring farmland supports another 12 per cent. The Salisbury Plain LIFE Project Partnership is trying to restore, promote, and protect the unique character and wildlife of the Porton range and Salisbury Plain. The project targets conservation management of chalk grassland and various Biodiversity Action Plan (BAP) species such as stone curlew, marsh fritillary butterfly (*Euphydryas aurinia*), and juniper (*Juniperus communis*).

• *Sparsely populated areas*. Some unprotected areas sustain their biodiversity because they are thinly populated. Mountains and deserts attract few people because they are generally inhospitable. Other areas are devoid of dense populations because they are dangerous border regions, like the demilitarized zone between North and South Korea. Areas owned by wealthy individuals and run as private estates, usually for hunting, often benefit from limited ecological impacts. A case in point is the old-growth forests maintained by royal families over hundreds of years in Europe.

• *Government-owned land*. Large tracts of government-owned land in many countries are designated for multiple uses. In the past, multiple uses have included logging, mining, grazing, wildlife management, and recreation. However, multiple-use lands are increasingly managed to protect species, communities, and ecosystems. In the south-eastern USA, longleaf pine (*Pinus palustris*) forest was traditionally managed for timber production. It is now managed to protect the endangered red-cockaded woodpecker (*P. borealis*) and the groves of old trees that the birds need for breeding.

Other conservation efforts that tackle populations, comprising communities across entire ecoregions, are also being explored. Options along these lines have been assessed and found feasible for conserving large and medium-sized mammals in the Cape Floristic Region, South Africa (Kerley *et al.* 2002). *Climate Change-integrated Conservation Strategies (CCSs)* take a similar approach, looking at the need to expand protected areas, to manage the surrounding matrix of unprotected areas, and attempt to coordinate efforts on a regional basis (e.g. Hannah *et al.* 2002a, b).

Ecosystem management

This is a philosophy and practice that seeks to reconcile human, economic, and ecosystem needs. It arose in the late 1980s and is advocated by many scientists and other people interested in the environment (e.g. Agee and Johnson 1988). Its ultimate aim is to enhance and ensure the diversity of species, communities, ecosystems, and landscapes. It is a fresh and emerging model of resource management. It covers a spectrum of approaches (see Yaffee 1999). A mild, technocentric version simply extends multiple use and sustained yield policies, and prosecutes a stewardship approach, in which the ecosystem is seen merely as a human life-support system (e.g. Kessler *et al.* 1992). In this view, public demands for habitat protection, recreation, and wildlife uses are simply seen as constraints to maximizing resource output (Cortner and Moote 1994). A more radical, ecocentric approach is to accept nature on its own terms, even where doing so means controlling incompatible human uses (e.g. Keiter and Boyce 1991). This extreme form of ecosystem management reflects a willingness to place environmental values, such as biodiversity and animal rights, and social and cultural values, such as the upholding of human rights, on an equal footing (Figure 18.4).

People exert a profound influence on ecological patterns and processes, and in turn ecological patterns and processes affect people. The connection between technology and the environment is well studied; the connection between social systems

and the environment is not. However, policies are tending to move away from the administrator-as-neutral-expert approach to policies that engender public deliberation and the discovery of shared values. Naturally, such extension of ecological matters to the social and political arena presents difficulties, though these may not be insuperable (e.g. Irland 1994). Ecosystem management accepts that human values must play a leading role in policy decision-making. Conservation strategies must take account of human needs and aspirations; and they must integrate ecosystem, economic, and social needs. The key players in ecosystem management are scientists, policy makers, managers, and the public. Indeed, the public, many of whom have a keen interest in environmental matters, are becoming more involved in ecosystem management as professionals recognize the legitimacy of claims that various groups make on natural resources. In Jervis Bay, Australia, the marine ecosystem is used by many existing and proposed conflicting interests (national park, tourism, urbanization, military training) (Ward and Jacoby 1992). Similarly, in the forests of the southwestern United States, ecosystem, economic, and social needs are considered in policy decision-making concerning ecology-based, multiple-use forest management (Kaufmann *et al.* 1994) (Figure 18.5).

RESTORING ECOSYSTEMS

What is restoration ecology?

Natural disturbing agencies and human actions damage many ecosystems. Vegetation succession often brings naturally damaged ecosystems back to their former state. *Restoration ecologists* seek to re-establish an indigenous, historical ecosystem at a site damaged or degraded by human activities. Restoration ecology, the science that deals with such restorations, originated in older technologies that sought to restore ecosystem functions known to have economic value. The older technological

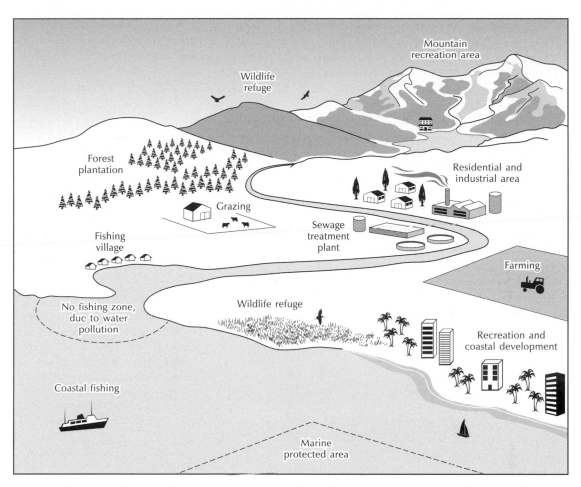

Figure 18.4 Interested parties that use an ecosystem and should have a say in its multipurpose management.
Source: Adapted from Miller (1996)

applications included wetland replication to prevent flooding, mine-site reclamation to prevent soil erosion, range management to ensure the productivity of grasses, and forest management for timber and amenity (see Gilbert and Anderson 1998). Often, these applications created simplified communities, or even communities that were not self-sustaining. The advent of biodiversity as a guiding principle in the ecological sciences shifted the emphasis in restoration ecology to the reestablishment of species assemblages and entire ecosystems. In addition, legal

requirements now often require businesses to restore the habitats that they have damaged. Of course, some ecosystems are 'too far gone' to be restored. Three situations severely curtail restoration efforts:

1 The source of damage still affects the ecosystem. This occurs in the degraded savannah woodland of Costa Rica where overgrazing by cattle is an ever-present ecosystem factor. Restoration of the savannahs is impossible while the cattle overgraze the land.

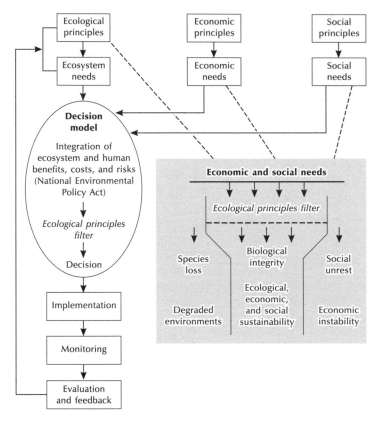

Figure 18.5 The integration of ecological, economic, and social needs in a decision-analysis model. Economic and social needs are tested against an 'ecological filter', which is shown in the shaded box. The aim is to determine economic and social actions that will produce the most desirable bridge between biological integrity and ecological, economic, and social sustainability. Bowing fully to economic and social needs would lead to species loss and environmental degradation. Bowing fully to ecological needs would lead to social unrest and economic instability. A compromise position allows the maintenance of biological integration while catering for economic and social needs. The result decision model leads to the implementation of an environmental policy. The effects of the policy are carefully monitored and evaluated. If the policy should fail to work as desired, then amendments can be made and the process started anew, until a satisfactory outcome is achieved.

Source: Adapted from Kaufmann *et al.* (1994)

2 The original species have disappeared over a large area. The establishment of cereal crops over the mid-western USA led to the elimination of prairie species over huge areas. Patches of land left uncultivated in the prairies do not revert to the original grassland community because the sources of seeds and animals, vital for recolonization to proceed, are too remote.

3 The original species cannot live at the site because the environmental conditions have changed and are no longer conducive to the original members of the community. Sites of old mines usually suffer from contamination by heavy metals and soils with poor structure and low nutrient status. Restoration of the original communities at such sites may take decades or even centuries.

Ways of restoring ecosystems

Approaches to restoring ecosystems are fourfold: no action, replacement, rehabilitation, and restoration (Figure 18.6). It is prudent to take *no action* where restoration is prohibitively costly, where earlier attempts at restoration failed, or where the ecosystem will recover unaided. The last case applies to abandoned fields in the eastern USA where forest usually returns within a few decades after abandonment (see p. 260). In some instances, a new productive ecosystem *replaces* a degraded ecosystem. Although the new ecosystem is not the same as the original ecosystem, it at least furnishes a biological community on a site and restores such important ecosystem functions as soil retention and flood control. An example is a productive pasture replacing a highly degraded forest. Like replacement, *rehabilitation* restores basic ecosystem functions, and additionally it brings back at least some of the original dominant species. An example is a tree plantation replacing a degraded forest. The practice of *restoration* reintroduces the original species, and particularly the original plant species, to recreate the original ecosystem. To achieve this goal, restoration ecologists must identify and reduce the degrading factors, and they must reinstate natural ecosystem processes to help repair the system.

Restored ecosystems

Examples of restored bottomland hardwood forests, tropical forests, and northern wetlands

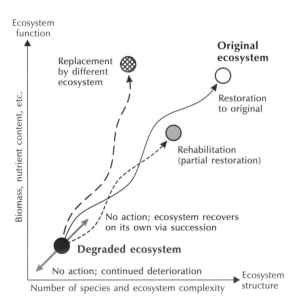

Figure 18.6 The repair of degraded ecosystems through replacement, rehabilitation, and restoration.
Source: Adapted from Bradshaw (1990)

will serve to show the scope of restoration ecology. However, before mentioning the success stories, it is fair to point out that not all restoration efforts succeed. Attempts to restore prairie grasslands in Illinois, USA, failed to recreate patterns of species found in remnant prairie communities (Sluis 2002).

Bottomland hardwood forests in Mississippi and Louisiana

Vast areas of bottomland hardwood forest once occupied the Mississippi Alluvial Valley, southern USA. The conversion of forest to agricultural land was made possible by flood control projects and led to three-quarters of the hardwood forests being lost. However, erosion and deposition cycles in this section of the Mississippi also destroy forest, instigating primary succession on consolidated river deposits. The first plants to colonize are eastern cottonwood (*Populus deltoides*) or black willow (*Salix nigra*). These are soon replaced by species that create riverfront forests, including American sycamore (*Planatus occidentalis*), sweet pecan (*Carya illinoensis*), green ash (*Fraxinus pennsylvanica*), American elm (*Ulmus americana*), and sugarberry (*Celtis laevigata*). In turn, wet oak forests, dominated by sweetgum (*Liquidambar styraciflua*) and oaks (*Quercus* spp.), characteristically replace the riverfront forests.

A study looked at 24 reforested stands, 2–10 years old, lying northeast of Tallulah, Louisiana (Twedt *et al.* 2002). These stands were planted on agricultural fields in Issaquena County, Mississippi, and Madison Parish, Louisiana. The trees planted were either predominantly oaks or eastern cottonwood. The focus of the study was the recolonization of the forest stands by birds over the 8-year period. In all, 48 species of bird were observed to hold territories, or parts of territories, in the reforested stands. Cottonwoods grew upwards fast (2–3 m per year) and created stands with a forest structure that supported a higher species richness of breeding birds than did the stands formed of the slower-growing oaks.

Cottonwood reforestation attracted red-winged blackbird (*Agelaius phoeniceus*) and northern bob-white (*Colinus virginianus*) when less than 4 years old, and 14 species of shrub-scrub birds, such as indigo bunting (*Passerina cyanea*), and early-successional forest birds, such as the warbling vireo (*Vireo gilvus*) when 5–9 years old. The oak stands up to 10 years old, on the other hand, attracted such grassland species as dickcissel (*Spiza americana*) and eastern meadowlark (*Sturnella magna*). Brown-headed cowbirds (*Molothrus ater*) were more common in old cottonwood stands, where 23 per cent of birds' nests were subject to parasitism by this species; the figures for the young (less than 4 years old) cottonwood stands and the oak stands were less than 3 per cent.

Given the benefits to the breeding birds, the recommendation was that the reforestation of bottomland hardwoods should involve planting a mixture of species with a large proportion of fast-growing early successional species such a cotton-woods. Subsequent silvicultural manipulation may then achieve the desired tree composition. Only where oak production is the key concern should reforestation with predominantly oak species be entertained.

Rehabilitating deforested areas in the Amazon Basin

Deforestation in the Amazon Basin continues apace. Studies of rehabilitation of the deforested areas are challenging, since there is little auto-ecological information about the species concerned. One study tested direct sowing as a rehabilitation technique (Camargo *et al.* 2002). Sites with differing degrees of disturbance were chosen: bare soil, pasture, and secondary and mature forests. Eleven native tree species were sown at each site. Germination and seedling survival was monitored throughout the following year. Germination varied with study site and species. All seedlings emerged, but the emergence rate varied between sites. Degree of site distur-bance had a positive effect on the seedling emer-

gence, the rates being lower on forest (12 per cent) and secondary forest sites (15 per cent) and higher on the pasture (23 per cent) and bare soils sites (33 per cent). Only seedlings of marindiba (*Buchenavia grandis*), piquiá (*Caryocar villosum*), periquiteira (*Cochlospermum orinoccense*), and marupá (*Simaruba amara*) emerged at all sites. In disturbed sites, the only survivor after one year with 45 per cent emergence was piquiá. Paricarana or fava arara tucupi (*Parkia multijuga*) also performed well. No pioneer species' seeds remained after a year. Larger seeds tended to survive better then smaller seeds. It was found that non-pioneer species with large seeds were best suited to direct sowing than species with small seeds. The recommendation was that a combination of direct sowing and seedling planting of piquiá and paricarana is the most suitable means for accelerating the rehabilitation of degraded areas.

Restored wetlands on Prince Edward Island, Canada

Since 1990, over 100 small wetlands on Prince Edward Island have been restored by dredging accumulated sediment and organic material from erosion to emulate pre-disturbance conditions (that is, open water and an extended flooded period) (Plate 18.1). A study set out to examine the use of 22 restored (i.e. dredged) wetlands by waterfowl pairs and broods compared with 24 reference wetlands (C. E. Stevens *et al.* 2003). The reference wetlands were mostly very shallow (20 cm) and dominated by emergent plants, including cattail (*Typha latifolia*), bulrush (*Scirpus* spp.), rushes (*Juncus* spp.), and sedges (*Carex* spp.). Ring-necked duck (*Aythya collaris*), gadwall (*Anas strepera*), green-winged teal (*A. crecca*), and American black duck (*A. rubripes*) all had more pairs and broods on the restored wetland sites than on the reference sites. Within the restored wetland sites, higher amounts of wetland and cattail (*Typha latifolia*) and easier access to freshwater rivers produced higher densities of

(a)

(b)

(c)

Plate 18.1 Restored wetlands on Prince Edward Island, Canada. (a) A newly restored wetland with high amounts of open water and unvegetated spoil following mechanical excavation of accumulated sediment and organic debris in the summer of 1993. (b) A restored wetland 3 years after excavation of accumulated sediment and organic debris to create open water. (c) A restored wetland 6 years after excavation of accumulated sediment and organic debris with near equal proportions of emergent vegetation cover (*Typha* spp.) and open water.

Photographs by Cam Stevens.

waterfowl pairs and greater species richness. Cattail cover also attracted more breeding pairs and more broods. Green-winged teal pairs occurred more frequently in restored wetlands with larger amounts of open water and with deeper water, and they seemed to use the sites for brood rearing. American black duck pairs used 86 per cent of the restored wetlands, probably as a stopover during overland or stream movements. Overall, the wetlands restoration appears to have been successful in increasing populations of green-winged teal and American black ducks. It had also benefited the green frog (*Rana pipiens*) and belted kingfisher (*Ceryle alcyon*) (C. E. Stevens *et al.* 2002).

SUMMARY

The conservation of communities and ecosystems is vital for the conservation of species. Worldwide, protected areas cover little more than a twentieth of the land area. The IUCN recognizes six categories of protected areas: strict nature reserves and wilderness areas, national parks, national monuments and landmarks, managed wildlife sanctuaries and nature reserves, protected landscapes and seascapes, and managed resource protected areas. In addition, there are several hundred biosphere reserves. Landscape ecology, and especially the theory of island biogeography, has informed the design and management of protected areas. Six themes in landscape ecology have proved useful – reserve number and size, reserve shape, reserve edges, conservation corridors, reserve functioning, and regional reserve settings. Over nine-tenths of the land area is unprotected. The management of species, communities, and ecosystems in unprotected areas is essential for the well-being of their counterparts in protected areas. Restoration ecology deals with the repair of damaged ecosystems. Restoration work may involve no action, replacement, rehabilitation, and restoration.

ESSAY QUESTIONS

1 To what extent has the theory of island biogeography aided the design and running of protected areas?

2 Why is the management of unprotected areas imperative in an effort to sustain biodiversity?

3 How successful are the practical applications of restoration ecology?

FURTHER READING

Bennett, A. F. (1999) *Linkages in the Landscape: The Role of Corridors and Connectivity in Wildlife Conservation*. Gland, Switzerland: IUCN (International Union for the Conservation of Nature).

An excellent source of information and ideas.

Gilbert, O. L. and Anderson, P. (1998) *Habitat Creation and Repair*. Oxford: Oxford University Press.

Not a beginner's book, but covers all aspects of restoration ecology.

Hambler, C. (2003) *Conservation*. Cambridge: Cambridge University Press.

A good place to start learning about conservation.

Jordan, W. R. III, Gilpin, M. E., and Aber, J. D. (eds) (1990) *Restoration Ecology: A Synthetic Approach to Ecological Research*. Cambridge: Cambridge University Press.

A valuable collection of essays.

Primack, R. B. (2000) *A Primer of Conservation Biology*, 2nd edn. Sunderland, MA: Sinauer Associates.

Highly readable and informative.

Shafer, C. L. (1990) *Nature Reserves: Island Theory and Conservation Practice*. Washington, DC, and London: Smithsonian Institution Press.

Good on the applications of island biogeography.

APPENDIX

THE GEOLOGICAL TIMESCALE

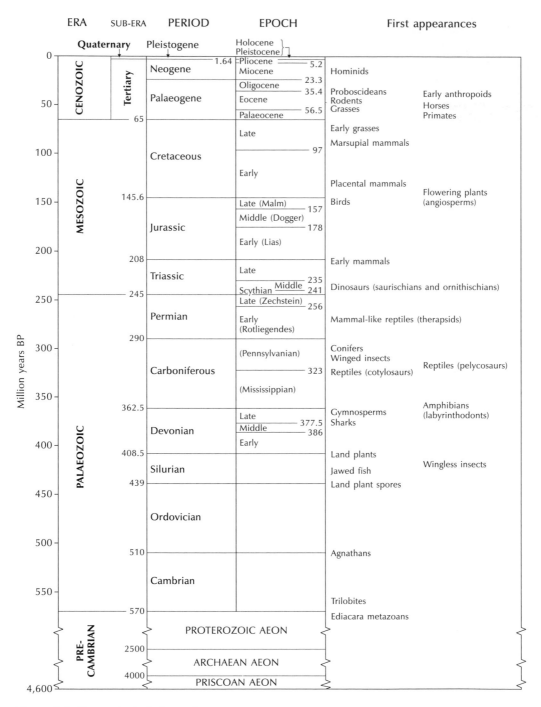

Figure A.1 The geological timescale.
Source: After Huggett (1997)

GLOSSARY

abiotic Characterized by the absence of life; inanimate.

acidophiles Organisms that love acidic conditions.

adaptation The adjustment or the process of adjustment by which characteristics of an entire organism, or some of its structures and functions, become better suited for life in a particular environment.

adaptive radiation The diversification of many species from a single founding species.

aerobic Depending on, or characterized by, the presence of oxygen.

algae Simple, unicellular or filamentous plants.

alkaliphiles (alkalophiles) Organisms that flourish in alkaline environments.

alleles Forms of a gene at a locus.

allelopathic Pertaining to the influence of plants upon each other through toxic products of metabolism – a sort of phytochemical warfare.

allogenic Originating from outside a community or ecosystem.

allopatric speciation The creation of new species through geographical isolation.

anaerobic Depending on, or characterized by, the absence of oxygen.

angiosperms Plants whose seeds develop in an ovary; flowering plants.

annuals Plants that germinate, grow, seed, and die in one season.

anthrax An infectious and often fatal disease caused by the bacterium *Bacillus anthracis*, common in cattle and sheep.

apical buds Buds located at the tips of shoots.

aquifers Subsurface geological formations containing water.

archipelagos Groups of islands, typically large groups.

aridity The state or degree of dryness.

atmosphere The gaseous envelope of the Earth, retained by the Earth's gravitational field.

autogenic Originating from inside a community or ecosystem.

autotrophs Organisms – green plants and some microorganisms – capable of making their own food from inorganic materials.

available moisture Precipitation less evaporation.

azonal soils Soils in which erosion and deposition dominate soil processes, as in soils formed in river alluvium and sand dunes.

bacteria Microorganisms, usually single-celled, that exist as free-living decomposers or parasites.

barriers Physical deterrents to the dispersal of organisms.

basal metabolic rate The rate of energy consumption by an organism at rest.

basophiles (basiphiles) Organisms that flourish in alkaline environments.

behaviour The reaction of organisms to given circumstances.

benthic Pertaining to the bottom of a water body.

biocides Poisons or other substances used to kill pests (and inadvertently other organisms).

bioclimates The climatic conditions that affect living things.

biocontrol The use of natural ecological interrelationships to control pest organisms.

biodiversity The diversity of species, genetic information, and habitats.

biogeochemical cycles The cycling of a mineral or organic chemical constituent through the biosphere; for example, the carbon cycle.

biological control (biocontrol) The use of natural ecological interrelationships to control pest organisms.

biomagnification The concentration of certain substances moving up a food chain.

biomantles The upper portions of soil that are worked by organisms.

biomass The mass of living material in a specified group of animals, or plants, or in a community, or in a unit area; usually expressed as a dry weight.

biomes Communities of animals and plants occupying a climatically uniform area on a continental scale.

biosphere All living beings.

biota All the animals (fauna) and plants (flora) living in an area or region.

biotic Pertaining to life; animate.

biotic potential The maximum rate of population growth that would occur in the absence of environmental constraints.

boreal forest A plant formation-type associated with cold-temperate climates (cool summers and long winters); also called taiga and coniferous evergreen forest. Spruces, firs, larches, and pines are the dominant plants.

bottlenecks Decreases in genetic diversity following substantial decreases in population size.

browsers Organisms that feed mainly on leaves and young shoots, especially those of shrubs and trees.

bryophytes Simple land plants with stems and leaves but no true roots or vascular tissue; for example, mosses, hornworts, and liverworts.

cacti (singular **cactus**) New World plants belonging to the family Cactaceae with thick, fleshy, and often prickly stems.

calcicoles (calciphiles) Plants that love soils rich in calcium.

calcifuges (calciphobes) Plants that hate soils rich in calcium.

capillary pressure The force of water in a capillary tube.

carrion Dead and decaying flesh.

carrying capacity The maximum population that an environment can support without environmental degradation occurring.

catenae The sequences of soils and vegetation found along a hillslope, from summit to valley bottom.

caviomorph rodents A suborder of the Rodentia.

Cenozoic era A slice of geological time spanning 65 million ago to the present; the youngest unit of the geological eras.

chasmophytes Plants that live in crevices.

chomophytes Plants that live on ledges.

chromosomes Threadlike structures of DNA that carry genes.

chronosequences Time sequences of vegetation or soils constructed by using sites of different age.

circulatory system The system of vessels through which blood is pumped by the heart.

clades Sets of species lineages arising from common ancestors.

cohorts Sets of individuals in a population born during the same period of time, e.g. during a particular year.

colonization The expansion of a species by establishment in a new area.

commensalism An interaction between two species in which one species (the commensal) benefits and the other species (the host) is unharmed.

community assembly The coming together of species to form a community.

competition An interaction between two species trying to share the same resource.

competitive exclusion principle The rule that two species with identical ecological requirements cannot coexist at the same place.

congeners Organisms belonging to the same genus as another or others.

coniferous Pertaining to trees of the Order Coniferales, commonly evergreen and bearing cones.

consumers Organisms that depend upon other organisms for their food.

consumption Community respiration.

continental drift The differential movement of continents caused by plate tectonic processes.

convergent evolution The evolution of similar forms from different common ancestors in different geographical regions, owing to the same selective pressures in the environment.

corollas The inner envelopes of flowers; they are made of fused or separate petals.

corridors Dispersal routes offering little or no resistance to migrating organisms.

Cretaceous period A slice of geological time spanning 144 million to 65 million years ago; the youngest unit of the Mesozoic era.

culling The act of removing or killing 'surplus' animals in a herd or flock.

cursorial Adapted for running.

cuticles Protective layers of cutin (a wax-like, water-repellent material) covering the epidermis (outermost layer of cells) of plants.

cycads Gymnosperms of the family Cycadaceae looking like palm trees but topped with compound, fern-like leaves.

deciduous Pertaining to perennial plants that lose their leaves during the adverse (cold or dry) season.

decomposers Organisms that help to break down organic matter.

demes Groups of interbreeding individuals; local populations.

density-dependent factors Causes of fecundity and mortality that become more effective as population density rises.

density-independent factors Causes of fecundity and mortality that act independently of population density; natural disasters are an example.

detritivores Organisms that comminute dead organic matter.

detritus Disintegrated matter.

diatoms Microscopic, unicellular, marine (planktonic) algae with skeletons composed of hydrous opaline silica.

dicotyledonous plant species Plants belonging to the Dicotyledoneae, one of the two major divisions of flowering plants. They are characterized by a pair of embryonic seed leaves that appear at germination.

diploids Cells or organisms that contain both sets of chromosome pairs.

disharmonious communities Communities of animals and plants adapted to a climate that has no modern counterpart.

dispersal The spread of organisms into new areas.

dispersal routes The paths followed by dispersing organisms.

distribution The geographical area occupied by a group of organisms (species, genus, family, etc.); the same as the geographical range.

drought A prolonged period with no rain.

ecological equivalents Species of different ancestry but with the same characteristics living in different biogeographical regions. Also called vicars.

ecosphere The global ecosystem – all life plus its life-support systems (air, water, and soil).

ecosystems Short for ecological systems – groups of organisms together with the physical environment with which they interact.

ecotones Transitional zones between two plant communities.

El Niño The appearance of warm water in the usually cold water regions off the coasts of Peru, Ecuador, and Chile.

endangered Pertaining to species facing extinction.

environments The biological (biotic) and physical (abiotic) surroundings that influence individuals, populations, and communities.

environmental degradation A temporary or permanent reduction in the ability of the environment to support the organisms that live in it.

environmental ethics A philosophy dealing with the ethics of the environment, including animal rights.

environmental factors Any of the biotic or abiotic components in the environment, such as heat, moisture, and nutrient levels.

environmental gradients Continuous changes in an environmental factor from one place to another; for example, a change from dry to wet conditions.

Eocene epoch A slice of geological time spanning 54.9 to 38 million years ago.

epiphytes Plants that grow on another plant or an object, using it for support but not nourishment.

estuarine Of, pertaining to, or found in an estuary.

evaporation The diffusion of water vapour into the atmosphere from sources of water exposed to the air (see evapotranspiration).

evapotranspiration Evaporation plus the water discharged into the atmosphere by plant transpiration.

evergreens Plants with foliage that stays green all year round.

extinction The demise of a species or any other taxon.

extirpation The local extinction of a species or any other taxon.

extremophiles Organisms that relish ultra-extreme conditions.

fauna All the animals living in an area or region.

fecundity Reproductive potential as measured by the number of eggs, sperms, or asexual structures produced.

filter routes Paths followed by dispersing organisms that, owing to the environment or geography, do not permit all species to pass.

flagship species Charismatic species that serve as a focal point for conservation campaigns.

flora All the plants living in an area or region.

food chains Simple expressions of feeding relationships in a community, starting with plants and ending with top carnivores.

food webs Networks of feeding relationships within a community.

frugivory Fruit eating.

Gaia hypothesis The idea that the chemical and physical conditions of the surface of the Earth, the atmosphere, and the oceans are actively controlled by life, for life.

gaps Openings in vegetation caused by disturbances both large and small.

genes Specific pieces of genetic information; the basic units of heredity carried and transmitted by the chromosomes.

genetic variability A measure of the amount of genetic diversity within a gene pool.

genomes The total genetic information contained in individuals (or organelles or cells).

genotypes The genetic make-up of individuals – the sums of their genes.

genera (singular **genus**) Taxonomic groups of lower rank than a family that consist of closely related species or, in extreme cases, of only one species.

geodiversity The diversity of the physical environment.

geographical ranges The areas occupied by a group of organisms (species, genus, family, etc.); the same as the distribution.

geosphere The solid Earth – core, mantle, and crust.

geothermal springs Springs of water made hot by the Earth's internal heat.

germination The time of sprouting in plants.

global extinction The total loss of a particular gene pool.

Gondwana A hypothetical, late Palaeozoic supercontinent lying chiefly in the southern hemisphere and comprising large parts of South America, Africa, India, Antarctica, and Australia.

grazers Organisms that feed on growing grasses and herbs.

gross primary productivity Productivity before respiration losses are accounted for.

guilds Groups of species living in the same community that bear similar forms, live in similar habitats, and have similar resource requirements.

habitats The places where organisms live.

habitat selection The selection of a particular habitat by an individual.

halophiles Plants (and other organisms) that live in salty conditions.

halophytes Plants that live in salty conditions.

heathland Tracts of open and uncultivated land supporting shrubby plants, especially the heathers (*Erica* spp).

heliophytes Plants growing best in full sunlight.

heliotropism The growth of a plant towards or away from sunlight.

helophytes Marsh plants.

heterogeneous Pertaining to an environment comprising a mosaic of dissimilar spatial elements; a non-uniform environment.

heterotrophs Organisms that obtain nourishment from the tissues of other organisms.

heterozygous Referring to a locus associated with more than one allele.

homeostasis A steady-state with inputs counterbalancing outputs.

homeotherms Animals that regulate their body temperature by mechanisms within their own bodies; endotherms or 'warm-blooded' animals (see poikilotherms).

homogeneous Pertaining to an environment comprising similar spatial elements; a uniform environment.

homozygous Referring to a locus associated with just one allele.

humification The process of humus formation.

hummock–hollow cycles Peatland cycles involving the infilling of a wet hollow by bog mosses until a hummock is formed and colonized by heather. The heather eventually degenerates and the process starts again.

humus An amorphous, colloidal substance produced by soil microorganisms transforming plant litter.

hurricanes Violent and sometimes devastating tropical cyclones.

hybridization Sexual reproduction between two different species.

hydrophytes Plants that grow wholly or partly submerged in water.

hydrosphere All the waters of the Earth.

hyperparasites Parasites that live on other parasites.

hyperthermophiles Organisms that thrive in ultra-hot conditions.

ice age A time when ice forms broad sheets in middle and high latitudes (often in conjunction with the widespread occurrence of sea ice and permafrost), and mountain glaciers form at all latitudes.

Ice Age An old term for the Quaternary glacial–interglacial sequence.

inbreeding Reproduction within a small population of related individuals, which often reduces fitness.

interspecific competition Competition among species.

intestinal endosymbionts Symbiotic organisms living in another organism's gut.

intraspecific competition Competition within a species.

intrazonal soils Soils whose formation is dominated by local factors of relief or substrate; an example is soils formed on limestone.

introduced species or **introductions** Species released outside their natural geographical range.

isolines Lines on a map joining points of equal value.

Jurassic period A slice of geological time spanning 213 million to 144 million years ago; the middle unit of the Mesozoic era.

keystone species Species that play a central role in a community or ecosystem, their removal having far-reaching effects.

land bridges Dry land connections between two landmasses that were previously separated by the sea.

Laurasia The ancient northern hemisphere landmass that broke away from Gondwana about 180 million year sago and subsequently split into North America, Greenland, Europe, and northern Asia.

leaf–area index The ratio of total leaf surface to ground surface. For example, a leaf area index of 2 would mean that, if you were to clip all the leaves hanging over 1 m^2 of ground, you would have 2 m^2 of leaf surface.

legumes Plants of the family Leguminosae, characteristically bearing pods that split in two to reveal seeds attached to one of the halves.

lichens Plants consisting of a fungus and an alga.

life-forms The characteristics of mature animals or plants.

life tables Tabulations of the complete mortality schedule of a population.

littorals Shallow water zones.

local extinction (extirpation) The loss of a species from a particular area.

loci (or singular **locus**) The sites or locations of genes on chromosomes.

macronutrients Chemical elements needed in large quantities by living things.

maquis A shrubby vegetation of evergreen small trees and bushes.

mass extinction Extinction episodes in which large numbers of species disappear.

meadows Grassland mown for hay.

mechanists Proponents of the view that all biological and ecological phenomena may be explained by the interaction of physical entities.

megafauna Large mammals, such as mammoths and sloths, contrasted with small mammals, such as rodents and insectivores.

megaherbivores Large browsers and grazers, such as elephants and rhinoceroses.

melanism A darkening of colour owing to increased amounts of black pigments.

mesophytes Plants flourishing under mesic (not too dry and not too wet) conditions.

Mesozoic era A slice of geological time spanning 248 million to 65 million years ago; comprises the Triassic, Jurassic, and Cretaceous periods.

metabolic processes The many physical and chemical process involved in the maintenance of life.

micronutrients Chemical elements required in small quantities by at least some living things.

microorganisms Organisms not visible to the unaided naked eye.

mineralization The release of nutrients from dead organic matter.

Miocene epoch A slice of geological time spanning 25 million to 5 million years ago.

monophyletic groups Sets of species descended from common ancestors.

mutations Changes in genes or chromosomes creating a genotype different from the parent.

mutualism An obligatory interaction between two species where both benefit.

natural selection The process by which the environmental factors sort and sift genetic variability and drive microevolution.

net primary productivity Gross primary productivity less respiration.

neutralism An interaction in which neither species is harmed or benefited.

neutrophiles Plants that like neutral conditions – not too acidic and not too alkaline.

nitrogen-fixing The conversion of atmospheric nitrogen into ammonium compounds by some bacteria.

nitrophiles Plants that live in nitrogen-rich environments.

nutrients Chemical elements required by at least some living things.

old-growth forests Virgin forests that have never been cut, or forests that have been undisturbed for a long time.

Oligocene epoch A slice of geological time spanning 38 million to 24.6 million years ago; the youngest unit of the Palaeogene period.

organelles Structures within a cell that specialize in particular functions; examples are mitochondria and chloroplasts.

orobiomes Biomes associated with the altitudinal climatic zones on mountains.

overwintering Spending the winter in a particular place.

Palaeocene epoch A slice of geological time spanning 65 million to 54.9 million years ago; the oldest unit of the Palaeogene period.

Palaeozoic era A slice of geological time spanning 590 million to 248 million years ago; comprises the Cambrian, Ordovician, Silurian, Devonian, Carboniferous, and Permian periods.

Pangaea The Triassic supercontinent comprising all the present continents.

parallel evolution (parallelism) The evolution of similar forms from the same common ancestor in different geographical regions, owing to the same selective pressures in the environment.

parapatric speciation The evolution of a new species in populations living next to each other.

pathogens Agents, such as bacteria or fungi, that cause a disease.

pedobiomes Areas of characteristic vegetation produced by distinctive soils.

pedosphere All the soils of the Earth.

peripatric speciation The evolution of a new species on the edge of a species range.

Permian period A slice of geological time spanning 286 million to 248 million years ago.

petrophytes Plants that grow on bare rock.

phenotypes The observed characteristics of species, resulting from the expression of the genotypes interacting with the environment.

phosphatiphiles Plants that live in phosphorus-rich environments.

photoperiod The duration of darkness and light.

photosynthesis The synthesis of carbohydrates from carbon dioxide and water by chlorophyll, using light as an energy source. Oxygen is a by-product.

phylogeography The reconstruction of biogeographical history using the modern geographical distribution of genetic characteristics.

phytomass The mass of living plant material in a community, or in a unit area; usually expressed as a dry weight.

phytoplankton Plant plankton: the plant community in marine and freshwater systems that floats free in the water and contains many species of algae and diatoms.

phytotoxic Poisonous to plants.

plant formations The vegetational equivalent of biomes, that is, communities of plants of like physiognomy (life-form) occupying climatically uniform areas on a continental scale.

Pleistocene epoch A slice of geological time spanning 2 million to 10,000 years ago; the older unit of the Pleistogene period (or Quaternary sub-era).

Pliocene epoch A slice of geological time spanning 5.2 million to 1.64 million years ago; the younger unit of the Neogene period.

poikilotherms Organisms whose body temperature is determined by the ambient temperature and who can control their body temperature only by taking advantage of sun and shade to heat up or to cool down; 'cold-blooded' animals (see homeotherms).

polymorphism Genetically controlled variation within a species population.

polyploids Organisms having twice, or more than twice, the number of chromosomes as either parent.

prairies Extensive areas of natural, dry grassland. Equivalent to steppes in Eurasia.

producers Organisms that obtain their energy from the physical environment.

profundal Pertaining to deep water zones.

propagules The stages in an organism's life cycle (a plant seed for instance), parts of an organism (such as plant cuttings), or groups of organisms (such as a male and a female rat of mating age) that can form a new colony.

protocooperation A non-obligatory interaction between two species where both benefit.

pscychrophiles Organisms that flourish at low temperatures.

races Animal or plant populations that differ from other populations of the same species in one or more hereditary characters.

radioisotopes Forms of a chemical element that undergo spontaneous radioactive decay.

rain shadows Regions with relatively low rainfall that are sheltered from rain-bearing winds by high ground.

rescue effect The saving of a small population from extinction's brink by the arrival of immigrants of the same species.

respiration A complex series of chemical reactions in all organisms by which energy is made available for use. The end products are carbon dioxide, water, and energy.

rhizomes Root-like stems, usually horizontal, growing underground with roots emerging from their lower surface and leaves or shoots from their upper surface.

salinity Saltiness.

saltatorial Adapted for hopping or jumping.

saprophages Organisms that feed on dead or decaying organic matter.

saprophytes Organisms, especially fungi and bacteria, that feed on dead or decaying organic matter.

saprovores Animals, mainly invertebrates, that feed on dead or decaying organic matter.

savannah Tropical grassland.

saxicolous Growing on or living among rocks.

scavengers Organisms feeding on dead animal flesh or other decaying organic material.

sciophytes Plants growing best in shade.

seasonality The degree of climatic contrast between summer and winter.

sessile Permanently attached or fixed; not free-moving.

soil Rock at, or near, the land surface that has been transformed by the biosphere.

solar radiation The total electromagnetic radiation emitted by the Sun. Electromagnetic radiation is energy propagated through space or through a material as an interaction between electric and magnetic waves; occurs at frequencies ranging from short wavelength, high frequency cosmic rays, through medium wavelength, medium frequency visible light, to long wavelength, low frequency radio waves.

speciation The process of species multiplication; the development of two or more genetically distinguishable species from a common ancestor.

species Reproductively isolated collections of interbreeding populations (demes).

species richness The number of species living in a particular area.

stasipatric speciation The evolution of a new species within a species range.

steppes Extensive areas of natural, dry grassland. Equivalent to prairies in North American terminology.

stomata (singular **stoma**) Epidermal pores in a leaf or stem through which air and water vapour pass.

sweepstakes routes Paths along which organisms may disperse, but very few will do so owing to the difficulties involved.

symbiotic Pertaining to two or more organisms of different species living together. In a narrow sense equivalent to mutualism.

sympatric speciation The evolution of a new species within the same area as the parent species.

taxa (singular **taxon**) Populations or groups of populations (taxonomic groups) that are distinct enough to be given distinguishing names and to be ranked in definite categories.

territories Areas defended by animals against other members of their species (and occasionally other species).

Tertiary period A slice of geological time spanning 65 million to 2 million years ago and the youngest unit of the Cenozoic era; now designated a sub-era.

theory of island biogeography A model that explains the species diversity of islands chiefly as a function of distance from the mainland and island area.

thermophiles Organisms that love hot conditions.

toposequences The sequences of soils and vegetation found along hillslopes, from summit to valley bottom.

trace elements Chemical elements used by an organism in minute quantities and vital to its physiology.

treelines The altitudinal and latitudinal limits of tree growth.

tree-throw The toppling of trees by strong winds.

Triassic period A slice of geological time spanning 248 million to 213 million years ago; the oldest unit of the Mesozoic era.

trophic Of, or pertaining to, nourishment or feeding.

tsunamis The scientific (and Japanese) name for tidal waves.

ungulates Hoofed mammals.

vascular plants Plants containing vascular tissue, the conducting system which enables water and minerals to move through them.

vegetation Collective plant life; the plants of an area.

vernalization The exposure of some plants (or their seeds) to a period of cold that allows them to flower or to flower earlier than usual.

vicariance The splitting apart of a species distribution by a geological or environmental event.

vicars Species of different ancestry but with the same characteristics living in different biogeographical regions. Also called ecological equivalents.

vitalists Proponents of the idea that some kind of living force resides in organisms.

volatilization The process of converting to vapour.

wetlands All the places in the world that are wet for at least part of the year and support characteristic soils and vegetation; they include marshes, swamps, and bogs.

xerophytes Plants adapted for life in a dry climate.

zonobiomes The nine, climatically defined, major community units of the Earth.

zoomass The mass of living animal material in a community, or in a unit area; usually expressed as a dry weight.

zooplankton Animal plankton; aquatic invertebrates living in sunlit waters of the hydrosphere.

REFERENCES

Ables, E. D. (1969) Home range studies of red foxes (*Vulpes vulpes*). *Journal of Mammalogy* 50, 108–20.

Agee, J. K. and Johnson, D. R. (1988) *Ecosystem Management for Parks and Wilderness*. Seattle, WA: University of Washington Press.

Ahlgren, I. F. and Ahlgren, C. E. (1960) Ecological effects of forest fires. *The Botanical Review* 26, 483–533.

Akçakaya, H. R., Burgman, M. A., and Ginzburg, L. R. (1999) *Applied Population Ecology: Principles and Computer Exercises Using RAMAS(r)EcoLab.*, 2nd edn. Sunderland, MA: Sinauer Associates.

Aldrich, J. W. and James, F. C. (1991) Ecogeographic variation in the American robin, *Turdus migratorius*. *The Auk* 108, 230–49.

Allainé, D., Graziani, L., and Coulon, J. (1998) Postweaning mass gain in juvenile alpine marmots *Marmota marmota*. *Oecologia* 113, 370–6.

Allainé, D., Rodrigue, I., Le Berre, M., and Ramousse, R. (1994) Habitat preferences of alpine marmots, *Marmota marmota*. *Canadian Journal of Zoology* 72, 2193–8.

Allen, J. A. (1877) The influence of physical conditions in the genesis of species. *Radical Review* 1, 108–40.

Alvarez, L. W., Alvarez, W., Asaro, F., and Michel, H. V. (1980) Extraterrestrial cause for the Cretaceous–Tertiary extinction. *Science* 208, 1095–108.

Andersen, S. T. (1966) Interglacial succession and lake development in Denmark. *Palaeobotanist* 15, 117–27.

Anderson, P. and Shimwell, D. W. (1981) *Wild Flowers and Other Plants of the Peak District: An Ecological Study*. Ashbourne, Derbyshire: Moorland Publishing.

Anderson, S. H., Mann, K., and Shugart, H. H. (1977) The effect of transmission-line corridors on bird populations. *American Midland Naturalist* 97, 216–21.

Arnold, E. N. (1994) Investigating the origins of performance advantage: adaptation, exaptation and lineage effects. In P. Eggleton and R. I. Vane-Wright (eds) *Phylogenetics and Ecology*, pp. 124–68. London: Academic Press.

Arnold, G. W., Steven, D. E., Weeldenburg, J. R., and Smith, E. A. (1993) Influences of remnant size, spacing pattern and connectivity on population boundaries and demography in euros *Macropus robustus* living in a fragmented landscape. *Biological Conservation* 64, 219–30.

Arnold, G. W., Weeldenburg, J. R., and Steven, D. E. (1991) Distribution and abundance of two species of kangaroo in remnants of native vegetation in the central wheatbelt of Western Australia and the role of native vegetation along road verges and fencelines as linkages. In D. A. Saunders and R. J. Hobbs (eds) *Nature Conservation 2: The Role of Corridors*, pp. 273–80. Chipping Norton, Australia: Surrey Beatty & Sons.

Arnold, H. R. (1993) *Atlas of Mammals In Britain* (Institute of Terrestrial Ecology Research Publication No. 6). London: Her Majesty's Stationery Office.

Atkinson, I. A. E. and Cameron, E. K. (1993) Human influence on the terrestrial biota and biotic communities of New Zealand. *Trends in Ecology and Evolution* 8, 447–51.

Avise, J. C. (2000) *Phylogeography: The History and Formation of Species*. Cambridge, MA: Harvard University Press.

Bach, R. (1950) Die Standorte jurassischer Buchenwaldgesellschaften mit besonderer Berücksichtigung der Böden (Humuskarbonatböden und Rendzinen). *Berichte der Schweizerischen Botanischen Gesellschaft* 60, 51–162.

Bailey, R. G. (1995) *Description of the Ecoregions of the United States*, 2nd edn, revised and enlarged (Miscellaneous Publication No. 1391). Washington, DC: United States Department of Agriculture, Forest Service.

Bailey, R. G. (1996) *Ecosystem Geography*, with a foreword by Jack Ward Thomas, Chief, USDA Forest Service. New York: Springer.

Baillie, M. G. L. (1994) Dendrochronology raises questions about the nature of the AD 536 dust-veil event. *The Holocene* 4, 212–17.

Baker, W. L. (1989) Landscape ecology and nature reserve design in the Boundary Waters Canoe Area, Minnesota. *Ecology* 70, 23–35.

Baker, W. L. (1992) The landscape ecology of large disturbances in the design and management of nature reserves. *Landscape Ecology* 7, 181–94.

Bakker, R. T. (1986) *The Dinosaur Heresies: A Revolutionary View of Dinosaurs*. Harlow, Essex: Longman Scientific & Technical.

Balling, R. C., Jr, Meyer, G. A., and Wells, S. G. (1992) Climate change in Yellowstone National Park: is the drought-related risk of wildfires increasing? *Climatic Change* 22, 35–45.

Balmford, A., Leader-Williams, N., and Green, M. J. B. (1995) Parks or arks: where to conserve threatened mammals? *Biodiversity and Conservation* 4, 595–607.

Barfield, C. S. and Stimac, J. L. (1980) Pest management: an entomological perspective. *BioScience* 30, 683–88.

Barnosky, A. D. (1985) Taphonomy and herd structure of the extinct Irish elk (*Megaloceras giganteus*). *Science* 228, 340–4.

Barnosky, A. D. (1986) 'Big game' extinction caused by late Pleistocene climatic change: Irish elk (*Megaloceras giganteus*) in Ireland. *Quaternary Research* 25, 128–35.

Barnosky, A. D. (1989) The late Pleistocene event as a paradigm for widespread mammal extinction. In S. K. Donovan (ed.) *Mass Extinctions: Processes and Evidence*, pp. 235–54. London: Belhaven Press.

Barry, J. C. and Flynn, L. L. (1990) Key biostratigraphic events in the Siwalik sequence. In E. H. Lindsay, V. Fahlbusch, and P. Mein (eds) *European Neogene Mammal Chronology*, pp. 557–71. New York: Plenum Press.

Barry, J. C., Johnson, N. M., Raza, S. M., and Jacobs, L. L. (1985) Neogene mammalian faunal change in southern Asia: correlations with climatic, tectonic, and eustatic events. *Geology* 13, 637–40.

Bartholomew, G. A. (1968) Body temperature and energy metabolism. In M. S. Gordon, G. A. Bartholomew, A. D. Grinnell, C. B. Jorgensen, and F. N. White (eds) *Animal Function: Principles and Adaptations*, pp. 290–354. New York: Macmillan.

Bartlein, P. J. and Prentice, I. C. (1989) Orbital variations, climate, and palaeoecology. *Trends in Ecology and Evolution* 4, 195–9.

Bartlett, D. H. (1992) Microbial life at high pressures. *Science Progress* 76, 479–96.

Barton, A. M. (1993) Factors controlling plant distributions: drought, competition, and fire in montane pines in Arizona. *Ecological Monographs* 63, 367–97.

Bascompte, J. and Sole, R. V. (1995) Rethinking complexity: modelling spatiotemporal dynamics in ecology. *Trends in Ecology and Evolution* 10, 361–6.

Bellamy, D. (1976) *Bellamy's Europe*. London: British Broadcasting Corporation.

Bennett, A. F. (1991) Roads, roadsides and wildlife conservation: a review. In D. A. Saunders and R. J. Hobbs (eds) *Nature Conservation 2: The Role of Corridors*, pp. 99–118. Chipping Norton, Australia: Surrey Beatty & Sons.

Bennett, A. F. (1999) *Linkages in the Landscape: The Role of Corridors and Connectivity in Wildlife Conservation*. Gland, Switzerland: IUCN (International Union for the Conservation of Nature).

Benton, M. J. (1985) Mass extinction among non-marine tetrapods. *Nature* 316, 811–14.

Benton, M. J. (1994) Palaeontological data and identifying mass extinctions. *Trends in Ecology and Evolution* 9, 181–5.

Benton, M. J. (2003) *When Life Nearly Died: The Greatest Mass Extinction of All Time*. London: Thames and Hudson.

Bergmann, C. (1847) Über die Verhältnisse der Wärmeökonomie der Thiere zu ihrer Grösse. *Göttinger Studien* 3, 595–708.

Berry, P. M., Dawson, T. P., Harrison, P. A., and Pearson, R. G. (2002) Modelling potential impacts of climate change on the bioclimatic envelope of species in Britain and Ireland. *Global Ecology & Biogeography* 11, 453–62.

Berry, P. M., Dawson, T. P., Harrison, P. A, Pearson, R., and Butt. N. (2003) The sensitivity and vulnerability of terrestrial habitats and species in Britain and Ireland to climate change. *Journal for Nature Conservation* 11, 15–23.

Billings, W. D. (1950) Vegetation and plant growth as affected by chemically altered rocks in the western Great Basin. *Ecology* 31, 62–74.

Billings, W. D. (1990) The mountain forests of North America and their environments. In C. B. Osmond, L. F. Pitelka, and G. M. Hidy (eds) *Plant Biology of the Basin and Range* (Ecological Studies, vol. 80), pp. 47–86. Berlin: Springer.

Birks, H. J. B. (1986) Late-Quaternary biotic changes in terrestrial and aquatic environments, with particular reference to north-west Europe. In B. E. Berglund (ed.) *Handbook of Holocene Palaeoecology and Palaeohydrology*, pp. 3–65. Chichester: John Wiley & Sons.

Birks, J. (1990) Feral mink and nature conservation. *British Wildlife* 1, 313–23.

Bjärvall, A. and Ullström, S. (1986) *The Mammals of Britain and Europe*, with a foreword by Ernest Neal, translated by David Christie. London and Sydney: Croom Helm.

Bloomer, J. P. and Bester, M. N. (1992) Control of feral cats on sub-Antarctic Marion Island, Indian Ocean. *Biological Conservation* 60, 211–19.

Bogaert, J. (2001) Size dependence of interior-to-edge ratios: size predominates shape. *Acta Biotheoretica* 49, 121–3.

Bogaert, J., Rousseau, R., Van Hecke, P., and Impens, I. (2000a) Alternative area–perimeter ratios for measurement of 2D shape compactness of habitats. *Applied Mathematics and Computation* 111, 71–85.

Bogaert, J., Salvador-Van Eysenrode, D., Van Hecke, P., Impens, I., and Ceulemans, R. (2001a) Land-cover change: quantification metrics for perforation using 2-D gap features. *Acta Biotheoretica* 49, 161–9.

Bogaert, J., Salvador-Van Eysenrode, D., Van Hecke, P., and Impens, I. (2001b) Geometrical considerations for evaluation of reserved design. *Web Ecology* 2, 65–70.

Bogaert, J., Salvador-Van Eysenrode, D., Impens, I., and Van Hecke, P. (2001c) The interior-to-edge breakpoint distance as a guideline for nature conservation policy. *Environmental Management* 27, 493–500.

Bogaert, J., Van Hecke, P., Salvador-Van Eysenrode, D., and Impens, I. (2000b) Landscape fragmentation assessment using a single measure. *Wildlife Society Bulletin* 28, 875–81.

Borhidi, A. (1991) *Phytogeography and Vegetation Ecology of Cuba*. Budapest: Akadémiai Kiadó.

Botkin, D. B. and Keller, E. A. (2003) *Environmental Science: Earth as a Living Planet*, 4th edn. New York: John Wiley & Sons.

Botkin, D. B., Woodby, D. A., and Nisbet, R. A. (1991) Kirtland's warbler habitats: a possible early indicator of climatic warming. *Biological Conservation* 56, 63–78.

Boursot, P., Auffray, J. C., Britton-Davidian, J., and Bonhomme, F. (1993) The evolution of the house mouse. *Annual Review of Ecology and Systematics* 24, 119–52.

Box, E. O. and Meentemeyer, V. (1991) Geographic modelling and modern ecology. In G. Esser and D. Overdieck (eds) *Modern Ecology: Basic and Applied Aspects*, pp. 773–804. Amsterdam: Elsevier.

Box, E. O., Crumpacker, D. W., and Hardin, E. D. (1993) A climatic model for location of plant species in Florida, USA. *Journal of Biogeography* 20, 629–44.

Bradshaw, A. D. (1990) The reclamation of derelict land and the ecology of ecosystems. In W. R. Jordan III, M. E. Gilpin, and J. D. Aber (eds) *Restoration Ecology: A Synthetic Approach to Ecological Research*, pp. 53–74. Cambridge: Cambridge University Press.

Brasier, M. D., Corfield, R. M., Derry, L. A., Rozanov, A. Yu., and Zhuravlev, A. Yu. (1994) Multiple $\delta^{13}C$ excursions spanning the Cambrian explosion to the Botomian crisis in Siberia. *Geology* 22, 455–8.

Briggs, J. C. (1995) *Global Biogeography* (Developments in Palaeontology and Stratigraphy No. 14). Amsterdam: Elsevier.

Bright, D. A., Dushenko, W. T., Grundy, S. L., and Reimer, K. J. (1995) Effects of local and distant contaminant sources: polychlorinated biphenyls and other organochlorines in bottom-dwelling animals from an Arctic estuary. *Science of the Total Environment* 160–1, 265–83.

Bright, P. W. (1998) Behaviour of specialist species in habitat corridors: arboreal dormice avoid corridor gaps. *Animal Behaviour* 56, 1485–90.

Brito, D. and de Souza Lima Figueiredo, M. (2003) Minimum viable population and conservation status of the Atlantic Forest spiny rat *Trinomys eliasi*. *Biological Conservation* 122, 153–8.

Brooks, R. R. (1987) *Serpentine and Its Vegetation: A Multidisciplinary Approach*. London and Sydney: Croom Helm.

Brothers, N. (1991) Albatross mortality and associated bait loss in the Japanese longline fishery in the Southern Ocean. *Biological Conservation* 55, 255–68.

Brower, L. P. (1969) Ecological chemistry. *Scientific American* 220, 22–9.

Brown, J. H. (1968) Adaptation to environmental temperature in two species of woodrats, *Neotoma cinerea* and *N. albigula*. *University of Michigan*

Museum of Zoology, Miscellaneous Publications 135, 1–48.

Brown, J. H. (1971) Mammals on montaintops: nonequilibrium insular biogeography. *The American Naturalist* 105, 467–78.

Brown, J. H. (1986) Two decades of interaction between the MacArthur–Wilson model and the complexities of mammalian distributions. *Biological Journal of the Linnean Society* 28, 231–51.

Brown, J. H. (1995) *Macroecology*. Chicago and London: The University of Chicago Press.

Brown, J. H. and Heske, E. J. (1990) Control of a desert–grassland transition by a keystone rodent guild. *Science* 250, 1705–7.

Brown, J. H. and Kodric-Brown, A. (1977) Turnover rates in insular biogeography: effect of immigration on extinction. *Ecology* 58, 445–9.

Brown, J. H. and Lomolino, M. V. (1998) *Biogeography*, 2nd edn. Sunderland, MA: Sinauer Associates.

Browne, J. (1983) *The Secular Ark: Studies in the History of Biogeography*. New Haven, CT: Yale University Press.

Budiansky, S. (1995) *Nature's Keepers: The New Science of Nature Management*. London: Weidenfeld & Nicolson.

Bunce, R. G. H., Barr, C. J., Gillespie, M. K., Howard, D. C., Scott, W. A., Smart, S. M., van de Poll, H. M., and Watkins, J. W. (1999) *Vegetation of the British Countryside: The Countryside Vegetation System* (ECOFACT, vol. 1). London: Merlewood Research Station, Institute of Terrestrial Ecology, Department of the Environment, Transport and the Regions (DETR).

Burel, F. and Baudry, J. (1990) Hedgerow network patterns and processes in France. In I. S. Zonneveld and R. T. T. Forman (eds) *Changing Landscapes: An Ecological Perspective*, pp. 99–120. New York: Springer.

Buringh, P. and Dudal, R. (1987) Agricultural land use in space and time. In M. G. Wolman and F. G. A. Fournier (eds) *Land Transformation in Agriculture* (SCOPE 32), pp. 9–44. Chichester: John Wiley & Sons.

Burke, A., Esler, K. J., Pienaar, E., and Barnard, P. (2003) Species richness and floristic relationships between mesas and their surroundings in southern African Nama Karoo. *Diversity and Distributions* 9, 43–53.

Burnham, C. P. and Mackney, D. (1968) Soils of Shropshire. *Field Studies* 2, 83–113.

Burt, W. H. (1943) Territoriality and home range concepts as applied to mammals. *Journal of Mammalogy* 24, 346–52.

Burton, R. (1896 edn) *The Anatomy of Melancholy*, edited by Rev. A. R. Shilleto, with an introduction by A. H. Bullen, 3 vols. London: George Bell & Sons.

Bush, M. B. and Whittaker, R. J. (1991) Krakatau: colonization patterns and hierarchies. *Journal of Biogeography* 18, 341–56.

Bush, M. B., Whittaker, R. J., and Partomihardjo, T. (1992) Forest development on Rakata, Panjang and Sertung: contemporary dynamics (1979–89). *GeoJournal* 28, 185–99.

Butler, D. R. (1992) The grizzly bear as an erosional agent in mountainous terrain. *Zeitschrift für Geomorphologie* NF 36, 179–89.

Butler, D. R. (1995) *Zoogeomorphology: Animals as Geomorphic Agents*. Cambridge: Cambridge University Press.

Cantlon, J. E. (1953) Vegetation and microclimates on north and south slopes of Cushetunk Mountain, New Jersey. *Ecological Monographs* 23, 241–70.

Carey, C. and Alexander, M. A. (2003) Climate change and amphibian declines: is there a link? *Diversity and Distributions* 9, 111–21.

Carmargo, J. L. C., Ferraz, I. D. K., and Imakawa, A. M. (2002) Rehabilitation of degraded areas of Central Amazonia using direct sowing of forest tree seeds. *Restoration Ecology* 10, 636–44.

Carrier, J.-A. and Beebee, T. J. C. (2003) Recent, substantial, and unexplained declines of the common toad *Bufo bufo* in lowland England. *Biological Conservation* 111, 395–9.

Carson, R. (1962) *Silent Spring*. Boston, MA: Houghton Mifflin.

Carter, S. P. and Bright, P. W. (2003) Reedbeds as refuges for water voles (*Arvicola terrestris*) from predation by introduced mink (*Mustela vison*). *Biological Conservation* 111, 371–6.

Case, T. J. and Bolger, D. T. (1991) The role of introduced species in shaping the distribution and abundance of island reptiles. *Evolutionary Ecology* 5, 272–90.

Caughley, G. (1970) Eruption of ungulate populations, with emphasis on the Himalayan thar in New Zealand. *Ecology* 51, 53–72.

Chapman, C. A., Balcomb, S. R., Gillespie, T. R., Skorupa, J. P., and Struhsaker, T. T. (2000) Long-term effects of logging on African primate communities: a 28-year comparison for Kibale National Park, Uganda. *Conservation Biology* 14, 207–17.

Chapman, N., Harris, S., and Stanford, A. (1994) Reeves' muntjac (*Muntiacus reevesi*) in Britain: their history, spread, habitat selection, and the role of human intervention in accelerating their dispersal. *Mammal Review* 24, 113–60.

Chapman, R. N. (1928) The quantitative analysis of environmental factors. *Ecology* 9, 111–22.

Choi, G. H. and Nuss, D. L. (1992) Hypovirulence of chestnut blight fungus conferred by an infectious viral cDNA. *Science* 257, 800–3.

Chown, S. L. and Smith, V. R. (1993) Climate change and the short-term impact of feral house mice at the sub-Antarctic Prince Edward Islands. *Oecologia* 96, 508–16.

Christensen, N. L., Agee, J. K., Brussard, P. F., Hughes, J., Knight, D. H., Minshall, G. W., Peek, J. M., Pyne, S. J., Swanson, F. J., Thomas, J. W., Wells, S., Williams, S. E., and Wright, H. A. (1989) Interpreting the Yellowstone fires of 1988. *BioScience* 39, 678–85.

Christian, K. A., Tracy, C. R., and Porter, W. P. (1983) Seasonal shifts in body temperature and use of microhabitats by the Galápagos land iguana (*Conolophus pallidus*). *Ecology* 64, 463–8.

Cifelli, R. L. (1993) Early Cretaceous mammals from North America and the evolution of marsupial dental characters. *Proceedings of the National Academy of Sciences USA* 90, 9413–16.

Clausen, J. (1965) Population studies of alpine and subalpine races of conifers and willows in the California high Sierra Nevada. *Evolution* 19, 56–68.

Clements, F. E. (1916) *Plant Succession: An Analysis of the Development of Vegetation* (Carnegie Institute of Washington Publication No. 242). Washington, DC: Carnegie Institute of Washington.

Clements, F. E. and Shelford, V. E. (1939) *Bio-Ecology*. New York: John Wiley & Sons.

Cloudsley-Thompson, J. L. (1975a) *Terrestrial Environments*. London: Croom Helm.

Cloudsley-Thompson, J. L. (1975b) *The Ecology of Oases*. Watford, Hertfordshire: Merrow.

Colbert, E. H. (1971) Tetrapods and continents. *The Quarterly Review of Biology* 46, 250–69.

Colinvaux, P. A., De Oliveira, P. E., Moreno, J. E., Miller, M. C., and Bush, M. B. (1996) A long pollen record from lowland Amazonia: forest and cooling in glacial times. *Science* 274, 85–8.

Collins, J. P. and Storfer, A. (2003) Global amphibian declines: sorting the hypotheses. *Diversity and Distributions* 9, 89–98.

Cone, C. D., Jr (1962) Thermal soaring of birds. *American Scientist* 50, 180–209.

Connell, J. H. (1961) The influence of interspecific competition and other factors on the distribution of the barnacle *Chthamalus stellatus. Ecology* 42, 710–23.

Conway, G. (1981) Man versus pests. In R. M. May (ed.) *Theoretical Ecology: Principles and Applications*, 2nd edn, pp. 356–86. Oxford: Blackwell Scientific Publications.

Coope, G. R. (1994) The response of insect faunas to glacial–interglacial climatic fluctuations. *Philosophical Transactions of the Royal Society of London* 344B, 19–26.

Cooper, W. S. (1923) The recent ecological history of Glacier Bay, Alaska. *Ecology* 6, 197.

Cooper, W. S. (1931) A third expedition to Glacier Bay, Alaska. *Ecology* 12, 61–95.

Cooper, W. S. (1939) A fourth expedition to Glacier Bay, Alaska. *Ecology* 20, 130–59.

Copper, P. (1994) Ancient reef ecosystem expansion and collapse. *Coral Reefs* 13, 3–11.

Corbett, A. L., Krannitz, P. G., and Aarssen, L. W. (1992) The influence of petals on reproductive success in the Arctic poppy (*Papaver radicatum*). *Canadian Journal of Botany* 70, 200–4.

Corbit, M., Marks, P. L., and Gardescu, S. (1999) Hedgerows as habitat corridors for forest herbs in central New York, USA. *Journal of Ecology* 87, 220–32.

Cortner, H. J. and Moote, M. A. (1994) Trends and issues in land and water resources management: setting the agenda for change. *Environmental Management* 18, 167–73.

Costanza, R., Sklar, F. H., and White, M. L. (1990) Modeling coastal landscape dynamics. *BioScience* 40, 91–107.

Cox, C. B. (1990) New geological theories and old biogeographical problems. *Journal of Biogeography* 17, 117–30.

Cox, C. B. (2001) The biogeographic regions reconsidered. *Journal of Biogeography* 28, 511–23.

Cox, C. B. and Moore, P. D. (1993) *Biogeography: An Ecological and Evolutionary Approach*, 5th edn. Oxford: Blackwell.

Cox, W. T. (1936) Snowshoe rabbit migration, tick infestation, and weather cycles. *Journal of Mammalogy* 17, 216–21.

Cracraft, J. (1973) Continental drift, palaeoclimatology and the evolution and biogeography of birds. *Journal of Zoology, London* 169, 455–545.

Cracraft, J. (1974) Phylogeny and evolution of ratite birds. *Ibis* 116, 494–521.

Cracraft, J. (1985) Biological diversification and its causes. *Annals of the Missouri Botanical Gardens* 72, 794–822.

Cramp, S. (ed.) (1988) *Handbook of the Birds of Europe, the Middle East and North Africa (The Birds of the Western Palaearctic). Volume 5: Tyrant Flycatchers to Thrushes*. Oxford: Oxford University Press.

Crawford, A. R. (1974) A greater Gondwanaland. *Science* 184, 1179–81.

Crawley, M. J. (1983) *Herbivory: The Dynamics of Animal–Plant Interactions* (Studies in Ecology, vol. 10). Oxford: Blackwell Scientific Publications.

Crocker, R. L. and Major, J. (1955) Soil development in relation to vegetation and surface age at Glacier Bay, Alaska. *Journal of Ecology* 43, 427–48.

Croizat, L. (1958) *Pangeography*, 2 vols. Caracas: Published by the author.

Croizat, L. (1964) *Space, Time, Form: The Biological Synthesis*. Caracas: Published by the author.

Crome, F. H. J. and Moore, L. A. (1990) Cassowaries in north-eastern Queensland: a report or a survey and a review and assessment of their status and conservation and management needs. *Australian Wildlife Management* 17, 369–85.

Crowell. K. L. (1986) A comparison of relict versus equilibrium models for insular mammals of the Gulf of Maine. *Biological Journal of the Linnean Society* 28, 37–64.

Crumpacker, D. W., Box, E. O., and Hardin, E. D. (2001a) Implications of climatic warming for conservation of native trees and shrubs in Florida. *Conservation Biology* 15, 1008–20.

Crumpacker, D. W., Box, E. O., and Hardin, E. D. (2001b) Temperate–subtropical transition areas for native woody plant species in Florida, USA: present locations, predicted changes under climatic warming, and implications for conservation. *Nature Areas Journal* 21, 136–48.

Crumpacker, D. W., Box, E. O., and Hardin, E. D. (2001c) Potential breakup of Florida plant communities as a result of climatic warming. *Florida Scientist* 64, 29–43.

Dale, V. H., Joyce, L. A., McNulty, S., Neilson, R. P., Ayres, M. P., Flannigan, M. D., Hanson, P. J., Irland, L. C., Lugo, A. E., Peterson, C. J., Simberloff, D., Swanson, F. J., Stocks, B. J., and Wotton, B. M. (2001) Climate change and forest disturbances. *BioScience* 51, 723–34.

Dansereau, P. (1957) *Biogeography: An Ecological Perspective*. New York: Ronald Press.

Darlington, P. J., Jr (1957) *Zoogeography: The Geographical Distribution of Animals*. New York: John Wiley & Sons.

Dayan, T., Simberloff, D., Tchernov, E., and Yom-Tov, Y. (1990) Feline canines: community-wide character displacement among small cats of Israel. *American Naturalist* 136, 39–60.

de Candolle, A.-P. (1820) Géographie botanique. In F. C. Levrault (ed.) *Dictionnaire des Sciences Naturelles*, vol. 18, pp. 359–436. Paris: Levrault.

Décamps, H. (2001) How a riparian landscape finds form and comes alive. *Landscape and Urban Planning* 57, 169–75.

Delcourt, H. R. and Delcourt, P. A. (1988) Quaternary landscape ecology: relevant scales in space and time. *Landscape Ecology* 2, 23–44.

Delcourt, H. R. and Delcourt, P. A. (1991) *Quaternary Ecology: A Palaeoecological Perspective*. London: Chapman & Hall.

Delcourt, H. R. and Delcourt, P. A. (1994) Postglacial rise and decline of *Ostrya virginiana* (Mill.) K. Koch and *Carpinus caroliniana* Walt. in eastern North America: predictable responses of forest species to cyclic changes in seasonality of climate. *Journal of Biogeography* 21, 137–50.

Delcourt, P. A. and Delcourt, H. R. (1987) *Long-term Forest Dynamics of the Temperate Zone: A Case Study of Late-Quaternary Forests in Eastern North America* (Ecological Studies, vol. 63). New York: Springer.

DeLucia, E. H. and Schlesinger, W. H. (1990) Ecophysiology of Great Basin and Sierra Nevada vegetation on contrasting soils. In C. B. Osmond, L. F. Pitelka, and G. M. Hidy (eds) *Plant Biology of the Basin and Range* (Ecological Studies, vol. 80), pp. 143–78. Berlin: Springer.

Desponts, M. and Payette, S. (1993) The Holocene dynamics of jack pine at its northern range limit in Quebec. *Journal of Ecology* 81, 719–27.

Diamond, J. M (1974) Colonization of exploded volcanic islands by birds: the supertramp strategy. *Science* 184, 803–6.

Diamond, J. M. (1987) Human use of world resources. *Nature* 328, 479–80.

Dieckmann, U. and Doebeli, M. (1999) On the origin of species by sympatric speciation. *Nature* 400, 354–7.

Dijak, W. D. and Thompson III, F. R. (2000) Landscape and edge effects on the distribution of mammalian predators in Missouri. *Journal of Wildlife Management* 64, 209–16.

Dixon, M. D. and Johnson, W. C. (1999) Riparian vegetation along the Middle Snake River, Idaho: zonation, geographical trends, and historical changes. *Great Basin Naturalist* 59, 18–34.

Dobler, G., Schneider, R., and Schweis, A. (1991) Die Invasion des Rauhfußbussards (*Buteo lagopus*) in Baden-Württemberg im Winter 1986/87. *Die Vogelwarte* 36, 1–18.

Dobzhansky, T., Ayala, F. J., Stebbins, G. L., and Valentine, J. W. (1977) *Evolution*. San Francisco, CA: W. H. Freeman and Co.

Dony, J. G. (1963) The expectation of plant records from prescribed areas. *Watsonia* 5, 377–85.

Doroff, A. M., Estes, J. A., Tinker, M. T., Burn, D. M., and Evans, T. J. (2003) Sea otter population declines in the Aleutian archipelago. *Journal of Mammalogy* 84, 55–64.

Downie, I. S., Butterfield, J. E. L., and Coulson, J. C. (1995) Habitat preferences of sub-montane spiders in northern England. *Ecography* 18, 51–61.

Drake, J. A. (1990) The mechanics of community assembly and succession. *Journal of Theoretical Biology* 147, 213–33.

Drury, W. H. and Nisbet, I. C. T. (1973) Succession. *Journal of the Arnold Arboretum* 54, 331–68.

Duggins, D. O. (1980) Kelp beds and sea otters: an experimental approach. *Ecology* 61, 447–53.

Dunbar, R. I. M. (1998) Impact of global warming on the distribution and survival of the gelada baboon: a modelling approach. *Global Change Biology* 4, 293–304.

Ebach, M. C. and Humphries, C. J. (2003) Ontology of biogeography. *Journal of Biogeography* 30, 959–62.

Ebach, M. C., Humphries, C. J., and Williams, D. M. (2002) Phylogenetic biogeography deconstructed. *Journal of Biogeography* 30, 1285–96.

Edenius, L. and Sjöberg, K. (1997) Distribution of birds in natural landscape mosaics of old-growth forests in northern Sweden: relations to habitat area and landscape context. *Ecography* 20, 425–31.

Ehleringer, J. R., Mooney, H. A., Rundel, P. W., Evans, R. D., Palma, B., and Delatorre, J. (1992) Lack of nitrogen cycling in the Atacama Desert. *Nature* 359, 316–18.

Ellenberg, H. (1986) *Vegetation Ecology of Central Europe*, 4th edn, translated by Gordon K. Strutt. Cambridge: Cambridge University Press.

Ellison, A. M., Gotelli, N. J., Brewer, J. S., Cochran-Stafira, D. L., Kneitel, J. M., Miller, T. E., Worley, A. C., and Zamora, R. (2003) The evolutionary ecology of carnivorous plants. *Advances in Ecological Research* 33, 1–74.

Elton, C. S. (1958) *The Ecology of Invasions by Animals and Plants*. London: Chapman & Hall.

Endler, J. A. (1977) *Geographic Variation, Speciation, and Clines* (Monographs in Population Biology No. 10). Princeton, NJ: Princeton University Press.

Erwin, T. L. (1979) A review of the natural history and evolution of ectoparisitoid relationships in carabid beetles. In T. L. Erwin, G. E. Ball, D. R. Whitehead, and A. L. Halpern (eds) *Carabid Beetles: Their Evolution, Natural History, and Classification*, pp. 479–84. The Hague: W. Junk.

Erwin, T. L. (1981) Taxon pulses, vicariance, and dispersal: an evolutionary synthesis illustrated by carabid beetles. In G. Nelson and D. E. Rosen (eds) *Vicariance Biogeography: A Critique* (Symposium of the Systematics Discussion Group of the American Museum of Natural History, 2–4 May 1979), pp. 159–87. New York: Columbia University Press.

Etherington, J. R. (1982) *Environment and Plant Ecology*, 2nd edn. Chichester: John Wiley & Sons.

Eyre, S. R. (1963) *Vegetation and Soils: A World Picture*. London: Edward Arnold.

Fagan, B. M. (1990) *The Journey from Eden: The Peopling of Our World*. London: Thames and Hudson.

Fastie, C. L. (1995) Causes and ecosystem consequences of multiple pathways of ecosystem succession at Glacier Bay, Alaska. *Ecology* 76, 1899–916.

Fayt, P. (1999) Available insect prey in bark patches selected by the three-toed woodpecker

Picoides tridactylus prior to reproduction. *Ornis Fennica* 76, 135–40.

Feeny, P. P (1970) Seasonal changes in oak leaf tannins and nutrients as a cause of spring feeding by winter moth caterpillars. *Ecology* 51, 565–81.

Fenneman, N. M. (1916) Physiographic divisions of the United States. *Annals of the Association of American Geographers* 6, 19–98.

Finney, H. R., Holowaychuk, N., and Heddleson, M. R. (1962) The influence of microclimate on the morphology of certain soils of the Allegheny Plateau of Ohio. *Soil Science Society of America Proceedings* 26, 287–92.

Fischer, A. G. (1981) Climatic oscillations in the biosphere. In M. H. Nitecki (ed.) *Biotic Crises in Ecological and Evolutionary Time*, pp. 103–31. New York: Academic Press.

Fischer, A. G. (1984) Biological innovations and the sedimentary record. In H. D. Holland and A. F. Trendall (eds) *Patterns of Change in Earth Evolution* (Dahlem Konferenzen 1984), pp. 145–57. Berlin: Springer.

Fischer, S. F., Poschlod, P., and Beinlich, B. (1996) Experimental studies on the dispersal of plants and animals on sheep in calcareous grasslands. *Journal of Applied Ecology* 33, 1206–22.

Flannery, T. F. (1994) *The Future Eaters: An Ecological History of the Australasian Lands and People*. New York: George Brazillier.

Flannery, T. F. (2001) *The Eternal Frontier: An Ecological History of North America and its Peoples*. Melbourne: Text Publishing.

Flannery, T. F., Rich, T. H., Turnbull, W. D., and Lundelius, E. L., Jr (1992) The Macropodoidea (Marsupialia) of the early Pliocene Hamilton local fauna, Victoria, Australia. *Fieldiana: Geology* (New Series No. 25). Chicago, IL: Field Museum of Natural History.

Flannigan, M. D., Stocks, B. J., and Wotton, B. M. (2000) Climate change and forest fires. *The Science of the Total Environment* 262, 221–9.

Fleagle, J. G. (1988) *Primate Adaptation and Evolution*. New York: Academic Press.

Fleagle, J. G. and Kay, R. F. (1997) Platyrrhines, catarrhines, and the fossil record. In W. G. Kinzey (ed.) *New World Primates: Ecology, Evolution, and Behavior*, pp. 3–23. New York: Aldine de Gruyter.

Flux, J. E. C. and Fullagar, P. J. (1992) World distribution of the rabbit *Oryctolagus cuniculus* on islands. *Mammal Review* 22, 151–205.

Flynn, L. J. and Jacobs, L. L. (1982) Effects of changing environments on Siwalik rodent faunas of northern Pakistan. *Palaeogeography, Palaeoclimatology, Palaeoecology* 38, 129–38.

Ford, M. J. (1982) *The Changing Climate: Responses of the Natural Fauna and Flora*. London: George Allen & Unwin.

Forman, R. T. T. (1990) Ecologically sustainable landscapes: the role of spatial configuration. In I. S. Zonneveld and R. T. T. Forman (eds) *Changing Landscapes: An Ecological Perspective*, pp. 261–78. New York: Springer.

Forman, R. T. T. (1995) *Land Mosaics: the Ecology of Landscapes and Regions*. Cambridge: Cambridge University Press.

Forman, R. T. T. and Godron, M. (1986) *Landscape Ecology*. New York: John Wiley & Sons.

Forman, R. T. T., Galli, A. E., and Leck, C. F. (1976) Forest size and avian diversity in New Jersey woodlots with some land use implications. *Oecologia* 26, 1–8.

Fortin, D. and Arnold, G. W. (1997) The influence of road verges on the use of nearby small shrubland remnants by birds in the central wheatbelt of Western Australia. *Wildlife Research* 24, 679–89.

Frankham, R., Ballou, J. D., Briscoe, D. A., and McInnes, K. H (illustrator) (2002) *Introduction to Conservation Genetics*. Cambridge: Cambridge University Press.

Fuccilo, D., Sears, L., and Stapleton, P. (1998) *Biodiversity in Trust: Conservation and Use of Plant Genetic Resources in CGIAR Centres.* New York: Cambridge University Press.

Funch, P. and Kristensen, R. M. (1995) Cycliophora is a new phylum with affinities to Entoprocta and Ectoprocta. *Nature* 378, 711–14.

Gates, D. M. (1980) *Biophysical Ecology.* New York: Springer.

Gates, D. M. (1993) *Climate Change and Its Biological Consequences.* Sunderland, MA: Sinauer Associates.

Gause, G. F. (1934) *The Struggle for Existence.* New York: Hafner.

Geist, V. (1987) Bergmann's Rule is invalid. *Canadian Journal of Zoology* 65, 1035–8.

Genelly, R. E. (1965) Ecology of the common mole-rat (*Cryptomys hottentotus*) in Rhodesia. *Journal of Mammalogy* 46, 647–65.

Gilbert, F. S. (1980) The equilibrium theory of island biogeography: fact or fiction? *Journal of Biogeography* 7, 209–35.

Gilbert, O. L. and Anderson, P. (1998) *Habitat Creation and Repair.* Oxford: Oxford University Press.

Givnish, T. J. and Sytsma, K. J. (1997) Preface. In T. J. Givnish and K. J. Sytsma (eds) *Molecular Evolution and Adaptive Radiation*, pp. xiii–xvii. Cambridge: Cambridge University Press.

Gloger, C. W. L. (1833) *Das Abändern der Vögel durch Einfluss des Klimas: Nach zoologischen, zunächst von dem europäischen Landvälgeln, entnommenen Beobachtungen dargestellt, mit den entsprechenden Erfahrungen bei den europäischen Säugthieren verglichen, und durch Tatsachen aus dem Gebiete der Physiologie, der Physik und der physischen Geographie erläutert.* Breslau: A. Schultz.

Godthelp, H. (1997) *Zyzomys rackhami* sp. nov. (Rodentia, Muridae): a rockrat from Pliocene Rackham's Roost Site, Riversleigh, northwestern Queensland. *Memoirs of Queensland Museum* 41, 329–33.

Goldspink, C. R. (1987) The growth, reproduction and mortality of an enclosed population of red deer (*Cervus elaphas*) in north-west England. *Journal of Zoology, London* 213, 23–44.

Good, R. (1974) *The Geography of the Flowering Plants*, 4th edn. London: Longman.

Gorman, M. L. (1979) *Island Ecology* (Outline Studies in Ecology). London: Chapman & Hall.

Gosling, L. M. and Baker, S. J. (1989) The eradication of muskrats and coypus from Britain. *Biological Journal of the Linnean Society* 38, 39–51.

Gotfryd, A. and Hansell, R. I. C. (1986) Prediction of bird-community metrics in urban woodlots. In J. Venner, M. L. Morrison, and C. J. Ralph (eds) *Wildlife 2000: Modeling Habitat Relationships of Terrestrial Vertebrates*, pp. 321–6. Madison, WI: University of Wisconsin Press.

Goudriaan, J. and Ketner, P. (1984) A simulation study for the global carbon cycle, including Man's impact on the biosphere. *Climatic Change* 6, 167–92.

Gould, S. J. (1974) The origin and function 'bizarre' structures: antler size and skull size in the 'Irish elk', *Megaloceras giganteus. Evolution* 28, 191–220.

Grace, J. (1987) Climatic tolerance and the distribution of plants. *New Phytologist* 106 (Supplement), 113–30.

Graham, J. B., Dudley, R., Aguilar, N. M., and Gans, C. (1995) Implications of the late Palaeozoic oxygen pulse for physiology and evolution. *Nature* 375, 117–20.

Graham, R. W. (1979) Paleoclimates and late Pleistocene faunal provinces in North America. In R. L. Humphrey and D. J. Stanford (eds) *Pre-Llano Cultures of the Americas: Paradoxes and Possibilities*, pp. 46–69. Washington, DC: Anthropological Society of Washington.

Graham, R. W. (1992) Late Pleistocene faunal changes as a guide to understanding effects of greenhouse warming on the mammalian fauna of North America. In R. L. Peters and T. E. Lovejoy (eds) *Global Warming and Biological Diversity*, pp. 76–87. New Haven, CT, and London: Yale University Press.

Graham, R. W. and Grimm, E. C. (1990) Effects of global climate change on the patterns of terrestrial biological communities. *Trends in Ecology and Evolution* 5, 289–92.

Graham, R. W. and Mead, J. I. (1987) Environmental fluctuations and evolution of mammalian faunas during the last deglaciation. In W. F. Ruddiman and H. E. Wright Jr (eds) *North American and Adjacent Oceans During the Last Deglaciation* (The Geology of North America, vol. K-3), pp. 371–402. Boulder, CO: The Geological Society of America.

Grant, P. R. (1986) *Ecology and Evolution of Darwin's Finches*. Princeton, NJ: Princeton University Press.

Grant, V. (1977) *Organismic Evolution*. San Francisco, CA: W. H. Freeman.

Grayson, D. K. (1987) Death by natural causes. *Natural History* (May), 8–13.

Grayson, D. K. (1991) Late Pleistocene mammalian extinction in North America: taxonomy, chronology, and explanation. *Journal of World Prehistory* 5, 193–231.

Green, D. M. (2003) The ecology of extinction: population fluctuation and decline in amphibians. *Biological Conservation* 11, 331–43.

Grime, J. P. (1977) Evidence for the existence of three primary strategies in plants and its relevance to ecological and evolutionary theory. *American Naturalist* 111, 1169–94.

Grime, J. P. (1989) The stress debate: symptom of impending synthesis?. *Biological Journal of the Linnean Society* 37, 3–17.

Grime, J. P., Hodgson, J. G., and Hunt, R. (1988) *Comparative Plant Ecology: A Functional Approach to Common British Species*. London: Unwin Hyman.

Grimm, E. C., Jacobson, G. L., Jr, Watts, W. A., Hansen, B. C. S., and Maasch, K. A. (1993) A 50,000-year record of climate oscillations from Florida and its temporal correlation with the Heinrich events. *Science* 261, 198–200.

Guilday, J. E. (1984) Pleistocene extinctions and environmental change: a case study of the Appalachians. In P. S. Martin and R. G. Klein (eds) *Quaternary Extinctions: A Prehistoric Revolution*, pp. 250–8. Tucson, AZ: The University of Arizona Press.

Gulland, F. M. D. (1992) The role of nematode parasites in Soay sheep (*Ovis aries* L.) mortality during a population crash. *Parasitology* 105, 493–503.

Guthrie, R. D. (1984) Mosaics, allelochemicals and nutrients: an ecological theory of Late Pleistocene megafaunal extinctions. In P. S. Martin and R. G. Klein (eds) *Pleistocene Extinctions: A Prehistoric Revolution*, pp. 259–98. Tucson, AZ: University of Arizona Press.

Gutiérrez, R. J. and Harrison, S. (1996) Applications of metapopulation theory to spotted owl management: a history and critique. In D. McCullough (ed.) *Metapopulations and Wildlife Conservation Management*, pp. 167–85. Covelo, CA: Island Press.

Haffer, J. (1969) Speciation in Amazonian forest birds. *Science* 165, 131–7.

Hafner, D. J. (1994) Pikas and permafrost: post-Wisconsin historical zoogeography of *Ochotona* in the southern Rocky Mountains, U.S.A. *Arctic and Alpine Research* 26, 375–82.

Hall, E. R. (1946) *Mammals of Nevada*. Berkeley, CA: University of California Press.

Halley, D. J. and Rosell, F. (2002) The beaver's reconquest of Eurasia: status, population development and management of a conservation success. *Mammal Review* 32, 153–78.

Hannah, L., Lohse, D., Hutchinson, C., Carr, J. L., and Lankerani, A. (1994) A preliminary inventory of human disturbance of world ecosystems. *Ambio* 23, 246–50.

Hannah, L., Midgley, G. F., Lovejoy, T., Bond, W. J., Bush, M. L., Scott, D., and Woodward, F. I. (2002a) Conservation of biodiversity in a changing climate. *Conservation Biology* 16, 11–15.

Hannah, L., Midgley, G. F., and Millar, D. (2002b) Climate change-integrated conservation strategies. *Global Ecology & Biogeography* 11, 485–95.

Hanski, I. (1982) Dynamics of regional distribution: the core and satellite species hypothesis. *Oikos* 38, 210–21.

Hanski, I. (1986) Population dynamics of shrews on small islands accord with the equilibrium model. *Biological Journal of the Linnean Society* 28, 23–36.

Hanski, I. and Henttonnen, H. (1996) Predation on competing rodent species: a simple explanation of complex patterns. *Journal of Animal Ecology* 65, 220–32.

Hanski, I. and Korpimäki, E. (1995) Microtine rodent dynamics in northern Europe: parameterized models for the predator–prey interaction. *Ecology* 76, 840–50.

Hanski, I., Foley, P., and Hassell, M. (1996a) Random walks in a metapopulation: how much density dependence is necessary for long-term persistence? *Journal of Animal Ecology* 65, 274–82.

Hanski, I., Moilanen, A., Pakkala, T., and Kuussaari, M. (1996b) The quantitative incidence function model and persistence of an endangered butterfly population. *Conservation Biology* 10, 587–90.

Hanski, I., Pakkala, T., Kuussaari, M., and Guangchun Lei (1995) Metapopulation persistence of an endangered butterfly in a fragmented landscape. *Oikos* 72, 21–8.

Hanski, I., Turchin, P., Korpimäki, E., and Henttonnen, H. (1993) Population oscillations of boreal rodents: regulation by mustelid predators leads to chaos. *Nature* 364, 232–5.

Hardin, G. (1960) The competitive exclusion principle. *Science* 131, 1292–97.

Harms, W. B. and Opdam, P. (1990) Woods as habitat patches for birds: application in landscape planning in The Netherlands. In I. S. Zonneveld and R. T. T. Forman (eds) *Changing Landscapes: An Ecological Perspective*, pp. 73–97. New York: Springer.

Harper, J. L. (1961) Approaches to the study of plant competition. *Symposia of the Society for Experimental Biology* 15, 1–39.

Harris, L. D. (1984) *The Fragmented Forest: Island Biogeography Theory and the Preservation of Biotic Diversity*, with a Foreword by Kenton R. Miller. Chicago and London: Chicago University Press.

Harris, P. (1993) Effects, constraints and the future of weed biocontrol. *Agriculture, Ecosystems and Environment* 46, 289–303.

Harrison, G. W. (1995) Comparing predator–prey models to Luckinbill's experiment with *Didinium* and *Paramecium*. *Ecology* 76, 357–74.

Harrison, P. A., Berry, P. M., and Dawson, T. P. (2001) *Climate Change and Nature Conservation in Britain and Ireland: Modelling Natural Resource Responses to Climate Change* (the MONARCH project). Oxford: UKCIP Technical Report.

Harrison, S. (1994) Metapopulations and conservation. In P. J. Edwards, R. M. May, and N. R. Webb (eds) *Large-Scale Ecology and Conservation Biology* (The 35th Symposium of the British Ecological Society with the Society for Conservation Biology, University of Southampton, 1993), pp. 111–28. Oxford: Blackwell Scientific Publications.

Hartman, G. (1994) Long-term population development of a reintroduced beaver (*Castor fiber*) population in Sweden. *Conservation Biology* 8, 713–17.

Haynes, G. (2002) The catastrophic extinction of North American mammoths and mastodons. *World Archaeology* 33, 391–416.

Heaney, L. R. (1986) Biogeography of mammals in S.E. Asia: estimates of rates of colonization, extinction and speciation. *Biological Journal of the Linnean Society* 28, 127–65.

Heard, D. C. and Ouellet, J. P. (1994) Dynamics of an introduced caribou population. *Arctic* 47, 88–95.

Heard, S. B. (1994) Pitcher-plant midges and mosquitoes: a processing chain commensalism. *Ecology* 75, 1647–60.

Hedges, S. B. (1989) Evolution and biogeography of West Indian frogs of the genus *Eleutherodactylus*: slow-evolving loci and the major groups. In C. A. Woods (ed.) *Biogeography of the West Indies: Past, Present, and Future*, pp. 305–70. Gainesville, FL: Sandhill Crane Press.

Henning, W. (1966) *Phylogenetic Systematics*. Urbana, IL: University of Illinois Press.

Heywood, V. H. (ed.) (1978) *Flowering Plants of the World*. Oxford: Oxford University Press.

Hill, J. M. and Knisley, C. B. (1992) Frugivory in the tiger beetle, *Cicindela repanda* (Coleoptera: Cicindelidae). *Coleopterist's Bulletin* 46, 306–10.

Hilligardt, M. (1993) Durchsetzungs- und Reproduktionsstrategien bei *Trifolium pallescens* Schreb. und *Trifolium thalii* Vill. II. Untersuchungen zur Populationsbiologie. *Flora (Jena)*, 188, 175–95.

Hinsley, S. A., Bellamy, P. E., Enoksson, B., Fry, G., Gabrielsen, L., McCollin, D., and Schotman, A. (1998) Geographical and land-use influences on bird species richness in small woods in agricultural landscapes. *Global Ecology and Biogeography Letters* 7, 125–35.

Hoffstetter, R. and Lavocat, R. (1970) Découverte dans le Déseadien de Bolivie de Genres Pentalophodentes Appuyant les Affinités Africaines des Rongeurs Caviomorphes. *Comptes Rendus, Académie des Sciences, Série D* 271, 172–5.

Horn, H. S. (1981) Succession. In R. M. May (ed.) *Theoretical Ecology: Principles and Applications*, 2nd edn, pp. 253–71. Oxford: Blackwell Scientific Publications.

Horovitz, I. and Meyer, A. (1997) Evolutionary trends in the ecology of New World monkeys inferred from a combined phylogenetic analysis of nuclear, mitochondrial, and morphological data. In T. J. Givnish and K. J. Sytsma (eds) *Molecular Evolution and Adaptive Radiation*, pp. 189–224. Cambridge: Cambridge University Press.

Horsley, G. A. (1966) Trees and shrubs. In B. L. Sage (ed.) *Northaw Great Wood: Its History and Natural History*, pp. 34–51. Hertford: Education Department of the Hertfordshire Country Council.

Hosking, J. R., Sullivan, P. R., and Welsby, S. M. (1994) Biological control of *Opuntia stricta* (Haw.) Haw. var. *stricta* using *Dactylopius opuntiae* (Cockerell) in an area of New South Wales, Australia, where *Cactoblastis cactorum* (Berg) is not a successful biological control agent. *Agriculture, Ecosystems and Environment* 48, 241–55.

Houde, P. W. (1988) Paleognathous birds from the early Tertiary of the northern hemisphere. *Publications of the Nuttall Ornithological Club* 22, 1–148.

Huffaker, C. B. (1958) Experimental studies on predation: dispersion factors and predator–prey oscillations. *Hilgardia* 27, 343–83.

Huffaker, C. B., Shea, K. P., and Herman, S. G. (1963) Experimental studies on predation: complex dispersion and levels of food in an acarine predator–prey interaction. *Hilgardia* 34, 305–30.

Huggett, R. J. (1980) *Systems Analysis in Geography*. Oxford: Clarendon Press.

Huggett, R. J. (1995) *Geoecology: An Evolutionary Approach*. London: Routledge.

Huggett, R. J. (1997) *Environmental Change: The Evolving Ecosphere*. London: Routledge.

Huggett, R. J. (1999) Ecosphere, biosphere, Gaia? What to call the global ecosystem. *Global Biogeography and Ecology* 8, 425–31.

Huggett, R. J. and Cheesman, J. E. (2002) *Topography and the Environment*. Harlow, Essex: Prentice Hall.

Humphries, C. J. (2000) Form, space and time; which come first? *Journal of Biogeography* 27, 11–15.

Hunt, W. G. and Selander, R. K. (1973) Biochemical genetics of hybridization in European house mice. *Heredity* 31, 11–33.

Huntley, B. (1990) European vegetation history: palaeovegetation maps from pollen data – 13,000 yr BP to present. *Journal of Quaternary Science* 5, 103–22.

Huntley, B. and Prentice, I. C. (1993) Holocene vegetation and climates of Europe. In H. E. Wright Jr, J. E. Kutzbach, T. Webb III, W. F. Ruddiman, F. A. Street-Perrott, and P. J. Bartlein (eds) *Global Climates Since the Last Glacial Maximum*, pp. 136–68. Minneapolis, MN, and London: University of Minnesota Press.

Hurlbert, S. H. and Archibald, J. D. (1995) No statistical support for sudden (or gradual) extinction of dinosaurs. *Geology* 23, 881–4.

Huxley, J. (1942) *Evolution: The Modern Synthesis*. London: George Allen & Unwin.

Hylander, L. D., Silva, E. C., Oliveira, L. J., Silva, S. A., Kuntze, E. K., and Silva, D. X. (1994) Mercury levels in Alto Pantanal: a screening study. *Ambio* 23, 478–84.

Illies, J. (1974) *Introduction to Zoogeography*, translated by W. D. Williams. London: Macmillan.

Irland, L. C. (1994) Getting from here to there: implementing ecosystem management on the ground. *Journal of Forestry* 92, 12–17.

Irving, L. (1966) Adaptations to cold. *Scientific American* 214, 94–101.

IUCN (1994) *Guidelines for Protected Area Management Categories*. IUCN, Gland, Switzerland and Cambridge, UK: CNPPA with the assistance of WCMC.

Iversen, J. (1944) *Viscum, Hedera* and *Ilex* as climate indicators. *Geologiska föreningens i Stockholm förhandlinger* 66, 463–83.

Iversen, J. (1958) The bearing of glacial and interglacial epochs on the formation and extinction of plant taxa. *Uppsala Universiteit Årsskrift* 6, 210–15.

Jablonski, D. (1986) Background and mass extinctions: the alternation of macroevolutionary regimes. *Science* 231, 129–33.

Jackson, R. M. and Raw, F. (1966) *Life in the Soil* (The Institute of Biology's Studies in Biology No. 2). London: Edward Arnold.

Jaffe, K., Michelangeli, F., Gonzalez, J. M., Miras, B., and Ruiz, M. C. (1992) Carnivory in pitcher plants of the genus *Heliamphora* (Sarraceniaceae). *New Phytologist* 122, 733–44.

James, F. C. (1970) Geographical size variation in birds and its relationship to climate. *Ecology* 51, 365–90.

James, F. C. (1991) Complementary descriptive and experimental studies of clinal variation in birds. *American Zoologist* 31, 694–706.

Jenny, H. (1980) *The Soil Resource: Origin and Behaviour* (Ecological Studies, vol. 37). New York: Springer.

Johnson, D. H. (1980) The comparison of usage and availability measurements for evaluating resource preference. *Ecology* 61, 65–71.

Johnson, D. L. (1980) Problems in the land vertebrate zoogeography of certain islands and the swimming powers of elephants. *Journal of Biogeography* 7, 383–98.

Johnson, D. L. (1989) Subsurface stone lines, stone zones, artifact-manuport layers and biomantles produced by bioturbation via pocket gophers (*Thomomys bottae*). *American Antiquity* 54, 370–89.

Johnson, D. L. (1990) Biomantle evolution and the redistribution of earth materials and artefacts. *Soil Science* 149, 84–102.

Johnson, M. P. and Simberloff, D. S. (1974) Environmental determinants of island species numbers in the British Isles. *Journal of Biogeography* 1, 149–54.

Johnston, R. F. and Selander, R. K. (1971) Evolution in the house sparrow. II. Adaptive differentiation in North American populations. *Evolution* 25, 1–28.

Joshi, S. (1991) Biological control of *Parthenium hysterophorus* L. (Asteraceae) by *Cassia uniflora* Mill (Leguminosae), in Bangalore, India. *Tropical Pest Management* 37, 182–4.

Juarez, K. M. and Marinho-Filho, J. (2002) Diet, habitat use, and home ranges of sympatric canids in central Brazil. *Journal of Mammalogy* 83, 925–33.

Karanth, K. U. and Stith, B. M. (1999) Prey depletion as a critical determinant of tiger population viability. In J. Seidensticker, S. Christie, and P. Jackson (eds) *Riding the Tiger: Tiger Conservation in Human-dominated Landscapes*, pp. 100–13. London: The Zoological Society of London; Cambridge: Cambridge University Press.

Karanth, K. U. and Sunquist, M. E. (1995) Prey selection by tiger, leopard and dhole in tropical forests. *Journal of Animal Ecology* 64: 439–50.

Kats, L. B. and Ferrer, R. P. (2003) Alien predators and amphibian decline: review of two decades of science and the transition to conservation. *Diversity and Distributions* 9, 99–110.

Kauffman, E. G. (1995) Global change leading to biodiversity crisis in a greenhouse world: the Cenomanian–Turonian (Cretaceous) mass extinction. In Board on Earth Sciences and Resources Commission on Geosciences, Environment, and Resources, National Research Council, *Effects of Past Global Change on Life*, pp. 47–71. Washington, DC: National Academy Press.

Kauffman, E. G. and Fagerstrom, J. A. (1993) The Phanerozoic evolution of reef diversity. In R. E. Ricklefs and D. Schulter (eds) *Species Diversity in Ecological Communities: Historical and Geographical Perspectives*, pp. 365–404. Chicago, IL: University of Chicago Press.

Kaufman, D. M. (1995) Diversity of New World mammals: universality of the latitudinal gradients of species and bauplans. *Journal of Mammalogy* 76, 322–34.

Kaufmann, M. R., Graham, R. T., Boyce, D. A., Jr, Moir, W. H., Perry, L., Reynolds, R. T., Bassett, R. L., Mehlhop, P., Edminster, C. B., Block, W. M., and Corn, P. S. (1994) *An Ecological Basis for Ecosystem Management* (General Technical Report RM – 246, United States Department of Agriculture, Forest Service, Rocky Mountain Forest and Range Experiment Station, Fort Collins, Colorado). Washington, DC: United States Government Printing Office.

Keiter, R. and Boyce, M. (1991) *The Greater Yellowstone Ecosystem*. New Haven, CT: Yale University Press.

Keller, G. (1989) Extended periods of extinctions across the Cretaceous/Tertiary boundary in planktonic foraminifera of continental shelf sections: implications for impact and volcanic theories. *Bulletin of the Geological Society of America* 101, 1408–19.

Kennerly, T. E., Jr (1964) Microenvironmental conditions of the pocket gopher burrow. *Texas Journal of Science* 14, 397–441.

Kenward, R. E. and Holm, J. L. (1993) On the replacement of the red squirrel in Britain: a phytotoxic explanation. *Proceedings of the Royal Society of London* 251B, 187–94.

Kerley, G. I. H., Pressey, R. L., Cowling, R. M., Boshoff, A. F., and Sims-Castley, R. (2003)

Options for the conservation of large and medium-sized mammals in the Cape Floral Province Region hotspot, South Africa. *Biological Conservation* 112, 169–90.

Kessel, B. (1953) Distribution and migration of the European starling in North America. *Condor* 55, 49–67.

Kessell, S. R. (1976) Gradient modeling: a new approach to fire modeling and wilderness resource management. *Environmental Management* 1, 39–48.

Kessell, S. R. (1979) *Gradient Modeling, Resource and Fire Management*. New York: Springer.

Kessler, W. B., Salwasser, H., Cartwright, C., Jr, and Caplan, J. (1992) New perspective for sustainable natural resources management. *Ecological Applications* 2, 221–5.

Kettlewell, H. B. D. (1973) *The Evolution of Melanism: The Study of a Recurring Necessity, with Special Reference to Industrial Melanism in Lepidoptera*. Oxford: Clarendon Press.

Kevan, P. G. (1975) Sun-tracking solar furnaces in high Arctic flowers: significance for pollination and insects. *Science* 189, 723–6.

King, C. M. (1990) Introduction. In C. M. King (ed.) *The Handbook of New Zealand Mammals*, pp. 3–21. Auckland: Oxford University Press.

King, D. W. (1966) The soils. In B. L. Sage (ed.) *Northaw Great Wood: Its History and Natural History*, pp. 26–33. Hertford: Education Department of the Hertfordshire Country Council.

King, J. E. and Saunders, J. J. (1984) Environmental insularity and the extinction of the American mastodon. In P. S. Martin and R. G. Klein (eds) *Quaternary Extinctions: A Prehistoric Revolution*, pp. 315–44. Tucson, AZ: The University of Arizona Press.

Kinnaird, M. F. and O'Brien, T. G. (1991) Viable populations for an endangered forest primate, the Tana River crested mangabey (*Cercocebus galeritus galeritus*). *Conservation Biology* 5, 203–13.

Kitayama, K., Mueller-Dombois, D., and Vitousek, P. M. (1995) Primary succession of Hawaiian montane rain forest on a chronosequence of eight lava flows. *Journal of Vegetation Science* 6, 211–22.

Kitchell, J. A. and Carr, T. R. (1985) Non-equilibrium model of diversification: faunal turnover dynamics. In J. W. Valentine (ed.) *Phanerozoic Diversity Patterns*, pp. 277–310. Princeton, NJ: Princeton University Press.

Kitchener, A. C. (1999) Tiger distribution, phenotypic variation and conservation issues. In J. Seidensticker, S. Christie, and P. Jackson (eds) *Riding the Tiger: Tiger Conservation in Human-dominated Landscapes*, pp. 19–39. London: The Zoological Society of London; Cambridge: Cambridge University Press.

Klein, R. G. (1986) Carnivore size and Quaternary climatic change in southern Africa. *Quaternary Research* 26, 153–70.

Knapp, P. A. (1992) Secondary plant succession and vegetation recovery in two western Great Basin Desert ghost towns. *Biological Conservation* 6, 81–9.

Kohn, D. D. and Walsh, D. M. (1994) Plant species richness: the effect of island size and habitat diversity. *Journal of Ecology* 82, 367–77.

Kormondy, E. J. (1996) *Concepts of Ecology*, 4th edn. Upper Saddle River, NJ: Prentice Hall.

Krauss, J., Steffan-Dewenter, I., and Tscharntke, T. (2003) How does landscape context contribute to effects of habitat fragmentation on diversity and population density of butterflies? *Journal of Biogeography* 30, 889–900.

Kupfer, J. A. (1995) Landscape ecology and biogeography. *Progress in Physical Geography* 19, 18–34.

Kurtén, B. (1969) Continental drift and evolution. *Scientific American* 220, 54–64.

Kurtén, B. and Anderson, E. (1980) *Pleistocene Mammals of North America*. New York: Columbia University Press.

Labandeira, C. and Sepkoski, J. J., Jr (1993) Insect diversity in the fossil record. *Science* 261, 310–15.

Lack, D. (1933) Habitat selection in birds with special references to the effects of afforestation on the Breckland avifauna. *Journal of Animal Ecology* 2, 239–62.

Lack, D. (1947) *Darwin's Finches*. Cambridge: Cambridge University Press.

Lack, D. (1970) Island birds. *Biotropica* 2, 29–31.

Lamberson, R. H., McKelvey, R., Noon, B. R., and Voss, C. (1992) A dynamic analysis of northern spotted owl viability in a fragmented forest landscape. *Conservation Biology* 6, 505–12.

Larcher, W. (1975) *Physiological Plant Ecology*, 1st edn. Berlin: Springer.

Larcher, W. (1995) *Physiological Plant Ecology: Ecophysiology and Stress Physiology of Functional Groups*, 3rd edn. Berlin: Springer.

Laurie, W. A. and Brown, D. (1990a) Population biology of marine iguanas (*Amblyrhynchus cristatus*). I. Changes in fecundity related to a population crash. *Journal of Animal Ecology* 59, 515–28.

Laurie, W. A. and Brown, D. (1990b) Population biology of marine iguanas (*Amblyrhynchus cristatus*). II. Changes in annual survival rates and the effects of size, sex, age and fecundity in a population crash. *Journal of Animal Ecology* 59, 529–44.

Laws, R. M. (1970) Elephants as agents of habitat and landscape change in East Africa. *Oikos* 21, 1–15.

Leggett, J. (1989) The biggest mass-extinction of them all. *New Scientist* 122 (1668), 62.

Lenihan, J. M. (1993) Ecological response surfaces for North American boreal tree species and their use in forest classification. *Journal of Vegetation Science* 4, 667–80.

Leslie, P. H. (1945) The use of matrices in certain population mathematics. *Biometrika* 33, 183–212.

Leslie, P. H. (1948) Some further notes on the use of matrices in population mathematics. *Biometrika* 35, 213–45.

Lessa, E. P., Van Valkenburgh, B., and Farina, R. A. (1997) Testing hypotheses of differential mammal extinctions subsequent to the Great American Biotic Interchange. *Palaeogeography, Palaeoclimatology, and Palaeoecology* 135, 157–62.

Leuschner, C. (1998) Vegetation an der Waldgrenze auf tropischen und subtropischen Inseln. *Geographische Rundschau* 50, 690–7.

Lever, C. (1979) *The Naturalized Animals of the British Isles*. London: Granada Publishing.

Levins, R. (1970) Extinction. In M. Gerstenhaber (ed.) *Some Mathematical Questions in Biology*, pp. 77–107. Providence, RI: American Mathematical Society.

Li, Y., Glime, J. M., and Liao, C. (1992) Responses of two interacting *Sphagnum* species to water level. *Journal of Bryology* 17, 59–70.

Liebig, J. (1840) *Organic Chemistry and its Application to Agriculture and Physiology*, English edn, edited by L. Playfair and W. Gregory. London: Taylor & Walton.

Ligon, J. D. (1978) Reproductive interdependence of piñon jays and piñon pines. *Ecological Monographs* 48, 111–26.

Lindemann, R. L. (1942) The trophic-dynamic aspect of ecology. *Ecology* 23, 399–418.

Lindenmayer, D. B. (1991) A note on the occupancy of nest trees by Leadbeater's possum in the montane ash forests of the Central Highlands of Victoria. *Victorian Naturalist* 108, 128–9.

Lindenmayer, D. B. (1997) Differences in the biology and ecology of arboreal marsupials in forests of southeastern Australia. *Journal of Mammalogy* 78, 1117–27.

Lindenmayer, D. B. (2000) Factors at multiple scales affecting distribution patterns and their implications for animal conservation: Leadbeater's possum as a case study. *Biodiversity and Conservation* 9, 15–35.

Lindenmayer, D. B. and Lacy, R. C. (1995) Metapopulation viability of Leadbeater's possum, *Gymnobelideus leadbeateri*, in fragmented old-growth forests. *Ecological Applications* 5, 164–82.

Lindenmayer, D. B., Cunningham, R. B., and Donnelly, C. F. (1993) The conservation of arboreal marsupials in the montane ash forests of the Central Highlands of Victoria, south-east Australia, IV: the presence and abundance or arboreal marsupials in retained linear habitats (wildlife corridors) within logged forest. *Biological Conservation* 66, 207–21.

Lindenmayer, D. B., Mackey, B. G., Cunningham, R. B., Donnelly, C. F., Mullen, I. C., McCarthy, M. A., and Gill, A. M. (2000) Factors affecting the presence of cool temperate rain forest tree myrtle (*Nothofagus cunninghamii*) in southeastern Australia: integrating climatic, terrain and disturbance predictors of distribution patterns. *Journal of Biogeography* 27, 1001–9.

Lindsey, C. C. (1966) Body size of poikilotherm vertebrates at different latitudes. *Evolution* 20, 456–65.

Linton, D. L. (1949) The delimitation of morphological regions. *Transactions of the Institute of British Geographers* 14, 86–7.

Lodge, D. M. (1993) Biological invasions: lessons for ecology. *Trends in Ecology and Evolution* 8, 133–7.

Lomolino, M. V. (2000a) Ecology's most general, yet protean pattern: the species–area relationship. *Journal of Biogeography* 27, 17–26.

Lomolino, M. V. (2000b) A species-based theory of insular biogeography. *Global Ecology & Biogeography* 9, 39–58.

Lomolino, M. V. and Weiser, M. D. (2001) Towards a more general species–area relationship: diversity on all islands, great and small. *Journal of Biogeography* 28, 431–45.

Lomolino, M. V., Brown, J. H., and Davis, R. (1989) Island biogeography of montane forest mammals in the American Southwest. *Ecology* 70, 180–94.

Losos, J. B. and Glor, R. E. (2003) Phylogenetic comparative methods and the geography of speciation. *Trends in Ecology and Evolution* 18, 220–7.

Losos, J. B. and Schluter, D. (2000) Analysis of an evolutionary species–area relationship. *Nature* 408, 847–50.

Lotka, A. J. (1925) *Elements of Physical Biology*, Baltimore, MD: Williams & Wilkins. Reprinted with corrections and bibliography as *Elements of Mathematical Biology*. New York: Dover, 1956.

Loughlin, T. R. and Miller, R. V. (1989) Growth of the northern fur seal colony on Bogoslof Island, Alaska. *Arctic* 42, 368–72.

Lovelock, J. E. (1979) *Gaia: A New Look at Life on Earth*. Oxford: Oxford University Press.

Lovelock, J. E. (1989) Geophysiology. *Transactions of the Royal Society of Edinburgh: Earth Sciences* 80, 169–75.

Luckinbill, L. S. (1973) Coexistence in laboratory populations of *Paramecium aurelia* and its predator *Didinium nasutum*. *Ecology* 54, 1320–7.

Luckinbill, L. S. (1974) The effects of space and enrichment on a predator–prey system. *Ecology* 55, 1142–7.

Lundelius, E. L., Jr (1987) The Pleistocene mammalian crisis: habitat destruction as an extinction mechanism. *AnthroQuest* 37, 13–14.

Lundelius, E. L., Jr, Graham, R. W., Anderson, E., Guilday, J., Holman, J. A., Steadman, D., and Webb, S. D. (1983) Terrestrial vertebrate faunas. In S. C. Porter (ed.) *Late-Quaternary Environments of the United States. Vol. 1. The Late Pleistocene*, pp. 311–53. London: Longman.

Luo, Z.-X., Ji, Q., Wible, J. R., and Yuan, C.-X. (2003) An early Cretaceous tribosphenic mammal and metatherian evolution. *Science* 302, 1934–40.

Lyell, C. (1863) *The Geological Evidences of the Antiquity of Man with Remarks on the Origin of Species by Variation*. London: John Murray.

Mabry, K. E., Dreelin, E. A., and Barrett, G. W. (2003) Influence of landscape elements on population densities and habitat use of three small-mammal species. *Journal of Mammalogy* 84, 20–5.

MacArthur, R. H. (1958) Population ecology of some warblers of northeastern coniferous forests. *Ecology* 39, 599–619.

MacArthur, R. H. and Wilson, E. O. (1963) An equilibrium theory of insular zoogeography. *Evolution* 17, 373–87.

MacArthur, R. H. and Wilson, E. O. (1967) *The Theory of Island Biogeography*. Princeton, NJ: Princeton University Press.

McCollin, D. (1993) Avian distribution patterns in a fragmented wooded landscape (North Humberside, U.K.): the role of between-patch and within-patch structure. *Global Ecology and Biogeography Letters* 3, 48–62.

McCollin, D. (1998) Forest edges and habitat selection in birds: a functional approach. *Ecography* 21, 247–60.

MacFadden, B. J. (1980) Rafting mammals or drifting islands?: biogeography of the Greater Antillean insectivores. *Journal of Biogeography* 7, 11–22.

McIntyre, S. and Barrett, G. W. (1992) Habitat variegation, an alternative to fragmentation. *Conservation Biology* 6, 146–7.

McKenna, M. C. (1973) Sweepstakes, filters, corridors, Noah's arks, and beached Viking funeral ships in palaeogeography. In D. H. Tarling and S. K. Runcorn (eds) *Implications of Continental Drift to the Earth Sciences, Volume 1*, pp. 295–308. London and New York, Academic Press.

McLaren, D. J. (1988) Detection and significance of mass killings. In N. J. McMillan, A. F. Embry, and D. J. Glass (eds) *Devonian of the World. Volume III: Paleontology, Palaeoecology and Biostratigraphy* (Proceedings of the Second International Symposium on the Devonian System, Calgary, Canada), pp. 1–7. Calgary, Canada: Canadian Society of Petroleum Geologists.

MacLulich, D. A. (1937) *Fluctuations in the Numbers of the Varying Hare* (Lepus americanus) (University of Toronto Studies, Biological Series, No. 43). Toronto, Canada: University of Toronto.

MacPhee, R. D. and Marx, P. A. (1997) The 40,000 year plague: humans, hyperdisease, and first-contact extincitons. In S. A. Goodman and B. D. Patterson (eds) *Natural Change and Human Impact in Madagascar*, pp. 169–217. Washington, DC: Smithsonian Institution Press.

Madigan, M. T. and Marrs, B. L. (1997) Extremophiles. *Scientific American* 276, 66–71.

Malhotra, A. and Thorpe, R. S. (1991) Experimental detection of rapid evolutionary response in natural lizard populations. *Nature* 353, 347–8.

Mallet, J. (2001) The speciation revolution. *Journal of Evolutionary Biology* 14, 887–8.

Mares, M. A. and Seine, R. H. (2000) The fauna of inselbergs. In S. Porembski and W. Barthlott (eds) *Inselbergs: Biotic Diversity of Isolated Rock Outcrops in Tropical and Temperate Regions* (Ecological Studies, vol. 146), pp. 483–91. Berlin: Springer.

Marshall, J. K. (1978) Factors limiting the survival of *Corynephorus canescens* (L.) Beauv. in Great Britain at the northern edge of its distribution. *Oikos* 19, 206–16.

Marshall, L. G. (1980) Marsupial paleobiogeography. In L. L. Jacobs (ed.) *Aspects of Vertebrate History: Essays in Honor of Edwin Harris Colbert*, pp. 345–86. Flagstaff, AZ: Museum of Northern Arizona Press.

Marshall, L. G. (1981a) The Great American Interchange: an invasion induced crisis for South American mammals. In M. H. Nitecki (ed.) *Biotic*

Crises in Ecological and Evolutionary Time, pp. 133–229. New York: Academic Press.

Marshall, L. G. (1981b) The Argentine connection. *Field Museum of Natural History Bulletin* (Chicago) 52, 16–25.

Marshall, L. G. (1994) The terror birds of South America. *Scientific American* 270, 64–9.

Martin, A. A. and Tyler, M. J. (1978) The introduction into western Australia of the frog *Limnodynastes tasmaniensis*. *Australian Zoologist* 19, 320–44.

Martin, P. S. (1984) Prehistoric overkill: the global model. In P. S. Martin and R. G. Klein (eds) *Quaternary Extinctions: A Prehistoric Revolution*, pp. 354–403. Tucson, AZ: The University of Arizona Press.

Martin, R., Rodriguez, A., and Delibes, M. (1995) Local feeding specialization by badgers (*Meles meles*) in a Mediterranean environment. *Oecologia* 101, 45–50.

May, R. M. (1976) Simple mathematical models with very complicated dynamics. *Nature* 261, 459–67.

May, R. M. (1981) Models for single populations. In R. M. May (ed.) *Theoretical Ecology: Principles and Applications*, 2nd edn, pp. 5–29. Oxford: Blackwell Scientific Publications.

Mayr, E. (1942) *Systematics and the Origin of Species*. New York: Columbia University Press.

Menge, B. A., Berlow, E. L., Blanchette, C. A., Navarrete, S. A., and Yamada, S. B. (1994) The keystone species concept: variation in interaction strength in a rocky intertidal habitat. *Ecological Monographs* 64, 249–86.

Merriam, C. H. (1894) Laws of temperature control of the geographic distribution of terrestrial animals and plants. *National Geographic Magazine* 6, 229–38.

Meyer, G. A., Wells, S. G., Balling, R. C., Jr, and Jull, A. J. T. (1992) Response of alluvial systems to fire and climate change in Yellowstone National Park. *Nature* 357, 147–50.

Midgley, G. F., Hannah, L., Millar, D., Rutherford, M. C., and Powrie, L. W. (2002) Assessing the vulnerability of species richness to anthropogenic climate change in a biodiversity hotspot. *Global Ecology & Biogeography* 11, 445–51.

Milchunas, D. G. and Lauenroth, W. K. (1993) Quantitative effects of grazing on vegetation and soils over a global range of environments. *Ecological Monographs* 63, 327–66.

Miller, K. R. (1996) *Balancing the Scales: Guidelines for Increasing Biodiversity's Chances Through Bioregional Management*. Washington, DC: World Resources Institute.

Mitchell-Jones, A. J., Amori, G., Bogdanowicz, W., Kryštufek, B., Reijnders, P. J. H., Spitzennberger, F., Stubbe, M., Thissen, J. B. M., Vohralík, V., and Zime, J. (1999). *The Atlas of European Mammals*. London: Academic Press.

Mittermeier, R. A., Myers, N., Gil, P. R., and Mittermeier, C. G. (1999) *Hotspots: Earth's Biologically Richest and Most Endangered Terrestrial Ecoregions*. Mexico City, Mexico: Cemex and Agrupación Sierra Madre.

Molfino, B., Heusser, L. H., and Woillard, G. M. (1984) Frequency components of a Grand Pile pollen record: evidence of precessional orbital forcing. In A. Berger, J. Imbrie, J. Hays, G. Kukla, and B. Saltzman (eds) *Milankovitch and Climate: Understanding the Response of Astronomical Forcing, Part 1* (Proceedings of the NATO Advanced Research Workshop on Milankovitch and Climate, Palisades, New York, 1982. NATO ASI Series C, Mathematical and Physical Sciences, vol. 126), pp. 392–404. Dordrecht: D. Reidel.

Moloney, C. L., Cooper, J., Ryan, P. G., and Siegfried, W. R. (1994) Use of a population model to assess the impact of longline fishing on wandering albatross *Diomedia exulans* populations. *Biological Conservation* 70, 195–203.

Morhardt, J. E. and Gates, D. M. (1974) Energy-exchange analysis of the Belding ground squirrel and its habitat. *Ecological Monographs* 44, 14–44.

Morkill, A. E. and Anderson, S. H. (1991) Effectiveness of marking powerlines to reduce sandhill crane collisions. *Wildlife Society Bulletin* 19, 442–9.

Moyle, P. B. and Williams, J. E. (1990) Biodiversity loss in the temperate zone: decline of the native fish fauna of California. *Conservation Biology* 4, 275–84.

Mueller-Dombois, D., Spatz, G., Conant, S., Tomich, P. Q., Radovsky, F. J., Tenorio, J. M., Gagné, W., Brennan, B. M., Mitchell, W. C., Springer, D., Samuelson, G. A., Gressit, J. L., Steffan, W. A., Paik, Y. K., Sung, K. C., Hardy, D. E., Delfinado, M. D., Fujii, D., Doty, M. S., Watson, L. J., Stoner, M. F., and Baker, G. E. (1981) Altitudinal distribution of organisms along an island mountain transect. In D. Mueller-Dombois, K. W. Bridges, and H. L. Carson (eds) *Island Ecosystems: Biological Organization in Selected Hawaiian Communities* (US/IBP Synthesis Series, vol. 15), pp. 77–180. Stroudsberg, PA, and Woods Hole, MA: Hutchinson Ross.

Mullin, B. H. (1998) The biology and management of purple loosestrife (*Lythrum salicaria*). *Weed Technology* 12, 397–401.

Murphy, D. D. and Freas, K. (1988) Habitat-based conservation: the case of the Amargosa vole. *Endangered Species Update* 5, 6.

Myklestad, Å. and Birks, H. J. B. (1993) A numerical analysis of the distribution patterns of *Salix* L. species in Europe. *Journal of Biogeography* 20, 1–32.

Naiman, R. J. and Décamps, H. (1997) The ecology of interfaces: riparian zones. *Annual Review of Ecology and Systematics* 28, 621–58.

Naiman, R. J., Johnson, C. A., and Kelley, J. C. (1988) Alteration of North American streams by beaver. *BioScience* 38, 753–62.

Naiman, R. J., Pinay, G., Johnson, C. A., and Pastor, J. (1994) Beaver influences on the long-term biogeochemical characteristics of boreal forest drainage networks. *Ecology* 75, 905–21.

Naqvi, S. M., Howell, R. D., and Sholas, M. (1993) Cadmium and lead residues in field-collected red swamp crayfish (*Procambarus clarkii*) and uptake by alligator weed, *Alternanthera philoxiroides*. *Journal of Environmental Science and Health* B28, 473–85.

Neilson, R. P. (1993a) Vegetation redistribution: a possible biosphere source of CO_2 during climatic change. *Water, Air, and Soil Pollution* 70, 659–73.

Neilson, R. P. (1993b) Transient ecotone response to climatic change: some conceptual and modelling approaches. *Ecological Applications* 3, 385–95.

Nelson, T. C. (1955) Chestnut replacement in the Southern Highlands. *Ecology* 36, 352–3.

Nevo, E. (1986) Mechanisms of adaptive speciation at the molecular and organismal levels. In S. Karlin and E. Nevo (eds) *Evolutionary Process and Theory*, pp. 438–74. New York: Academic Press.

Newmark, W. D. (1987) A land-bridge island perspective on mammalian extinctions in western North American parks. *Nature* 325, 430–2.

Newton, I. (1972) *Finches*. London: Collins.

Nicholls, A. O. and McKenzie, N. J. (1994) Environmental control of the local-scale distribution of funnel ants, *Aphaenogaster longiceps*. *Memoirs of the Queensland Museum* 36, 165–72.

Niklas, K. J., Tiffney, B. H., and Knoll, A. H. (1983) Patterns in vascular land plant diversification. *Nature* 303, 614–16.

Noon, B. R. and Franklin, A. B. (2002) Scientific research and the spotted owl (*Strix occidentalis*): opportunities for major contributions to avian population ecology. *The Auk* 119, 311–20.

Nores, M. (1999) An alternative hypothesis for the origin of Amazonian bird diversity. *Journal of Biogeography* 26, 475–85.

Noss, R. F. (1983) A regional landscape approach to maintain diversity. *BioScience* 33, 700–6.

Noss, R. F. (1987) Corridors in real landscapes: a reply to Simberloff and Cox. *Conservation Biology* 1, 159–64.

Noss, R. F. and Cooperrider, A. Y. (1994) *Saving Nature's Legacy: Protecting and Restoring Biodiversity.* Washington, DC: Island Press.

Odum, H. T. (1971) *Environment, Power, and Society.* New York: John Wiley & Sons.

Orphanides, G. M. (1993) Control of *Saissetia oleae* (Hom.: Coccidae) in Cyprus through establishment of *Metaphycus bartletti* and *M. helvolus* (Hym.: Encyrtidae). *Entomophaga* 38, 235–9.

Osburn, R. C., Dublin, L. I., Shimer, H. W., and Lull, R. S. (1903) Adaptation to aquatic, arboreal, fossorial, and cursorial habits in mammals. *American Naturalist* 37, 651–65, 731–6, 819–25; 38, 322–32.

Ouin, A., Paillat, G., Butet, A., and Burel, F. (2000) Spatial dynamics of wood mouse (*Apodemus sylvaticus*) in an agricultural landscape under intensive use in the Mont Saint Michel Bay (France). *Agriculture, Ecosystems and Environment* 78, 159–65.

Overpeck, J. T., Rind, D., and Goldberg, R. (1990) Climate-induced changes in forest disturbance and vegetation. *Nature* 343, 51–3.

Owen, R. (1846) *A History of British Fossil Mammals and Birds.* London: John van Voorst.

Owen-Smith, R. N. (1987) Pleistocene extinctions: the pivotal role of megaherbivores. *Paleobiology* 13, 351–62.

Owen-Smith, R. N. (1988) *Megaherbivores: The Influence of Very Large Body Size on Ecology.* Cambridge: Cambridge University Press.

Owen-Smith, R. N. (1989) Megafaunal extinctions: the conservation message from 11,000 years BP. *Conservation Biology* 3, 405–12.

Paine, R. T. (1974) Intertidal community structure: experimental studies on the relationship between a dominant competitor and its principle predator. *Oecologia* 15, 93–120.

Parker, A., Holden, A. N. G., and Tomley, A. J. (1994) Host specificity testing and assessment of the pathogenicity of the rust, *Puccinia abrupta* var. *partheniicola*, as a biological control agent of parthenium weed (*Parthenium hysterophorus*). *Plant Pathology* 43, 1–16.

Parkes, J. P. (1993) Feral goats: designing solutions for a designer pest. *New Zealand Journal of Ecology* 17, 71–83.

Parmesan, C. (1996) Climate and species' range. *Nature* 382, 765–6.

Pastor, J. R. and Post, W. M. (1988) Response of northern forests to CO_2-induced climatic change. *Nature* 334, 55–8.

Patterson, B. D. (1984) Mammalian extinction and biogeography in the Southern Rocky Mountains. In M. H. Nitecki (ed.) *Extinctions*, pp. 247–93. Chicago: University of Chicago Press.

Patterson, B. D. (1987) The principle of nested subsets and its implications for biological conservation. *Conservation Biology* 1, 323–34.

Patterson, B. D. (1991) The integral role of biogeographic theory in the conservation of tropical rain forest diversity. In M. A. Mares and D. J. Schmidly (eds) *Latin American Mammalogy: History, Biodiversity, and Conservation*, pp. 125–49. Norman, OK: University of Oklahoma Press.

Patterson, B. D. and Atmar, W. (1986) Nested subsets and the structure of insular mammalian faunas and archipelagos. *Biological Journal of the Linnean Society* 28, 65–82.

Patterson, B. D. and Brown, J. H. (1991) Regionally nested patterns of species composition in granivorous rodent assemblages. *Journal of Biogeography* 18, 395–402.

Patterson, C. (1982) Cladistics and classification. In J. Cherfas (ed.) *Darwin Up to Date* (A New Scientist Guide), pp. 35–9. London: IPC Magazines.

Payette, S. (1993) The range limit of boreal tree species in Quebec–Labrador: an ecological and palaeoecological interpretation. *Review of Palaeobotany and Palynology* 79, 7–30.

Pearl, R. (1928) *The Rate of Living*, New York: Alfred Knopf.

Pearlstine, L., McKellar, H., and Kitchens, W. (1985) Modelling the impacts of a river diversion on bottomland forest communities in the Santee River floodplain, South Carolina. *Ecological Modelling* 29, 283–302.

Pears, N. (1985) *Basic Biogeography*, 2nd edn. Harlow: Longman.

Pedlar, J. H., Fahrig, L., and Merriam, H. G. (1997) Raccoon habitat use at 2 spatial scales. *Journal of Wildlife Management* 61, 109–12.

Pérez, F. L. (1987) Soil moisture and the upper altitudinal limit of giant paramo rosette. *Journal of Biogeography* 14, 173–86.

Pérez, F. L. (1989) Some effects of giant Andean stem-rosettes on ground microclimate, and their ecological significance. *International Journal of Biometeorology* 33, 131–5.

Pérez, F. L. (1991) Soil moisture and the distribution of giant Andean rosettes on talus slopes of a desert paramo. *Climate Research* 1, 217–31.

Perfit, M. R. and Williams, E. E. (1989) Geological constraints and biological reintroductions in the evolution of the Caribbean Sea and its islands. In C. A. Woods (ed.) *Biogeography of the West Indies: Past, Present, and Future*, pp. 47–102. Gainesville, FL: Sandhill Crane Press.

Perring, F. H. and Walters, S. M. (1962) *Atlas of the British Flora*. London: Nelson.

Perrins, C. (ed.) (1990) *The Illustrated Encyclopaedia of Birds: The Definitive Guide to Birds of the World*. London: Headline Publishing.

Peters, R. L. (1992a) Conservation of biological diversity in the face of climatic change. In R. L. Peters and T. E. Lovejoy (eds) *Global Warming and Biological Diversity*, pp. 15–30. New Haven, CT, and London: Yale University Press.

Peters, R. L. (1992b) Introduction. In R. L. Peters and T. E. Lovejoy (eds) *Global Warming and Biological Diversity*, pp. 3–14. New Haven, CT, and London: Yale University Press.

Phillipson, J. (1966) *Ecological Energetics* (The Institute of Biology's Studies in Biology, No. 1). London: Edward Arnold.

Pickett, S. T. A. and Thompson, J. N. (1978) Patch dynamics and the design of nature reserves. *Biological Conservation* 13, 23–37.

Pielou, E. C. (1979) *Biogeography*. New York: John Wiley & Sons.

Pigott, C. D. (1974) The response of plants to climate and climatic change. In F. H. Perring (ed.) *The Flora of a Changing Britain*, pp. 32–44. London: Classey.

Pigott, C. D. (1981) Nature of seed sterility and natural regeneration of *Tilia cordata* near its northern limit in Finland. *Annales Botanici Fennici* 18, 255–63.

Pigott, C. D. and Huntley, J. P. (1981) Factors controlling the distribution of *Tilia cordata* at the northern limits of its geographical range. III. Nature and causes of seed sterility. *New Phytologist* 87, 817–39.

Pimm, S. L. (1991) *Balance of Nature? Ecological Issues in the Conservation of Species and Communities*. Chicago, IL: Chicago University Press.

Pither, J. (2002) Climate tolerance and interspecific variation in geographic range size. *Proceedings of the Royal Society of London* 270B, 475–81.

Poiani, K. A. and Johnson, W. C. (1991) Global warming and prairie wetlands. *BioScience* 41, 611–18.

Poiani, K. A., Richter, B. D., Anderson, M. G., and Richter, H. E. (2000) Biodiversity conservation at multiple scales: functional sites, landscapes, and networks. *BioScience* 50, 133–46.

Polischuk, S. C., Letcher, R. J., Norstrom, R. J., and Ramsay, M. A. (1995) Preliminary results of fasting on the kinetics of organochlorines in polar bears (*Ursus maritimus*). *Science of the Total Environment* 160–1, 465–72.

Pond, D. (1992) Protective-commensal mutualism between the queen scallop *Chlamys opercularis* (Linnaeus) and the encrusting sponge *Suberites*. *Journal of Molluscan Studies* 58, 127–34.

Porembski, S., Becker, U., and Seine, R. (2000a) Islands on islands: habitats on inselbergs. In S. Porembski and W. Barthlott (eds) *Inselbergs: Biotic Diversity of Isolated Rock Outcrops in Tropical and Temperate Regions* (Ecological Studies, vol. 146), pp. 49–67. Berlin: Springer.

Porembski, S., Seine, R., and Barthlott, W. (2000b) Factors controlling species richness of inselbergs. In S. Porembski and W. Barthlott (eds) *Inselbergs: Biotic Diversity of Isolated Rock Outcrops in Tropical and Temperate Regions* (Ecological Studies, vol. 146), pp. 451–81. Berlin: Springer.

Pounds, J. A. (1990) Disappearing gold. *BBC Wildlife* 8, 812–17.

Pounds, J. A. and Crump, M. L. (1994) Amphibian declines and climate disturbance: the case of the golden toad and the harlequin frog. *Conservation Biology* 8, 72–85.

Pregill, G. K. (1981) An appraisal of the vicariance hypothesis of Caribbean biogeography and its application to West Indian terrestrial vertebrates. *Systematic Zoology* 30, 147–55.

Preston, C. D., Pearman, D. A., and Dines, T. D. (2002) *New Atlas of the British and Irish Flora: An Atlas of the Vascular Plants of Britain, Ireland, the Isle of Man and the Channel Islands*. Oxford: Oxford University Press.

Preston, F. W. (1962) The canonical distribution of commonness and rarity. *Ecology* 43, 185–215, 410–32.

Priddel, D. and Wheeler, R. (1994) Mortality of captive-raised malleefowl, *Leipoa ocellata*, released into a mallee remnant within the wheat-belt of New South Wales. *Wildlife Research* 21, 543–52.

Prins, H. H. T. and Van der Jeugd, H. P. (1993) Herbivore population crashes and woodland structure in East Africa. *Journal of Ecology* 81, 305–14.

Puigcerver, M., Gallego, S., Rodriguez-Teijeiro, J. D., and Senar, J. C. (1992) Survival and mean life span of the quail *Coturnix c. coturnix*. *Bird Study* 39, 120–3.

Purdue, J. R. (1980) Clinal variations in some mammals during the Holocene in Missouri. *Quaternary Research* 13, 242–58.

Purdue, J. R. (1989) Changes during the Holocene in the size of the white-tailed deer (*Odocoileus virginianus*) from central Illinois. *Quaternary Research* 32, 307–16.

Quade, J. and Cerling, T. E. (1995) Expansion of C_4 grasses in the Late Miocene of northern Pakistan: evidence from stable isotopes in paleosols. *Palaeogeography, Palaeoclimatology, Palaeoecology* 115, 91–116.

Rabb, G. B. (1994) The changing roles of zoological parks in conserving biological diversity. *American Zoologist* 34, 159–64.

Rabinowitz, A. R. and Walker, S. R. (1991) The carnivore community in a dry tropical forest mosaic in Huai Kha Khaeng Wildlife Sanctuary, Thailand. *Journal of Tropical Ecology* 7, 37–47.

Rahbek, C. (1995) The elevational gradient of species richness: a uniform pattern? *Ecography* 18, 200–5.

Rampino, M. R. and Caldeira, K. (1993) Major episodes of geologic change: correlations, time structure and possible causes. *Earth and Planetary Science Letters* 114, 215–27.

Rapoport, E. H. (1982) *Aerography: Geographical Strategies of Species*. Oxford: Pergamon Press.

Raunkiaer, C. (1934) *The Life Forms of Plants and Statistical Plant Geography, Being the Collected Papers of C. Raunkiaer*, translated by H. Gilbert-Carter and A. G. Tansley. Oxford: Clarendon Press.

Raup, D. M. and Sepkoski, J. J. (1982) Mass extinctions in the marine fossil record. *Science* 215, 1501–3.

Raup, H. M. (1981) Physical disturbance in the life of plants. In M. H. Nitecki (ed.) *Biotic Crises in Ecological and Evolutionary Time*, pp. 39–52. New York: Academic Press.

Raven, P. H. (1963) Amphitropical relationships in the floras of North and South America. *The Quarterly Review of Biology* 38, 151–77.

Ray, P. M. and Alexander, W. E. (1966) Photoperiodic adaptation to latitude in *Xanthium strumasium. American Journal of Botany* 53, 806–16.

Reddish, P. (1996) *Spirits of the Jaguar: The Natural History and Ancient Civilizations of the Caribbean and Central America*. London: BBC Books.

Reh, W. (1989) Investigations into the influence of roads on the genetic structure of populations of the common frog *Rana temporaria*. In T. E. S. Langton (ed.) *Amphibians and Roads*, pp. 101–3. Shefford, Bedfordshire: ACO Polymer Products.

Reid, D. G., Herrero, S. M., and Code, T. E. (1988) River otters as agents of water loss from beaver ponds. *Journal of Mammalogy* 69, 100–7.

Reid, W. V. and Miller, K. R. (1989) *Keeping Options Alive: The Scientific Basis for Conserving Biodiversity*. Washington, DC: World Resources Institute.

Revelle, R. (1984) The effects of population on renewable resources. *Population Studies* 90, 223–40.

Rhodes II, R. S. (1984) Paleoecological and regional paleoclimatic implications of the Farmdalian Craigmile and Woodfordian Waubonsie mammalian local faunas, southwestern Iowa. *Illinois State Museum Report of Investigations* 40, 1–51.

Ripple, W. J., Johnson, D. H., Hershey, K. T., and Meslow, E. C. (1991) Old-growth and mature forests near spotted owl nests in western Oregon. *Journal of Wildlife Management* 55, 316–18.

Roberts, D. L., Cooper, R. J., and Petit, L. J. (2000) Use of premontane moist forest and shade coffee agroecosystems by army ants in Western Panama. *Conservation Biology* 14, 192–9.

Robinson, J, M. (1990) Lignin, land plants and fungi: biological evolution affecting Phanerozoic oxygen balance. *Geology* 15, 607–10.

Robinson, J. M. (1991) Phanerozoic atmospheric reconstructions: a terrestrial perspective. *Palaeogeography, Palaeoclimatology, Palaeoecology (Global and Planetary Change Section)* 97, 51–62.

Rodríguez de la Fuente, F. (1975) *Animals of South America* (World of Wildlife Series), English language version by John Gilbert. London: Orbis Publishing.

Rohde, K. (1992) Latitudinal gradients in species diversity: the search for the primary cause. *Oikos* 65, 514–27.

Romer, A. S. (1966) *Vertebrate Paleontology*, 3rd edn. Chicago, IL, and London: Chicago University Press.

Root, R. B. (1967) The niche exploitation pattern of the blue-gray gnatcatcher. *Ecological Monographs* 37, 317–50.

Root, T. L. (1988a) Environmental factors associated with avian distributional boundaries. *Journal of Biogeography* 15, 489–505.

Root, T. L. (1988b) Energy constraints on avian distributions and abundances. *Ecology* 69, 330–9.

Root, T. L. and Schneider, S. H. (1993) Can large-scale climatic models be linked with multiscale ecological studies? *Conservation Biology* 7, 256–70.

Rorison, I. H., Sutton, F., and Hunt, R. (1986) Local climate, topography and plant growth in Lathkill Dale NNR. I. A twelve-year summary of solar radiation and temperature. *Plant, Cell, and Environment* 9, 49–56.

Rosenzweig, M. L. (1992) Species diversity gradients: we know more and less than we thought. *Journal of Mammalogy* 73, 715–30.

Rosenzweig, M. L. (1995) *Species Diversity in Space and Time*. Cambridge: Cambridge University Press.

Rossignol-Strick, M. and Planchais, N. (1989) Climate patterns revealed by pollen and oxygen isotope records of a Tyrrhenian sea core. *Nature* 342, 413–16.

Rudge, M. R. (1990) Feral goat. In C. M. King (ed.) *The Handbook of New Zealand Mammals*, pp. 406–23. Auckland: Oxford University Press.

Rupp, T. S., Chapin, F. S., III, and Starfield, A. M. (2000a) Response of subarctic vegetation to transient climatic change on the Seward Peninsula in north-west Alaska. *Global Change Biology* 6, 541–55.

Rupp, T. S., Starfield, A. M., and Chapin, F. S., III (2000b) A frame-based spatially explicit model of subarctic vegetation response to climatic change: a comparison with a point model. *Landscape Ecology* 15, 383–400.

Rykiel, E. J., Jr, Coulson, R. N., Sharpe, P. J. H., Allen, T. F. H., and Flamm, R. O. (1988) Disturbance propagation by bark beetles as an episodic landscape phenomenon. *Landscape Ecology* 1, 129–39.

Sage, B. L. (ed.) (1966a) *Northaw Great Wood: Its History and Natural History*. Hertford: Education Department of the Hertfordshire Country Council.

Sage, B. L. (1966b) Geology. In B. L. Sage (ed.) *Northaw Great Wood: Its History and Natural History*, pp. 22–5. Hertford: Education Department of the Hertfordshire Country Council.

Sakai, A. (1970) Freezing resistance in willows from different climates. *Ecology* 51, 485–91.

Sakai, A. and Otsuka, K. (1970) Freezing resistance of alpine plants. *Ecology* 51, 665–71.

Salisbury, E. J. (1926) The geographical distribution of plants in relation to climatic factors. *Geographical Journal* 57, 312–35.

Sallabanks, R. (1992) Fruit fate, frugivory, and fruit characteristics: a study of the hawthorn, *Crataegus monogyna* (Rosaceae). *Oecologia* 91, 296–304.

Saloman, M. (2002) A revised cline theory that can be used for quantified analyses of evolutionary processes without parapatric speciation. *Journal of Biogeography* 29, 509–17.

Samways, M. J. (1989) Climate diagrams and biological control: an example from the areography of the ladybird *Chilocorus nigritus* (Fabricius, 1798) (Insecta, Coleoptera, Coccinellidae). *Journal of Biogeography* 16, 345–51.

Sanders, N. J., Moss, J., and Wagner, D. (2003) Patterns of ant species richness along elevational gradients in an arid ecosystem. *Global Ecology & Biogeography* 12, 93–102.

Saracco, J. F. and Collazo, J. A. (1999) Carolina bottomland hardwood forest. *Wilson Bulletin* 111, 541–9.

Sarre, S. (1995) Size and structure of populations of *Oedura reticulata* (Reptilia: Gekkonidae) in woodland remnants: implications for the future regional distribution of a currently common species. *Australian Journal of Ecology* 20, 288–98.

Sauer, J. D. (1969) Oceanic islands and biogeographic theory: a review. *Geographical Review* 59, 582–93.

Saunders, H. (1889) *An Illustrated Manual of British Birds*. London: Gurney & Jackson.

Schaetzl, R. J., Burns, S. F., Johnson, D. L., and Small, T. W. (1989a) Tree uprooting: a review of impacts on forest ecology. *Vegetatio* 79, 165–76.

Schaetzl, R. J., Johnson, D. L., Burns, S. F., and Small, T. W. (1989b) Tree uprooting: review of terminology, process, and environmental implications. *Canadian Journal of Forest Research* 19, 1–11.

Schennum, W. E. and Willey, R. B. (1979) A geographical analysis of quantitative morphological variation in the grasshopper *Arphia conspersa*. *Evolution* 33, 64–84.

Schlesinger, W. H., DeLucia, E. H., and Billings, W. D. (1989) Nutrient-use efficiency of woody plants on contrasting soils in the western Great Basin, Nevada. *Ecology* 70, 105–13.

Schmidt-Nielsen, K. and Schmidt-Nielsen, B. (1953) The desert rat. *Scientific American* 189, 73–8.

Schnitzler, A. (1994) European alluvial hardwood forests of large floodplains. *Journal of Biogeography* 21, 605–23.

Schultz, J. (1995) *The Ecozones of the World: The Ecological Divisions of the Geosphere.* Hamburg: Springer.

Schwalbe, C. P., Mastro, V. C., and Hansen, R. W. (1991) Prospects for genetic control of the gypsy moth. *Forest Ecology and Management* 39, 163–71.

Sclater, P. L. (1858) On the general geographical distribution of the members of the class Aves. *Journal of the Linnean Society, Zoology* 2, 130–45.

Sengonca, C., Uygun, N., Kersting, U., and Ulusoy, M. R. (1993) Successful colonization of *Eretmocerus debachi* (Hym.: Aphelinidae) in the eastern Mediterranean citrus region of Turkey. *Entomophaga* 38, 383–90.

Senut, B., Pickford, M., and Dauphin, Y. (1995) Découverte d'œufs de type "Aepyornithoide" dans le Miocène inférieur de Namibie. *Comptes Rendus, Académie des Sciences, Série II: Sciences de la Terre et des Planètes* 320, 71–6.

Sepkoski, J. J., Jr (1989) Periodicity in extinction and the problem of catastrophism in the history of life. *Journal of the Geological Society, London* 146, 7–19.

Sepkoski, J. J., Jr, Bambach, R. K., Raup, D. M., and Valentine, J. W. (1981) Phanerozoic marine diversity and the fossil record. *Nature* 293, 435–7.

Sexton, W. T., Szaro, R. C., Johnson, N. C., and Malik, A. J. (eds) (1999) *Ecological Stewardship: A Common Reference for Ecosystem Management*, 3 vols. New York: Elsevier Science Ltd.

Shafer, C. L. (1990) *Nature Reserves: Island Theory and Conservation Practice.* Washington, DC, and London: Smithsonian Institution Press.

Shafer, C. L. (1997) Terrestrial nature reserve design at the urban/rural interface. In M. W. Schwartz (ed.) *Conservation in Highly Fragmented Landscapes*, pp. 345–78. New York: Chapman & Hall.

Shaffer, M. L. (1981) Minimum population sizes for species conservation. *BioScience* 31, 131–4.

Sheenan, P. M. and Russell, D. A. (1994) Faunal change following the Cretaceous–Tertiary impact: using paleontological data to assess the hazard of impacts. In T. Gehrels (ed.), with the editorial assistance of M. S. Matthews and A. M. Schumann, *Hazards Due to Comets and Asteroids*, pp. 879–93. Tucson, AZ, and London: The University of Arizona Press.

Sheenan, P. M., Fastovsky, D. E., Hoffman, D. E., Berghaus, R. G., and Gabriel, D. L. (1991) Sudden extinction of the dinosaurs: latest Cretaceous, upper Great Plains, U.S.A. *Science* 254, 835–9.

Shelford, V. E. (1911) Physiological animal geography. *Journal of Morphology* 22, 551–618.

Shimwell, D. W. (1971) *Description and Classification of Vegetation.* London: Sidgwick & Jackson.

Shorten, M. (1954) *Squirrels.* London: Collins.

Shugart, H. H. and Noble, J. R. (1981) A computer model of succession and fire response of the high altitude *Eucalyptus* forest of the Brindabella Range, Australian Capital Territory. *Australian Journal of Ecology* 6, 149–64.

Signor, P. W. (1994) Biodiversity in geological time. *American Zoologist* 43, 23–32.

Simberloff, D. and Dayan, T. (1991) The guild concept and the structure of ecological communities. *Annual Review of Ecology and Systematics* 22, 115–43.

Simberloff, D. S. and Wilson, E. O. (1970) Experimental zoogeography of islands: a two-year record of colonization. *Ecology* 51, 934–7.

Simmons, I. G. (1979) *Biogeography: Natural and Cultural*. London: Edward Arnold.

Simpson, G. G. (1940) Mammals and land bridges. *Journal of the Washington Academy of Science* 30, 137–63.

Simpson, G. G. (1953) *Evolution and Geography: An Essay on Historical Biogeography with Special Reference to Mammals* (Condon Lectures). Eugene, OR: Oregon State System of Higher Education.

Simpson, G. G. (1961) Historical zoogeography of Australian mammals. *Evolution* 16, 413–46.

Simpson, G. G. (1980) *Splendid Isolation: The Curious History of South American Mammals*. New Haven, CT, and London: Yale University Press.

Sinsch, U. (1992) Structure and dynamic of a natterjack toad metapopulation (*Bufo calamita*). *Oecologia* 90, 489–99.

Sjögren, P. (1991) Extinction and isolation gradients in metapopulations: the case of the pool frog (*Rana lessonae*). *Biological Journal of the Linnean Society* 42, 135–47.

Sluis, W. J. (2002) Patterns of species richness and composition in re-created grassland. *Restoration Ecology* 10, 677–84.

Smith, A. G., Smith, D. G., and Funnell, B. M. (1994) *Atlas of Mesozoic and Cenozoic Coastlines*. Cambridge: Cambridge University Press.

Smith, A. T. (1974) The distribution and dispersal of pikas: consequences of insular population structure. *Ecology* 55, 1112–19.

Smith, A. T. (1980) Temporal changes on insular populations of the pika (*Ochotona princeps*). *Ecology* 61, 8–13.

Smith, C. H. (1983) A system of world mammal faunal regions. I. Logical and statistical derivation of the regions. *Journal of Biogeography* 10, 455–66.

Smith, F. D. M., May, R. M., and Harvey, P. H. (1994) Geographical ranges of Australian mammals. *Journal of Animal Ecology* 63, 441–50.

Smith, J. M. (1974) *Models in Ecology*. London: Cambridge University Press.

Solomon, A. M. (1986) Transient response of forests to CO_2-induced climatic change: simulation modeling experiments in eastern North America. *Oecologia* 68, 567–79.

Soulé, M. (1990) The onslaught of alien species and other challenges in the coming decades. *Conservation Biology* 4, 233–9.

Soulé, M. E. and Gilpin, M. E. (1991) The theory of wildlife corridor capability. In D. A. Saunders and R. J. Hobbs (eds) *Nature Conservation 2: The Role of Corridors*, pp. 3–8. Chipping Norton, Australia: Surrey Beatty & Sons.

Spence, D. H. N. (1970) Scottish serpentine vegetation. *Oikos* 21, 22–31.

Spencer, J. W. and Kirby, K. J. (1992) An inventory of ancient woodland for England and Wales. *Biological Conservation* 62, 77–93.

Spicer, R. A. (1993) Palaeoecology, past climate systems, and C_3/C_4 photosynthesis. *Chemosphere* 27, 947–78.

Springer, M. S., Kirsch, J. A. W., and Case, J. A. (1997) The chronicle of marsupial evolution. In T. J. Givnish and K. J. Sytsma (eds) *Molecular Evolution and Adaptive Radiation*, pp. 129–61. Cambridge: Cambridge University Press.

Stanley, S. M. (1988a) Paleozoic mass extinctions: shared patterns suggest global cooling as a common cause. *American Journal of Science* 288, 334–52.

Stanley, S. M. (1988b) Climatic cooling and mass extinction of Paleozoic reef communities. *Palaios* 3, 228–32.

Stanley, S. M. (1990) Delayed recovery and the spacing of major extinctions. *Paleobiology* 16, 401–14.

Stanley, S. M. (1992) An ecological theory for the origin of *Homo*. *Paleobiology* 18, 237–57.

Stanley, S. M. (1995) Climatic forcing and the origin of the human genus. In Board on Earth Sciences and Resources Commission on Geosciences, Environment, and Resources, National Research Council, *Effects of Past Global Change on Life*, pp. 233–43. Washington, DC: National Academy Press.

Stanley, S. M. and Yang, X. (1994) A double mass extinction at the end of the Paleozoic era. *Science* 266, 1340–4.

Stanton, M. L. and Galen, C. (1989) Consequences of flower heliotropism for reproduction in an alpine buttercup (*Ranunculus adoneus*). *Oecologia* 78, 477–85.

Stephens, E. P. (1956) The uprooting of trees, a forest process. *Soil Science Society of America Proceedings* 20, 113–16.

Stevens, C. E., Diamond, A. W., and Gabor, T. S. (2002) Anuran call surveys on small wetlands in Prince Edward Island, Canada restored by dredging of sediments. *Wetlands* 22, 90–9.

Stevens, C. E., Gabor, T. S. and Diamond, A. W. (2003) Use of restored small wetlands by breeding waterfowl in Prince Edward Island, Canada. *Restoration Ecology* 11, 3–12.

Stevens, G. C. (1989) The latitudinal gradient in geographical range: how so many species coexist on the tropics. *The American Naturalist* 133, 240–56.

Stevens, G. C. (1992) The elevational gradient in altitudinal range: an extension of Rapoport's latitudinal rule to altitude. *The American Naturalist* 140, 893–911.

Stoltz, J. F., Botkin, D. B., Dastoor, M. N. (1989) The integral biosphere. In M. B. Rambler, L. Margulis, and R. Fester (eds) *Global Ecology: Towards a Science of the Biosphere*, pp. 31–49. San Diego, CA: Academic Press.

Storer, R. W. (1966) Sexual dimorphism and food habits in three North American accipiters. *The Auk* 83, 423–6.

Storfer, A. (2003) Amphibian declines: future directions. *Diversity and Distributions* 9, 151–63.

Stoutjesdijk, P. and Barkman, J. J. (1992) *Microclimate, Vegetation and Fauna*. Uppsala, Sweden: Opulus Press.

Strahan, R. (ed.) (1995) *Mammals of Australia*, 2nd edn. Washington, DC: Smithsonian Institution Press.

Stuart, A. J. (1991) Mammalian extinctions in the Late Pleistocene of northern Eurasia and North America. *Biological Reviews* 66, 453–62.

Stuart, A. J. (1993) The failure of evolution: Late Quaternary mammalian extinctions in the Holarctic. *Quaternary International* 19, 1–7.

Sukachev, V. N. and Dylis, N. V. (1964) *Fundamentals of Forest Biogeocoenology*, translated by J. M. Maclennan. Edinburgh and London: Oliver & Boyd.

Tagawa, H. (1992) Primary succession and the effect of first arrivals on subsequent development of forest types. *GeoJournal* 28, 175–83.

Takhtajan, A. L. (1986) *The Floristic Regions of the World*, translated by T. Crovello. Berkeley, CA: The University of California Press.

Tambussi, C. P., Noriega, J. I., Gaździcki, A., Tatur, A., Reguero, M. A., and Vizcaino, S. F. (1994) Ratite bird from the Paleogene La Meseta Formation, Seymour Island, Antarctica. *Polish Polar Research* 15, 15–20.

Tansley, A. G. (1935) The use and abuse of vegetational concepts and terms. *Ecology* 16, 284–307.

Tansley, A. G. (1939) *The British Isles and Their Vegetation*. Cambridge: Cambridge University Press.

Tassy, P. and Debruyne, R. (2001) The timing of early Elephantinae differentiation: the palaeontological record with a short comment on molecular data. In J. Shoshani (Chair) Workshop: Elephantidae: origine ed evoluzione (Primo Congresso Internazionale La Terra delgi Elefanti Roma 16–20 ottobre; Campidoglio, Sala della Protomoteca, CNR, Piazzale Aldo Moro), pp. 685–7. Available on-line at http//www.cq.rm.cnr.it/elephants2001/T6.htm.

Tattersall, I. (1993) Madagascar's lemurs. *Scientific American* 268(1), 90–7.

Taulman, J. F. and Robbins, L. W. (1996) Recent range expansion and distributional limits of the nine-banded armadillo (*Dasypus novemcinctus*) in the United States. *Journal of Biogeography* 23, 635–48.

Taylor, C. R. (1969) The eland and the oryx. *Scientific American* 220, 89–95.

Terbough, J. (1988) The big things that run the world: a sequel to E. O. Wilson. *Conservation Biology* 2, 402–3.

Terborgh, J. and Weske, J. S. (1975) The role of competition in the distribution of Andean birds. *Ecology* 56, 562–76.

Tewari, J. C., Rikhari, H. C., and Singh, S. P. (1989) Compositional and structural features of certain tree stands along an elevational gradient in central Himalaya. *Vegetatio* 85, 107–20.

Thoday, J. M. (1972) Disruptive selection. *Proceedings of the Royal Society of London*, 182B, 109–43.

Thomas, C. D. and Jones, T. M. (1993) Partial recovery of a skipper butterfly (*Hesperia comma*) from population refuges: lessons for conservation in a fragmented landscape. *Journal of Animal Ecology* 62, 472–81.

Thomas, C. D., Bodsworth, E. J., Wilson, R. J., Simmons, A. D., Davies, Z. G., Musche, M., and Conradt, L. (2001) Ecological and evolutionary processes at expanding range margins. *Nature* 411, 577–81.

Thomas, D. J., Tracey, B., Marshall, H., and Norstrom, R. J. (1992) Arctic terrestrial ecosystem contamination. *Science of the Total Environment* 122, 135–64.

Thompson, D. Q., Stuckey, R. L., and Thompson, E. B. (1987) *Spread, Impact, and Control of Purple Loosestrife* (Lythrum salicaria) *in North American Wetlands*. Washington, DC: United States Department of Agriculture, Fish, and Wildlife Service.

Thuiller, W. (2003) BIOMOD: optimising predictions of species distributions and projecting potential future shifts under global change. *Global Change Biology* 9, 1353–62.

Thulin, C.-G. (2003) The distribution of mountain hares *Lepus timidus* in Europe: a challenge from brown hares *L. europaeus. Mammal Review* 33, 29–42.

Tilman, D. (1988) *Plant Strategies and the Dynamics and Structure of Plant Communities* (Monographs in Population Biology 26). Princeton, NJ: Princeton University Press.

Tilman, D. (1994) Competition and biodiversity in spatially structured habitats. *Ecology* 75, 2–16.

Tilman, D. and Wedin, D. (1991) Oscillations and chaos in the dynamics of a perennial grass. *Nature* 353, 653–5.

Tivy, J. (1992) *Biogeography: A Study of Plants in the Ecosphere*, 3rd edn. Edinburgh: Oliver & Boyd.

Tucker, M. E. and Benton, M. J. (1982) Triassic environments, climates and reptile evolution. *Palaeogeography, Palaeoclimatology, Palaeoecology* 40, 361–79.

Turkry, K. and Bogaert, J. (2000) Fragmentation and the regional decree regarding nature conservancy and the natural environment. In R. Ceulemans, J. Bogaert, G. Deckmyn, and I. Nijs (eds) *Topics in Ecology: Structure and Function in Plants and Ecosystems*, pp. 61–9. Wilrijk, Belgium: University of Antwerp.

Turner, M. G. and Stratton, S. P. (1987) Fire, grazing, and the landscape heterogeneity of a Georgia barrier island. In M. G. Turner (ed.) *Landscape Heterogeneity and Disturbance* (Ecological Studies, vol. 64), pp. 85–101. New York: Springer.

Twedt, D. J., Wilson, R. R., Henne-Kerr, J. K., and Grosshuesch, D. A. (2002) Avian response to bottomland hardwood reforestation: the first 10 years. *Restoration Ecology* 10, 645–55.

University of Liverpool Botanic Gardens (Undated) *Plants from China Collected by George Forrest*. Ness, Cheshire: The University of Liverpool Botanic Gardens.

Usher, M. B. (1972) Developments in the Leslie matrix model. In J. N. R. Jeffers (ed.) *Mathematical Models in Ecology*, pp. 29–60. Oxford: Blackwell Scientific Publications.

Usher, M. B., Brown, A. C., and Bedford, S. E. (1992) Plant species richness in farm woodlands. *Forestry* 65, 1–13.

Utida, S. (1957) Population fluctuation, an experimental and theoretical approach. *Cold Spring Harbor Symposium in Quantitative Biology* 22, 139–51.

Valentine, J. W. (1989) Phanerozoic marine faunas and the stability of the Earth system. *Palaeogeography, Palaeoclimatology, Palaeoecology (Global and Planetary Change Section)* 75, 137–55.

Valentine, J. W., Foin, T. C., and Peart, D. (1978) A provincial model of Phanerozoic marine diversity. *Paleobiology* 4, 55–66.

Vannote, R. L., Minshall, G. W., Cummins, K. W., Sedell, J. R., and Cushing, C. E. (1980) The river continuum concept. *Canadian Journal of Fisheries and Aquatic Sciences* 37, 130–7.

Van Valkenburgh, B. and Janis, C. M. (1993) Historical diversity patterns in North American large herbivores and carnivores. In R. E. Ricklefs and D. Schluter (eds) *Species Diversity in Ecological Communities: Historical and Geographical Perspectives*,

pp. 330–40. Chicago, IL: University of Chicago Press.

van Wilgen, B. W., Richardson, D. M., Kruger, F. J., and van Hensbergen, H. J. (eds) (1992) *Fire in South African Mountain Fynbos: Ecosystem, Community and Species Response at Swartboskloof* (Ecological Studies, vol. 93). New York: Springer.

Varley, G. C. (1970) The concept of energy flow applied to a woodland community. In A. Watson (ed.) *Animal Populations in Relation to Their Food Resources* (A Symposium of the British Ecological Society, Aberdeen 24–28 March 1969), pp. 389–405. Oxford and Edinburgh: Blackwell Scientific Publications.

Vaughan, T. A. (1954) Mammals of the San Gabriel Mountains of California. *University of Kansas Publications, Museum of Natural History* 7, 513–82.

Vaughan, T. A. (1978) *Mammalogy*, 2nd edn. Philadelphia, PA: W. B. Saunders.

Veena, T. and Lokesha, R. (1993) Association of drongos with myna flocks: are drongos benefited?. *Journal of Biosciences, Indian Academy of Sciences* 18, 111–19.

Venables, L. S. V. and Venables U. M. (1955) *Birds and Mammals of Shetland*. Edinburgh and London: Oliver & Boyd.

Verboom, J., Schotman, A., Opdam, P., and Metz, J. A. J. (1991) European nuthatch metapopulations in a fragmented agricultural landscape. *Oikos* 61, 149–56.

Via, S. (2001) Sympatric speciation in animals: the ugly duckling grows up. *Trends in Ecology and Evolution* 16, 381–90.

Virolainen, K. M., Suomi, T., Suhonen, J., and Kuitunen, M. (1998) Conservation of vascular plants in single large and several small mires: species richness, rarity and taxonomic diversity. *Journal of Applied Ecology* 35, 700–7.

Vitousek, P. M., Ehrlich, P. R., Ehrlich, A. H., and Matson, P. A. (1986) Human appropriation of the products of photosynthesis. *BioScience* 36, 368–73.

Vogiatzakis, I. N., Griffiths, G. H., and Mannion, A. M. (2003) Environmental factors and vegetation composition, Lefka Ori massif, Crete, S. Aegean. *Global Ecology & Biogeography* 12, 121–46.

Volterra, V. (1926) Fluctuations in the abundance of a species considered mathematically. *Nature* 188, 558–60.

Volterra, V. (1928) Variations and fluctuations of the number of individuals in animal species living together. *Journal du Conseil International pour l'Exploration de la Mer* 3, 3–51.

Volterra, V. (1931) *Leçons sur la Théorie mathématique de la Lutte pour la Vie*, edited by Marcel Brelot. Paris: Gauthier-Villars.

Vrba, E. S., Denton, G. H., and Prentice, M. L. (1989) Climatic influences on early hominid behaviour. *Ossa* 14, 127–56.

Wagner, J. and Reichegger, B. (1997) Phenology and seed development of the alpine sedges *Carex curvula* and *Carex firma* in response to contrasting topoclimates. *Arctic and Alpine Research* 219, 291–9.

Wallace, A. R. (1876) *The Geographical Distribution of Animals; with A Study of the Relations of Living and Extinct Faunas as Elucidating the Past Changes of the Earth's Surface*, 2 vols. London: Macmillan.

Wallace A. R. (1880) *Island Life*. London: Macmillan.

Walter, H. (1985) *Vegetation of the Earth and Ecological Systems of the Geo-Biosphere*, 3rd revised and enlarged edn, translated from the 5th revised German edn by O. Muise. Berlin: Springer.

Walter, H. and Breckle, S.-W. (1985) *Ecological Systems of the Geobiosphere. Vol. 1. Ecological Principles in Global Perspective*, translated by S. Gruber. Berlin: Springer.

Walter, H. and Lieth, H. (1960–7) *Klimadiagramm–Weltatlas*. Jena: Gustav Fischer.

Wang, K., Geldsetzer, H. H. J., and Krouse, H. R. (1994) Permian–Triassic extinction: organic $\delta^{13}C$ evidence from British Columbia, Canada. *Geology* 22, 580–4.

Ward, J. P. and Anderson, S. H. (1988) Influences of cliffs on wildlife communities in southcentral Wyoming. *Journal of Wildlife Management* 52, 673–8.

Ward, T. J. and Jacoby, C. A. (1992) A strategy for assessment and management of marine ecosystems: baseline and monitoring studies in Jervis Bay, a temperate Australian embayment. *Marine Pollution Bulletin* 25, 163–71.

Wardle, D. A., Nicholson, K. S., Rahman, A. (1993) Influence of plant age on the allelopathic potential of nodding thistle (*Carduus nutans* L.) against pasture grasses and legumes. *Weed Research* 33, 69–78.

Watt, A. S. (1947) Pattern and process in the plant community. *Journal of Ecology* 35, 1–22.

Watts, W. A. and Hansen, B. C. S. (1994) Pre-Holocene and Holocene pollen records of vegetation history from the Florida peninsula and their climatic implications. *Palaeogeography, Palaeoclimatology, Palaeoecology* 109, 163–76.

Webb, N. R. and Thomas, J. A. (1994) Conserving insect habitats in heathland biotopes: a question of scale. In P. J. Edwards, R. M. May, and N. R. Webb (eds) *Large-Scale Ecology and Conservation Biology* (The 35th Symposium of the British Ecological Society with the Society for Conservation Biology, University of Southampton, 1993), pp. 129–51. Oxford: Blackwell Scientific Publications.

Webb, S. D. (1983) The rise and fall of the late Miocene ungulate fauna in North America. In M. H. Nitecki (ed.) *Coevolution*, pp. 267–306. Chicago, IL, and London: The University of Chicago Press.

Webb, S. D. (1984) Ten million years of mammal extinctions in North America. In P. S. Martin and R. G. Klein (eds) *Quaternary Extinctions: A Prehistoric Revolution*, pp. 189–210. Tucson, AZ: The University of Arizona Press.

Webb, S. D. (1991) Ecogeography and the Great American Interchange. *Paleobiology* 17, 266–80.

Webb, S. D. and Opdyke, N. D. (1995) Global climatic influence on Cenozoic land mammal faunas. In Board on Earth Sciences and Resources Commission on Geosciences, Environment, and Resources, National Research Council, *Effects of Past Global Change on Life*, pp. 184–208. Washington, DC: National Academy Press.

Weber, M. G. and Flannigan, M. D. (1997) Canadian boreal forest ecosystem structure and function in a changing climate: impact on fire regimes. *Environmental Reviews* (National Research Council of Canada) 5, 145–66.

Weiser, C. J. (1970) Cold resistance and injury in woody plants. *Science* 169, 1269–78.

Wendland, W. M., Benn, A., Semken, H. A., Jr (1987) Evaluation of climatic changes on the North American Great Plains determined from faunal evidence. In R. W. Graham and H. A. Semken Jr (eds) *Late Quaternary Mammalian Biogeography and Environment of the Great Plains and Prairies* (Scientific Papers 22), pp. 460–73. Springfield, IL: Illinois State Museum.

Western, D. (1989) Conservation without parks: wildlife in the rural landscape. In D. Western and M. Pearl (eds) *Conservation for the Twenty-first Century*, pp. 158–65. Oxford: Oxford University Press.

Whelan, R. J. (1995) *The Ecology of Fire*. Cambridge: Cambridge University Press.

White, M. J. D. (1978) *Modes of Speciation*. San Francisco, CA: W. H. Freeman.

Whitmore, T. C. (1975) *Tropical Rain Forests of the Far East*. Oxford: Oxford University Press.

Whittaker, J. B. (2001) Impacts and responses at population level of herbivorous insects to elevated CO_2. *European Journal of Entomology* 96, 149–56.

Whittaker, R. H. (1953) A consideration of climax theory: the climax as a population and pattern. *Ecological Monographs* 23, 41–78.

Whittaker, R. H. (1954) The ecology of serpentine soils. IV. The vegetational response to serpentine soils. *Ecology* 35, 275–88.

Whittaker, R. J. and Bush, M. B. (1993) Dispersal and establishment of tropical forest assemblages, Krakatoa, Indonesia. In J. Miles and D. W. H. Walton (eds) *Primary Succession on Land* (The British Ecological Society, Special Publication No. 12), pp. 147–60. Oxford: Blackwell Scientific Publications.

Whittaker, R. J. and Jones, S. H. (1994) Structure in re-building insular ecosystems: an empirically derived model. *Oikos* 69, 524–30.

Whittaker, R. J., Bush, M. B., and Richards, K. (1989) Plant recolonization and vegetation succession on the Krakatau Islands, Indonesia. *Ecological Monographs* 59, 59–123.

Whittaker, R. J., Bush, M. B., Asquith, N. M., and Richards, K. (1992) Ecological aspects of plant colonisation of the Krakatau Islands. *GeoJournal* 28, 201–11.

Whittlesey, D. (1954) The regional concept and the regional method. In P. E. James and C. F. Jones (eds) *American Geography: Inventory and Prospect*, pp. 19–68. Syracuse, NY: Syracuse University Press.

Williams, E. E. (1989) Old problems and new opportunities in West Indian Biogeography. In C. A. Woods (ed.) *Biogeography of the West Indies: Past, Present, and Future*, pp. 1–46. Gainesville, FL: Sandhill Crane Press.

Williams, M. (1996) European expansion and land cover transformation. In I. Douglas, R. J. Huggett, and M. E. Robinson (eds) *Companion*

Encyclopedia of Geography: The Environment and Humankind, pp. 182–205. London and New York: Routledge.

Williams, M. E. (1994) Catastrophic versus noncatastrophic extinction of the dinosaurs: testing, falsifiability, and the burden of proof. *Journal of Paleontology* 68, 183–90.

Williamson, M. (1981) *Island Populations*. Oxford: Oxford University Press.

Willis, K. J. and Whittaker, R. J. (2000) The refugial debate. *Science* 287, 1406–7.

Willis, K. J., Rudner, E., and Sümegi, P. (2000) The full-glacial forests of central and southeastern Europe. *Quaternary Research* 53, 203–13.

Wilson, E. O. (1992) *The Diversity of Life*. Cambridge, MA: The Belknap Press of Harvard University Press.

Wilson, R., Allen-Gil, S., Griffin, D., and Landers, D. (1995) Organochlorine contaminants in fish from an Arctic lake in Alaska, USA. *Science of the Total Environment* 160–1, 511–19.

Woods, C. A. (ed.) (1989) *Biogeography of the West Indies: Past, Present, and Future*. Gainesville, FL: Sandhill Crane Press.

Woodward, F. I. (1992) A review of the effects of climate on vegetation: ranges, competition, and composition. In R. L. Peters and T. E. Lovejoy (eds) *Global Warming and Biological Diversity*, pp. 105–23. New Haven, CT, and London: Yale University Press.

Woodwell, G. M. (1967) Toxic substances and ecological cycles. *Scientific American* 216 (March), 24–31.

Wright, S. J., Gompper, M. E., and DeLeon, B. (1994) Are large predators keystone species in Neotropical forests? The evidence from Barro Colorado Island. *Oikos* 71, 279–94.

Wu, C.-I. (2001) The genic view of the process of speciation. *Journal of Evolutionary Biology* 14, 851–65.

Yaffee, S. L. (1999) Three faces of ecosystem management. *Conservation Biology* 13, 713–25.

Yom-Tov, Y. (1993) Does the rock hyrax, *Procavia capensis*, conform with Bergmann's rule? *Zoological Journal of the Linnean Society* 108, 171–7.

Yom-Tov, Y. (2001) Global warming and body mass decline in Israeli passerine birds. *Proceedings of the Royal Society of London* 268B, 947–52.

Zackrisson, O. and Nilsson, M. C. (1992) Allelopathic effects by *Empetrum hermaphroditum* on seed germination of two boreal tree species. *Canadian Journal of Forest Research* 22, 1310–19.

Zhou, Z. and Zheng, S. (2003) The missing link in Gingko evolution. *Nature* 423, 821–2.

Ziswiler, V. (1967) *Extinct and Vanishing Animals: A Biology of Extinction and Survival*, revised English edn by Fred and Pille Bunnell (The Heidelberg Science Library, vol. 2). London: The English Universities Press.

INDEX

NB Italicized numbers indicate mentions within Figures or Tables; emboldened numbers indicate mentions within Boxes.